Geophysical Continua presents a systematic treatment of deformation in the Earth from seismic to geologic time scales, and demonstrates the linkages between different aspects of the Earth's interior that are often treated separately.

A unified treatment of solids and fluids is developed to include thermodynamics and electrodynamics, in order to cover the full range of tools needed to understand the interior of the globe. A close link is made between microscopic and macroscopic properties manifested through elastic, viscoelastic and fluid rheologies, and their influence on deformation. Following a treatment of geological deformation, a global perspective is taken on lithospheric and mantle properties, seismology, mantle convection, the core and Earth's dynamo. The emphasis throughout the book is on relating geophysical observations to interpretations of earth processes. Physical principles and mathematical descriptions are developed that can be applied to a broad spectrum of geodynamic problems.

Incorporating illustrative examples and an introduction to modern computational techniques, this textbook is designed for graduate-level courses in geophysics and geodynamics. It is also a useful reference for practising Earth Scientists. Supporting resources for this book, including exercises and full-colour versions of figures, are available at www.cambridge.org/9780521865531.

BRIAN KENNETT is Director and Distinguished Professor of Seismology at the Research School of Earth Sciences in The Australian National University. Professor Kennett's research interests are directed towards understanding the structure of the Earth through seismological observations. He is the recipient of the 2006 Murchison Medal of the Geological Society of London, and the 2007 Gutenberg Medal of the European Geosciences Union, and he is a Fellow of the Royal Society of London. Professor Kennett is the author of three other books for Cambridge University Press: *Seismic Wave Propagation in Stratified Media* (1983), *The Seismic Wavefield: Introduction and Theoretical Development* (2001), and *The Seismic Wavefield: Interpretation of Seismograms on Regional and Global Scales* (2002).

HANS-PETER BUNGE is Professor and Chair of Geophysics at the Department of Earth and Environmental Sciences, University of Munich, and is Head of the Munich Geo-Center. Prior to his Munich appointment, he spent 5 years on the faculty at Princeton University. Professor Bunge's research interests lie in the application of high performance computing to problems of Earth and planetary evolution, including core, mantle and lithospheric dynamics. A member of the Bavarian Academy of Sciences, Bunge is also President of the Geodynamics Division of the European Geosciences Union (EGU).

Geophysical Continua

Deformation in the Earth's Interior

B.L.N. KENNETT
Research School of Earth Sciences, The Australian National University

H.-P. BUNGE
Department of Geosciences, Ludwig Maximilians University, Munich

CAMBRIDGE
UNIVERSITY PRESS

CAMBRIDGE
UNIVERSITY PRESS

University Printing House, Cambridge CB2 8BS, United Kingdom

One Liberty Plaza, 20th Floor, New York, NY 10006, USA

477 Williamstown Road, Port Melbourne, VIC 3207, Australia

314-321, 3rd Floor, Plot 3, Splendor Forum, Jasola District Centre, New Delhi - 110025, India

79 Anson Road, #06-04/06, Singapore 079906

Cambridge University Press is part of the University of Cambridge.

It furthers the University's mission by disseminating knowledge in the pursuit of education, learning and research at the highest international levels of excellence.

www.cambridge.org
Information on this title: www.cambridge.org/9781108462730

First published 2008
First paperback edition 2018

A catalogue record for this publication is available from the British Library

ISBN 978-0-521-86553-1 Hardback
ISBN 978-1-108-46273-0 Paperback

Contents

Preface

Geophysical Continua is designed to present a systematic treatment of deformation in the Earth from seismic to geologic time scales. In this way we demonstrate the linkages between different aspects of the Earth's interior that are commonly treated separately. We provide a coherent presentation of non-linear continuum mechanics with a uniform notation, and then specialise to the needs of particular topics such as elastic, viscoelastic and fluid behaviour. We include the concepts of continuum thermodynamics and link to the properties of material under pressure in the deep interior of the Earth, and also provide the continuum electrodynamics needed for conducting fluids such as the Earth's core.

Following an introduction to continuum methods and the structure of the Earth, Part I of the book takes the development of continuum techniques to the level where they can be applied to the diverse aspects of Earth structure and dynamics in Part II. At many levels there is a close relation between microscopic properties and macroscopic consequences such as effective rheology, and so Part II opens with a discussion of the relation of phenomena at the atomic scale to continuum properties. We follow this with a treatment of geological deformation at the grain and outcrop scale. In the subsequent chapters we emphasise the physical principles that allow understanding of Earth processes, taking a global perspective towards lithospheric and mantle properties, seismology, mantle convection, the core and Earth's dynamo. We make links to experimental results and seismological observations to provide insight into geodynamic interpretations.

The material in the book has evolved over a considerable time period and has benefited from interactions with many students in Cambridge, Canberra, Princeton and Munich. Particular thanks go to the participants in the Geodynamics Seminar in Munich in 2005, which helped to refine Part I and the discussion of the lithosphere in Part II.

In a work of this complexity covering many topics with their own specific notation it is difficult to avoid reusing symbols. Nevertheless we have have tried to sustain a unified notation throughout the whole book and to minimise multiple use.

We have had stimulating discussions with Jason Morgan, John Suppe and Geoff Davies over a wide range of topics. Gerd Steinle-Neumann provided very helpful input on mineral properties and *ab initio* calculations, and Stephen Cox provided valuable insight into the relation of continuum mechanics and structural geology.

Special thanks go to the Alexander von Humboldt Foundation for the Research Award to Brian Kennett that led to the collaboration on this volume.

Acknowledgements

We are grateful to the many people who have gone to trouble to provide us with figures, in particular: A. Barnhoorn, G. Batt, J. Besse, C. Bina, S. Cox, J. Dawson, E. Debayle, U. Faul, A. Fichtner, S. Fishwick, J. Fitz Gerald, E. Garnero, A. Gorbatov, B. Goleby, O. Heibach, M. Heintz, G. Houseman, R. Holme, G. Iaffeldano, M. Ishii, A. Jackson, I. Jackson, J. Jackson, M. Jessell, J. Kung, P. Lorinczi, S. Micklethwaite, M. Miller, D. Mueller, A. Piazzoni, K. Priestley, M. Sandiford, W. Spakman, B. Steinberger, J. Suppe, F. Takahashi, and K. Yoshizawa.

1

Introduction

The development of quantitative methods for the study of the Earth rests firmly on the application of physical techniques to the properties of materials without recourse to the details of atomic level structure. This has formed the basis of seismological methods for investigating the internal structure of the Earth, and for modelling of mantle convection through fluid flow. The deformation behaviour of materials is inextricably tied to microscopic properties such as the elasticity of individual crystals and processes such as the movement of dislocations. In the continuum representation such microscopic behaviour is encapsulated in the description of the rheology of the material through the connection between stress and strain (or strain rate).

Different classes of behaviour are needed to describe the diverse aspects of the Earth both in depth and as a function of time. For example, in the context of the rapid passage of a seismic wave the lithosphere may behave elastically, but under the sustained load of a major ice sheet will deform and interact with the deeper parts of the Earth. When the ice sheet melts at the end of an ice age, the lithosphere recovers and the pattern of post-glacial uplift can be followed through raised beaches, as in Scandinavia.

The Earth's core is a fluid and its motions create the internal magnetic field of the Earth through a complex dynamo interaction between fluid flow and electromagnetic interactions. The changes in the magnetic field at the surface on time scales of a few tens of years are an indirect manifestation of the activity in the core. By contrast, the time scales for large-scale flow in the silicate mantle are literally geological, and have helped to frame the configuration of the planet as we know it.

We can link together the many different facets of Earth behaviour through the development of a common base of continuum mechanics before branching into the features needed to provide a detailed description of specific classes of behaviour. We start therefore by setting the scene for the continuum representation. We then review the structure of the Earth and the different types of mechanical behaviour that occur in different regions, and examine some of the ways in which information at the microscopic level is exploited to infer the properties of the Earth through both experimental and computational studies.

1.1 Continuum properties

A familiar example of the concept of a continuum comes from the behaviour of fluids, but we can use the same approach to describe solids, glasses and other more general substances that have short-term elastic and long-term fluid responses. The behaviour of such continua can then be established by using the conservation laws for linear and angular momentum and energy, coupled to explicit descriptions of the relationship between the stress, describing the force system within the material, and the strain, which summarises the deformation.

We adopt the viewpoint of continuum mechanics and thus ignore all the fine detail of atomic level structure and assume that, for sufficiently large samples,:

- the highly discontinuous structure of real materials can be replaced by a smoothed hypothetical continuum; and
- every portion of this continuum, however small, exhibits the macroscopic physical properties of the bulk material.

In any branch of continuum mechanics, the field variables (such as density, displacement, velocity) are conceptual constructs. They are taken to be defined at all points of the imagined continuum and their values are calculated via axiomatic rules of procedure.

The continuum model breaks down over distances comparable to interatomic spacing (in solids about 10^{-10} m). Nonetheless the *average* of a field variable over a small but *finite* region is meaningful. Such an average can, in principle, be compared directly to its nominal counterpart found by experiment – which will itself represent an average of a kind taken over a region containing many atoms, because of the physical size of any measuring probe.

For solids the continuum model is valid in this sense down to a scale of order 10^{-8} m, which is the side of a cube containing a million or so atoms.

Further, when field variables change slowly with position at a microscopic level $\sim 10^{-6}$ m, their averages over such volumes (10^{-20} m^3 say) differ insignificantly from their centroidal values. In this case pointwise values can be compared directly to observations.

Within the continuum we take the behaviour to be determined by
a) conservation of mass;
b) linear momentum balance: the rate of change of total linear momentum is equal to the sum of the external forces; and
c) angular momentum balance.
The continuum hypothesis enables us to apply these laws on a local as well as a global scale.

1.1.1 Deformation and strain

If we take a solid cube and subject it to some deformation, the most obvious change in external characteristics will be a modification of its shape. The specification of this deformation is thus a geometrical problem, that may be carried out from two different viewpoints:
a) with respect to the undeformed state (Lagrangian), or
b) with respect to the deformed state (Eulerian).
Locally, the mapping from the deformed to the undeformed state can be assumed to be linear and described by a differential relation, which is a combination of pure stretch (a rescaling of each coordinate) and a pure rotation.

The mechanical effects of the deformation are confined to the stretch and it is convenient to characterise this by a *strain* measure. For example, for a wire under load the strain ϵ would be the relative extension, i.e.,

$$\epsilon = \frac{\text{change in length}}{\text{initial length}}, \tag{1.1.1}$$

The generalisation of this idea requires us to introduce a *strain tensor* at each point of the continuum to allow for the three-dimensional nature of deformation.

1.1.2 The stress field

Within a deformed continuum a force system acts. If we were able to cut the continuum in the neighbourhood of a point, we would find a force acting on a cut surface which would depend on the inclination of the surface and is not necessarily perpendicular to the surface (Figure 1.1).

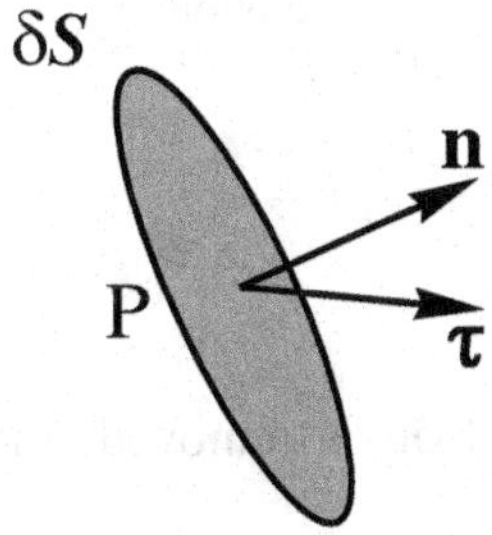

Figure 1.1. The force vector $\boldsymbol{\tau}$ acting on an internal surface specified by the vector normal $\mathbf{n}$ will normally not align with $\mathbf{n}$.

This force system can be described by introducing a *stress tensor* σ at each point, whose components describe the loading characteristics, and from which the force vector $\boldsymbol{\tau}$ can be found for a surface with arbitary normal $\mathbf{n}$.

For a loaded wire, the stress σ would just be the force per unit area.

1.1.3 Constitutive relations

The specification of the stress and strain states of a body is insufficient to describe its full behaviour, we need in addition to link these two fields. This is achieved by introducing a *constitutive relation*, which prescribes the response of the continuum to arbitrary loading and thus defines the connection between the stress and strain tensors for the particular material.

At best, a mathematical expression provides an approximation to the actual behaviour of the material. But, as we shall see, we can simulate the behaviour of a wide class of media by using different mathematical forms.

We shall assume that the forces acting at a point depend on the *local* geometry of deformation and its history, and possibly also on the history of the local temperature. This concept is termed the *principle of local action*, and is designed to exclude 'action at a distance' for stress and strain.

Solids

Solids are a familiar part of the Earth through the behaviour of the outer layers, which exhibit a range of behaviours depending on time scale and loading.

We can illustrate the range of behaviour with the simple case of extension of a wire under loading. The tensile stress σ and tensile strain ϵ are then typically related as shown in Figure 1.2.

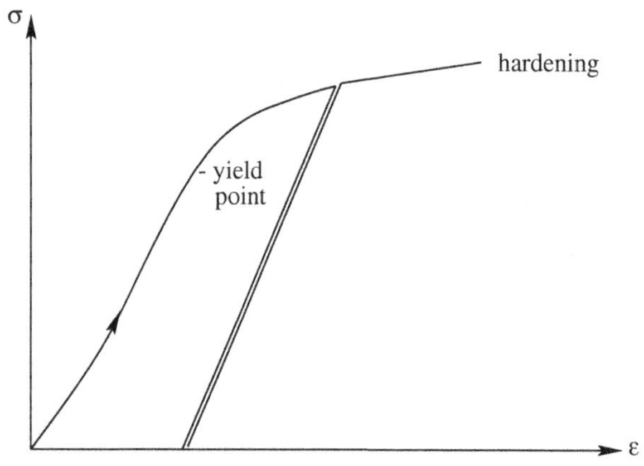

Figure 1.2. Behaviour of a wire under load

Elasticity

If the wire returns to its original configuration when the load is removed, the behaviour is said to be elastic:
(i) linear elasticity $\sigma = \mathsf{E}\epsilon$ – usually valid for small strains;
(ii) non-linear elasticity $\sigma = \mathsf{f}(\epsilon)$ – important for rubber-like materials, but not significant for the Earth.

Plasticity

Once the yield point is exceeded, permanent deformation occurs and there is no unique stress–strain curve, but a unique $d\sigma - d\epsilon$ relation. As a result of microscopic processes the yield stress rises with increasing strain, a phenomenon known as work hardening. Plastic flow is important for the movement of ice, e.g., in glacier flow.

Viscoelasticity (rate-dependent behaviour)

Materials may creep and show slow long-term deformation, e.g., plastics and metals at elevated temperatures. Such behaviour also seems to be appropriate to the Earth, e.g., the slow uplift of Fennoscandia in response to the removal of the loading of the glacial ice sheets.

Elementary models of viscoelastic behaviour can be built up from two basic building blocks: the *elastic spring* for which

$$\sigma = m\,\epsilon, \tag{1.1.2}$$

and the *viscous dashpot* for which

$$\sigma = \eta\,\dot{\epsilon}. \tag{1.1.3}$$

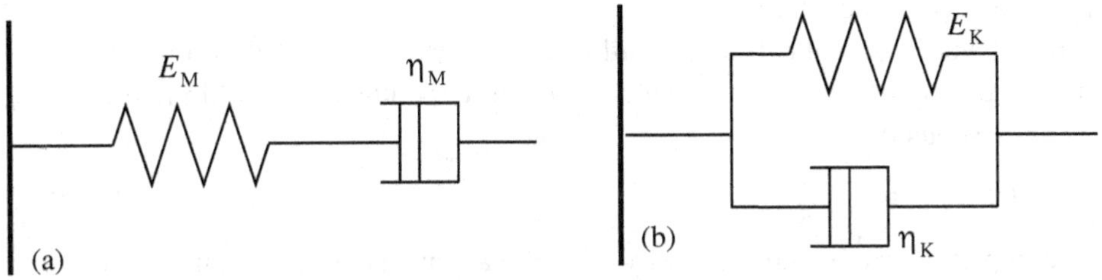

Figure 1.3. Mechanical models for linear viscoelastic behaviour combining a spring and viscous dashpot: (a) Maxwell model, (b) Kelvin–Voigt model.

The stress-strain relations depend on how these elements are combined.

(i) Maxwell model
The spring and dashpot are placed in series (Figure 1.3a) so that

$$\dot{\sigma} = E_M(\dot{\epsilon} + \epsilon/\tau_M); \tag{1.1.4}$$

this allows for instantaneous elasticity and represents a crude description of a fluid. The constitutive relation can be integrated using, e.g., Laplace transform methods and we find

$$\sigma(t) = E_M\Big(\epsilon(t) + \int^t dt'\,\epsilon(t')\exp[-(t-t')/\tau_M]\Big), \tag{1.1.5}$$

so the stress state depends on the history of strain.

(ii) Kelvin–Voigt model
The spring and dashpot are placed in parallel (Figure 1.3b) and so

$$\sigma = E_K(\dot{\epsilon} + \epsilon/\tau_K), \tag{1.1.6}$$

which displays long-term elasticity. For the initial condition $\epsilon = 0$ at $t = 0$ and constant stress σ_0, the evolution of strain in the Kelvin–Voigt model is

$$\epsilon = \frac{\sigma_0}{2E_K}\left[1 - \exp\left(-\frac{t}{\tau_K}\right)\right], \tag{1.1.7}$$

so that the viscous damping is not relevant on long time scales.

More complex models can be generated, but all have the same characteristic that the stress depends on the time history of deformation.

Fluids

The simplest constitutive equation encountered in continuum mechanics is that for an ideal fluid, where the pressure field p is isotropic and depends on density and temperature

$$\sigma = -p(\rho, T), \tag{1.1.8}$$

where ρ is the density, and T is the absolute temperature. If the fluid is incompressible ρ is a constant.

The next level of complication is to allow the pressure to depend on the flow of the fluid. The simplest such form includes a linear dependence on strain rate $\dot{\epsilon}$ – a Newtonian viscous fluid:

$$\sigma = -p(\rho, T) + \eta\dot{\epsilon}. \tag{1.1.9}$$

Further complexity can be introduced by allowing a non-linear dependence of stress on strain rate, as may be required for the flow of glacier ice.

1.2 Earth processes

The Earth displays a broad spectrum of continuum properties varying with both depth and time. A dominant influence is the effect of pressure with increasing depth, so that properties of materials change as phase transitions in minerals accommodate closer packed structures. Along with the pressure the temperature increases, so we need to deal with the properties of materials at conditions that are not simple to reproduce under laboratory conditions.

The nature of the deformation processes within the Earth depends strongly on the frequency of excitation. At high frequencies appropriate to the passage of seismic waves the dominant contribution is elastic, with some seismic attenuation that can be represented with a small linear viscoelastic component. However, as the frequency decreases and the period lengthens viscous flow effects become more prominent, so that elastic contributions can be ignored in the study of mantle convection. This transition in behaviour is illustrated in Figure 1.4, and is indicative of a very complex rheology for the interior of the Earth as different facets of material behaviour become important in different frequency bands. The observed behaviour reflects competing influences at the microscopic level, and

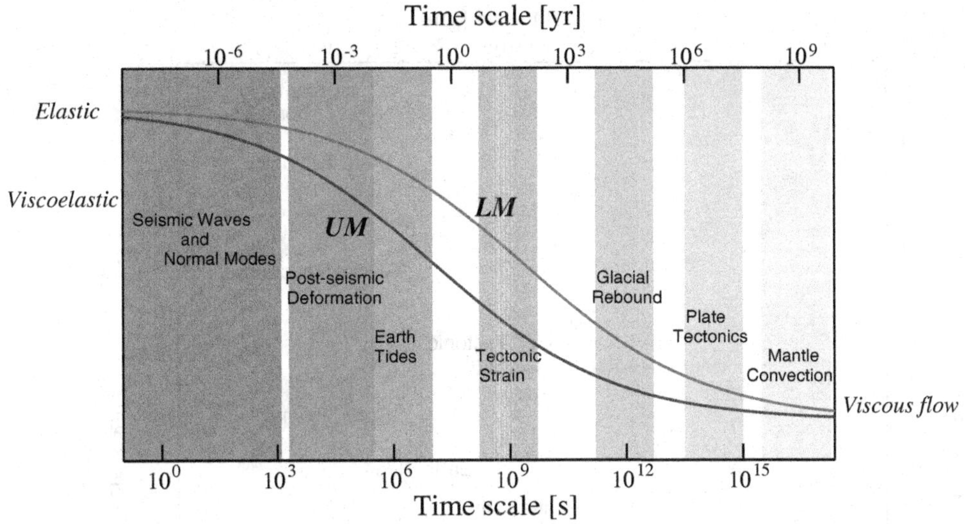

Figure 1.4. Spectrum of Earth deformation processes indicating the transition from viscoelastic to fully viscous behaviour as the frequency decreases. The upper curve refers to the lower mantle (*LM*), and the lower curve to the upper mantle (*UM*), indicating the differences in viscosity and deformation history.

varies significantly with depth as indicated by the two indicative curves (*UM*, *LM*) in Figure 1.4 for the transition from near elastic behaviour to fully viscous flow behaviour, representing the states for the upper and lower mantle.

Further, the various classes of deformation occur over a very wide range of spatial scales (Figure 1.5). As a result, a variety of different techniques is needed to examine the behaviour from seismological to geodetic through to geological observations. There is increasing overlap in seismic and space geodetic methods for studying the processes associated with earthquake sources that has led to new insights for fault behaviour. Some phenomena, such as the continuing recovery of the Earth from glacial loading, can be studied using multiple techniques that provide direct constraints on rheological properties.

Our aim is to integrate understanding of continuum properties and processes with the nature of the Earth itself, and to show how the broad range of terrestial phenomena can be understood within a common framework. We therefore now turn our attention to the structure of the Earth and the classes of geodynamic and deformation processes that shape the planet we live on.

In Part I that follows, we embark on a more detailed examination of the development of continuum methods, in a uniform treatment encompassing solid, fluid and intermediate behaviour. Then in Part II we address specific Earth issues building on the continuum framework

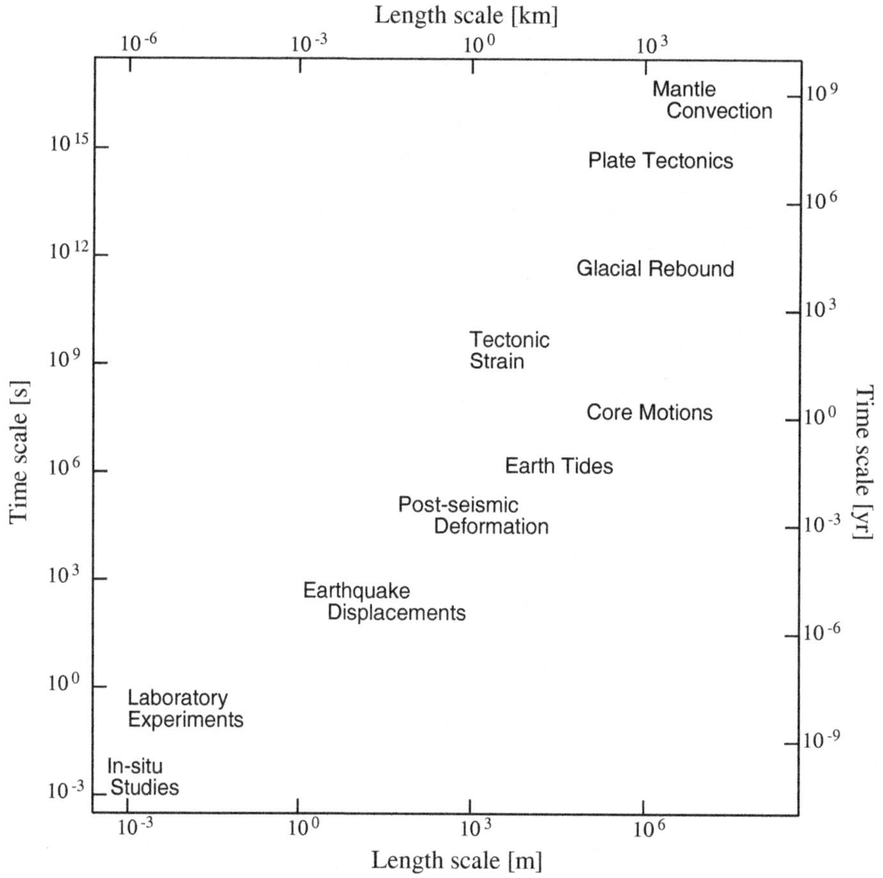

Figure 1.5. Temporal and spatial spectrum of Earth deformation processes.

1.3 Elements of Earth structure

Much of our knowledge of the interior of the Earth comes from the analysis of seismological data, notably the times of passage of seismic body waves at high frequencies ($\approx$ 1 Hz) and the behaviour of the free oscillations of the Earth at lower frequencies (0.03 – 3 mHz). Such studies provide information both on the dominant radial variations in physical properties, and on the three-dimensional variations in the solid parts of the Earth. Important additional constraints are provided by the mass and moments of inertia of the Earth, which can be deduced from satellite observations. The moments of inertia are too low for the Earth to have uniform density, there has to be a concentration of mass towards the centre that can be identified with the seismologically defined *core*.

The resulting picture of the dominant structure of the Earth is presented in Figure 1.6. The figure of the Earth is close to an oblate spheroid with a flattening of 0.003356. The radius to the pole is 6357 km and the equatorial radius is 6378 km,

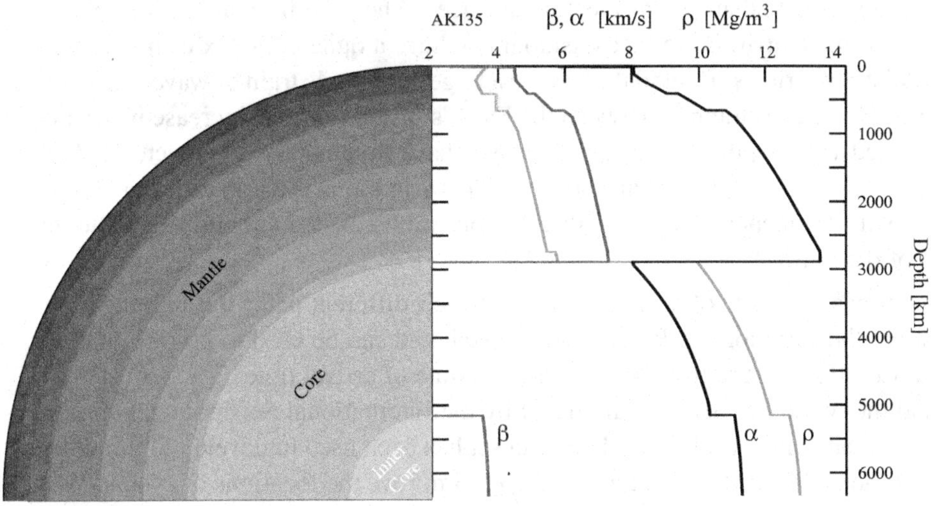

Figure 1.6. The major divisions of the radial structure of the Earth linked to the radial reference Earth model AK135, seismic wave speeds α (*P*), β (*S*): Kennett et al. (1995); density ρ: Montagner & Kennett (1996). The gradations in tone in the Earth's mantle indicate the presence of discontinuities at 410 and 660 km depth, and the presence of the D″ near the core–mantle boundary.

but for most purposes a spherical model of the Earth with a mean radius of 6371 km is adequate. Thus reference models for internal structure in which the physical properties depend on radius can be used. Three-dimensional variations can then be described by deviations from a suitable reference model.

Beneath the thin crustal shell lies the silicate mantle which extends to a depth of 2890 km. The mantle is separated from the metallic core by a major change of material properties that has a profound effect on global seismic wave propagation. The outer core behaves as a fluid at seismic frequencies and does not allow the passage of shear waves, while the inner core appears to be solid.

The existence of a discontinuity at the base of the crust was found by Mohorovičić in the analysis of the Kupatal earthquake of 1909 from only a limited number of records from permanent seismic stations. Knowledge of crustal structure from seismic methods has developed substantially in past decades through the use of controlled sources, e.g., explosions. Indeed most of the information on the oceanic crust comes from such work. The continental crust varies in thickness from around 20 km in rift zones to 70 km under the Tibetan Plateau. Typical values are close to 35 km. The oceanic crust is thinner, with a basalt pile about 7 km thick whose structure changes somewhat with the age of the oceanic crust.

Earthquakes and man-made sources generate two types of seismic waves that propagate through the Earth. The earliest arriving (*P*) wave has longitudinal motion; the second (*S*) wave has particle motion perpendicular to the path. In the Earth the direct *P* and *S* waves are accompanied by multiple reflections and conversions, particulary from the free surface. These additional seismic phases follow the main arrivals, so that seismograms have a quite complex character with many distinct arrivals. Behind the *S* wave a large-amplitude train of waves builds up from surface waves trapped between the Earth's surface and the increase in seismic wavespeed with depth. These surface waves have dominantly *S* character and are most prominent for shallow earthquakes. The variation in the properties of surface waves with frequency provides valuable constraints on the structure of the outer parts of the Earth.

The times of arrival of seismic phases on their different paths through the globe constrain the variations in *P* and *S* wavespeed, and can be used to produce models of the variation with radius. A very large volume of arrival time data from stations around the world has been accumulated by the International Seismological Centre and is available in digital form. This data set has been used to develop high-quality travel-time tables, that can in turn be used to improve the locations of events. With reprocessing of the arrival times to improve locations and the identification of the picks for later seismic phases, a set of observations of the relation between travel time and epicentral distance have been produced for a wide range of phases. The reference model AK135 of Kennett et al. (1995) for both *P* and *S* wave speeds, illustrated in figure 1.6, gives a good fit to the travel times of mantle and core phases. The reprocessed data set and the AK135 reference model have formed the basis of much recent work on high-resolution travel-time tomography to determine three-dimensional variations in seismic wavespeed.

The need for a core at depth with greatly reduced seismic wave speeds was recognised at the end of the nineteenth century by Oldham in his analysis of the great Assam earthquake of 1890, because of a zone without distinct *P* arrivals (a 'shadow zone' in *PKP*). By 1914 Gutenburg had obtained an estimate for the radius of the core which is quite close to the current value. The presence of the inner core was inferred by Inge Lehmann in 1932 from careful analysis of arrivals within the shadow zone (*PKiKP*), which had to be reflected from some substructure within the core.

The mantle shows considerable variation in seismic properties with depth, with strong gradients in seismic wavespeed in the top 800 km. The presence of distinct structure in the upper mantle was recognised by Jeffreys in the 1930's from the change in the slope of the travel time as a function of distance from events near 20°. Detailed analysis at seismic arrays in the late 1960s provided evidence for significant discontinuities in the upper mantle. Subsequent studies have demonstrated the global presence of discontinuities near 410 and 660 km depth, but also significant variations in seismic structure within the upper mantle (for a review see Nolet et al., 1994).

The use of the times of arrival of seismic phases enables the construction of models for *P* and *S* wavespeed, but more information is needed to provide a full model for Earth structure. The density distribution in the Earth has to be inferred from indirect observations and the main constraints come from the mass and moment of inertia. The mean density of the Earth can be reconciled with the moment of inertia if there is a concentration of mass towards the centre of the Earth; which can be associated with a major density jump going from the mantle into the outer core and a smaller density contrast at the boundary between the inner and outer cores (Bullen, 1975).

With successful observations of the free oscillations of the Earth following the great Chilean earthquake of 1960, additional information on both the seismic wave speeds and the density could be extracted from the frequencies of oscillation. Fortunately the inversion of the frequencies of the free oscillations for a spherically symmetric reference model provides independent constraints on the P wavespeed structure in the outer core. Even with the additional information from the normal modes the controls on the density distribution are not strong (Kennett, 1998), and additional assumptions such as an adiabatic state in the core and lower mantle have often been employed to produce a full model.

The reference model PREM of Dziewonski & Anderson (1981) combined the free-oscillation and travel-time information available at the time. A parametric representation of structure was employed in terms of simple mathematical functions to aid the inversion; thus a single cubic was used for seismic wavespeed in the outer core and again for most of the lower mantle. The PREM model forms the basis of much current global seismology using quantitative exploitation of seismic waveforms at longer periods (e.g., Dahlen & Tromp, 1998).

In order to reconcile the information derived from the free oscillations of the Earth and the travel time of seismic phases, it is necessary to take account of the influence of *anelastic attenuation* within the Earth. A consequence of the energy loss of seismic energy due to attenuation is a small variation in the seismic wave speeds with frequency, so that waves with frequencies of 0.01 Hz (at the upper limit of free-oscillation observations) travel slightly slower than the 1 Hz waves typical of the short-period observations used in travel-time studies. The differences in the apparent wavespeeds between travel-time analysis and free-oscillation results thus provides constraints on the attenuation distribution with depth. The density and attenuation model shown in figure 1.6 was derived by Montagner & Kennett (1996) to satisfy a broad set of global information with a common structure based on the wavespeed profiles of the AK135 model of Kennett et al. (1995).

The process of subduction brings the cold oceanic lithosphere into the upper mantle and locally there are large contrasts in seismic wave speeds, well imaged by detailed seismic tomography, that extend down to at least 660 km and in some zones even deeper. Remnant subducted material can have a significant presence in some regions, e.g., above the 660 km discontinuity in the north-west Pacific and in the zone from 660 down to 1100 km beneath Indonesia.

1.3.1 Mantle

The nature of the structure of the silicate mantle varies with depth and it is convenient to divide the mantle up into four major zones (e.g., Jackson & Ridgen, 1998)

Upper Mantle (depth $z < 350$ km), with a high degree of variability in seismic wavespeed (exceeding $\pm$ 4%) and relatively strong attenuation in many locations.

Transition Zone ($350 < z < 800$ km), including significant discontinuities in *P* and *S* wavespeeds and generally high velocity gradients with depth.

Lower Mantle ($800 < z < 2600$ km) with a smooth variation of seismic wavespeeds with depth that is consistent with adiabatic compression of a chemically homogeneous material.

D″ layer ($2600 < z < 2900$ km) with a significant change in velocity gradient and evidence for strong lateral variability and attenuation.

As the pressure increases with depth, there are phase transformations in the silicate minerals of the minerals as the oxygen coordination varies to accommodate denser packing. The two major discontinuities in seismic wavespeeds near depths of 410 and 660 km are controlled by such phase transitions The changes in seismic wave speed across these two discontinuities occur over just a few kilometres, and they are seen in both short-period and long-period observations. Other minor discontinuities have been proposed, but only one near 520 km appears to have some global presence in long-period stacks, although it is not seen in short-period data. This 520 km transition may occur over an extended zone, e.g., 30–50 km, so that it still appears sharp for long-period waves with wavelengths of 100 km or more. A broad ranging review of the interpretation of seismological models for the transition zone and their reconciliation with information from mineral physics is provided by Jackson & Ridgen (1998).

Frequently a definition for the lower mantle is adopted that begins below the 660 km discontinuity. However, strong gradients in seismic wavespeeds persist to depths of the order of 800 km and it seems appropriate to retain this region within the transition zone. There is increasing evidence for localised sharp transitions in seismic properties at depth around 900 km that appear to be related to the penetration of subducted material into the lower mantle.

Between 800 km and 2600 km, the lower mantle has, on average, relatively simple properties which would be consistent with the adiabatic compression of a mineral assemblage of constant chemical composition and phase. Although tomographic studies image some level of three-dimensional structure in this region the variability is much less than in the upper part of the mantle or near the base of the mantle.

The D″ layer from 2600 km to the core–mantle boundary has a distinctive character. The nature of seismic wavespeed distribution changes significantly with a sharp drop in the average velocity gradient. There is a strong increase in the level of wavespeed heterogeneity near the core–mantle boundary compared with the rest of the lower mantle. The base of the Earth's mantle is a complex zone with widespread indications of heterogeneity on many scales, discontinuities of variable character, and shear-wave anisotropy (e.g., Gurnis et al., 1998; Kennett, 2002). The results of seismic tomography give a consistent picture of the long-wavelength structure of the D″ region: there are zones of markedly lower *S* wavespeed in the central Pacific and southern Africa, whereas the Pacific is ringed by relatively fast wavespeeds that may represent a 'slab graveyard' arising from past subduction. A discordance between *P* and *S* wave results suggests the presence of chemical heterogeneity rather than just the effect of temperature (e.g., Masters et al., 2000).

1.3.2 Core

The core–mantle boundary at about 2890 km depth marks a substantial change in physical properties associated with a transition from the silicate mantle to the metallic core (see Figure 1.6). There is a significant jump in density, and a dramatic drop in *P* wavespeed from 13.7 to 8.0 km/s. The major change in wavespeed arises from the absence of shear strength in the fluid outer core, so that the *P* wave speed depends just on the bulk modulus and density. No shear waves can be transmitted through the outer core.

The process of core formation requires the segregation of heavy iron-rich components in the early stages of the accretion of the Earth (e.g., O'Neill & Palme, 1998). The core is believed to be largely composed of an iron–nickel alloy, but its density requires the presence of some lighter elemental components. A wide variety of candidates has been proposed for the light components, but it is difficult to satisfy the geochemical constraints on the nature of the bulk composition of the Earth.

The inner core appears to be solid and formed by crystallisation of material from the outer core, but it is possible that it could include some entrained fluid in the top 100 km or so. The shear wave speed for the inner core inferred from free-oscillation studies is very low and the ratio of *P* to *S* wavespeeds is comparable to that of a slurry-like material at normal pressures. The structure of the inner core is both anisotropic and shows three-dimensional variation (e.g., Creager, 1999). There is also some evidence to suggest that the central part of the inner core may have distinct properties from the rest (Ishii & Dziewonski, 2003), but this region is very difficult to sample adequately.

The fluid outer core is conducting and motions within the core create a self-sustaining dynamo which generates the main component of the magnetic field at the surface of the Earth. The dominant component of the geomagnetic field is dipolar but with significant secondary components. Careful analysis of the historic record of the variation of the magnetic field has led to a picture of the evolution of the flow in the outer part of the core (e.g., Bloxham & Gubbins, 1989). The presence of the inner core may well be important for the action of the dynamo, and electromagnetic coupling between the inner and outer cores could give rise to differential rotation between the two parts of the core (Glatzmaier & Roberts, 1996). Efforts have been made to detect this differential rotation using the time history of different classes of seismic observations but the results are currently inconclusive.

1.4 The state of the Earth

The complexity of the processes within the Earth giving rise to the presence of three-dimensional structure is indicated in Figure 1.7. We discuss many of these processes in Part II.

Heterogeneity in the mantle appears to occur on a wide range of scale lengths, from the kilometre level (or smaller) indicated by the scattering of seismic waves to thousands of kilometres in large-scale mantle convection. The mantle in Figure 1.7 is shown with large-scale convective motions (large arrows), primarily driven by subduction of dense, cold oceanic lithosphere (darker outer layer, and dark slabs). The different configurations reflect conditions in various subduction zones; including the possibility of stagnant slabs on top of the 660 km discontinuity, penetration into the lower mantle and ultimately cumulation at the core–mantle boundary. Such downwelling needs to be matched by a return flow of hotter material, this is most likely to be localised plume-like features which tend to entrain mantle material in their ascent towards the surface. Plumes which traverse the whole mantle are expected to form near or above the hottest deep regions, possibly guided by topographical features in the structure near the core-mantle boundary. The dominant upper mantle phase boundaries near 410 and 660 km depth are expected to be deflected by thermal effects or chemical heterogeneity (e.g., slabs and plumes). Other boundaries have also been detected but might not be global (e.g., the 220 and 520 km discontinuities, dashed).

The dominant lower mantle mineral structure, magnesium-silicate perovskite, is predicted to transform to a denser phase, post-perovskite (ppv), in the lowermost few hundred kilometres of the mantle (D''). If slab material is also dominated by perovskite chemistry, then subducted material may independently transform to ppv (white dashed lines near D'' in slabs). The pressure–temperature behaviour of the phase transition has yet to be fully established and is likely to be noticeably influenced by minor components. Complex structure exists near the core–mantle boundary. Large scale features with lowered seismic wavespeed are indicated by seismic tomography that are inferred to have higher density and are likely to be chemically distinct from the rest of the mantle. These dense thermo-chemical piles (DTCP in Figure 1.7) may be reservoirs of incompatible elements and act as foci for large-scale return flow in the overlying mantle. Seismological studies characterize significant reductions in shear velocity in such regions, which may well be the hottest zones in the lowermost mantle, and thus related to partially molten material

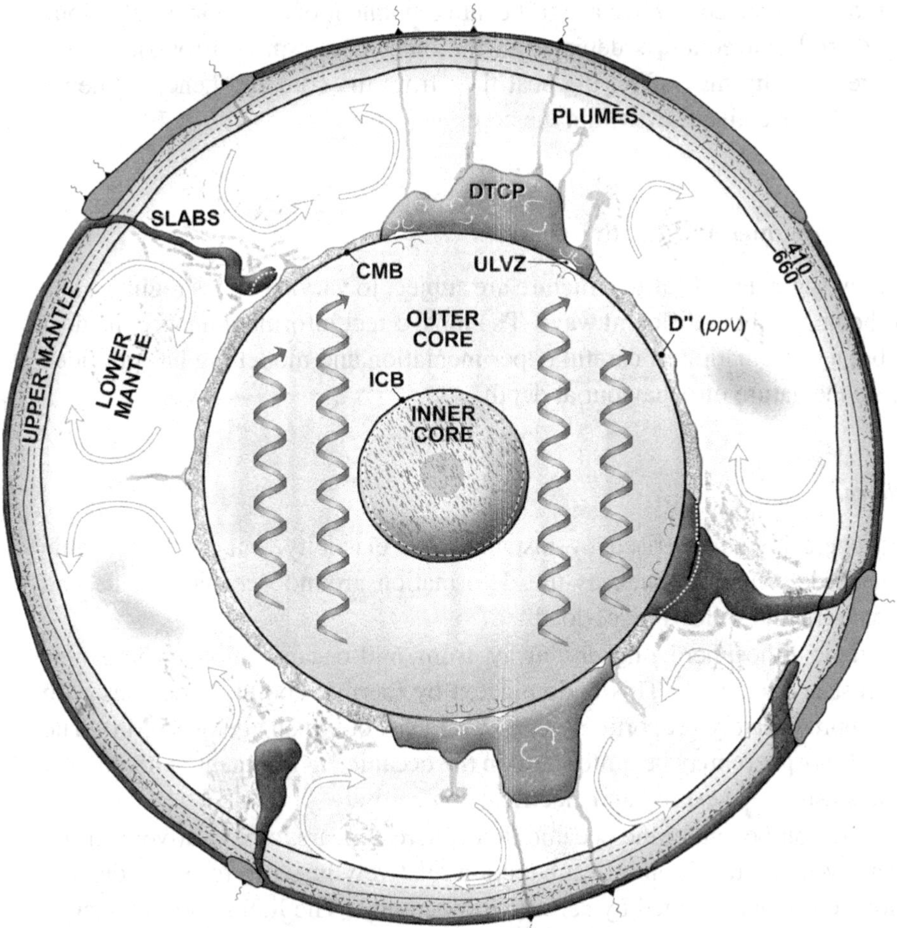

Figure 1.7. Schematic cross-section of the Earth's interior indicating a range of processes that have been indicated by recent studies [courtesy of E. Garnero].

that comprises ultra-low velocity zones (ULVZ) right at the core–mantle boundary (CMB).

Abundant evidence now exists for seismic wavespeed anisotropy (stippled or grainy areas in Figure 1.7) near the major boundary layers in the mantle: in the top few hundred kilometres below the surface, and in the lowermost few hundred kilometres of the mantle (the D″ region). The inner core is also anisotropic in its seismic properties, and has been characterized as having a fast propagation direction aligned similar to, but slightly offset from the Earth's rotation axis. The 100–200 km immediately below the inner core boundary (ICB) appears to have much reduced anisotropy compared with the rest. The innermost inner core may have its own unique subdivision (slightly darker shading).

The convective motions in the conducting outer core that give rise to the geodynamo are expected to have a significant component of columnnar behaviour. This Taylor roll convection is depicted in the outer core (spiral arrows). Lower mantle heterogeneity may affect the heat flow from the core and hence influence the pattern of convective flow within the core.

Deformation regimes and Earth dynamics

The different segments of Earth structure are subject to varying stress regimes, and respond rheologically in different ways. The most direct information is for the near surface, but a combination of careful experimentation and modelling has provided insight into the nature of behaviour at depth.

Lithosphere:

The lithosphere is characterised by instantaneous elasticity, but is also capable of long-term deformation, such as the deformation around oceanic islands and post-glacial rebound following ice-load.

The oceanic lithosphere thickens away from mid-ocean regions where new oceanic crust is generated. This is dominated by thermal cooling processes with thickness approximately proportional to $t^{1/2}$ (at least out to an age of 85 Ma). The base of the lithosphere may be quite sharp in the oceanic environment, with distinct changes in seismic wavespeed and electrical conductivity.

The mantle component of the oceanic lithosphere appears to be relatively strong since it survives the transition into subduction relatively intact to form the distinct subduction zones well-imaged by seismic tomography. The lithosphere is bent as it descends into the subduction zone and this produces shallow earthquakes near the trench. Earthquakes are generally concentrated near the top of the subducting plate close to the division between the former oceanic crust and mantle component. However, in some subduction zones such as northern Japan there is a second deeper zone of earthquakes near the centre of the subducting material.

The relative uniformity of the oceanic lithosphere is in striking contrast to the complexity of the continental environment, where the crust reflects a complex amalgamation of units dating back 3 Ga or more. Lithospheric properties are somewhat variable, but the lithosphere is significantly thinner ($<$ 120 km) under Phanerozoic belts than for the Precambrian. The resilience of the ancient components beneath the shield is achieved because they are underlain by slightly lowered densities in the lithospheric mantle; this material is highly refractory (and hence difficult to melt), but is intrinsically weak if stretched. The base of the lithosphere is only locally sharp.

The crustal component of the lithosphere is the most accessible and exhibits a range of character. In the near surface the materials are relatively brittle, but plastic deformation becomes more significant with depth. As a result earthquakes occur predominantly in the top 15 km above the brittle–ductile transition.

Asthenosphere:

Beneath the lithosphere in the upper mantle lies the asthenosphere that is more susceptible to shorter-term deformation and thus can sustain flow.

The asthenosphere generally has lowered shear wavespeed, enhanced attenuation of seismic wavespeeds and lowered apparent viscosity. These properties were originally ascribed to the presence of partial melt, but recent studies suggest that enhanced water content could produce the requisite change in physical properties. The rate of change of elastic moduli and attenuation increase significantly with temperature, and for temperatures above 1200 K the effects are noticeable even though there is no actual melt.

Seismological studies of the properties of shear waves and surface waves indicate the presence of anisotropy in mantle materials, manifested either by differences in the arrival times of shear waves of different polarisation or by angular variations in the apparent propagation speed of surface waves. The shear-wave-splitting measurements do not allow localisation of the source of anisotropy and there has been considerable debate as to whether the observations are best explained by 'frozen' anisotropy in the lithosphere reflecting past deformation or current asthenospheric flow.

Transition zone:

The properties of the transition zone are dominated by the influences of the various phase transformations in the silicate minerals of the mantle. The dominant influence comes from the transformations of olivine, but the minor minerals can play a significant role in modifying behaviour. Further, many nominally anhydrous minerals appear to be capable of incorporating significant amounts of water in their crystalline lattices, and the presence of water at depth may have a strong local influence on the behaviour of materials.

Lower mantle

The dominant mineral in the lower mantle is ferro-magnesian perovskite [$(Fe,Mg)SiO_3$] with an admixture of magnesiowustite [$(Fe,Mg)O$] and much smaller amounts of calcium- and aluminium-bearing minerals, which nevertheless may have an important influence on the seismic properties. A small fraction of the lower mantle is occupied by material that has arrived through the action of past subduction. There are relatively coherent sheet-like features as beneath the Americas, associated with the extinct Farallon plate. Elsewhere, such as in the Indonesian region, there is ponding of material down to 1000–1100 km depth. Distinct, but enigmatic, wavespeed anomalies occur to substantial depth (1800–2000 km beneath present-day Australia) but have no connection to subduction in the last 120 Ma.

This seismic evidence provides a major argument for the presence of some form of whole-mantle convection, even though some classes of geochemical information favour some degree of segregation between the upper and lower mantle.

No major phase transition occurs within the lower mantle, but there is a possibility of a change of iron partitioning with depth associated with a spin-state transition in magnesiowustite. The consequences of such subtle changes in density on convective processes have yet to be explored.

Core–mantle Boundary zone – D'':

This region just above the core–mantle boundary is highly heterogeneous on both large and small scales. The recent discovery of a post-perovskite phase transition (see, e.g., Murakami et al., 2004) provides a possible mechanism for explaining the presence of seismic discontinuities. However, the constraints on the pressure and temperature characteristics of this transition are still not tight enough for us to be confident that such a transition actually occurs within the silicate mantle. With large scale chemical heterogeneity suggested by seismic tomography, the regimes in the regions with lowered wavespeeds beneath southern Africa and the Pacific that appear to be related to major upwellings may well differ from the rest of the D'' layer.

Outer core:

The outer core is a conducting fluid with a complex pattern of flow, and is the seat of the internal magnetic field of the Earth. Direct evidence for variation in the core comes from the variations in the magnetic field at the Earth's surface, first recognised through an apparent westerly drift of the magnetic pole. Careful work on reconstructing the magnetic field patterns over the last few centuries (e.g., Bloxham & Gubbins, 1989) has been exploited to map flow patterns at the top of the core. There is not quite sufficient information to make a direct mapping, but different classes of approximation give similar results.

We have little information on the way in which the deeper parts of the outer core behave, although the analysis of the free oscillations of the Earth suggests that the overall behaviour is very close to an adiabatic state. The convective motions in the internal dynamo induce small, and time-varying, fluctuations about this state.

Inner core:

The crystallisation of the solid inner core provides substantial energy that is available to drive the flows in the outer core. The assymmetry and anisotropy of the seismic properties of the inner core suggests that the formation of crystalline material is not uniform over the surface and may reflect a rather complex pattern of growth.

Part I

Continuum Mechanics in Geophysics

2

Description of Deformation

In this Part we will introduce the concepts of continuum mechanics, starting with the description of the geometry of deformation and the notion of strain. We introduce the force field within the continuum through the stress tensor and then link it to the rheological properties of the medium through the appropriate constitutive equations. The treatment is based on the concepts of finite deformation and the results are derived in a general fashion so that the links between the descriptions of solids, fluids, and intermediate properties have a common basis. Having established the general results, we specialise to the important special cases of small deformation in the treatment of linearised elasticity and viscoelasticity. The materials deep within the Earth exist under states of both high pressures and high temperatures so we examine the way in which we can provide a suitable description that can tie to both laboratory experiments and seismological observations. We then treat the evolution of flow in a viscous fluid and the introduction of non-dimensional variables; we present some simple examples including the description of the onset of convection. We bring this Part to a close by bringing together the differential representations of the conservation of mass, momentum and energy with the necessary boundary conditions. The active core of the Earth produces the internal magnetic field that we perceive at the surface, so we need to be able to consider the interaction of continua with the electromagnetic field to describe both the highly conducting core fluid and the much lower conductivity of the silicate mantle. A section is therefore devoted to the development of continuum electrodynamics and comparisons with the simpler cases discussed in the earlier chapters.

2.1 Geometry of deformation

The pattern of deformation within a medium can be described by the geometry imposed by the change to the medium which can be recognised through the behaviour of points, lines and volumes. Such a description of deformation can be based on the transformation from the reference state to the current deformed state, or alternatively by relating the deformed state back to the reference configuration from which it was derived. This distinction between a viewpoint based on the initial (reference) configuration often called a *material description*, and the alternative *spatial description* based on the current state plays a important role in the way that different aspects of the properties of the continuum are studied.

After an arbitrary deformation of a material continuum, the amounts of compression (or expansion) and distortion of material vary with position throughout the continuum.

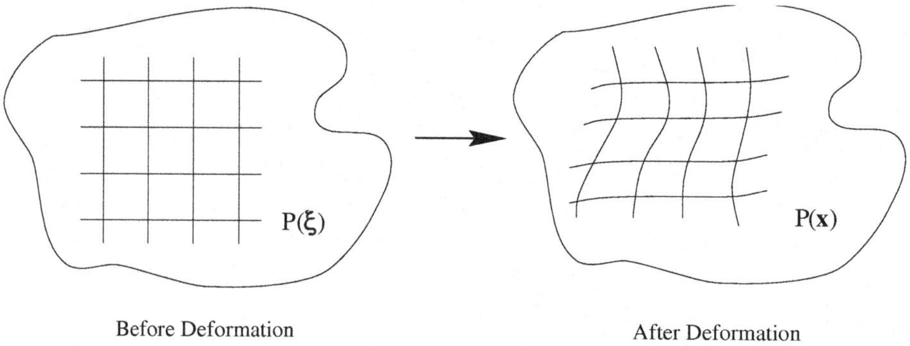

Figure 2.1. Representation of the deformation of a continuum by the relation of a grid in the current state derived from a simple configuration in the reference state.

We need to look at the deformation on a local basis and so examine the geometrical aspects in the neighbourhood of a point P. We consider any two configurations of a material continuum. One of these is then taken as a *reference state* relative to which the deformation in the other is assessed.

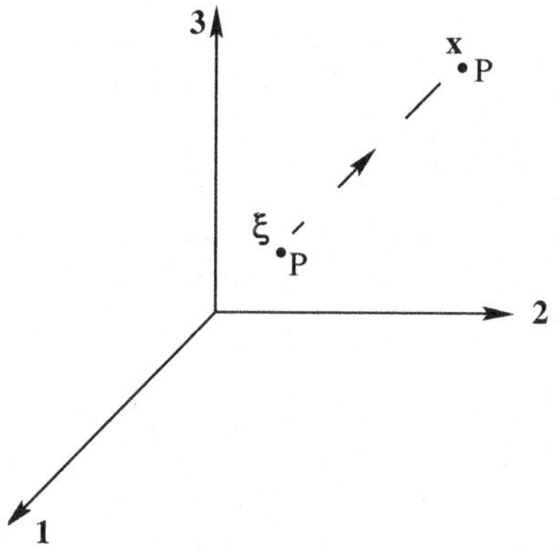

Figure 2.2. The relation of a point P in the reference state $\boldsymbol{\xi}$ and the current, deformed state $\mathbf{x}$.

We take a set of rectangular background axes and use these to specify the coordinates of a material point P (Figure 2.2):
i) in the reference state $\boldsymbol{\xi} \equiv (\xi_1, \xi_2, \xi_3)$,
ii) in the deformed state $\mathbf{x} \equiv (x_1, x_2, x_3)$.

The nature of the deformation from the reference state to the current, deformed state is specified by knowing

$$\mathbf{x} = \mathbf{x}(\boldsymbol{\xi}, t), \tag{2.1.1}$$

as a function of $\mathbf{x}$, or alternatively $\boldsymbol{\xi}$. When the functions $\mathbf{x}(\boldsymbol{\xi}, t)$ are linear, the deformation is said to be *homogeneous*; in this case planes remain planes and lines remain lines.

2.1.1 Deformation of a vector element

We can describe the local properties of the deformation, even when it varies with position, by looking at the way in which a vector element transforms between the reference and current states (Figure 2.3).

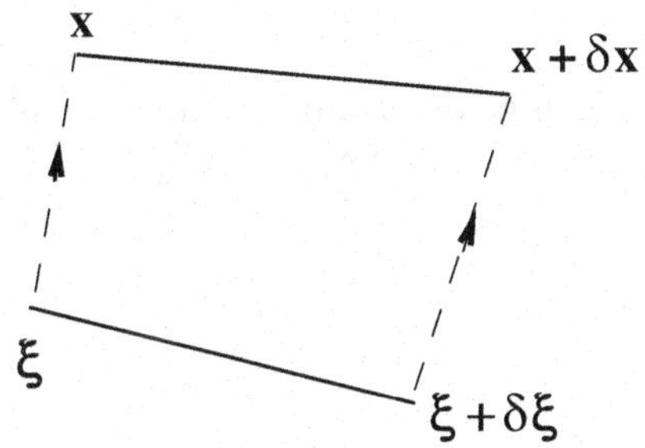

Figure 2.3. Transformation of a vector element between the reference and deformed states.

In general, near the point P, if $\boldsymbol{\xi} + \delta\boldsymbol{\xi} \to \mathbf{x} + \delta\mathbf{x}$ and $\mathbf{x}(\boldsymbol{\xi}, t)$ is differentiable,

$$dx_i = \left(\frac{\partial x_i}{\partial \xi_j}\right) d\xi_j, \tag{2.1.2}$$

or, symbolically,

$$d\mathbf{x} = \mathsf{F}(\boldsymbol{\xi}, t) d\boldsymbol{\xi}. \tag{2.1.3}$$

The matrix $\mathsf{F} \equiv (\partial x_i/\partial \xi_j)$ is called the local deformation gradient, and for a (1,1) mapping the Jacobian $J = \det \mathsf{F} \neq 0$.

Note: under change of background coordinates F transforms like a second-rank tensor. However, since F relates vectors in two different spaces it is strictly a *two-point tensor*.

The inverse transformation from the vector element in the current state to the corresponding element in the reference state is given by

$$d\boldsymbol{\xi} = \mathsf{F}^{-1}(\boldsymbol{\xi}, t) d\mathbf{x}, \quad \text{with } F^{-1}_{kl} = \partial \xi_k/\partial x_l. \tag{2.1.4}$$

The local deformation gradient F plays an important role in summarising the nature of deformation. The combination $\mathsf{F}^T\mathsf{F}$ is the metric for the deformed state relative to the reference state. $(\mathsf{F}\mathsf{F}^T)^{-1}$ is the corresponding metric for the inverse transformation. The deviations of these metric tensors from the unit diagonal tensor provide measures of strain.

2.1.2 Successive deformations

The result of successive deformations is to compound the effects of the two transformations, so that the total deformation gradient between the reference and final state is the product of the deformation gradients for the successive stages of deformation.

If $\mathbf{x} = \mathbf{x}(\mathbf{y})$ with deformation gradient $\mathsf{F}_1 = \partial\mathbf{x}/\partial\mathbf{y}$, and $\mathbf{y} = \mathbf{y}(\boldsymbol{\xi})$ with deformation gradient $\mathsf{F}_2 = \partial\mathbf{y}/\partial\boldsymbol{\xi}$, then $\mathbf{x} = \mathbf{x}(\mathbf{y}(\boldsymbol{\xi})) = \mathbf{x}(\boldsymbol{\xi})$ with deformation gradient $\mathsf{F} = \partial\mathbf{x}/\partial\boldsymbol{\xi}$, where

$$\mathsf{F} = \mathsf{F}_1\mathsf{F}_2 \qquad \text{(matrix multiplication).} \tag{2.1.5}$$

2.1.3 Deformation of an element of volume

The way in which an element of volume deforms can be determined by looking at the transformation of a local triad of vector elements (Figure 2.4).

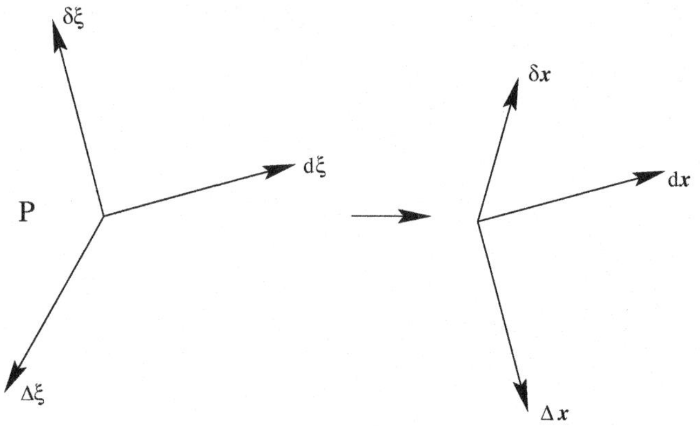

Figure 2.4. Transformation of a triad of vector elements from the reference to the current state.

The volume is given by the scalar triple product

$$\begin{aligned} dV &= [d\mathbf{x}, \delta\mathbf{x}, \Delta\mathbf{x}], \\ &= [\mathsf{F}\, d\boldsymbol{\xi}, \mathsf{F}\,\delta\boldsymbol{\xi}, \mathsf{F}\,\Delta\boldsymbol{\xi}], \\ &= \det \mathsf{F}[d\boldsymbol{\xi}, \delta\boldsymbol{\xi}, \Delta\boldsymbol{\xi}] = \det \mathsf{F}\, dV_0 = \mathsf{J}\, dV_0, \end{aligned} \tag{2.1.6}$$

i.e., as would be expected the scaling is by the Jacobian of the transformation. The condition $\mathsf{J} = \det \mathsf{F} \neq 0$ prevents physically infeasible operations such as kinking, tearing or inversion during the deformation.

If we think in terms of a fixed volume, (2.1.6) specifies the density after transformation. Since a mass element $\mathrm{d}m = \rho\,\mathrm{d}V = \rho_0\,\mathrm{d}V_0$, the density in the current state is related to that in the reference state by

$$\rho(\mathbf{x}) = J^{-1}\rho_0(\boldsymbol{\xi}) = (\det \mathsf{F})^{-1}\rho_0(\boldsymbol{\xi}). \tag{2.1.7}$$

2.1.4 Deformation of an element of area

We can use the result for the deformation of a volume element to derive the equivalent result for an element of area by considering the transformation of a skew cylinder.

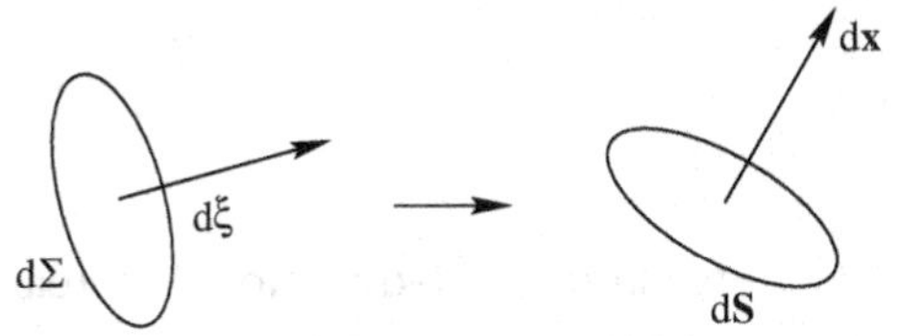

Figure 2.5. Deformation of a skew cylinder from the reference to the current state.

Consider a skew cylinder with base $\mathbf{d\Sigma}$, generators $\mathrm{d}\boldsymbol{\xi}$ (Figure 2.5). Under the deformation described by F the volume will transform to

$$\mathrm{d}V = \mathbf{dx}\cdot\mathbf{dS} = \mathbf{dx}^{\mathsf{T}}\,\mathbf{dS}. \tag{2.1.8}$$

Thus,

$$\mathrm{d}V = \mathbf{dx}^{\mathsf{T}}\,\mathbf{dS} = J\,\mathrm{d}V_0 = \det\mathsf{F}\,\mathrm{d}\boldsymbol{\xi}^{\mathsf{T}}\,\mathbf{d\Sigma}, \tag{2.1.9}$$

but $\mathrm{d}\boldsymbol{\xi} = \mathsf{F}^{-1}\mathbf{dx}$, and so

$$\mathbf{dx}^{\mathsf{T}}\,\mathbf{dS} = \mathbf{dx}^{\mathsf{T}}\,\det\mathsf{F}(\mathsf{F}^{-1})^{\mathsf{T}}\,\mathbf{d\Sigma}. \tag{2.1.10}$$

This relation must be true for any $\mathbf{dx}$ and so for any area we require

$$\mathbf{dS} = \det\mathsf{F}\,\mathsf{F}^{-\mathsf{T}}\,\mathbf{d\Sigma}. \tag{2.1.11}$$

Note: $\mathsf{F}^{-\mathsf{T}} \equiv (\mathsf{F}^{\mathsf{T}})^{-1} \equiv (\mathsf{F}^{-1})^{\mathsf{T}}$.

2.1.5 Homogeneous deformation

Since only the ratios of differentials occur in (2.1.2), such relations at any point P are formally the same as those representing *homogeneous* deformation relative to P

$$\mathbf{x} = \mathsf{F}\boldsymbol{\xi} \tag{2.1.12}$$

with F independent of $\boldsymbol{\xi}$. This mapping may be regarded as between the positions of either a **particle** (material point) or a **fibre** (material line segment). The deformation gradient F is uniquely determined by the mappings of any three non-coplanar fibres and this provides a convenient way to find F experimentally.

Some simple deformations:

(a) Dilatation: with a simple rescaling of the coordinates

$$x_i = \lambda \xi_i, \qquad i = 1,2,3 \tag{2.1.13}$$

the deformation gradient is simply $\mathsf{F} = \lambda \mathbf{I}$, where $\mathbf{I}$ is the unit matrix, and the volume change $J = \lambda^3$.

(b) Extension with lateral contraction or expansion: Suppose that

$$x_1 = \lambda_1 \xi_1, \quad x_2 = \lambda_\perp \xi_2, \quad x_3 = \lambda_\perp \xi_3, \tag{2.1.14}$$

then the deformation gradient F is given by

$$\mathsf{F} = \begin{pmatrix} \lambda_1 & 0 & 0 \\ 0 & \lambda_\perp & 0 \\ 0 & 0 & \lambda_\perp \end{pmatrix}, \quad \text{with} \quad J = \lambda_1 \lambda_\perp^2. \tag{2.1.15}$$

If $\lambda_1 > 1$ the deformation is a uniform extension in the 1-direction, whereas if $\lambda_1 < 1$ we have a uniform contraction. $\lambda_\perp$ measures the lateral contraction ($\lambda_\perp < 1$), or extension ($\lambda_\perp > 1$), in the 2–3 plane. For an *incompressible* material there can be no volume change, and so $J = \lambda_1 \lambda_\perp^2 = 1$ with the result that an extension in the 1-direction must be accompanied by a lateral contraction in the 2–3 plane.

(c) General extension: when the extensions along the three coordinate axes are different so that

$$x_i = \lambda_i \xi_i, \qquad i = 1,2,3 \tag{2.1.16}$$

the deformation gradient $\mathsf{F} = \text{diag}\{\lambda_1, \lambda_2, \lambda_3\}$, with $J = \lambda_1 \lambda_2 \lambda_3$. If $\lambda_1 = \lambda_2^{-1}$ and $\lambda_3 = 1$ there will be no change of volume, areas are preserved in planes normal to the x_3 direction, and the deformation is a *pure shear*.

EXAMPLE: DEFORMATION OF CRYSTALS

The unit cell of a crystal can be represented through a vector triad $\boldsymbol{\alpha}_i$, where the vectors are not required to be orthogonal.

Introduce the *reciprocal triad* $\boldsymbol{\beta}_j$ such that $\boldsymbol{\alpha}_i^T \boldsymbol{\beta}_j = \boldsymbol{\alpha}_i . \boldsymbol{\beta}_j = \delta_{ij}$. Show that $\boldsymbol{\alpha}_p \boldsymbol{\beta}_p^T = \mathbf{I}$ where $\mathbf{I}$ is the unit matrix.

Under homogeneous deformation the triad $\boldsymbol{\alpha}_i$ is transformed into the triad $\mathbf{a}_i$; show that the deformation gradient tensor F can be represented as $\mathsf{F} = \mathbf{a}_p \boldsymbol{\beta}_p^T$.

The reciprocal triad can be found by considering vector products, e.g., $\boldsymbol{\beta}_1 = (\mathbf{a}_2 \times \mathbf{a}_3)/(\mathbf{a}_1 . \mathbf{a}_2 \times \mathbf{a}_3)$.
These vectors have the required property that $\boldsymbol{\alpha}_i^T \boldsymbol{\beta}_j = \boldsymbol{\alpha}_i . \boldsymbol{\beta}_j = \delta_{ij}$.
Then $(\boldsymbol{\alpha}_p \boldsymbol{\beta}_p^T)\boldsymbol{\alpha}_q = \boldsymbol{\alpha}_p (\boldsymbol{\beta}_p^T \boldsymbol{\alpha}_q) = \boldsymbol{\alpha}_p \delta_{pq} = \boldsymbol{\alpha}_q$ and, since a similar relation holds for any linear combination of the $\boldsymbol{\alpha}_i$, we require $\boldsymbol{\alpha}_p \boldsymbol{\beta}_p^T = \mathbf{I}$.
Under deformation $\boldsymbol{\alpha}_i \to \mathbf{a}_i$, so $\mathbf{a}_i = \mathsf{F}\boldsymbol{\alpha}_i$ and hence $\mathbf{a}_p \boldsymbol{\beta}_p^T = \mathsf{F}\boldsymbol{\alpha}_p \boldsymbol{\beta}_p^T = \mathsf{F}$. •

Consider a crystal lattice whose elementary cell is a unit cube. A *lattice-plane* is defined to have intercepts ν_i^{-1} $(i = 1, 2, 3)$ on the cube edges from node 0, where ν_i are positive or negative integers. Suppose the crystal is deformed so that each cell becomes a parallelepiped with edges $\mathbf{a}_i$, and let a *reciprocal lattice* be defined with cell edges $\mathbf{b}_j$ such that $\mathbf{a}_i.\mathbf{b}_j = \delta_{ij}$. Show that, under arbitrary deformation, the plane remains perpendicular to an embedded direction in the reciprocal lattice, while the distance of the plane from 0 varies as $|\nu_j \mathbf{b}_j|^{-1}$.

We have initially a cubic basis $\boldsymbol{\alpha}_1, \boldsymbol{\alpha}_2, \boldsymbol{\alpha}_3$ with reciprocal triad $\boldsymbol{\beta}_1, \boldsymbol{\beta}_2, \boldsymbol{\beta}_3$. The lattice plane with intercepts ν_i^{-1} has the equation

$$\boldsymbol{\xi} = \frac{(1 - l - m)}{\nu_1}\boldsymbol{\alpha}_1 + \frac{l}{\nu_2}\boldsymbol{\alpha}_2 + \frac{m}{\nu_3}\boldsymbol{\alpha}_3,$$

and the normal satisfying

$$\boldsymbol{\nu} \cdot \left(\frac{1}{\nu_1}\boldsymbol{\alpha}_1 - \frac{1}{\nu_2}\boldsymbol{\alpha}_2\right) = \boldsymbol{\nu} \cdot \left(\frac{1}{\nu_1}\boldsymbol{\alpha}_1 - \frac{1}{\nu_3}\boldsymbol{\alpha}_3\right) = 0$$

i.e., $\boldsymbol{\xi} \cdot \boldsymbol{\nu} = 1$, is

$$\boldsymbol{\nu} = \nu_1 \boldsymbol{\beta}_1 + \nu_2 \boldsymbol{\beta}_2 + \nu_3 \boldsymbol{\beta}_3.$$

After homogeneous deformation the orthogonal triad $\boldsymbol{\alpha}_i$ is mapped to a vector triad $\mathbf{a}_i$ that will represent a parallepiped. Using an embedded basis, the equation of the plane will transform to

$$\mathbf{x} = \frac{1 - l - m}{\nu_1}\mathbf{a}_1 + \frac{l}{\nu_2}\mathbf{a}_2 + \frac{m}{\nu_3}\mathbf{a}_3,$$

and the perpendicular to the embedded directions $\mathbf{x} \cdot \mathbf{n} = 1$ is given by

$$\mathbf{n} = \nu_1 \mathbf{b}_1 + \nu_2 \mathbf{b}_2 + \nu_3 \mathbf{b}_3,$$

where $\mathbf{b}_i$ is the reciprocal triad to $\mathbf{a}_i$.
The distance to the plane varies as $1/|\mathbf{n}| = 1/|\nu_j \mathbf{b}_j|$. •

The reciprocal lattice is important in crystallography and controls the X-ray diffraction patterns from crystals.

2.2 Strain

The mechanical effects of deformation can be separated into a part which affects the length of fibres and a rotation component. The rotation can be regarded as occurring either before or after the stretch, and this leads to different algebraic representations.

The differential deformation in the neighbourhood of $\boldsymbol{\xi}$ is defined by $d\mathbf{x} = \mathbf{F}\, d\boldsymbol{\xi}$, and we now explore how this affects the length and orientation of a line segment (fibre).

2.2.1 Stretch

For a fibre in the direction $\boldsymbol{\nu}$,

$$d\boldsymbol{\xi} = |d\boldsymbol{\xi}|\boldsymbol{\nu}, \tag{2.2.1}$$

we introduce the stretch

$$\lambda_{(\nu)} = \frac{|\mathbf{dx}|}{|\mathbf{d\xi}|} = \frac{\text{final length}}{\text{initial length}}. \tag{2.2.2}$$

The length of the line segment in the deformed state

$$|\mathbf{dx}|^2 = \mathbf{dx}^{\mathsf{T}}\,\mathbf{dx} = \mathbf{d\xi}^{\mathsf{T}}\,(\mathsf{F}^{\mathsf{T}}\mathsf{F})\mathbf{d\xi}. \tag{2.2.3}$$

Consider also the transformation of two vectors such that

$$\mathbf{dx} = \mathsf{F}\,\mathbf{d\xi}, \quad \mathbf{dy} = \mathsf{F}\,\mathbf{d\eta}. \tag{2.2.4}$$

The scalar product in the current state is

$$\mathbf{dx}\cdot\mathbf{dy} = \mathbf{dx}^{\mathsf{T}}\,\mathbf{dy} = \mathbf{d\xi}^{\mathsf{T}}(\mathsf{F}^{\mathsf{T}}\mathsf{F})\mathbf{d\eta}; \tag{2.2.5}$$

the final angle between the vectors in the deformed state is $\mathbf{dx}\cdot\mathbf{dy}/(|\mathbf{dx}||\mathbf{dy}|)$. Thus $\mathsf{F}^{\mathsf{T}}\mathsf{F}$ is the metric for the deformed state relative to the reference state. Lengths and angles are unchanged if and only if F is proper orthogonal (i.e. a rotation).

The *stretch* for the direction $\boldsymbol{\nu}$ is to be found from

$$(\lambda_{(\nu)})^2 = \frac{|\mathbf{dx}|^2}{|\mathbf{d\xi}|^2} = \frac{\mathbf{d\xi}^{\mathsf{T}}(\mathsf{F}^{\mathsf{T}}\mathsf{F})\mathbf{dx}}{|\mathbf{d\xi}|^2} = \boldsymbol{\nu}^{\mathsf{T}}(\mathsf{F}^{\mathsf{T}}\mathsf{F})\boldsymbol{\nu}. \tag{2.2.6}$$

A corresponding form can be found in terms of the deformed configuration. The length of the line segment in the reference state is

$$\begin{aligned} |\mathbf{d\xi}|^2 &= \mathbf{d\xi}^{\mathsf{T}}\,\mathbf{d\xi} = \mathbf{dx}^{\mathsf{T}}(\mathsf{F}^{-1})^{\mathsf{T}}\,\mathsf{F}^{-1}\,\mathbf{dx} \\ &= \mathbf{dx}^{\mathsf{T}}(\mathsf{F}\mathsf{F}^{\mathsf{T}})^{-1}\,\mathbf{dx}. \end{aligned} \tag{2.2.7}$$

Thus, if $\mathbf{dx}$ lies in the direction $\mathbf{n}$,

$$(\lambda_{(\nu)})^2 = [\mathbf{n}^{\mathsf{T}}(\mathsf{F}\mathsf{F}^{\mathsf{T}})^{-1}\mathbf{n}]^{-1}. \tag{2.2.8}$$

2.2.2 Principal fibres and principal stretches

If we consider the eigenvectors $\mathbf{d\xi}_r$ of the local metric tensor $\mathsf{F}^{\mathsf{T}}\mathsf{F}$ we have

$$\mathsf{F}^{\mathsf{T}}\mathsf{F}\,\mathbf{d\xi}_r = \lambda_r^2\,\mathbf{d\xi}_r, \tag{2.2.9}$$

and the eigenvalues λ_r^2 satisfy $\det|\mathsf{F}^{\mathsf{T}}\mathsf{F} - \lambda_r^2\mathbf{I}| = 0$, where $\mathbf{I}$ is the unit tensor.

In the reference state $\mathbf{d\xi}^{\mathsf{T}}\,\mathsf{F}^{\mathsf{T}}\mathsf{F}\,\mathbf{d\xi} = \textit{const}$ is an ellipsoid (the Lagrangian ellipsoid) with mutually orthogonal principal axes $\pm\mathbf{d\xi}_r$. Under deformation this ellipsoid transforms to a sphere (see Figure 2.6) – for a principal fibre from (2.2.3):

$$|\mathbf{dx}_r|^2 = \mathbf{d\xi}_r^{\mathsf{T}}(\mathsf{F}^{\mathsf{T}}\mathsf{F})\mathbf{d\xi}_r = \lambda_r^2|\mathbf{d\xi}_r|^2, \tag{2.2.10}$$

and thus the λ_r are the principal stretches. The triad of principal fibres $\mathbf{d\xi}_r$ is termed the Lagrangian triad (or *material, referential* triad).

If we premultiply the eigenrelation (2.2.9) by F we obtain

$$\mathsf{F}\mathsf{F}^{\mathsf{T}}\mathsf{F}\,\mathbf{d\xi}_r = \lambda_r^2\mathsf{F}\,\mathbf{d\xi}_r, \tag{2.2.11}$$

and so

$$\mathsf{FF}^{\mathsf{T}}\,\mathbf{dx}_r = \lambda_r^2\,\mathbf{dx}_r, \tag{2.2.12}$$

so that the principal fibres are also finally orthogonal. The $\mathbf{dx}_r$ form a triad (the Eulerian or *spatial* triad) of the principal axes of an ellipsoid $\mathbf{dx}_r^{\mathsf{T}}(\mathsf{FF}^{\mathsf{T}})^{-1}\,\mathbf{dx}_r =$ *const* which results from the deformation of a sphere in the reference state (Figure 2.6).

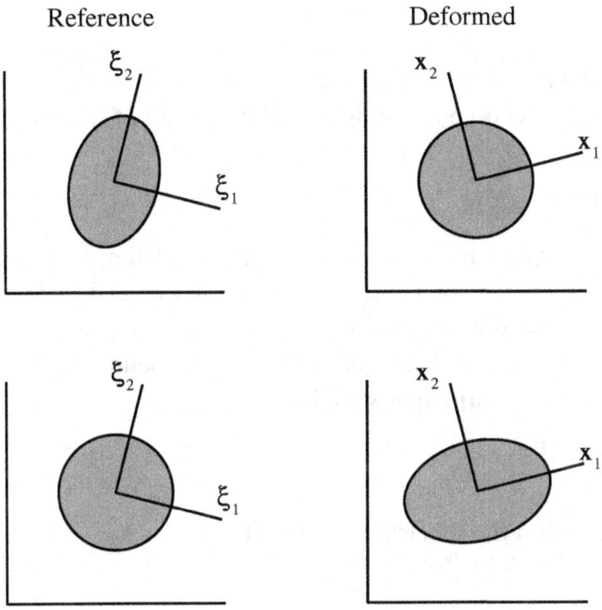

Figure 2.6. Relation of Lagrangian (material) and Eulerian (spatial) triads.

The Lagrangian triad (2.2.9) and the Eulerian triad (2.2.12) generally have different orientations as illustrated in Figure 2.6.

2.2.3 The decomposition theorem

We define the stretch tensor U to be coaxial with the Lagrangian triad of F, that is to have the same principal axes, and to have eigenvalues λ_r. The array of background components U can be constructed by the spectral formula

$$\mathsf{U} = \sum_r \lambda_r \hat{\boldsymbol{\xi}}_r \hat{\boldsymbol{\xi}}_r^{\mathsf{T}}, \qquad |\hat{\boldsymbol{\xi}}_r| = 1. \tag{2.2.13}$$

Considered as a mapping, U acts as a *pure strain*: it stretches the Lagrangian triad of F just like F itself, but maintains the orientation of the triad. Hence F can differ from U only by a final rotation

$$\mathsf{F} = \mathsf{RU}. \tag{2.2.14}$$

In the context of matrix algebra this is an example of a *polar decomposition*.

Algebraically: since F^TF is coaxial with U and has eigenvalues λ_r^2, $F^TF = U^2$. When F is given, the equation $F^TF = U^2$ can be solved uniquely for a positive definite symmetric U. Then R defined by (2.2.14) is automatically proper orthogonal

$$F^TF = U^TR^TRU = UR^TRU = U^2, \tag{2.2.15}$$

and so the rotation matrix R must satisfy

$$R^TR = I, \tag{2.2.16}$$

and $\det R > 0$ since $\det F > 0$ and $\det U > 0$.

There is also a similar decomposition with first a rotation of the principal fibres, and then a stretching:

$$F = RU = (RUR^T)R = VR. \tag{2.2.17}$$

$V = RUR^T$ is the array of background components of the pure strain that is coaxial with the Eulerian triad of F and has principal values λ_r. $V^2 = FF^T$ is the combination that we need in comparing the deformed to the reference state.

Note: The deformation gradient F can be specified independently of coordinate frame by using the triads and principal stretches

$$F = \sum_r \hat{x}_r \hat{\xi}_r^T, \quad |\hat{x}_r|, |\hat{\xi}_r| = 1, \tag{2.2.18}$$

which may be derived from the spectral form for U.

2.2.4 Pure rotation

A rotation by an angle θ about an axis $\mathbf{n}$ transforms a vector $\boldsymbol{\xi}$ to

$$\mathbf{x} = \boldsymbol{\xi} + \sin\theta(\mathbf{n}\times\boldsymbol{\xi}) + (1-\cos\theta)\mathbf{n}\times(\mathbf{n}\times\boldsymbol{\xi}). \tag{2.2.19}$$

We now introduce the skew matrix $N(\equiv \mathbf{n}\times)$ associated with $\mathbf{n}$, $N_{ik} = \epsilon_{ijk}n_j$, where the alternating symbol ϵ_{ijk} takes the value 1 if $\{i, j, k\}$ is an even permutation of $\{1,2,3\}$, the value -1 if $\{i, j, k\}$ is an odd permutation of $\{1,2,3\}$, and the value zero otherwise.

The rotation R can thus be represented as

$$R = I + N\sin\theta + N^2(1-\cos\theta). \tag{2.2.20}$$

Now, $N^2 = \mathbf{n}\mathbf{n}^T - 1$, $N\mathbf{n} = 0$ and so $N^3 = -N$, $N^4 = -N^2$. This enables us to write R in the form

$$\begin{aligned} R = \exp(N\theta) &= I + N\sum_r (-1)^r \frac{\theta^{2r+1}}{(2r+1)!} + N^2 \sum_r (-1)^r \frac{\theta^{2r}}{(2r)!}, \\ &= I + \sum_r N^r \frac{\theta^r}{r!}. \end{aligned} \tag{2.2.21}$$

Once $\mathbf{R}$ is given, the angle θ may be computed from

$$\operatorname{tr}\mathbf{R} = 1 + 2\cos\theta, \tag{2.2.22}$$

and then $\mathbf{N}$ and $\mathbf{n}$ can be found from

$$\mathbf{N}\sin\theta = \tfrac{1}{2}(\mathbf{R} - \mathbf{R}^{\mathrm{T}}). \tag{2.2.23}$$

EXAMPLE: DEFORMATION GRADIENT AND STRETCH MATRIX

The deformation gradient F is specified by

$$\mathsf{F} = \frac{1}{9}\begin{pmatrix} 7 & -5 & 2 \\ 4 & 4 & -1 \\ -5 & 7 & 2 \end{pmatrix}.$$

Show that its Lagrangian triad is coaxial with $(0,0,1)$, $(1,1,0)$, $(1,-1,0)$ and find the principal stretches. Hence construct U and U^{-1} and by constructing $\mathbf{R} = \mathsf{F}\mathsf{U}^{-1}$, verify that it is a rotation of $\pi/3$ about $(1,1,1)$. Finally show that the Eulerian triad is coaxial with $(2,-1,2)$, $(1,4,1)$, $(1,0,-1)$.

The stretch matrix is given by

$$\mathsf{U}^2 = \mathsf{F}^{\mathrm{T}}\mathsf{F} = \frac{1}{81}\begin{pmatrix} 90 & -54 & 0 \\ -54 & 90 & 0 \\ 0 & 0 & 9 \end{pmatrix},$$

The Lagrangian triad is determined by

$$\begin{aligned} 10\xi_1 - 6\xi_2 &= 9\lambda^2\xi_1 \\ -6\xi_1 + 10\xi_2 &= 9\lambda^2\xi_2 \\ \xi_3 &= 9\lambda^2\xi_3 \end{aligned}$$

and is thus coaxial with $(1,1,0)$, $(1,-1,0)$, $(0,0,1)$
with principal stretches $\frac{2}{3}$, $\frac{4}{3}$, $\frac{1}{3}$.
We can construct the stretch matrix U and its inverse U^{-1} from

$$\mathsf{U} = \lambda_1\hat{\boldsymbol{\xi}}^{(1)}\hat{\boldsymbol{\xi}}^{(1)\mathrm{T}} + \lambda_2\hat{\boldsymbol{\xi}}^{(2)}\hat{\boldsymbol{\xi}}^{(2)\mathrm{T}} + \lambda_3\hat{\boldsymbol{\xi}}^{(3)}\hat{\boldsymbol{\xi}}^{(3)\mathrm{T}},$$

and

$$\mathsf{U}^{-1} = \lambda_1^{-1}\hat{\boldsymbol{\xi}}^{(1)}\hat{\boldsymbol{\xi}}^{(1)\mathrm{T}} + \lambda_2^{-1}\hat{\boldsymbol{\xi}}^{(2)}\hat{\boldsymbol{\xi}}^{(2)\mathrm{T}} + \lambda_3^{-1}\hat{\boldsymbol{\xi}}^{(3)}\hat{\boldsymbol{\xi}}^{(3)\mathrm{T}},$$

so that

$$\mathsf{U} = \frac{1}{3}\begin{pmatrix} 3 & -1 & 0 \\ -1 & 3 & 0 \\ 0 & 0 & 1 \end{pmatrix}, \qquad \mathsf{U} = 3\begin{pmatrix} \frac{3}{8} & \frac{1}{8} & 0 \\ \frac{1}{8} & \frac{3}{8} & 0 \\ 0 & 0 & 1 \end{pmatrix}.$$

The rotation

$$\mathbf{R} = \mathsf{F}\mathsf{U}^{-1} = \frac{1}{3}\begin{pmatrix} 2 & -1 & 2 \\ 2 & 2 & -1 \\ -1 & 2 & 2 \end{pmatrix}.$$

Now $\mathrm{tr}\mathbf{R} = 2 = 1 + 2\cos\theta$ and so $\cos\theta = \frac{1}{2}$, hence $\theta = \pi/3$. Further

$$\mathbf{R}\begin{pmatrix}1\\1\\1\end{pmatrix} = \begin{pmatrix}1\\1\\1\end{pmatrix},$$

so $(1,1,1)$ is the rotation axis.

To obtain the Eulerian triad we operate on the Lagrangian triad with $\mathbf{R}$,

$$\mathbf{R}\begin{pmatrix}1 & 1 & 0\\ 1 & -1 & 0\\ 0 & 0 & 1\end{pmatrix} = \frac{1}{3}\begin{pmatrix}1 & 3 & 2\\ 4 & 0 & -1\\ 1 & -3 & 2\end{pmatrix}$$

and is coaxial with $(1,4,1)$, $(1,0,-1)$, $(2,-1,2)$. •

2.2.5 *Tensor measures of strain*

We have seen that the change in shape of any cell can be calculated when we know which three fibres in the initial configuration are principal, and by how much each is stretched. Moreover, this is the minimal information for this purpose.

In this sense a suitable strain measure can be found by looking at the change in the length of $\mathbf{d\xi}$, using (2.2.3) and (2.2.7)

$$|\mathbf{dx}|^2 - |\mathbf{d\xi}|^2 = \mathbf{d\xi}^{\mathsf{T}}\,\mathsf{F}^{\mathsf{T}}\mathsf{F}\,\mathbf{d\xi} - \mathbf{d\xi}^{\mathsf{T}}\,\mathbf{d\xi} = 2\,\mathbf{d\xi}^{\mathsf{T}}\,\mathsf{E}\,\mathbf{d\xi}, \tag{2.2.24}$$

$$= \mathbf{dx}^{\mathsf{T}}\,\mathbf{dx} - \mathbf{dx}^{\mathsf{T}}(\mathsf{F}\mathsf{F}^{\mathsf{T}})^{-1}\,\mathbf{dx} = 2\,\mathbf{dx}^{\mathsf{T}}\,\mathbf{e}\,\mathbf{dx}. \tag{2.2.25}$$

Here we have introduced the *Green strain*,

$$\mathsf{E} = \tfrac{1}{2}(\mathsf{F}^{\mathsf{T}}\mathsf{F} - \mathbf{I}) = \tfrac{1}{2}(\mathsf{U}^2 - \mathbf{I}), \tag{2.2.26}$$

and the *Cauchy strain*,

$$\mathbf{e} = \tfrac{1}{2}(\mathbf{I} - (\mathsf{F}\mathsf{F}^{\mathsf{T}})^{-1}) = \tfrac{1}{2}(\mathbf{I} - \mathbf{V}^{-2}), \tag{2.2.27}$$

associated with some specified background frame.

The components E of the *Green strain* generate a tensor whose axes define the reference directions of the Lagrangian principal fibres and whose principal values are

$$\tfrac{1}{2}(\lambda_r^2 - 1) \quad r = 1,2,3. \tag{2.2.28}$$

This strain measure is *objective* (i.e. unaffected by any final rotation) and it vanishes when, and only when, the shape is unchanged.

The normalising factor $\frac{1}{2}$ is included so that when the strain is infinitesimal, the principal values of the measure are just the fractional changes in length of the principal fibres; since, if $\lambda_r = 1 + \epsilon$,

$$\tfrac{1}{2}[(1+\lambda)^2 - 1] \approx \epsilon, \tag{2.2.29}$$

when ϵ is a first-order quantity.

We have comparable properties for the *Cauchy strain* $\mathbf{e}$ in terms of the Eulerian triad, where the principal axes lie along the Eulerian principal fibres and the principal values are

$$\tfrac{1}{2}(1 - \lambda_r^{-2}) \qquad r = 1, 2, 3. \tag{2.2.30}$$

Once again for infinitesimal strain the principal values reduce to the fractional changes of length of the principal fibres.

Displacement representations:
We introduce the displacement $\mathbf{u}$ between the deformed and reference configurations

$$\mathbf{u} = \mathbf{x} - \boldsymbol{\xi}, \tag{2.2.31}$$

and its gradients,

$$\frac{\partial \mathbf{u}}{\partial \boldsymbol{\xi}} = \mathsf{F} - \mathbf{I}, \quad \mathsf{F} = \mathbf{I} + \frac{\partial \mathbf{u}}{\partial \boldsymbol{\xi}}, \tag{2.2.32}$$

$$\frac{\partial \mathbf{u}}{\partial \mathbf{x}} = \mathbf{I} - \mathsf{F}^{-1}, \quad \mathsf{F}^{-1} = \mathbf{I} - \frac{\partial \mathbf{u}}{\partial \mathbf{x}}. \tag{2.2.33}$$

In terms of these displacement gradients the Green strain tensor is given by

$$\begin{aligned} 2\mathsf{E} = \mathsf{F}\mathsf{F}^T - \mathbf{I} &= \left[\mathbf{I} + \frac{\partial \mathbf{u}}{\partial \boldsymbol{\xi}}\right]\left[\mathbf{I} + \frac{\partial \mathbf{u}}{\partial \boldsymbol{\xi}}\right]^T - \mathbf{I}, \\ &= \left[\frac{\partial \mathbf{u}}{\partial \boldsymbol{\xi}}\right] + \left[\frac{\partial \mathbf{u}}{\partial \boldsymbol{\xi}}\right]^T + \left[\frac{\partial \mathbf{u}}{\partial \boldsymbol{\xi}}\right]\left[\frac{\partial \mathbf{u}}{\partial \boldsymbol{\xi}}\right]^T, \end{aligned} \tag{2.2.34}$$

and the Cauchy strain by

$$\begin{aligned} 2\mathsf{e} = \mathbf{I} - \mathsf{F}^{-T}\mathsf{F}^{-1} &= \mathbf{I} - \left[\mathbf{I} - \frac{\partial \mathbf{u}}{\partial \mathbf{x}}\right]^T\left[\mathbf{I} - \frac{\partial \mathbf{u}}{\partial \mathbf{x}}\right], \\ &= \left[\frac{\partial \mathbf{u}}{\partial \mathbf{x}}\right] + \left[\frac{\partial \mathbf{u}}{\partial \mathbf{x}}\right]^T - \left[\frac{\partial \mathbf{u}}{\partial \mathbf{x}}\right]\left[\frac{\partial \mathbf{u}}{\partial \mathbf{x}}\right]^T. \end{aligned} \tag{2.2.35}$$

The cartesian components of the strain tensors are therefore given by
i) Green strain tensor E:

$$\mathsf{E}_{ij} = \frac{1}{2}\left[\frac{\partial u_i}{\partial \xi_j} + \frac{\partial u_j}{\partial \xi_i} + \frac{\partial u_k}{\partial \xi_i}\frac{\partial u_k}{\partial \xi_j}\right], \tag{2.2.36}$$

and
ii) Cauchy strain tensor e:

$$e_{ij} = \frac{1}{2}\left[\frac{\partial u_i}{\partial x_j} + \frac{\partial u_j}{\partial x_i} - \frac{\partial u_k}{\partial x_i}\frac{\partial u_k}{\partial x_j}\right]. \tag{2.2.37}$$

When the displacements and displacement gradients are sufficiently small, the distinction between the two small-strain definitions is usually ignored, since the second-order terms are then unimportant.

2.3 Plane deformation

Introduce a unit vector **n**, and then in a deformation with displacement

$$\mathbf{u} = \mathbf{x} - \boldsymbol{\xi} = \gamma\mathbf{m}(\mathbf{n} \cdot \boldsymbol{\xi}) = \gamma\mathbf{m}(\mathbf{n}^T\boldsymbol{\xi}), \tag{2.3.1}$$

particles are not displaced in the *basal plane* $\mathbf{n}^T\boldsymbol{\xi} = 0$ irrespective of the direction of the unit vector **m**. The deformation gradient

$$\mathsf{F} = \mathbf{I} + \gamma\mathbf{m}\mathbf{n}^T, \tag{2.3.2}$$

and the amount of deformation scales with the height of $\boldsymbol{\xi}$ above the basal plane $(\mathbf{n}.\boldsymbol{\xi})$ (Figure 2.7).

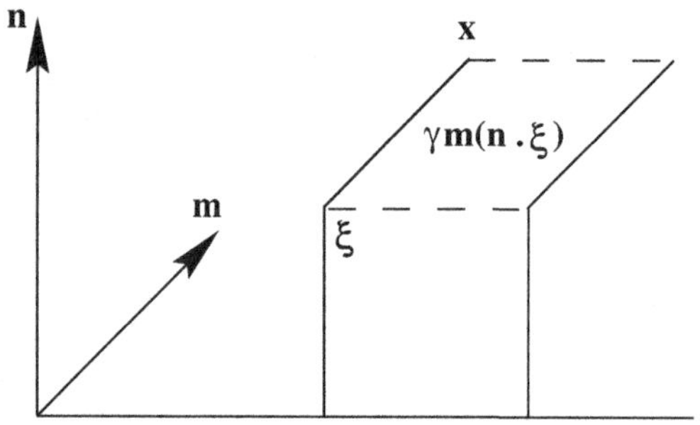

Figure 2.7. Representation of plane deformation in the direction **m** from a basal plane with normal **n**.

Shear

An equivoluminal plane deformation is called a shear, the principal stretches are then such that

$$\lambda_3 = 1 \quad \text{with} \quad \lambda_1\lambda_2 = 1. \tag{2.3.3}$$

If $\mathbf{m}^T\mathbf{n} = 0$ the translation components in (2.3.1) are parallel to the basal plane and volume is conserved. This is *simple shear*

$$\mathsf{F} = \mathbf{I} + \gamma\mathbf{m}\mathbf{n}^T, \quad \mathbf{m}^T\mathbf{n} = 0. \tag{2.3.4}$$

γ is called the amount of shear and a plane $\mathbf{n}^T\boldsymbol{\xi} = const$ a shearing plane.

The simple shear process has the effect of rotating the principal strain axes, as illustrated in Figure 2.8, and sets of initially perpendicular lines will be sheared and end up at an oblique angle.

In contrast for the pure shear situation in which there is simultaneous lengthening and perpendicular shortening, such that $\lambda_1\lambda_2 = 1$, the principal axes of strain remain fixed (Figure 2.8). Any set of perpendicular lines inclined to the principal axes will suffer shear strain and be no longer orthogonal after deformation.

In each case the two-dimensional area and three-dimensional volume will remain unchanged in the homogeneous shear.

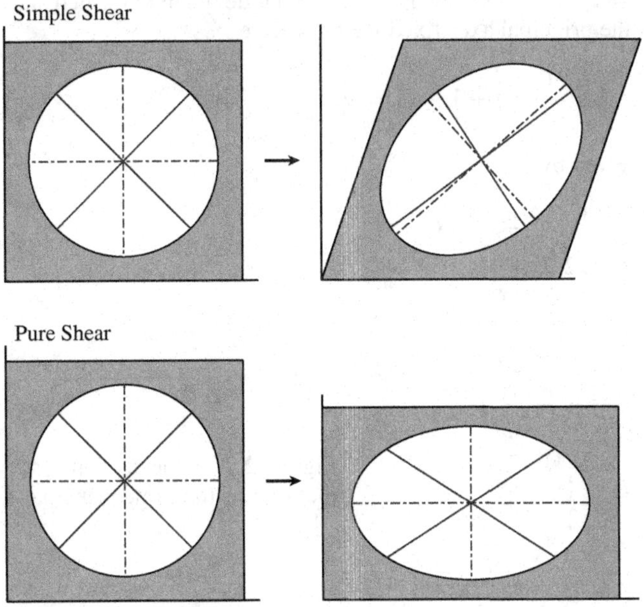

Figure 2.8. Illustration of the action of a simple shear and a pure shear process. In each homogeneous deformation the area of a body is preserved. The principal strain axes shown by chain dotted lines rotate for simple shear, but remain fixed for pure shear. The solid reference lines are shear strained in each case and are no longer orthogonal after deformation.

EXAMPLE: A SHEAR PROCESS

Consider a type of shearing

$$x_1 = s\xi_1 + t\xi_2/s, \quad x_2 = \xi_2/s, \quad x_3 = \xi_3, \tag{2.3.5}$$

with $s \geq 1, t \geq 0$ ($s \equiv 1$ is a path of simple shear). Show that any *resultant* simple shear can be obtained by a path of this type in (s, t) space such that the same fibres are principal at *every* stage.

We have a homogeneous deformation gradient

$$\mathsf{F} = \begin{pmatrix} s & t/s \\ 0 & 1/s \end{pmatrix}, \quad \det \mathsf{F} = 1.$$

We require the principal axes of

$$\mathsf{F}^T\mathsf{F} = \begin{pmatrix} s^2 & t \\ t & (t^2+1)/s^2 \end{pmatrix}$$

to be fixed. For a matrix of the form

$$\begin{pmatrix} a & c \\ c & b \end{pmatrix},$$

the inclinations of the principal axes depend only on $(a - b)/c$.

Because we require the final state to be a simple shear the trajectory in (s, t) space has to pass through $s = 1, t = \gamma$ and $s = 1, t = 0$ to include the undeformed state. To keep the inclination of the principal axes fixed we must have

$$\frac{1}{t}\left(\frac{t^2+1}{s^2} - s^2\right) = \frac{1}{\gamma}\left(\gamma^2 + 1 - 1\right) = \gamma,$$

and so the path is given by

$$s^4 + \gamma s^2 t - t^2 = 1, \quad 0 \le t \le \gamma,$$

as illustrated in Figure 2.9.

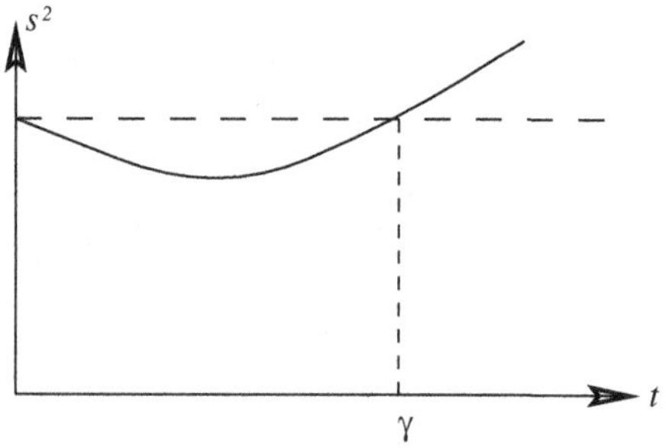

Figure 2.9. Trajectory in (s,t) space that retains the same principal fibres.

2.4 Motion

We can follow the motion of a particle by tracking its position relative to its reference position $\boldsymbol{\xi}$ (a Lagrangian viewpoint), or alternatively we may follow the particle directly in an Eulerian viewpoint. In a similar way we can represent a field $q^L(\boldsymbol{\xi}, t)$ in terms of the Lagrangian frame so that q^L follows the particle initially at $\boldsymbol{\xi}$. Or, alternatively, we can work in terms of the current configuration of the medium with an Eulerian description $q^E(\mathbf{x}, t)$ in terms of a fixed position $\mathbf{x}$. Because both representations link the value of q for the particle labelled by $\boldsymbol{\xi}$ which is at $\mathbf{x}$ at time t:

$$q^E(\mathbf{x}(\boldsymbol{\xi}, t), t) = q^L(\boldsymbol{\xi}, t). \tag{2.4.1}$$

When we follow a given particle $\boldsymbol{\xi}$ (a Lagrangian viewpoint) then the time rate of change of interest is

$$\left.\frac{\partial}{\partial t}\right]_{\xi}, \tag{2.4.2}$$

but, using the chain rule, we find that the 'material time derivative'

$$\frac{D}{Dt}\psi \equiv \left.\frac{\partial}{\partial t}\right]_{\xi}\psi = \left.\frac{\partial}{\partial t}\right]_{\mathbf{x}}\psi + \left.\frac{\partial}{\partial t}\right]_{\xi} x_i \frac{\partial}{\partial x_i}\psi, \tag{2.4.3}$$

in terms of an Eulerian decomposition.

We define the *velocity field* corresponding to a particle initially at $\boldsymbol{\xi}$ as

$$\mathbf{v} = \left.\frac{\partial}{\partial t}\right]_{\xi} \mathbf{x}(\boldsymbol{\xi}, t) \equiv \frac{D}{Dt}\mathbf{x}. \tag{2.4.4}$$

The *acceleration field* is then

$$\mathbf{f} = \left.\frac{\partial}{\partial t}\right]_{\xi} \mathbf{v}(\boldsymbol{\xi}, t) \equiv \frac{D}{Dt}\mathbf{v}, \tag{2.4.5}$$

and in the Eulerian viewpoint

$$\mathbf{f} = \left.\frac{\partial}{\partial t}\right]_{\mathbf{x}} \mathbf{v}(\mathbf{x}, t) + \left[\mathbf{v} \cdot \frac{\partial}{\partial \mathbf{x}}\right] \mathbf{v}, \tag{2.4.6}$$

where we have written $\partial/\partial\mathbf{x}$ for the gradient operator in the Eulerian configuration (for fluids this is the most common viewpoint to use).

In general, then the mobile derivative following the particles is

$$\frac{D}{Dt} \equiv \left.\frac{\partial}{\partial t}\right]_{\mathbf{x}} + \left[\mathbf{v} \cdot \frac{\partial}{\partial \mathbf{x}}\right]. \tag{2.4.7}$$

If we consider a line element $d\mathbf{x} = \mathsf{F}\, d\boldsymbol{\xi}$, its material derivative is

$$\frac{D}{Dt}d\mathbf{x} = d\boldsymbol{\xi} \cdot \frac{\partial}{\partial \boldsymbol{\xi}}\mathbf{v}^{L} = \left(d\mathbf{x} \cdot \frac{\partial}{\partial \mathbf{x}}\right)\mathbf{v}, \tag{2.4.8}$$

a result that can also be derived from geometrical considerations.

EXAMPLE: TIME-DEPENDENT DEFORMATION

A body undergoes the deformation specified by

$$x_1 = \xi_1(1 + a^2t^2), \quad x_2 = \xi_2, \quad x^3 = \xi_3.$$

Find the displacement and velocity in both the material and the spatial description.

The displacement $\mathbf{u} = \mathbf{x} - \boldsymbol{\xi}$ in the material description is

$$u_1 = \xi_1 a^2 t^2, \quad u_2 = 0, \quad u_3 = 0,$$

and in the spatial description

$$u_1 = x_1 a^2 t^2/(1 + a^2t^2), \quad u_2 = 0, \quad u_3 = 0.$$

The velocity, $\partial\mathbf{u}/\partial t)_{\xi}$ in the material description is

$$v_1 = 2\xi_1 a^2 t, \quad v_2 = 0, \quad v_3 = 0,$$

and in the spatial description is

$$v_1 = 2x_1 a^2 t/(1 + a^2t^2), \quad v_2 = 0, \quad v_3 = 0.$$

2.5 The continuity equation

Consider a volume V_0 in the undeformed state and monitor its state during deformation. The conservation of mass requires that the integral

$$M_0 = \int_{V_0} \rho_0(\boldsymbol{\xi})\, dV_0 \tag{2.5.1}$$

should remain invariant. In terms of the deformed state, the mass

$$M_0 = \int_{V} \rho(\mathbf{x}, t)\, dV = \int_{V_0} \rho(\mathbf{x}, t)\, J(t)\, dV_0, \tag{2.5.2}$$

where we have used (2.1.6). If we now equate the two expressions for M_0 we have the expression for the conservation of mass in the Lagrangian representation

$$\int_{V_0} [\rho_0(\boldsymbol{\xi}) - \rho(\mathbf{x}, t)J(t)]\, dV_0 = 0. \tag{2.5.3}$$

This relation must be true for any volume V_0 and so the integrand must vanish, which gives an alternative derivation of the density relation deduced above in section 2.1.3.

Now the density in the undeformed state is a constant whatever the deformation, and so

$$\left.\frac{\partial}{\partial t}\right]_{\xi} \rho_0(\boldsymbol{\xi}) = \left.\frac{\partial}{\partial t}\right]_{\xi} \rho(\mathbf{x}, t)J(t) = \frac{D}{Dt}\rho(\mathbf{x}, t)J(t) = 0. \tag{2.5.4}$$

The material time derivative of density can therefore be found from

$$\frac{D}{Dt}\rho(\mathbf{x}, t) + \rho(\mathbf{x}, t)J^{-1}(t)\frac{D}{Dt}J(t) = 0. \tag{2.5.5}$$

From the properties of the determinant of the deformation gradient

$$\frac{D}{Dt}J(t) = \left.\frac{\partial}{\partial t}\right]_{\xi} J(t) = J(t)\frac{\partial}{\partial \mathbf{x}} \cdot \mathbf{v}(\mathbf{x}, t). \tag{2.5.6}$$

in terms of the velocity field $\mathbf{v}(\mathbf{x}, t)$. The material time derivative of the density is thus determined by

$$\frac{D}{Dt}\rho(\mathbf{x}, t) + \rho(\mathbf{x}, t)\frac{\partial}{\partial \mathbf{x}} \cdot \mathbf{v}(\mathbf{x}, t) = 0. \tag{2.5.7}$$

When we make use of the representation (2.4.7) for the material time derivative we can write the Eulerian form of the equation of mass conservation in the form

$$\left.\frac{\partial}{\partial t}\right]_{\mathbf{x}} \rho(\mathbf{x}, t) + \frac{\partial}{\partial \mathbf{x}} \cdot [\rho(\mathbf{x}, t)\mathbf{v}(\mathbf{x}, t)] = 0, \tag{2.5.8}$$

which is known as the continuity equation. The relation (2.5.8) can also be deduced by considering a fixed volume in space and then allowing for mass flux across the boundary.

Material derivative of a volume integral

Consider some physical quantity Ψ (scalar, vector or tensor) associated with the continuum so that ψ is the quantity per unit mass. The material rate of change of Ψ for a fixed material volume V with surface S and outward normal $\hat{\mathbf{n}}$ must be equated to the rate of change of Ψ for the particles instantaneously within V and the net flux of Ψ into V across the surface S so that

$$\frac{d}{dt}\int_V \Psi \, dV = \int_V \frac{\partial}{\partial t}\Psi \, dV + \int_S \Psi \mathbf{v} \cdot \hat{\mathbf{n}} \, dS. \tag{2.5.9}$$

If we now represent Ψ as $\psi\rho$ we have

$$\frac{d}{dt}\int_V \psi\rho \, dV = \int_V \frac{\partial}{\partial t}(\psi\rho) \, dV + \int_S \psi\rho\mathbf{v} \cdot \hat{\mathbf{n}} \, dS. \tag{2.5.10}$$

We apply the divergence theorem to the surface integral in (2.5.10), and then

$$\begin{aligned}\frac{d}{dt}\int_V \psi\rho \, dV &= \int_V \left\{\frac{\partial}{\partial t}(\psi\rho) + \frac{\partial}{\partial \mathbf{x}}(\psi\rho\mathbf{v})\right\} dV \\ &= \int_V \left\{\psi\left[\frac{\partial}{\partial t}\rho + \frac{\partial}{\partial \mathbf{x}}(\rho\mathbf{v})\right] + \rho\left[\frac{\partial}{\partial t}\psi + \mathbf{v}\cdot\frac{\partial}{\partial \mathbf{x}}\psi)\right]\right\} dV. \end{aligned} \tag{2.5.11}$$

From the continuity equation (2.5.8) the first contribution to the volume integral on the right-hand side of (2.5.11) vanishes, and the second can be recognised as $\rho \, D\psi/Dt$. Hence

$$\frac{d}{dt}\int_V \rho\psi \, dV = \int_V \rho \frac{D}{Dt}\psi \, dV; \tag{2.5.12}$$

for $\psi = 1$ we recover the intergal form of the continuity condition (2.5.8).

Appendix: Properties of the determinant of the deformation gradient

In terms of the deformation gradient $\mathbf{F} = \partial\mathbf{x}/\partial\boldsymbol{\xi}$, $F_{ij} = \partial x_i/\partial\xi_j$, the determinant

$$\begin{aligned} J = \det \mathbf{F} &= \tfrac{1}{6}\epsilon_{ijk}\epsilon_{pqr}F_{pi}F_{qj}F_{rk} \qquad (2A.1)\\ &= \tfrac{1}{6}\epsilon_{ijk}\epsilon_{pqr}\frac{\partial x_p}{\partial\xi_i}\frac{\partial x_q}{\partial\xi_j}\frac{\partial x_r}{\partial\xi_k}. \qquad (2A.2)\end{aligned}$$

Now, from the orthogonality relation

$$\frac{\partial x_i}{\partial\xi_j}\frac{\partial\xi_j}{\partial x_k} = \delta_{ik}, \tag{2A.3}$$

we can apply Cramer's rule for the solution of linear equations to obtain

$$\frac{\partial\xi_j}{\partial x_k} = \frac{1}{2J}\epsilon_{jmp}\epsilon_{kqr}\frac{\partial x_p}{\partial\xi_i}\frac{\partial x_q}{\partial\xi_j}, \tag{2A.4}$$

and thus we can recognise that

$$\frac{\partial J}{\partial F_{ij}} = \frac{\partial J}{\partial(\partial x_i/\partial\xi_j)} = J\frac{\partial\xi_j}{\partial x_i}. \tag{2A.5}$$

We can therefore establish the important relations

$$\frac{\partial}{\partial x_k}\left(\frac{1}{J}\frac{\partial x_j}{\partial \xi_k}\right) = 0, \tag{2A.6}$$

and

$$\left.\frac{\partial}{\partial t}\right]_{\xi} J = J\frac{\partial \xi_j}{\partial x_i}.\left.\frac{\partial}{\partial t}\right]_{\xi}\frac{\partial x_i}{\partial \xi_j} = J\frac{\partial}{\partial x_i}\left.\frac{\partial}{\partial t}\right]_{\xi} x_i = J\frac{\partial}{\partial \mathbf{x}}.\mathbf{v}\,, \tag{2A.7}$$

which is used in (2.5.6).

3

The Stress-Field Concept

The stress tensor provides a description of the force distribution through the continuum and is most readily treated in the deformed state (an Eulerian treatment). We suppress the details of complex interatomic forces and make the simplifying assumption that the effect of all the forces acting on a given surface may be represented through a single vector field defined over the surface.

3.1 Traction and stress

Consider any point P of a continuum and any infinitesimal plane element δS with P as centroid. Let $\mathbf{x}$ be the position vector of P, and $\mathbf{n}$ the unit normal to δS (in some consistent sense).

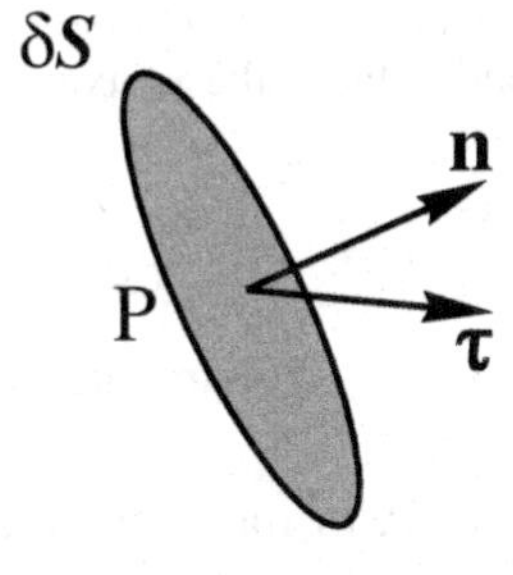

Figure 3.1. The traction $\boldsymbol{\tau}$ acting on an element with normal $\mathbf{n}$.

Then the mechanical influence exerted across $\delta\mathbf{S}$ **by** the material on the side to which $\mathbf{n}$ points is postulated to be represented by a force

$$\boldsymbol{\tau}(\mathbf{n}, \mathbf{x})\delta\mathbf{S} \qquad (3.1.1)$$

through P. The negative of this force is exerted across $\delta\mathbf{S}$ **on** this material. $\boldsymbol{\tau}$ is called the *traction vector* and has the dimensions of force per unit area.

Note: a mutual couple could also be introduced across $\delta\mathbf{S}$, but the need for it has never been indisputably established in any particular situation.

The properties of traction can be determined from a hypothesis due substantially to Euler: the vector fields of force and mass-acceleration in a continuum have equal linear and rotational resultants over any part of the continuum.

Consider a volume V, bounded by a surface S with outward normal $\mathbf{n}$, and dm a generic mass element in V, and set

$\mathbf{f}$ – local acceleration,
$\mathbf{g}$ – local body force per unit mass (e.g., gravitational or fictitious),
$\boldsymbol{\mu}$ – local body moment per unit mass (e.g., electrostatic or magnetic).

Then we can write the equations of conservation of linear and angular momentum in the form

$$\int_S \boldsymbol{\tau}(\mathbf{x})\, dS + \int_V \mathbf{g}(\mathbf{x})\, dm = \int_V \mathbf{f}(\mathbf{x})\, dm, \tag{3.1.2}$$

$$\int_S \mathbf{x} \times \boldsymbol{\tau}(\mathbf{x})\, dS + \int_V \mathbf{x} \times \mathbf{g}(\mathbf{x})\, dm + \int_V \boldsymbol{\mu}(\mathbf{x})\, dm = \int_V \mathbf{x} \times \mathbf{f}(\mathbf{x})\, dm. \tag{3.1.3}$$

Tensor character of stress

At a point P consider a plane element perpendicular to the background axis x_i. Let $\boldsymbol{\sigma}_i$ denote the vector traction exerted over this element by the material situated on its positive side (Figure 3.2).

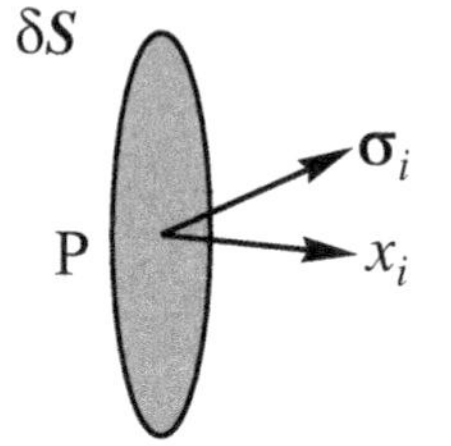

Figure 3.2. Traction $\boldsymbol{\sigma}_i$ acting on an element perpendicular to the x_i axis.

The components σ_{ij} of $\boldsymbol{\sigma}_i$ constitute the *stress-matrix*. σ_{11}, σ_{22}, σ_{33} are called the *normal components* and the others *shear components*.

The stress matrix fully specifies the mechanical state at P, in the sense that it determines the traction $\boldsymbol{\tau}$ for any $\mathbf{n}$. Consider an infinitesimal tetrahedron at P with three faces parallel to the coordinate planes and the fourth with outward normal $\mathbf{n}$ and area ϵ^2 (Figure 3.3).

We now apply the equation (3.1.2) for the linear force resultant to this volume. Suppose that $\mathbf{n}$ falls in the first octant; the facial areas are then

$$(n_1, n_2, n_3, 1)\epsilon^2. \tag{3.1.4}$$

The surface integral over the tetrahedron is thus

$$\int_S \boldsymbol{\tau}\, dS = (\boldsymbol{\tau} - n_i \boldsymbol{\sigma}_i)\epsilon^2 + O(\epsilon^3) = \int_V (\mathbf{f} - \mathbf{g})\, dm. \tag{3.1.5}$$

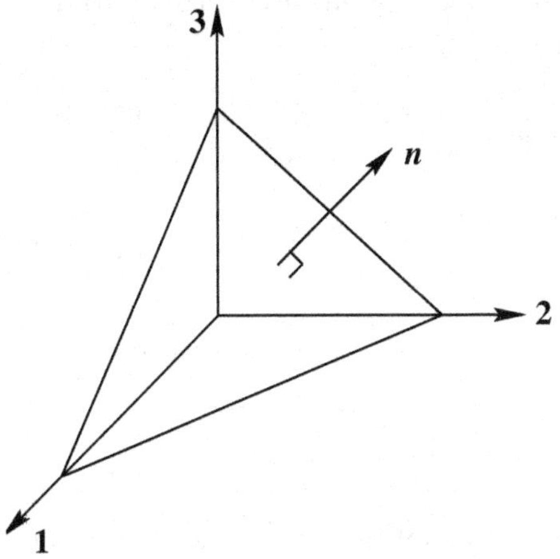

Figure 3.3. Resultant forces on an elementary tetrahedron

The same value is obtained when $\mathbf{n}$ falls in any other octant, or when $\mathbf{n}$ is perpendicular to one axis (in which case the tetrahedron is replaced by a triangular prism).

Now the mass integrals over the small volume are of order $O(\epsilon^3)$ because we can take $\mathbf{f}$, $\mathbf{g}$ constant within the volume, and so, as $\epsilon \to 0$,

$$\int_S \boldsymbol{\tau}\, dS \to 0, \tag{3.1.6}$$

Thus

$$\boldsymbol{\tau}(\mathbf{n}) = n_i \boldsymbol{\sigma}_i, \quad \text{i.e.} \quad \tau_j(\mathbf{n}) = n_i \sigma_{ij}. \tag{3.1.7}$$

which is known as *Cauchy's relation.*

We note that (3.1.7) also implies that σ_{ij} transforms like a second-rank tensor. Consider a new set of rectangular axes indicated by stars and set l_{pi} to be the direction cosine between the pth starred axis and the ith old axis (Figure 3.4).

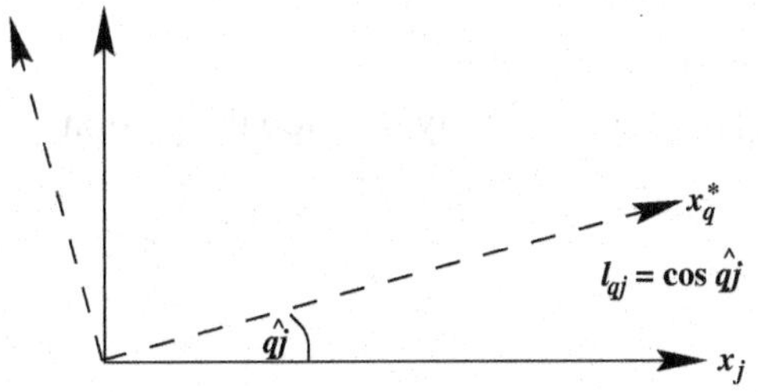

Figure 3.4. Rotation of coordinate axes

Then

$$\sigma^*_{pq} = \sigma_{ij} l_{pi} l_{qj}, \tag{3.1.8}$$

since

$$\tau^*_q = l_{qj} \tau_j, \quad \boldsymbol{\sigma}_i = l_{pi} \boldsymbol{\sigma}^*_p, \quad \text{as} \quad l_{ip} l_{jp} = \delta_{ij}. \tag{3.1.9}$$

Alternatively, in terms of the rotation R which carries the unstarred axes into the starred ones

$$R_{ip} = l_{pi}, \tag{3.1.10}$$

and

$$\sigma^* = R^T \sigma R. \tag{3.1.11}$$

3.2 Local equations of linear motion

We rewrite (3.1.2) in component form

$$\int_S \tau_j(\mathbf{x})\, dS = \int_S \sigma_{ij}(\mathbf{x}) n_i(\mathbf{x})\, dS = \int_V [f_j(\mathbf{x}) - g_j(\mathbf{x})]\, dm, \tag{3.2.1}$$

and then use the tensor divergence theorem

$$\int_S \sigma_{ij}(\mathbf{x}) n_i(\mathbf{x})\, dS = \int_V \frac{\partial}{\partial x_i} \sigma_{ij}(\mathbf{x})\, dV = \int_V \frac{\partial}{\partial x_i} \sigma_{ij}(\mathbf{x}) \frac{dm}{\rho(\mathbf{x})}, \tag{3.2.2}$$

to represent the effects of the surface tractions though a volume integral. Here $\rho(\mathbf{x})$ is the local density and the stress tensor σ_{ij} is assumed to be continuous and differentiable within V.

From (3.2.1) and (3.2.2) we have the relation

$$\int_V \frac{\partial}{\partial x_i} \sigma_{ij}(\mathbf{x}) \frac{dm}{\rho(\mathbf{x})} = \int_V [f_j(\mathbf{x}) - g_j(\mathbf{x})]\, dm, \tag{3.2.3}$$

and since the region V is arbitrary

$$\frac{\partial}{\partial x_i} \sigma_{ij}(\mathbf{x}) + \rho g_j(\mathbf{x}) = \rho f_j(\mathbf{x}). \tag{3.2.4}$$

We see that it is the *stress gradient*, in conjunction with the body force, that determines the local acceleration.

If $\partial \sigma_{ij}/\partial x_i = 0$, the stress field is said to be *self-equilibrated*

Note: the local equations (3.1.7) [$\tau_j = \sigma_{ij} n_i$] and (3.2.4) imply the global equation (3.1.2).

3.2.1 Symmetry of the stress tensor

With the aid of the tensor divergence theorem the component form of the conservation of angular momentum equation (3.1.3) can be written as

$$\int_S \epsilon_{ijk} x_j \tau_k\, dS = \int_V \epsilon_{ijk} \frac{\partial}{\partial x_l} (x_j \sigma_{lk}) \frac{dm}{\rho}, \tag{3.2.5}$$

$$= -\int_V \mu_i\, dm + \int_V \epsilon_{ijk} x_j [f_k - g_k]\, dm. \tag{3.2.6}$$

From the equation of motion (3.2.4)

$$\int_V \epsilon_{ijk}\left[\frac{\partial}{\partial x_l}(x_j\sigma_{lk}) - x_j\frac{\partial}{\partial x_l}\sigma_{lk}\right]\frac{dm}{\rho} = -\int_V \mu_i\,dm, \tag{3.2.7}$$

and so,

$$\int_V \epsilon_{ijk}\delta_{jl}\sigma_{lk}\,dm = \int_V \epsilon_{ijk}\sigma_{jk}\,dm = -\rho\int_V \mu_i\,dm. \tag{3.2.8}$$

Since the region V is again arbitrary

$$\epsilon_{ijk}\sigma_{jk}(\mathbf{x}) = -\mu_i(\mathbf{x}), \tag{3.2.9}$$

and we may recover (3.1.3) from these local equations.

We will subsequently suppose that *body moments are absent* ($\mu_i \equiv 0$) and then

$$\sigma_{jk}(\mathbf{x}) = \sigma_{kj}(\mathbf{x}), \tag{3.2.10}$$

and the stress matrix is *symmetric.*

EXAMPLE: STRESS FIELDS

Show that the stress field $\sigma_{ij} = p\delta_{ij} + \rho v_i v_j$ is self-equilibrated, where p, ρ and $\mathbf{v}$ are the pressure, density and velocity in a steady flow of some inviscid compressible fluid.

The gradient of the stress tensor $\boldsymbol{\sigma}$

$$\begin{aligned}\frac{\partial\sigma_{ij}}{\partial x_j} &= \frac{\partial p}{\partial x_j} + \rho v_j\frac{\partial v_i}{\partial x_j} + \rho v_i\frac{\partial v_j}{\partial x_j} + v_i v_j\frac{\partial\rho}{\partial x_j}\\ &= [\boldsymbol{\nabla}p + \rho(\mathbf{v}\cdot\boldsymbol{\nabla})\mathbf{v}]_i + v_i[\boldsymbol{\nabla}\cdot(\rho\mathbf{v})].\end{aligned}$$

In steady flow,

$$\boldsymbol{\nabla}p + \rho(\mathbf{v}\cdot\boldsymbol{\nabla})\mathbf{v} = 0,$$

and from the continuity equation

$$\frac{\partial\rho}{\partial t} + \boldsymbol{\nabla}\cdot(\rho\mathbf{v}) = 0,$$

but since flow is steady $\partial\rho/\partial t = 0$. Thus

$$\frac{\partial\sigma_{ij}}{\partial x_j} = 0,$$

and so the stress field is self-equilibrated.•

Show that the stress field $\sigma_{ij} = v_i v_j - \frac{1}{2}\mathbf{v}^2\delta_{ij}$ exerts tractions on any closed surface which are statically equivalent overall to body forces $\mathbf{v}(\boldsymbol{\nabla}\cdot\mathbf{v})$ per unit volume, when $\mathbf{v}$ is any irrotational field (i.e. $\boldsymbol{\nabla}\times\mathbf{v} = 0$).

The gradient of the stress tensor $\boldsymbol{\sigma}$

$$\frac{\partial \sigma_{ij}}{\partial x_j} = v_i \frac{\partial v_j}{\partial x_j} + v_j \frac{\partial v_i}{\partial x_j} - v_k \frac{\partial v_k}{\partial x_i}$$
$$= v_i \frac{\partial v_j}{\partial x_j} + v_j \left(\frac{\partial v_i}{\partial x_j} - \frac{\partial v_j}{\partial x_i} \right)$$
$$= v_j \boldsymbol{\nabla} \cdot \mathbf{v},$$

since $\boldsymbol{\nabla} \times \mathbf{v} = 0$ and we can recognise the bracketed term as a component of $\boldsymbol{\nabla} \times \mathbf{v}$. In a static scenario, we have from (3.2.1) and (3.2.2)

$$\int_V \frac{\partial \sigma_{ij}}{\partial x_j} \, dV = \int_S \tau_i(\mathbf{x}) dS = -\int_V g_i \, dV$$

and the body force per unit volume $\mathbf{g}$ is $\mathbf{v}(\boldsymbol{\nabla} \cdot \mathbf{v})$ •

3.2.2 Stress jumps (continuity conditions)

A discontinuity in material properties across an internal interface generally causes a discontinuity in the stress tensor.

Such interfaces are present in simple laminates like plywood and in composite solids of any kind. Well known composites are rocks, reinforced concrete, glass-fibre resins (used in the hull of boats), cermets – sintered mixtures of ceramic inclusions in a metallic matrix (used in cutting tools), and ceramic fibre-reinforced materials (used in turbine blades).

Major internal discontinuities in the Earth occur at the base of the crust, where the Mohorovičić discontinuity can in places be very sharp, and at the phase boundaries in the mantle transition zone. Important fluid–solid discontinuities are at the core–mantle boundary between the silicate mantle and the metallic fluid in the outer core and the transition from the fluid outer core to the solid inner core.

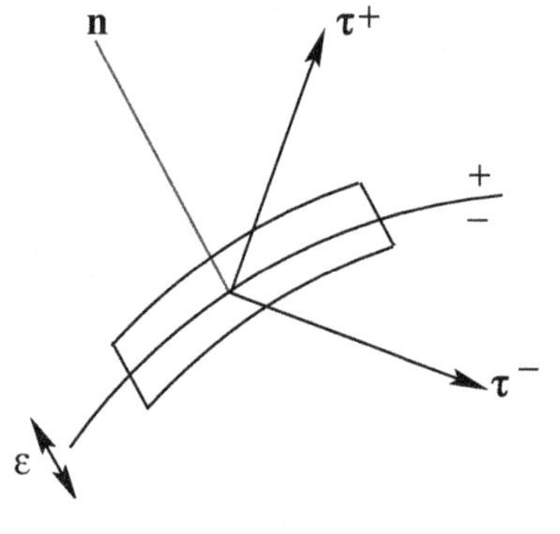

Figure 3.5. Interface conditions on traction

Possible jumps in the stress components at an interface are, however, restricted by the inviolable continuity of the traction vector acting on the interface

$$[\tau_j]_-^+ = n_i [\sigma_{ij}]_-^+ = 0, \tag{3.2.11}$$

where $\mathbf{n}$ is the interface normal. We may prove continuity of traction by applying (3.1.2) to a vanishingly thin disc enclosing an interfacial area A (Figure 3.5),

$$\int_A \boldsymbol{\tau}^+ dA^+ + \int_A \boldsymbol{\tau}^- dA^- + O(\epsilon) = \mathbf{0}, \tag{3.2.12}$$

where the $O(\epsilon)$ term arises from tractions on the edge, body forces and finite mass-accelerations (we exclude idealised 'shocks' with a jump in velocity).

In the limit as $\epsilon \to 0$

$$\boldsymbol{\tau}^+(\mathbf{n}) + \boldsymbol{\tau}^-(-\mathbf{n}) = \mathbf{0}, \quad \text{i.e.} \quad [\boldsymbol{\tau}(\mathbf{n})]_-^+ = \mathbf{0}. \tag{3.2.13}$$

Since (3.2.11) imposes only three constraints on six independent components of σ_{ij}, the *stress tensor* is not necessarily continuous at an interface. Consider a local coordinate scheme with x_1, x_2 in the tangent plane and x_3 normal: then $\boldsymbol{\tau}$ in (3.2.11) becomes $\boldsymbol{\sigma}_3$ and so

$$[\sigma_{31}] = [\sigma_{32}] = [\sigma_{33}] = 0; \tag{3.2.14}$$

consequently $[\sigma_{13}] = [\sigma_{23}] = 0$, but we cannot require $[\sigma_{11}]$, $[\sigma_{22}]$, $[\sigma_{12}]$ to be zero, simply from (3.2.11).

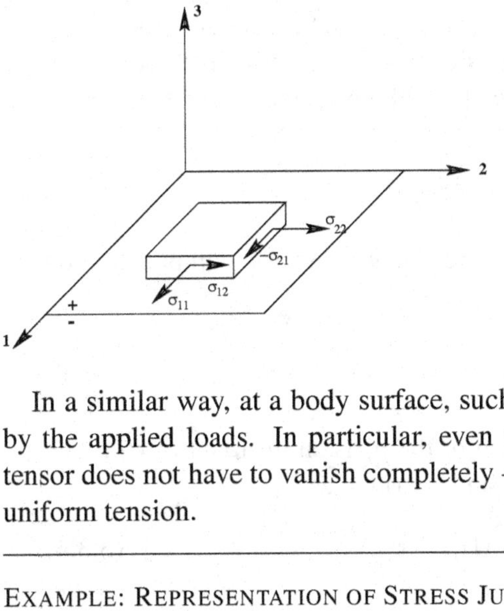

Figure 3.6. Unconstrained elements of the stress tensor from the condition of continuity of traction at an interface.

In a similar way, at a body surface, such 'interior' tractions are not constrained by the applied loads. In particular, even where the load is zero, the local stress tensor does not have to vanish completely – such a case is provided by a rod under uniform tension.

EXAMPLE: REPRESENTATION OF STRESS JUMPS

Show that every statically admissable jump in the stress tensor is of the type

$$(\delta_{ir} - n_i n_r)(\delta_{js} - n_j n_s) p_{rs}$$

for some symmetric p_{rs}, where $\mathbf{n}$ is the unit normal to the jump surface.

Set the stress jump $\hat{p}_{ij} = (\delta_{ik} - n_i n_k)(\delta_{jl} - n_j n_l) p_{kl}$. Consider the normal component of the associated traction

$$\begin{aligned} \hat{p}_{ij} n_j &= (\delta_{ik} - n_i n_k)(n_l - |n^2| n_l) p_{kl} \\ &= 0, \qquad \text{since } |n| = 1. \end{aligned}$$

The traction in a direction **m** orthogonal to the normal **n**, with $\mathbf{m}\cdot\mathbf{n} = 0$, would be

$$\begin{aligned}\hat{p}_{ij}m_j &= (\delta_{ik} - n_i n_k)(m_l - n_j m_j n_l)p_{kl} \\ &= (\delta_{ik} - n_i n_k)m_l p_{kl}, \qquad \text{as } \mathbf{m}\cdot\mathbf{n} = 0 \\ &= m_j p_{ij} - (n_k m_l p_{kl})n_i.\end{aligned}$$

Thus the traction jump in the normal direction is zero, and in orthogonal directions is non-zero as required for a statically admissable stress jump. •

3.3 Principal basis for stress

The traction vector $\boldsymbol{\tau}(\mathbf{n})$ over a plane element is purely normal when $\boldsymbol{\tau} = \sigma\mathbf{n}$ for some σ. Thus since $\tau_j = \sigma_{ij}n_i$, we have

$$(\sigma_{ij} - \sigma\delta_{ij})n_i = 0, \tag{3.3.1}$$

The eigenvectors $\pm\mathbf{n}_r$ $(r = 1, 2, 3)$ are called the *principal axes of stress* and the eigenvalues σ_r given by the roots of $\det(\sigma_{ij} - \sigma\delta_{ij}) = 0$ are called the *principal stresses.*

Since σ_{ij} is symmetric, the principal stresses σ_r are real and the axes are orthogonal. Each σ_r acts in opposite directions on opposite faces of a 'principal' box. The vectors $\mathbf{n}_r$ can also be thought of geometrically as the principal axes of the stress quadric surface

$$\sigma_{ij}x_i x_j = \textit{const.} \tag{3.3.2}$$

When the principal axes and stresses are known, stress components on background axes are given by the spectral formula

$$\boldsymbol{\sigma} = \sum_r \sigma_r n_r n_r^T. \tag{3.3.3}$$

Note: any quantity which is solely a function of the σ_r is an invariant of σ_{ij}. The cubic equation for the principal stresses σ_r can be written as

$$\det(\sigma_{ij} - \sigma\delta_{ij}) = -\sigma^3 + I_\sigma\sigma^2 - II_\sigma\sigma + III_\sigma = 0 \tag{3.3.4}$$

in terms of the invariants

$$\begin{aligned}I_\sigma &= \sigma_1 + \sigma_2 + \sigma_3 = \sigma_{kk} \\ II_\sigma &= \sigma_1\sigma_2 + \sigma_2\sigma_3 + \sigma_3\sigma_1 = \tfrac{1}{2}[\sigma_{ii}\sigma_{jj} - \sigma_{ij}\sigma_{ji}] \\ III_\sigma &= \sigma_1\sigma_2\sigma_3 = \det\boldsymbol{\sigma}.\end{aligned} \tag{3.3.5}$$

On occasion one may need the stress tensor components with respect to a particular set of axes, and it is often easier to use the general tensor transformation rule (3.1.8) from the principal basis, rather than use the spectral form (3.3.3).

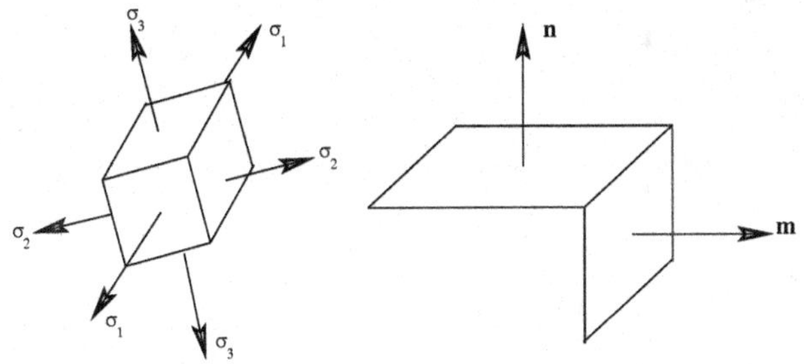

Figure 3.7. Transformation from the principal basis for stress to normal and shear tractions on a surface.

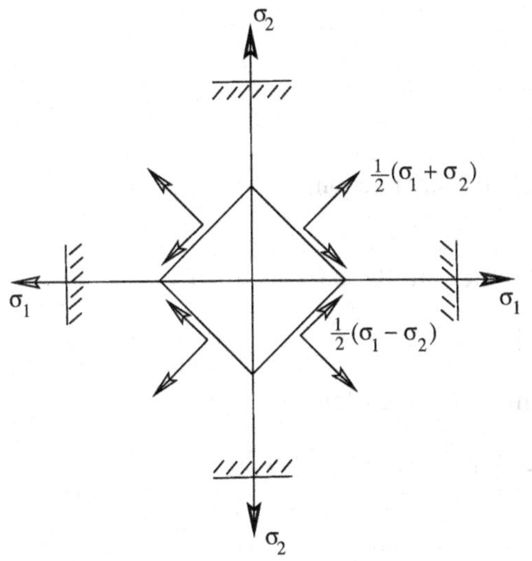

Figure 3.8. Configuration with maximum shear stress.

If, in particular, we seek the normal component of traction on an element with normal **n**, and the shear component in a direction **m** with $\mathbf{m} \cdot \mathbf{n} = 0, |\mathbf{m}| = 1$, (Figure 3.7), then we should use

$$\boldsymbol{\tau}(\mathbf{n}) = (n_1\sigma_1, n_2\sigma_2, n_3\sigma_3), \tag{3.3.6}$$

making use of the relation $\boldsymbol{\tau} = n_i\sigma_i$, with $\mathbf{n} = (n_1, n_2, n_3)$ in a principal basis. Then

$$\mathbf{n} \cdot \boldsymbol{\tau}(\mathbf{n}) = n_1^2\sigma_1 + n_2^2\sigma_2 + n_3^2\sigma_3, \tag{3.3.7}$$

$$\mathbf{m} \cdot \boldsymbol{\tau}(\mathbf{n}) = m_1n_1\sigma_1 + m_2n_2\sigma_2 + m_3n_3\sigma_3 = \mathbf{n} \cdot \boldsymbol{\tau}(\mathbf{m}). \tag{3.3.8}$$

From (3.3.8), any normal stress lies between the algebraically least and greatest principal stresses.

Stress circle

Consider the tractions on a box with one pair of faces perpendicular to a principal axis. Then the traction on this pair is purely normal, σ_3, and the tractions on the other faces have no components in that direction.

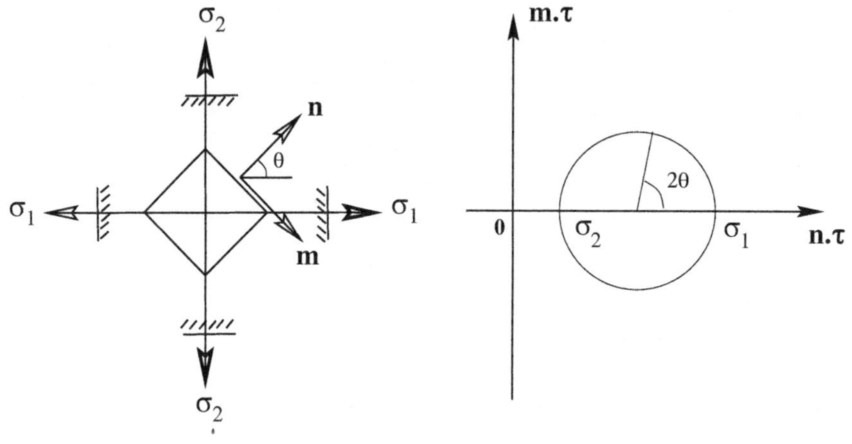

Figure 3.9. Stress-circle construction for normal and shear tractions.

Consider the normal in the $\widehat{12}$ plane at an angle θ to the σ_1 axis, then

$$\mathbf{n} = (\cos\theta, \sin\theta, 0)\mathbf{m} = (\sin\theta, -\cos\theta, 0) \tag{3.3.9}$$

so that the normal and shear stress components in the $\widehat{12}$ plane are

$$\begin{aligned} \mathbf{n}\cdot\boldsymbol{\tau} &= \cos^2\theta\,\sigma_1 + \sin^2\theta\sigma_2 = \tfrac{1}{2}(\sigma_1+\sigma_2) + \tfrac{1}{2}(\sigma_1-\sigma_2)\cos 2\theta \\ \mathbf{m}\cdot\boldsymbol{\tau} &= \tfrac{1}{2}(\sigma_1-\sigma_2)\sin 2\theta \end{aligned} \tag{3.3.10}$$

As θ is varied from 0 to π, the locus of $\mathbf{m}\cdot\boldsymbol{\tau}$ versus $\mathbf{n}\cdot\boldsymbol{\tau}$ is a circle: centre $\frac{1}{2}(\sigma_1+\sigma_2)$, radius $\frac{1}{2}(\sigma_1-\sigma_2)$, where we have assumed that σ_1 is greater than σ_2 (Figure 3.9).

The shear stress is largest when $\theta = \pi/4$

$$(\mathbf{m}\cdot\boldsymbol{\tau})_{\max} = \tfrac{1}{2}(\sigma_1-\sigma_2) \tag{3.3.11}$$

and then the normal tractions are equal to $\frac{1}{2}(\sigma_1+\sigma_2)$ on all four faces as illustrated in Figure 3.8.

EXAMPLE: PRINCIPAL STRESSES

A cylinder whose axis is parallel to the x_3-axis and whose normal cross-section is the square $-a \leq x_1 \leq a$, $-a \leq x_2 \leq a$ is subjected to torsion by couples acting over its ends $x_3 = 0$ and $x_3 = L$. The stress components are given by $\sigma_{13} = \partial\psi/\partial x_2$, $\sigma_{23} = -\partial\psi/\partial x_1$, $\sigma_{11} = \sigma_{12} = \sigma_{22} = \sigma_{33} = 0$, where $\psi = \psi(x_1, x_2)$.
(a) Show that this stress tensor is self-equilibrated;
(b) show that the difference between the maximum and minimum stress components is $2[(\partial\psi/\partial x_1)^2 + (\partial\psi/\partial x_2)^2]^{1/2}$, and find the principal axis which corresponds to the zero principal component;
(c) for the special case $\psi = (x_1^2 - a^2)(x_2^2 - a^2)$ show that the lateral surfaces are free from traction and that the couple acting on each end face is $32a^6/9$.

The field is self-equilibrated because

$$\frac{\partial \sigma_{13}}{\partial x_1} + \frac{\partial \sigma_{23}}{\partial x_2} = \frac{\partial^2 \psi}{\partial x_2 \partial x_1} - \frac{\partial^2 \psi}{\partial x_1 \partial x_2} = 0$$

and all other components of the gradient of the stress tensor are zero.
Now set $a = \partial\psi/\partial x_2$, $b = -\partial\psi/\partial x_1$, then the principal stresses satisfy

$$\det \begin{pmatrix} \sigma & 0 & a \\ 0 & \sigma & -b \\ a & -b & \sigma \end{pmatrix} = 0,$$

and so $\sigma(\sigma^2 - b^2 - a^2) = 0$, with solutions $\sigma = 0, \pm(b^2 + a^2)^{1/2}$.
Thus the difference between the minimum and maximum principal stresses is $2(b^2 + a^2)^{1/2}$, i.e., $[(\partial\psi/\partial x_1)^2 + (\partial\psi/\partial x_2)^2]^{1/2}$.
For the choice $\psi = (x_1^2 - a^2)(x_2^2 - a^2)$,

$$\sigma_{13} = \frac{\partial \psi}{\partial x_2} = 2x_2(x_1^2 - a^2),$$

$$\sigma_{23} = -\frac{\partial \psi}{\partial x_1} = -2x_1(x_2^2 - a^2),$$

and so σ_{13} vanishes when $x_1 = \pm a$, and σ_{23} vanishes when $x_2 = \pm a$. Hence the lateral faces of the cylinder are free from traction.
The couple on the end is given by

$$-\int_{end} \left(x_1 \frac{\partial \psi}{\partial x_1} + x_2 \frac{\partial \psi}{\partial x_2} \right) dS$$

$$= -\int_a^a dx_1 \int_a^a dx_2 [2x_1^2(x_2^2 - a^2) + 2x_2^2(x_1^2 - a^2)] = \frac{32a^6}{9}. \quad \bullet$$

3.4 Virtual work rate principle

We can establish a useful set of integral relations for the interaction of the stress field with a test field $\mathbf{v}^*$, which reduce to the conservation of energy when $\mathbf{v}^*$ is the actual velocity field.

Consider material in a bounded region of volume V and surface S, subject to surface tractions $\boldsymbol{\tau}$ and body forces $\mathbf{g}$, with a local acceleration field $\mathbf{f}$.

The entire body must satisfy the equations of linear and angular motion

$$\int_S \boldsymbol{\tau}\, dS + \int_V \mathbf{g}\, dm = \int_V \mathbf{f}\, dm, \tag{3.4.1}$$

$$\int_S \mathbf{x} \times \boldsymbol{\tau}\, dS + \int_V \mathbf{x} \times \mathbf{g}\, dm = \int_V \mathbf{x} \times \mathbf{f}\, dm. \tag{3.4.2}$$

A stress field σ_{ij} is said to be compatible with these constraints if it possesses the properties we have just established:

$$\begin{aligned} &\sigma_{ij} \text{ is symmetric,} \\ &\frac{\partial}{\partial x_i}\sigma_{ij} = \rho(f_j - g_j) \quad \text{in } V, \\ &n_i\sigma_{ij} = \tau_j \quad \text{on } S, \\ &n_i[\sigma_{ij}]_-^+ = 0 \quad \text{on any jump surface } J. \end{aligned} \tag{3.4.3}$$

For a piecewise continuous and piecewise continuously differentiable stress field σ_{ij} and a field $\mathbf{v}^*$ which is continuous and piecewise continuously differentiable:

$$\int_S \boldsymbol{\tau} \cdot \mathbf{v}^* \, dS + \int_V \mathbf{g} \cdot \mathbf{v}^* \, dm = \int_V \sigma_{ij}\frac{\partial}{\partial x_i}v_j^* \, dV + \int_V \mathbf{f} \cdot \mathbf{v}^* \, dm. \tag{3.4.4}$$

Usually we would take $\mathbf{v}^*$ to be a velocity or an infinitesimal displacement. When $\mathbf{v}$ is velocity then as we shall see later (3.4.4) corresponds to the balance of externally applied and internal work rates.

PROOF

$$\begin{aligned} \int_S \boldsymbol{\tau}.\mathbf{v}^* \, dS &= \int_S n_i\sigma_{ij}v_j^* \, dS \\ &= \int_V \frac{\partial}{\partial x_i}(\sigma_{ij}v_j^*)dV + \int_J n_i[\sigma_{ij}]_-^+ v_j^* \, dJ \\ &= \int_V \left(v_j^*\frac{\partial}{\partial x_i}\sigma_{ij} + \sigma_{ij}\frac{\partial}{\partial x_i}v_j^*\right) dV \\ &= \int_V \sigma_{ij}\frac{\partial}{\partial x_i}v_j^* \, dV + \int_V v_j^*\rho(f_j - g_j)\, dV, \end{aligned}$$

using the conditions (3.4.3). •

Converse principle

If σ_{ij} is a symmetric tensor field such that (3.4.4) holds for arbitrary $\mathbf{v}^*$, then σ_{ij} is compatible with the constraints and conforms to (3.1.2) and (3.1.3).

Now

$$\int_S n_i\sigma_{ij}v_j^* \, dS = \int_V \frac{\partial}{\partial x_i}(\sigma_{ij}v_j^*)\, dV + \int_J n_i[\sigma_{ij}]_-^+ v_j^* \, dJ \tag{3.4.5}$$

and subtracting this from (3.4.4) we finally obtain

$$\int_S (\tau_j - n_i\sigma_{ij})v_j^* \, dS + \int_J n_i[\sigma_{ij}]_-^+ v_j^* \, dJ + \int_V \left[\frac{\partial}{\partial x_i}\sigma_{ij} - \rho(f_j - g_j)\right] v_j^* \, dV \tag{3.4.6}$$

and, since v_j^* is arbitrary, each integrand must vanish. We are therefore able to recover (3.4.3) and hence (3.1.2) and (3.1.3).

3.5 Stress from a Lagrangian viewpoint

The stress tensor σ_{ij} enables us to calculate traction vectors that act on an element of area in the current (deformed) configuration. However, in some circumstances we would like to be able to relate the stress back to the reference (undeformed configuration).

From (2.1.11) we can relate an element of area in the deformed state dS to the equivalent element in the reference state by

$$\mathbf{dS} = \det \mathsf{F}\, \mathsf{F}^{-\mathsf{T}}\, \mathbf{d\Sigma}, \tag{3.5.1}$$

which we can rewrite in terms of the normal vectors to the surface elements as

$$n_j\, dS = J\, F_{pj}^{-1} \nu_p\, d\Sigma. \tag{3.5.2}$$

The components of the traction vector $\boldsymbol{\tau}$ acting on dS can therefore be expressed as:

$$\tau_i\, dS = \sigma_{ij} n_j\, dS = J\, F_{pj}^{-1} \sigma_{ij} \nu_p\, d\Sigma = \sigma_{pi}^{PK} \nu_p\, d\Sigma. \tag{3.5.3}$$

The first Piola–Kirchhoff stress tensor

$$\sigma_{pi}^{PK} = J\, F_{pj}^{-1} \sigma_{ij}, \quad \boldsymbol{\sigma}^{PK} = J\, \mathsf{F}^{-1} \boldsymbol{\sigma}, \tag{3.5.4}$$

is not symmetric, but from the symmetry of $\boldsymbol{\sigma}$ we find that $[\boldsymbol{\sigma}^{PK}]^{\mathsf{T}} \mathsf{F}^{\mathsf{T}} = \boldsymbol{\sigma}^{PK} \mathsf{F}$. The first Piola–Kirchhoff stress tensor is a two-point tensor since it links the current and reference states. A second Piola–Kirchhoff tensor can be defined that acts solely in the reference state

$$\boldsymbol{\sigma}^{SK} = J\, \mathsf{F}^{-1} \boldsymbol{\sigma} \mathsf{F}^{-\mathsf{T}} = \boldsymbol{\sigma}^{PK} \mathsf{F}^{-\mathsf{T}}; \quad \boldsymbol{\sigma}^{PK} = \boldsymbol{\sigma}^{SK} \mathsf{F}^{\mathsf{T}}. \tag{3.5.5}$$

This second, symmetric tensor $\boldsymbol{\sigma}^{SK}$ plays an important role in the development of elastic properties in terms of a Lagrangian strain-energy (see section 4.3). The stress tensor $\boldsymbol{\sigma}$ in the deformed state can be recovered from

$$J\, \boldsymbol{\sigma} = \mathsf{F} \boldsymbol{\sigma}^{SK} \mathsf{F}^{\mathsf{T}}. \tag{3.5.6}$$

4
Constitutive Relations

The balance laws for linear and angular momentum (3.1.2), (3.1.3) do not make any reference to the properties of the body. The full specification of the mechanical and thermal properties of a material requires further information as to the relation between stress and strain, and thus to the relation between the geometry of deformation and the resulting stress field. This information is provided by a constitutive equation which gives a mathematical representation of the functional dependence between stress and strain, designed to provide agreement with the observed behaviour. The behaviour of real materials is both diverse and complex, and so most constitutive equations are designed to capture the most important features of the mechanical behaviour of a material in a particular situation. The mathematical formulation may well be an approximation to the macroscopic response induced by a specific set of atomic scale processes such as, e.g., the movement of dislocations in plastic deformation.

Mathematical and physical models of constitutive relations are most effective when dealing with a single material, since it is difficult to capture the full behaviour of a composite with a single relation. However, although geological materials are composed of an aggregate of minerals, many aspects of the behaviour of rocks can be captured through fairly simple representations with temperature and pressure dependence.

4.1 Constitutive relation requirements

A constitutive relation takes the form

$$\text{stress} = \text{functional(deformation)} \tag{4.1.1}$$

and the possible behaviour is restricted by a set of postulates which are designed to reconcile the mathematical and physical behaviour.

(a) Determinism:

> The stress at time t depends only on the history of the deformation up to t and not on future values.

(b) Local Action:

The stress at ξ at some time t depends only on the deformation in the neighbourhood of ξ

(c) Material Objectivity:

Constitutive equations are form invariant with respect to rigid motions of the spatial frame of reference.

(d) Material Invariance (Symmetry)

Constitutive equations are form invariant with respect to a symmetry group. For *fluids* the symmetry group is the full unimodular group. For *solids* the symmetry group is restricted to a group of orthogonal transformations of the material coordinates.

It is only at this stage that we meet any distinction between fluids and solids. The first principle (a) allows for material with memory as in viscoelastic behaviour, e.g., plastics, the Earth's upper mantle.

4.1.1 Simple materials

A important class of material behaviour is described by *simple materials* for which the stress depends on the local deformation through a functional in time:

$$\sigma(\xi, t) = f(F^t, \xi, t), \tag{4.1.2}$$

where F^t denotes the history up to time t of the deformation gradient $F(\xi, t)$ and f is a tensor functional in time alone.

In this case the principles (a) and (b) are satisfied automatically. In order to meet the requirement of material objectivity (c) we consider the effect of a rigid rotation. We rotate the spatial frame by $\mathbf{Q}^T$, so that

$$\mathbf{x}^* = \mathbf{Q}\mathbf{x}, \quad F^* = \mathbf{Q}F. \tag{4.1.3}$$

The stress transforms as a second-rank tensor under this rotation:

$$\sigma^* = \mathbf{Q}\sigma\mathbf{Q}^T. \tag{4.1.4}$$

After rotation we must have the same functional form for the dependence of stress on deformation as in (4.1.2) i.e.,

$$\sigma^*(t) = f(F^{*t}, \xi, t), \tag{4.1.5}$$

and thus

$$\mathbf{Q}\, f(F^t)\, \mathbf{Q}^T = f([\mathbf{Q}F^t]), \qquad \text{for all } \mathbf{Q}. \tag{4.1.6}$$

The condition for invariance under rotation (4.1.6) thus imposes a constraint on the possible form of the functional f.

We can proceed further by making the polar decomposition $\mathsf{F} = \mathsf{RU}$. Since (4.1.6) is an identity in $\mathbf{Q}$, choose $\mathbf{Q} = \mathbf{R}^{\mathsf{T}}$. Then

$$\mathbf{R}^{\mathsf{T}}\mathsf{f}(\mathsf{F}^{t},\boldsymbol{\xi},t)\mathbf{R} = \mathsf{f}([\mathbf{R}^{\mathsf{T}}\mathsf{F}]^{t},\boldsymbol{\xi},t) = \mathsf{f}(\mathsf{U}^{t},\boldsymbol{\xi},t), \tag{4.1.7}$$

and so, from the orthogonality of $\mathbf{R}$, we have

$$\mathsf{f}(\mathsf{F}^{t},\boldsymbol{\xi},t) = \mathbf{R}\,\mathsf{f}(\mathsf{U}^{t},\boldsymbol{\xi},t)\mathbf{R}^{\mathsf{T}}, \tag{4.1.8}$$

which satisfies (a), (b) and (c) with

$$\sigma(\boldsymbol{\xi},t) = \mathbf{R}\,\mathsf{f}(\mathsf{U}^{t},\boldsymbol{\xi},t)\,\mathbf{R}^{\mathsf{T}}. \tag{4.1.9}$$

The stress therefore depends on the history of deformation only through the stretch U or equivalently on any strain measure such as

$$\mathsf{E} = \tfrac{1}{2}(\mathsf{F}^{\mathsf{T}}\mathsf{F} - \mathbf{I}) = \tfrac{1}{2}(\mathsf{U}^{2} - \mathbf{I}). \tag{4.1.10}$$

If we write

$$\mathsf{f}(\mathsf{U}^{t}) = \mathsf{U}\,\mathsf{g}(\mathsf{E}^{t})\,\mathsf{U}^{\mathsf{T}}, \tag{4.1.11}$$

in terms of a new functional g, then the stress tensor

$$\sigma = \mathsf{RU}\,\mathsf{g}(\mathsf{E}^{t})\,\mathsf{U}^{\mathsf{T}}\mathsf{R}^{\mathsf{T}} = \mathsf{F}\,\mathsf{g}(\mathsf{E}^{t})\,\mathsf{F}^{\mathsf{T}}. \tag{4.1.12}$$

We may therefore express the stress behaviour for the simple material in the form

$$\sigma(\boldsymbol{\xi},t) = \mathsf{F}\,\mathsf{g}(\mathsf{E}^{t},\boldsymbol{\xi},t)\,\mathsf{F}^{\mathsf{T}}. \tag{4.1.13}$$

in terms of a functional of strain and time.

4.1.2 Material symmetry

When we use the decomposition $\mathsf{F} = \mathsf{RU}$, we represent the form of the deformation in terms of first stretching an element and then rotating it. If the material possesses some intrinsic symmetry, rotations within the *symmetry group* of the material applied before the deformation RU will leave the resulting stress the same.

Suppose the material has symmetry with respect to the orthogonal transformation P, such that $\mathsf{PP}^{\mathsf{T}} = \mathbf{I}$, then states FP and F will be indistinguishable. We therefore require that the stress representation

$$\sigma = \mathsf{f}(\mathsf{F}) = \mathsf{f}(\mathsf{FP}), \tag{4.1.14}$$

for all F and any $\mathbf{P}$ in the symmetry group of the material. If the body is *isotropic* then (4.1.14) holds for any orthogonal matrix $\mathbf{P}$.

With the representation of the stress (4.1.13) as a function of strain and time, the symmetry property (4.1.14) gives

$$\mathsf{Fg}(\mathsf{E})\mathsf{F}^{\mathsf{T}} = \mathsf{FPg}(\mathsf{P}^{\mathsf{T}}\mathsf{EP})\mathsf{P}^{\mathsf{T}}\mathsf{F}^{\mathsf{T}}, \tag{4.1.15}$$

since $E = \frac{1}{2}(F^T F - I)$. As a result we find that the functional g needs to have the property,

$$P^T g(E) P = g(P^T E P), \tag{4.1.16}$$

as an expression of the required symmetries.

4.1.3 Functional dependence

We can describe different classes of material by the functional dependence appearing in the constitutive relation. For viscoelastic materials we need the full dependence on the history of deformation, but for many materials such effects are not important.

We can characterise a *fluid* by the assumption that the deviatoric part of the stress depends on the current strain rate $\dot{\epsilon} = \partial \mathbf{v}/\partial \mathbf{x}$, i.e.,

$$\sigma = -p\mathbf{I} + f(\dot{\epsilon}, \rho), \tag{4.1.17}$$

where $\mathbf{I}$ is the unit tensor and p is the pressure. If the dependence on $\dot{\epsilon}$ is a linear function, we have a Newtonian viscous fluid.

For solid materials, the deformation itself is a more natural variable than strain rate. Thus elastic behaviour can be characterised by

$$\text{stress} = \text{function(deformation)}, \tag{4.1.18}$$

$$\text{i.e.} \quad \sigma = f(F), \tag{4.1.19}$$

and so from (4.1.9) $\sigma = \mathbf{R}\, f(\mathbf{U})\, \mathbf{R}^T$.

This approach, Cauchy elasticity, may be used with material symmetry arguments to find the form of the function f. However, a slightly more restrictive development based on the idea of a work function (due to Green) seems to be in close accord with experimental viewpoints, and we will adopt this approach in our subsequent development.

4.2 Energy balance

We cannot separate the mechanical aspects of deformation from thermal and other changes and have therefore to introduce the concepts of continuum thermodynamics. In many applications thermal effects are small and are often neglected, but we should be aware how they may enter into the consideration of the deformation of a medium.

As in section 3.4 we consider a volume V with surface S (and outward normal $\hat{\mathbf{n}}$) subject to surface tractions τ and body forces $\mathbf{g}$. The work rate of the body and surface forces for a velocity field $\mathbf{v}$ is then

$$\frac{d}{dt} Y = \int_S \tau \cdot \mathbf{v}\, dS + \int_V \mathbf{g} \cdot \mathbf{v}\, dm. \tag{4.2.1}$$

We can recast the right-hand side of (4.2.1) in terms of the acceleration field $\mathbf{f}$ and the stress tensor σ_{ij} by using the virtual work-rate principle (3.4.4) with $\mathbf{v}^*$ set equal to the actual velocity $\mathbf{v}$,

$$\frac{d}{dt}Y = \int_V \sigma_{ij}\frac{\partial}{\partial x_i}v_j\,dV + \int_V \mathbf{f}\cdot\mathbf{v}\,dm. \tag{4.2.2}$$

We can recognise $\int_V \mathbf{f}\cdot\mathbf{v}\,dm$ as the rate of change of kinetic energy

$$\int_V \mathbf{f}\cdot\mathbf{v}\,dm = \frac{d}{dt}\int_V \tfrac{1}{2}\mathbf{v}^2\,dm = \frac{d}{dt}\int_V \tfrac{1}{2}\rho\mathbf{v}^2\,dV = \int_V \rho\frac{D}{Dt}(\tfrac{1}{2}\mathbf{v}^2)dV, \tag{4.2.3}$$

where we have employed (2.5.12) to recast the material derivative of the volume integral.

The rate of increase of thermal energy for a material volume is the resultant of the heat production by internal sources, h per unit volume, and the heat conducted across the surface S:

$$\frac{d}{dt}H = \int_V h\,dV - \int_S \mathbf{q}\cdot\hat{\mathbf{n}}\,dS. \tag{4.2.4}$$

$\mathbf{q}$ is the heat flux vector which can be related to the temperature gradient by Fourier's law of heat conduction

$$\mathbf{q} = -k\frac{\partial}{\partial\mathbf{x}}T, \tag{4.2.5}$$

where k is the thermal conductivity of the material and T is the temperature. We can recast the expression for the rate of change of the thermal energy in terms of a volume integral by using the divergence theorem

$$\frac{d}{dt}H = \int_V \left[h + \frac{\partial}{\partial\mathbf{x}}\cdot\left(k\frac{\partial}{\partial\mathbf{x}}T\right)\right]\,dV. \tag{4.2.6}$$

Conservation of energy requires that the gain in the internal and kinetic energy balance the sum of the work rate by external forces and the rate of increase of thermal energy. In terms of the internal energy density U per unit mass, we obtain

$$\frac{d}{dt}\int_V \rho(U + \tfrac{1}{2}\mathbf{v}^2)dV = \frac{d}{dt}(H + Y). \tag{4.2.7}$$

From (2.5.12) we can rewrite the left-hand side of (4.2.7) in terms of an integral over the material derivative of the specific energy in the continuum,

$$\frac{d}{dt}\int_V \rho(U + \tfrac{1}{2}\mathbf{v}^2)dV = \int_V \rho\frac{D}{Dt}(U + \tfrac{1}{2}\mathbf{v}^2)dV = \frac{dH}{dt} + \frac{dY}{dt}. \tag{4.2.8}$$

Now from (4.2.2) and (4.2.3) we can express the work rate of the various forces as

$$\frac{dY}{dt} = \int_V \sigma_{ij}\frac{\partial}{\partial x_i}v_j\,dV + \int_V \rho\frac{D}{Dt}(\tfrac{1}{2}\mathbf{v}^2)dV. \tag{4.2.9}$$

With some rearrangement of (4.2.8) using (4.2.9) and the expression (4.2.6) for the rate of change of thermal energy, we can now find an expression for the volume integral of the material rate of change of the specific internal energy,

$$\int_V \rho \frac{D}{Dt} U \, dV = \int_V \left[h + \frac{\partial}{\partial \mathbf{x}} \cdot \left(k \frac{\partial}{\partial \mathbf{x}} T \right) + \sigma_{ij} \frac{\partial}{\partial x_i} v_j \right] dV. \tag{4.2.10}$$

Since the volume is arbitrary we may equate the integrands on the two sides of this equation and so the local form of the *principle of conservation of energy* is

$$\rho \frac{D}{Dt} U = h + \frac{\partial}{\partial \mathbf{x}} \cdot \left(k \frac{\partial}{\partial \mathbf{x}} T \right) + \sigma_{ij} \frac{\partial}{\partial x_i} v_j. \tag{4.2.11}$$

Neglect of thermal effects

The internal heat generation h from, e.g., radioactive materials in the Earth occurs sufficiently slowly that it is not important except for very long-term deformation.

The heat transport contribution to the rate of change of internal energy,

$$\frac{\partial}{\partial \mathbf{x}} \cdot \left(k \frac{\partial}{\partial \mathbf{x}} T \right), \tag{4.2.12}$$

can be neglected in discussions of deformation in two simplifying circumstances:

(i) Static Problems: if the time scale over which the deformation is occurring is sufficiently long, heat flow will occur to equalise temperature and conditions will be *isothermal.* Such is the case in static deformation in the laboratory or in the long-term deformation of Earth materials.

(ii) Dynamic Problems: if the time scale of deformation is short enough, the thermal state has no time to adjust to the disturbance and the effect is conditions comparable to thermal isolation. This is an *adiabatic* or *isentropic* state, appropriate to most elastic wave propagation, but not, e.g., for rubber-like materials. For seismic waves in the Earth with periods of around a second and wavelengths of a few kilometres, the thermal time constant is many thousands of years and so the *adiabatic* assumption is assured. However for small specimens of the same material around 5 mm in size, the thermal time scale is much shorter (about 20 seconds) and the adiabatic assumption requires high-frequency waves (1 kHz or greater).

The form of the internal energy function will differ in the two cases, and so the mechanical response to deformation will be different. For intermediate time scales, the roles of the deformation and thermal behaviour are closely linked and need to be treated with continuum thermodynamics.

4.3 Elastic materials

In those cases where thermal effects can be neglected, we can equate the rate of change of the energy of deformation (the *stress power*) to the deformation term in (4.2.2):

$$\int_{V_0} \frac{D}{Dt} W \, dV_0 = \int_V \sigma_{ij} \frac{\partial}{\partial x_i} v_j \, dV, \tag{4.3.1}$$

where W is called the work density or strain energy.

We can recast the left-hand side of (4.3.1) in terms of the current volume V by using the relation of dV to dV_0 from (2.1.6). We find

$$\int_V (\det F)^{-1} \frac{D}{Dt} W \, dV = \int_V \sigma_{ij} \frac{\partial}{\partial x_i} v_j \, dV, \tag{4.3.2}$$

but once again the volume V is arbitrary and so

$$(\det F)^{-1} \frac{D}{Dt} W = \sigma_{ij} \frac{\partial}{\partial x_i} v_j. \tag{4.3.3}$$

We can find an alternative expression for the mobile derivative DW/Dt by employing the chain rule in terms of the deformation gradient;

$$\frac{D}{Dt} W = \frac{\partial W}{\partial F_{jk}} \frac{D}{Dt} F_{jk} = \frac{\partial W}{\partial F_{jk}} \frac{D}{Dt} \left(\frac{\partial x_j}{\partial \xi_k} \right), \tag{4.3.4}$$

$$= \frac{\partial W}{\partial F_{jk}} \frac{\partial v_j}{\partial \xi_k} = \frac{\partial W}{\partial F_{jk}} \frac{\partial x_i}{\partial \xi_k} \frac{\partial v_j}{\partial x_i}. \tag{4.3.5}$$

With this representation for DW/Dt inserted in (4.3.3) we have

$$(\det F)^{-1} \frac{\partial W}{\partial F_{jk}} F_{ik} \frac{\partial v_j}{\partial x_i} = \sigma_{ij} \frac{\partial v_j}{\partial x_i}, \tag{4.3.6}$$

and, since this is true for an arbitrary velocity field,

$$\det F \, \sigma_{ij} = F_{ik} \frac{\partial W}{\partial F_{jk}}. \tag{4.3.7}$$

We already know that the stress $\boldsymbol{\sigma}$ depends only on the stretch $\mathbf{U}$ (or strain $\mathbf{E}$), we therefore recast (4.3.7) in terms of the Green strain $\mathbf{E}$. The elements of the Green strain are

$$2E_{kl} = F_{jk} F_{jl} - \delta_{kl}, \tag{4.3.8}$$

and we can express the increment in the strain energy as,

$$dW = \frac{\partial W}{\partial E_{kl}} \tfrac{1}{2} (F_{jk} \, dF_{jl} + F_{jl} \, dF_{jk}) = \frac{\partial W}{\partial E_{kl}} F_{jl} \, dF_{jk} = \frac{\partial W}{\partial F_{jk}} \, dF_{jk}. \tag{4.3.9}$$

The representation (4.3.7) for the stress tensor can therefore be written in the form,

$$\det F \, \sigma_{ij} = F_{ik} F_{jl} \frac{\partial W}{\partial E_{kl}}, \tag{4.3.10}$$

or, symbolically, as,

$$\det \mathsf{F}\,\sigma = \mathsf{F}\frac{\partial W}{\partial \mathsf{E}}\mathsf{F}^{\mathrm{T}} = \mathbf{R}\left[\mathsf{U}\frac{\partial W}{\partial \mathsf{E}}\mathsf{U}\right]\mathbf{R}^{\mathrm{T}}. \tag{4.3.11}$$

We can now identify (4.3.11) as having precisely the form (4.1.13) with the functional $\mathsf{g}(\mathsf{E})$ identified with the tensor derivative $\partial W/\partial \mathsf{E}$. Further, from (3.5.6), we can recognise $\mathsf{g}(\mathsf{E}) = \partial W/\partial \mathsf{E}$ as the second Piola–Kirchhoff tensor in the reference state.

From the symmetry properties (4.1.16) we can deduce

$$\frac{\partial W}{\partial \mathsf{E}}(\mathbf{P}^{\mathrm{T}}\mathsf{E}\mathbf{P}) = \mathbf{P}^{\mathrm{T}}\frac{\partial W}{\partial \mathsf{E}}\mathbf{P}. \tag{4.3.12}$$

for all orthogonal $\mathbf{P}$ in the symmetry group of the elastic medium.

Material symmetry

The strain energy function W is required to be invariant under the symmetry group of the material $\{\mathbf{P}\}$, so that

$$W(\mathsf{F}) \equiv W(\mathsf{F}\mathbf{P}), \tag{4.3.13}$$

and since a rigid post-rotation makes no difference

$$W(\mathbf{P}^{\mathrm{T}}\mathsf{F}\mathbf{P}) = W(\mathsf{F}) = W(\mathsf{U}), \tag{4.3.14}$$

and thus W depends only on the stretch matrix U.

4.4 Isotropic elastic material

If the material properties are unaffected by any rotation, the solid is said to be *isotropic*. In these circumstances there are some simplifications in the form of the relations between stress and strain.

4.4.1 Effect of rotation

For all rotations $\mathbf{Q}$, from (4.3.14),

$$W(\mathbf{Q}\mathsf{F}\mathbf{Q}^{\mathrm{T}}) = W(\mathsf{F}) = W(\mathsf{U}) = W(\mathbf{Q}^{\mathrm{T}}\mathsf{U}\mathbf{Q}). \tag{4.4.1}$$

If we choose $\mathbf{Q}$ as the rotation from background axes to the Lagrangian triad, then

$$\mathbf{Q}^{\mathrm{T}}\mathsf{F}\mathbf{Q} = \mathrm{diag}\{\lambda_1, \lambda_2, \lambda_3\} \tag{4.4.2}$$

in some ordering of the principal stretches.

Thus for any isotropic solid, $W(\mathsf{F})$ is a symmetric function of the principal stretches, since it is unchanged by a rotation of $\pi/2$ about the x_3 principal axis which takes the indices $1 \rightarrow 2$ etc.

4.4.2 Coaxiality of the Cauchy stress tensor and the Eulerian triad

From the symmetry property of stress (4.1.16) and the stress formula (4.3.11) we have the relation (4.3.12)

$$\frac{\partial W}{\partial E}(Q^{T}EQ) = Q^{T}\frac{\partial W}{\partial E}Q, \tag{4.4.3}$$

which now must hold for all rotations Q.

A consequence of isotropy is that $\partial W/\partial E$ and E are coaxial. The tensor E is symmetric with principal axes along the Lagrangian triad and principal values $\frac{1}{2}(\lambda_r^2 - 1)$, $r = 1, 2, 3$. In this principal basis consider a rotation P of π about the 3-axis: which takes $1 \rightarrow -1$, and $2 \rightarrow -2$, then $P^{T}EP = E$, and so from (4.3.13)

$$\frac{\partial W}{\partial E}(P^{T}EP) = P^{T}\frac{\partial W}{\partial E}P. \tag{4.4.4}$$

We write $g(E) = \partial W/\partial E$, and thus $g = P^{T}gP$. Since P is orthogonal $Pg = gP$. The effect of the rotation by π about the 3-axis is then that the elements g_{13}, g_{23} would change sign, but this is not allowed by the symmetry operation and hence

$$g_{13} = g_{23} = 0. \tag{4.4.5}$$

Similarly we may show that all off-diagonal terms vanish, i.e., with respect to the Lagrangian triad

$$g(E) = \frac{\partial W}{\partial E} = \mathrm{diag}\{g_1, g_2, g_3\}, \tag{4.4.6}$$

and so $\partial W/\partial E$ and E are coaxial. This is in fact a general theorem for an isotropic tensor function W.

Now from the stress relation (4.3.11) we have

$$\det F\,\sigma = R\left[U\frac{\partial W}{\partial E}U\right]R^{T}, \tag{4.4.7}$$

and so

$$R^{T}\sigma R = U\frac{\partial W}{\partial E}U\,(\det F)^{-1}. \tag{4.4.8}$$

With the Lagrangian triad as basis the right-hand side of (4.4.8) is diagonal. The action of the rotation R is to take the Lagrangian into the Eulerian triad and so, *for an isotropic solid*, the principal axes of σ lie along the Eulerian triad. Thus the Cauchy stress tensor and the Eulerian triad are coaxial in the case of isotropy.

4.4.3 Principal stresses

For a pure strain deformation, $F_{ij} = U_{ij}$, and referred to the Eulerian triad $F_{ij} = \mathrm{diag}\{\lambda_1, \lambda_2, \lambda_3\}$. From (4.3.7) the principal stress σ_r is given by

$$\det F\,\sigma_r = \lambda_1\lambda_2\lambda_3\,\sigma_r = \lambda_r\frac{\partial W}{\partial \lambda_r} \quad \text{(no sum)}. \tag{4.4.9}$$

Suppose the work function is given as $W(\alpha, \beta, \gamma)$ with $\alpha = \log(\lambda_1\lambda_2\lambda_3)$, $\beta = \lambda_1 + \lambda_2 + \lambda_3$, $\gamma = \frac{1}{2}(\lambda_1^2 + \lambda_2^2 + \lambda_3^2)$. Then the principal stress

$$\sigma_r = \frac{1}{\lambda_1\lambda_2\lambda_3}\left[\frac{\partial W}{\partial \alpha} + \frac{\partial W}{\partial \beta}\lambda_r + \frac{\partial W}{\partial \gamma}\lambda_r^2\right], \tag{4.4.10}$$

and since σ is coaxial with the Eulerian triad, λ_r are the principal values of $\mathbf{R}\mathbf{U}\mathbf{R}^T$, since $\mathbf{F} = \mathbf{R}\mathbf{U} = (\mathbf{R}\mathbf{U}\mathbf{R}^T)\mathbf{R}$. Further $\det \mathbf{F} = \lambda_1\lambda_2\lambda_3$ so we can equate principal values,

$$\sigma_r = (\det \mathbf{F})^{-1}\mathbf{R}\left[\frac{\partial W}{\partial \alpha}\mathbf{I} + \frac{\partial W}{\partial \beta}\mathbf{U} + \frac{\partial W}{\partial \gamma}\mathbf{U}^2\right]_r \mathbf{R}^T. \tag{4.4.11}$$

Transferring to general axes we have

$$\sigma = (\det \mathbf{F})^{-1}\mathbf{R}\left[\frac{\partial W}{\partial \alpha}\mathbf{I} + \frac{\partial W}{\partial \beta}\mathbf{U} + \frac{\partial W}{\partial \gamma}\mathbf{U}^2\right]\mathbf{R}^T \tag{4.4.12}$$

$$= (\det \mathbf{F})^{-1}\left[\frac{\partial W}{\partial \alpha}\mathbf{I} + \frac{\partial W}{\partial \beta}(\mathbf{F}\mathbf{F}^T)^{1/2} + \frac{\partial W}{\partial \gamma}(\mathbf{F}\mathbf{F}^T)\right], \tag{4.4.13}$$

which, as we would expect, depends on $\mathbf{F}\mathbf{F}^T$.

4.4.4 Some isotropic work functions

The simplest type of work function is $W = \phi(\lambda_1\lambda_2\lambda_3)$ that depends only on change of volume, for which from (4.4.9) each of the principal stresses $\sigma_r = \phi'(\lambda_1\lambda_2\lambda_3)$, and so the stress tensor is purely hydrostatic.

A second type of work function is provided by $W = \psi(\lambda_1) + \psi(\lambda_2) + \psi(\lambda_3)$ for which the principal stresses are $\sigma_r = \lambda_r\psi'(\lambda_r)/(\lambda_1\lambda_2\lambda_3)$. However this second type of work function has the defect that it predicts no change in lateral dimensions in a tension or compression test

$$\sigma_1 \neq 0, \quad \sigma_2 = \sigma_3 = 0, \quad \Rightarrow \quad \lambda_1 = \lambda_2 = 1. \tag{4.4.14}$$

To remedy this problem we combine the two forms for the work function to give

$$W = \phi(\lambda_1\lambda_2\lambda_3) + \psi(\lambda_1) + \psi(\lambda_2) + \psi(\lambda_3), \tag{4.4.15}$$

for which the principal stresses are

$$\sigma_r = \phi'(\lambda_1\lambda_2\lambda_3) + \lambda_r\psi'(\lambda_r)/(\lambda_1\lambda_2\lambda_3). \tag{4.4.16}$$

Such a representation for the work function is suitable for small deformations in most materials. For large elastic deformation it provides an adequate description of the static deformation of vulcanised or synthetic rubbers (i.e. random assemblages or cross-linked, long-chain molecules) for dimensional changes of a factor of 10 or so. Large elastic deformations are also feasible in rocks at high temperatures.

Near the reference state we can consider incremental deformation by setting

$\lambda_r = 1 + e_r$, where e_r is a first-order quantity and can denote any normalised measure of strain. Then the principal stresses are given by

$$\begin{aligned}\sigma_r &= [\phi''(1) - \psi'(1)](e_1 + e_2 + e_3) \\ &\quad + [\psi''(1) + \psi'(1)]e_r + [\phi'(1) + \psi'(1)] + O(e_r^2).\end{aligned} \tag{4.4.17}$$

If the ground state is stress-free, so that $\phi'(1) + \psi'(1) = 0$, then the expressions for the principal stress simplify somewhat:

$$\sigma_r = [\phi''(1) + \phi'(1)](e_1 + e_2 + e_3) + [\psi''(1) + \psi'(1)]e_r + O(e_r^2). \tag{4.4.18}$$

We may identify the elastic moduli in the ground state by comparison with a general linearised development (see Section 5.2.1).

Note that we may also make an expansion $\lambda_r = \lambda_{r0} + e_r$ about a distorted state but then the moduli for incremental deformation need not be isotropic (see Section 6.4).

4.5 Fluids

We have so far considered elastic solids for which the stress depends on strain. We can characterise a fluid by a constitutive equation in which stress depends on the rate of strain. A suitable constitutive relation is thus

$$\sigma_{ij} = s_{ij}(\partial v_p/\partial x_q, \rho, T), \tag{4.5.1}$$

in terms of the velocity gradient $\partial v_p/\partial x_q$, density ρ and temperature T.

We introduce the rate of deformation tensor

$$D_{ij} = \tfrac{1}{2}\left(\frac{\partial v_i}{\partial x_j} + \frac{\partial v_j}{\partial x_i}\right), \tag{4.5.2}$$

which represents the symmetric part of the velocity gradient in the Eulerian configuration. The corresponding anti-symmetric part

$$\varpi_{ij} = \tfrac{1}{2}\left(\frac{\partial v_i}{\partial x_j} - \frac{\partial v_j}{\partial x_i}\right), \tag{4.5.3}$$

is called the *spin or vorticity tensor.*

We will now impose the requirement on the fluid that the stress σ is independent of any superimposed rigid body rotation. We rewrite (4.5.1) in the form

$$\sigma = \mathbf{s}(\mathbf{D}, \boldsymbol{\varpi}, \rho, T) \tag{4.5.4}$$

to separate the dependence on the symmetric and anti-symmetric parts of the velocity gradient. Suppose we have a deformation process

$$\mathbf{x} = \mathbf{x}(\boldsymbol{\xi}, t), \qquad \mathbf{v} = \mathbf{v}(\mathbf{x}, t), \tag{4.5.5}$$

and superpose a time-dependent rigid rotation so that in the new flow

$$\bar{\mathbf{x}} = \mathbf{Q}(t)\mathbf{x}(\boldsymbol{\xi}, t), \tag{4.5.6}$$

where $Q(t)$ is an orthogonal rotation tensor. The associated velocity

$$\bar{\mathbf{v}} = \frac{D}{Dt}\bar{\mathbf{x}} = \dot{\mathbf{Q}}\mathbf{x} + \mathbf{Q}\dot{\mathbf{x}} = \dot{\mathbf{Q}}\mathbf{x} + \mathbf{Q}\mathbf{v}, \tag{4.5.7}$$

and the velocity gradient

$$\frac{\partial \bar{\mathbf{v}}}{\partial \bar{\mathbf{x}}} = \frac{\partial \bar{\mathbf{v}}}{\partial \mathbf{x}}\frac{\partial \mathbf{x}}{\partial \bar{\mathbf{x}}} = \left(\dot{\mathbf{Q}} + \mathbf{Q}\frac{\partial \mathbf{v}}{\partial \mathbf{x}}\right)\mathbf{Q}^{\mathsf{T}}. \tag{4.5.8}$$

The rate of strain tensor for the new flow is given by

$$\bar{\mathbf{D}} = \tfrac{1}{2}\left(\frac{\partial \bar{\mathbf{v}}}{\partial \bar{\mathbf{x}}} + \frac{\partial \bar{\mathbf{v}}}{\partial \bar{\mathbf{x}}}^{\mathsf{T}}\right) = \tfrac{1}{2}(\dot{\mathbf{Q}}\mathbf{Q}^{\mathsf{T}} + \mathbf{Q}\dot{\mathbf{Q}}^{\mathsf{T}}) + \mathbf{Q}\mathbf{D}\mathbf{Q}^{\mathsf{T}}, \tag{4.5.9}$$

but, since $\mathbf{Q}\mathbf{Q}^{\mathsf{T}} = \mathbf{I}$, the first term vanishes and thus

$$\bar{\mathbf{D}} = \mathbf{Q}\mathbf{D}\mathbf{Q}^{\mathsf{T}}. \tag{4.5.10}$$

The spin tensor for the new flow

$$\begin{aligned}\bar{\boldsymbol{\varpi}} &= \tfrac{1}{2}\left(\frac{\partial \bar{\mathbf{v}}}{\partial \bar{\mathbf{x}}} - \frac{\partial \bar{\mathbf{v}}}{\partial \bar{\mathbf{x}}}^{\mathsf{T}}\right) \\ &= \tfrac{1}{2}(\dot{\mathbf{Q}}\mathbf{Q}^{\mathsf{T}} - \mathbf{Q}\dot{\mathbf{Q}}^{\mathsf{T}}) + \mathbf{Q}\boldsymbol{\varpi}\mathbf{Q}^{\mathsf{T}} = \mathbf{Q}(\mathbf{Q}^{\mathsf{T}}\dot{\mathbf{Q}} + \boldsymbol{\varpi})\mathbf{Q}^{\mathsf{T}}.\end{aligned} \tag{4.5.11}$$

If σ is the stress from the original deformation process, the stress associated with the new flow including the time-dependent rotation is

$$\bar{\sigma} = \mathbf{Q}\sigma\mathbf{Q}^{\mathsf{T}}, \tag{4.5.12}$$

and in terms of $\bar{\mathbf{D}}$, $\bar{\boldsymbol{\varpi}}$ we can express $\bar{\sigma}$ as

$$\bar{\sigma} = \mathbf{s}(\bar{\mathbf{D}}, \bar{\boldsymbol{\varpi}}, \rho, T). \tag{4.5.13}$$

Now, equating the representations for $\bar{\sigma}$ from (4.5.12) with (4.5.1) and (4.5.13), using (4.5.10), (4.5.11), we require

$$\mathbf{s}(\mathbf{Q}\mathbf{D}\mathbf{Q}^{\mathsf{T}}, \mathbf{Q}(\mathbf{Q}^{\mathsf{T}}\dot{\mathbf{Q}} + \boldsymbol{\varpi})\mathbf{Q}^{\mathsf{T}}, \rho, T) = \mathbf{Q}\mathbf{s}(\mathbf{D}, \boldsymbol{\varpi}, \rho, T)\mathbf{Q}^{\mathsf{T}}, \tag{4.5.14}$$

for all rotations $\mathbf{Q}$. Suppose now that $\mathbf{Q} = I$, $\dot{\mathbf{Q}} \neq 0$, then

$$\mathbf{s}(\mathbf{D}, \dot{\mathbf{Q}} + \boldsymbol{\varpi}, \rho, T) = \mathbf{s}(\mathbf{D}, \boldsymbol{\varpi}, \rho, T), \tag{4.5.15}$$

and so $\mathbf{s}$ must be independent of the spin tensor $\boldsymbol{\varpi}$. We can therefore reduce (4.5.14) to the form

$$\mathbf{s}(\mathbf{Q}\mathbf{D}\mathbf{Q}^{\mathsf{T}}, \rho, T) = \mathbf{Q}\mathbf{s}(\mathbf{D}, \rho, T)\mathbf{Q}^{\mathsf{T}}, \tag{4.5.16}$$

for all rotations $\mathbf{Q}$.

Following the arguments of Section 4.4.2, the isotropy property (4.5.16) requires that $\sigma = \mathbf{s}(\mathbf{D}, \rho, T)$ and $\mathbf{D}$ are coaxial. The most general form for σ is therefore

$$\sigma = a\mathbf{I} + b\mathbf{D} + c\mathbf{D}^2. \tag{4.5.17}$$

where a, b, c are functions of ρ and T and the invariants of $\mathbf{D}$: $\operatorname{tr}\mathbf{D}$, $\det\mathbf{D}$, $\frac{1}{2}[\operatorname{tr}\mathbf{D}^2 - (\operatorname{tr}\mathbf{D})^2]$, which are symmetric functions of the principal values d_1, d_2, d_3.

In the absence of motion $\mathbf{D} \equiv 0$ and

$$\sigma = \mathbf{s}(0, \rho, T) = -p(\rho, T)\mathbf{I}, \tag{4.5.18}$$

where $p(\rho, T)$ is the hydrostatic pressure.

The classical Newtonian viscous fluid is a special case of (4.5.17) in which the stress is linear in the velocity gradient

$$\sigma_{ij} = -p(\rho, T)\delta_{ij} + b_{ijkl}D_{kl}. \tag{4.5.19}$$

The b_{ijkl} are the components of a fourth-order isotropic tensor with $i \leftrightarrow j$, $k \leftrightarrow l$ symmetry so

$$b_{ijkl} = \zeta\delta_{ij}\delta_{kl} + \eta(\delta_{ik}\delta_{jl} + \delta_{il}\delta_{jk}), \tag{4.5.20}$$

and thus we can write

$$\sigma_{ij} = [-p(\rho, T) + \zeta(\rho, T)]\delta_{ij} + 2\eta(\rho, T)D_{ij}. \tag{4.5.21}$$

For a dilatational flow from the origin

$$\begin{aligned} &\mathbf{v} = d\mathbf{r}, \qquad d_{ij} = d\delta_{ij}, \\ &\sigma_{ij} = [-p(\rho, T) + (\zeta + \tfrac{2}{3}\eta)]\delta_{ij}. \end{aligned} \tag{4.5.22}$$

and so $\zeta + \frac{2}{3}\eta$ is called the *coefficient of bulk viscosity* (and is often assumed to be zero).

For a shear flow

$$\begin{aligned} &v_1 = dx_2, \qquad v_2 = v_3 = 0, \\ &d_{12} = d_{21} = \tfrac{1}{2}d, \quad \text{all other} \quad d_{ij} = 0, \end{aligned} \tag{4.5.23}$$

and

$$\sigma_{12} = \sigma_{21} = \eta d, \quad \sigma_{11} = \sigma_{22} = \sigma_{33} = -p, \quad \sigma_{23} = \sigma_{13} = 0, \tag{4.5.24}$$

so that η is known as the coefficient of *shear viscosity* (more commonly just *viscosity*).

The explicit form of the stress tensor when the bulk viscosity vanishes (*a Stokes fluid*) is

$$\sigma_{ij} = -p\delta_{ij} + \eta\left(\frac{\partial v_i}{\partial x_j} + \frac{\partial v_j}{\partial x_i}\right) - \frac{2}{3}\eta\frac{\partial v_k}{\partial x_k}\delta_{ij}, \tag{4.5.25}$$

and this gives an excellent description of many fluids e.g. water, air.

The additional dependence on D^2 in (4.5.17) is insufficient to provide an adequate description of the flow of many composite materials, e.g., blood, polymeric liquids which deviate from linear dependence on rate of strain, these materials have a memory of deformation which can be described using viscoelastic constitutive laws.

4.6 Viscoelasticity

We can characterise a solid by a stress-state which depends on strain and a fluid by a stress-state which depends on the rate of strain. The intermediate situation in which stress depends on the strain and strain-rate via the history of deformation covers a wide variety of behaviour described as viscoelastic.

We will confine our attention to situations in which the strains are small so that the relation between stress and strain is linear. Even this development of *linear viscoelasticity* will allow the description of quite complex behaviour.

We assume that the stress tensor σ depends on the Cauchy strain tensor e by a linear transformation

$$\sigma(\mathbf{x}, t) = \mathbf{C}\{\mathbf{e}^t\}, \tag{4.6.1}$$

where, as in Chapter 4, $\mathbf{e}^t$ denotes the history of strain up to time t. This linear relation is assumed to be single-valued and to possess a unique inverse relation

$$e(\mathbf{x}, t) = \mathbf{J}\{\sigma^t\}, \tag{4.6.2}$$

Because the relation is linear $\mathbf{C}$, $\mathbf{J}$ represent tensor functions, e.g.,

$$\sigma_{ij}(\mathbf{x}, t) = C_{ijpq}\{e^t_{pq}\}, \tag{4.6.3}$$

with ij, pq symmetry associated with the symmetries of the stress and strain tensor.

Models of viscoelastic behaviour

One of the most useful models that can be built from the mechanical elements introduced in Section 1.1.3 is the *standard linear solid* consisting of a spring in series with a spring and dashpot combination.

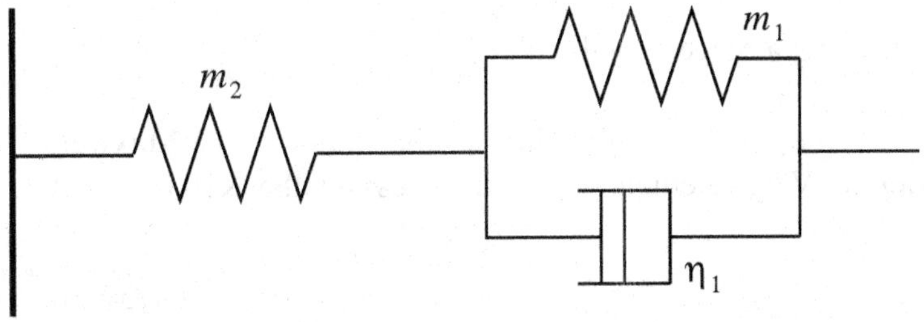

Figure 4.1. Mechanical model for a standard linear solid, built of springs and dashpot elements.

If the strain in the parallel spring and dashpot is e_1 then the stress is

$$\sigma = m_1 e_1 + \eta_1 \dot{e}_1 = m_2 e_2, \tag{4.6.4}$$

since this must also balance the strain e_2 in the second spring. The total strain is

$$e = e_1 + e_2, \tag{4.6.5}$$

and thus

$$\mathfrak{m}_1\mathfrak{m}_2 e + \eta_1\mathfrak{m}_2\dot{e} = (\mathfrak{m}_1 + \mathfrak{m}_2)\sigma + \eta_1\dot{\sigma}. \tag{4.6.6}$$

We may rewrite this relation as

$$\dot{\sigma} + \frac{\sigma}{\tau_R} = \mathfrak{m}_2\left(\dot{e} + \frac{e}{\tau_C}\right), \tag{4.6.7}$$

and the characteristic time scales for relaxation (τ_R) and creep (τ_C) are given by

$$\tau_R = \frac{\eta_1}{\mathfrak{m}_1 + \mathfrak{m}_2}, \quad \tau_C = \frac{\eta_1}{\mathfrak{m}_1}. \tag{4.6.8}$$

This model gives a three parameter description ($\mathfrak{m}_2$, τ_R, τ_C) of a viscoelastic material and covers all the mechanical features of such solids. For very rapid deformations the terms in $\dot{\sigma}$ and $\dot{e}$ are important and the material behaves like an elastic solid with dynamic modulus $\mathfrak{m}_2$; for very slow deformation the behaviour is again elastic with a static modulus $\tau_R\mathfrak{m}_2/\tau_C$.

A Burgers material provides a further generalisation of allowed behaviour by including the possibility of long-term viscosity, with the introduction of a further dashpot. The Burgers model thus consists of a Maxwell element and a Kelvin–Voigt element in series.

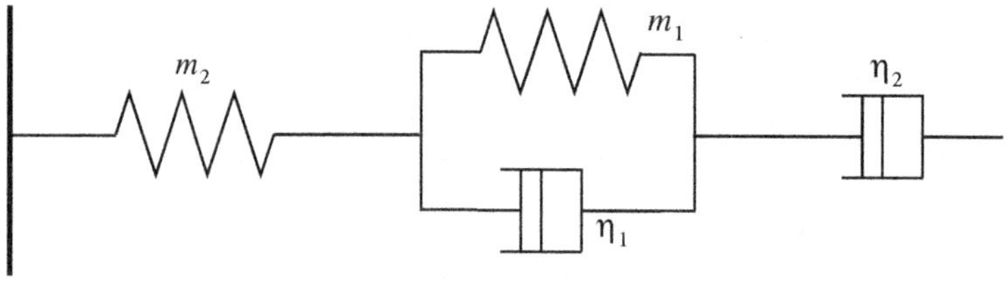

Figure 4.2. Mechanical model for a Burgers solid.

The total strain $\epsilon = \epsilon_1 + \epsilon_2$ is the combination of the strains in the Maxwell element ϵ_1 and the Kelvin–Voigt element ϵ_2. The stress in the Kelvin–Voigt element is

$$\sigma = \mathfrak{m}_1(\epsilon - \epsilon_2) + \eta_1(\dot{\epsilon} - \dot{\epsilon}_2), \tag{4.6.9}$$

and for the Maxwell element

$$\epsilon_2 = \frac{\dot{\sigma}}{\mathfrak{m}_2} + \frac{\sigma}{\eta_2}. \tag{4.6.10}$$

Now, by eliminating ϵ_2 between (4.6.9) and (4.6.10), we can derive the constitutive relation for the Burgers model in the form

$$\mathfrak{m}_2\ddot{\epsilon} + \frac{\mathfrak{m}_1\mathfrak{m}_2}{\eta_2}\dot{\epsilon} = \ddot{\sigma} + \left(\frac{\mathfrak{m}_1 + \mathfrak{m}_2}{\eta_1} + \frac{\mathfrak{m}_2}{\eta_2}\right)\dot{\sigma} + \frac{\mathfrak{m}_1\mathfrak{m}_2}{\eta_1\eta_2}\sigma, \tag{4.6.11}$$

which we can rewrite as

$$m_2\ddot{\epsilon} + \frac{m_1}{\tau_M}\dot{\epsilon} = \ddot{\sigma} + \left(\frac{1}{\tau_R} + \frac{1}{\tau_M}\right)\dot{\sigma} + \frac{1}{\tau_C\tau_M}\sigma. \tag{4.6.12}$$

where we have introduced the Maxwell time $\tau_M = \eta_2/m_2$, in addition to the characteristic time scales for relaxation (τ_R) and creep (τ_C) introduced above.

Under constant stress σ_0, the evolution of strain for the Burgers model takes the form

$$\epsilon = \frac{\sigma_0}{m_2} + \frac{\sigma_0}{m_1}\left[1 - \exp\left(-\frac{t}{\tau_C}\right)\right] + \frac{\sigma_0}{\eta_2}t. \tag{4.6.13}$$

The strain relation (4.6.13) separates rather neatly into an initial elastic response from the first spring element, continuing creep from the Kelvin–Voigt element and viscous flow from the final dashpot. The flexibility of the Burgers model means that it can provide a useful representation of a wide range of behaviour.

4.7 Plasticity and flow

Although a linear dependence of stress on strain-rate provides a reasonable description of many fluids, it is less representative of the slow deformation of solids. At high temperature and low stresses, polycrystalline silicate minerals frequently show a power-law dependence of strain rate on stress

$$\dot{\epsilon} = A\sigma^n, \qquad \text{with } n > 1. \tag{4.7.1}$$

A similar relation has also been used to describe glacier flow. This simple power-law dependence reflects the macroscopic effects of plastic deformation and rearrangement within the polycrystalline materials.

The influence of weak non-linearity will be most pronounced when comparing deformation at different time scales, since the effective viscosity will depend on strain rate and hence the frequency of deformation. For example, although the long-term effects of glacial rebound following the melting of the major ice sheets of the last glacial can be quite well represented using a Newtonian viscosity for the Earth's mantle, the effective viscosity deduced from modern high-precision studies using satellite based geodesy is slightly different. The difference is sufficient to suggest that weakly non-linear viscosity is present in the upper mantle.

EXAMPLE: NON-LINEAR VISCOSITY

The behaviour of certain viscous fluids can be modelled by the constitutive equation

$$\sigma_{ij} = -p\delta_{ij} + 2\eta(K_2)D_{ij}$$

where $K_2 = 2D_{ij}D_{ij}$, for rate-of-strain tensor D_{ij}, and $\eta(K_2) = kK_2^{(n-1)/2}$, where k and n are positive constants ($n = 1$ corresponds to a Newtonian fluid).

Such a power-law fluid undergoes a simple shearing flow between two large parallel plates a distance h apart, such that one plate is held fixed and the other moves with a constant speed U in its plane. Show that the shearing force per unit area on the plates is $k(U/h)^n$ and that the apparent viscosity is $k(U/h)^{n-1}$ as a function of the shear rate U/h.

Take cartesian axes such that the 1-axis is directed along the flow, and the 3-axis across flow. The fluid will be in stationary contact with the plates so that $v_1(0) = 0$, $v_1(h) = U$. The linear velocity field $\mathbf{v} = (Uz/h, 0, 0)$ The rate-of-deformation tensor D_{ij} will only have non-zero components $D_{13} = D_{31} = U/2h$, so that $K_2 = 2D_{ij}D_{ij} = U^2/h^2$. Thus $\eta(K_2) = k(U/h)^{n-1}$ and

$$\sigma_{ij} = -p\delta_{ij} + k(U/h)^n;$$

hence

$$\sigma_{13} = \sigma_{13} = k(U/h)^n, \qquad \sigma_{11} = \sigma_{22} = \sigma_{33} = p.$$

The traction on the top plate is $\tau_j = \sigma_{3j}$ and so

$$\tau_1 = k(U/h)^n, \quad \tau_2 = 0, \quad \tau_3 = -p.$$

The magnitude of the shearing force per unit area is therefore $k(U/h)^n$. The apparent viscosity $\bar{\eta}$ such that

$$\sigma_{ij} = -p\delta_{ij} + \bar{\eta}D_{ij},$$

is $\bar{\eta} = k(U/h)^{n-1}$. •

5

Linearised Elasticity and Viscoelasticity

The only known continua which are capable of 'finite' elastic strain and which have also comparable resistance to shear and compression are the new strong fibres or whiskers of ceramics based on carbon, boron etc.

In common metals the elastic range of strain is only *infinitesimal* and may be treated by a linearised theory. In the Earth as well, small additional disturbances can be treated by linearising about an existing stress state.

5.1 Linearisation of deformation

The assumption we shall make in subsequent work is that the strain is small, i.e., the difference $\mathsf{U} - \mathbf{I}$ is first order. However, for bodies such as slender columns or thin panels, the rotation can be large even though the strain is small, and in this case we have to retain rotation terms R. In general, we can take $\mathsf{R} - \mathbf{I}$ to also be small, and work with a fully linearised theory.

In terms of particle displacement

$$\mathbf{u} = \mathbf{x} - \boldsymbol{\xi} = \mathsf{A}\boldsymbol{\xi}, \quad \mathsf{A} = \mathsf{F} - \mathbf{I}, \tag{5.1.1}$$

where A is the displacement gradient. To first order in A, the Green strain tensor

$$\mathsf{E} = \tfrac{1}{2}(\mathsf{F}^{\mathsf{T}}\mathsf{F} - \mathbf{I}) = \tfrac{1}{2}(\mathsf{A} + \mathsf{A}^{\mathsf{T}}) + \cdots, \tag{5.1.2}$$

and also the incremental stretch

$$(\mathsf{U} - \mathbf{I}) = \tfrac{1}{2}(\mathsf{A} + \mathsf{A}^{\mathsf{T}}) + \cdots. \tag{5.1.3}$$

Now, from (5.1.1) we can express the displacement gradient as

$$\mathsf{A} = \mathsf{R}\mathsf{U} - \mathbf{I} = (\mathsf{R} - \mathbf{I}) + (\mathsf{U} - \mathbf{I}) + \cdots, \tag{5.1.4}$$

so that we have a representation for the rotation R

$$(\mathsf{R} - \mathbf{I}) = \tfrac{1}{2}(\mathsf{A} - \mathsf{A}^{\mathsf{T}}) + \cdots, \tag{5.1.5}$$

but for a first-order rotation we also have

$$(\mathsf{R} - \mathbf{I}) = \theta\mathsf{N} = \theta\,\mathbf{n}\times. \tag{5.1.6}$$

To first order the Green strain tensor elements are given by

$$E_{ij} = \tfrac{1}{2}\left(\frac{\partial u_i}{\partial \xi_j} + \frac{\partial u_j}{\partial \xi_i}\right) + \cdots, \tag{5.1.7}$$

and the distinction between E_{ij} and the Cauchy strain tensor e_{ij} is negligible. The fractional dilatation

$$E_{kk} = \frac{\partial u_k}{\partial \xi_k} + \cdots = \operatorname{div}\mathbf{u} + \cdots . \tag{5.1.8}$$

The rotation yields

$$\theta\,\mathbf{n} = \tfrac{1}{2}\boldsymbol{\nabla}\times\mathbf{u} + \cdots = \tfrac{1}{2}\operatorname{curl}\mathbf{u} + \cdots . \tag{5.1.9}$$

In this linearised theory we will use $\mathbf{x}$ rather than $\boldsymbol{\xi}$ to denote the reference position of a particle.

5.2 The elastic constitutive relation

We make an expansion of the strain energy function about the reference state

$$2W(\mathsf{E}) = a + b_{ij}E_{ij} + c_{ijkl}E_{ij}E_{kl} + O(\mathsf{E}^3), \tag{5.2.1}$$

and since E_{ij} is symmetric we can arrange $i \leftrightarrow j$ and $k \leftrightarrow l$ symmetry in the coefficients and also $ij \leftrightarrow kl$ symmetry.

$$b_{ij} = b_{ji}, \quad c_{ijkl} = c_{jikl} = c_{ijlk} = c_{klij}. \tag{5.2.2}$$

If the reference state is stress-free $b_{ij} = 0$. In the Earth, for example, there is a large hydrostatic pressure field due to self-gravitation and then $\tfrac{1}{2}b_{ij} = -p\delta_{ij}$. *Note:* the fact that earthquakes occur means that there must also be non-hydrostatic stresses representable by the second term in (5.2.1).

Now from the relation between the stress tensor and the strain derivative of the strain energy, (4.3.11), we have

$$\det\mathsf{F}\,\boldsymbol{\sigma} = \mathsf{F}\frac{\partial W}{\partial \mathsf{E}}\mathsf{F}^{\mathsf{T}}. \tag{5.2.3}$$

Thus, in the absence of non-hydrostatic pre-stress, when E is first order but $\mathbf{R}$ is unrestricted

$$\boldsymbol{\sigma} = \mathbf{R}\frac{\partial W}{\partial \mathsf{E}}\mathbf{R}^{\mathsf{T}} + O(\mathsf{E}^2), \tag{5.2.4}$$

as noted above this semi-linearised relation is needed for slender bodies.

For full linearity $\mathbf{A} - \mathbf{I}$ is first order, and

$$\boldsymbol{\sigma} = \frac{\partial W}{\partial \mathsf{E}} + O(\mathsf{E}^2), \tag{5.2.5}$$

and so the component representation of the stress tensor, using (5.1.1), becomes

$$\sigma_{ij} = \tfrac{1}{2}b_{ij} + c_{ijkl}E_{kl} + O(\mathsf{E}^2). \tag{5.2.6}$$

For a stress-free ground state (or working in terms of incremental stresses) we have the *generalised form of Hooke's Law*

$$\sigma_{ij} = c_{ijkl} e_{kl} + O(e^2), \qquad W' = \tfrac{1}{2}\sigma_{ij} e_{ij} + O(e^3), \tag{5.2.7}$$

where we have used the traditional notation for the strain, in view of the equivalence of the Cauchy and Green strain definitions for small deformation.

In this full anisotropic case the fourth order tensor of elastic moduli c_{ijkl} is restricted by the $i \leftrightarrow j$, $k \leftrightarrow l$ and $ij \leftrightarrow kl$ symmetries to 21 independent components. The presence of material symmetry further reduces the number of independent moduli, e.g. for hexagonal symmetry there are five moduli.

5.2.1 Isotropic response

If the material is isotropic then c_{ijkl} must be an isotropic tensor subject to the required symmetries in permutation of indices, and so

$$c_{ijkl} = \lambda \delta_{ij} \delta_{kl} + \mu(\delta_{ik}\delta_{jl} + \delta_{il}\delta_{jk}). \tag{5.2.8}$$

Thus the stress–strain relation is

$$\sigma_{ij} = \lambda \delta_{ij} e_{kk} + 2\mu e_{ij}; \tag{5.2.9}$$

λ and μ are known as the Lamé moduli.

The inverse relation to (5.2.9) is most conveniently expressed in terms of another set of constants

$$E e_{ij} = (1+\upsilon)\sigma_{ij} - \upsilon \sigma_{kk} \delta_{ij}, \tag{5.2.10}$$

where E is Young's modulus and υ is Poisson's ratio.

5.2.2 Nature of moduli

The interpretation of these moduli can be obtained by looking at simple deformations.

(a) Consider a *purely volumetric strain*

$$\begin{aligned} &e_{ij} = 0, \quad i \neq j, \\ &e_{11} = e_{22} = e_{33} = e, \end{aligned} \tag{5.2.11}$$

then the stress components are

$$\sigma_{ij} = 0, \quad i \neq j, \qquad \text{and} \quad \sigma_{11} = 3\kappa e, \ \text{etc.} \tag{5.2.12}$$

$\kappa = \lambda + \frac{2}{3}\mu$ is called the *bulk modulus* and we have a hydrostatic stress field.

(b) If we consider a *simple shear deformation*

$$x_1 = \xi_1 + \gamma \xi_2, \quad x_2 = \xi_2, \quad x_3 = \xi_3, \tag{5.2.13}$$

the only non-zero infinitesimal strain component is $e_{12} = \frac{1}{2}\gamma$ and the corresponding stress $\sigma_{12} = \mu\gamma$, with all other $\sigma_{ij} = 0$. μ is called the *shear or rigidity modulus*. In general, if the dilatation $\sum_k e_{kk}$ vanishes $\sigma_{ij} = 2\mu e_{ij}$.

(c) Consider *uniaxial tension* $\sigma_{11} = \sigma_1, \sigma_{22} = \sigma_{33} = 0, \sigma_{ij} = 0, \;\; i \neq j$, then

$$\begin{aligned} e_{11} &= \sigma_1/E, \\ e_{22} &= -\upsilon\sigma_1/E, \qquad e_{ij} = 0, \quad i \neq j, \\ e_{33} &= -\upsilon\sigma_1/E, \end{aligned} \tag{5.2.14}$$

so υ is the quotient of transverse contraction and longitudinal extension.

In terms of κ and μ the work function can be written as

$$W = \tfrac{1}{2}\kappa e_{kk}^2 + \mu(e_{ij} - \tfrac{1}{3}e_{kk}\delta_{ij})(e_{ij} - \tfrac{1}{3}e_{qq}\delta_{ij}), \tag{5.2.15}$$

and this will be positive definite if $\kappa > 0$, $\mu > 0$.

5.2.3 Interrelations between moduli

The interrelations between the moduli may be obtained by comparing the descriptions of different deformations

$$\sigma_{kk} = 3\kappa e_{kk}, \quad e_{kk} = \frac{1-2\upsilon}{E}\sigma_{kk}, \tag{5.2.16}$$

and

$$\begin{aligned} 2\mu e_{ij} &= \sigma_{ij} - \frac{\lambda}{3\lambda + 2\mu}\sigma_{kk}\delta_{ij} \\ Ee_{ij} &= (1+\upsilon)\sigma_{ij} - \upsilon\sigma_{kk}\delta_{ij}. \end{aligned} \tag{5.2.17}$$

On combining the results of (5.2.16) and (5.2.17) we obtain

$$\begin{aligned} \frac{E}{1+\upsilon} &= 2\mu, \qquad \frac{E}{1-2\upsilon} = 3\kappa, \\ \frac{3}{E} &= \frac{1}{\mu} + \frac{1}{3\kappa}, \qquad 2\upsilon = \frac{3\kappa - 2\mu}{3\kappa + \mu}. \end{aligned} \tag{5.2.18}$$

For positive-definite W we note that $\kappa > 0$ with $\mu > 0$ is equivalent to $E > 0$ with $-1 < \upsilon < \frac{1}{2}$. Normally υ is positive since there is transverse contraction under compression.

The limiting case of incompressibility (or second-order volume change) is obtained by letting $\kappa \to \infty$ with μ fixed so that $E \to 3\mu$, and $\upsilon \to \frac{1}{2}$.

The notation we have used for the elastic moduli is common in elasticity and seismology. However, in the engineering and materials physics literature the notation conventions are somewhat different: the bulk modulus κ is designated K and the shear modulus μ is represented by G.

5.2.4 An example of linearisation

Suppose we have a work function specified in terms of principal stretches as in (4.4.16), so that

$$W = \phi(\lambda_1\lambda_2\lambda_3) + \sum_r \psi(\lambda_r), \tag{5.2.19}$$

with

$$\phi = \frac{c}{mn}[(\lambda_1\lambda_2\lambda_3)^{-mn} - 1], \quad \psi(\lambda_r) = \frac{c}{m}[\lambda_r^m - 1], \tag{5.2.20}$$

for which the principal stress is

$$\sigma_r = c[\lambda_r^m - \nu^{-mn}]/\nu \quad \text{with} \quad \nu = \lambda_1\lambda_2\lambda_3. \tag{5.2.21}$$

Near the reference state

$$\sigma_r = mc[n(e_1 + e_2 + e_3) + e_r] + O(e^2). \tag{5.2.22}$$

More general W may be constructed by summing over various (c, m) pairs keeping n fixed.

Comparing with the isotropic constitutive relation in the principal frame

$$\sigma_r = 2\mu e_r + \lambda(e_1 + e_2 + e_3), \tag{5.2.23}$$

we can identify the shear modulus $\mu = \frac{1}{2}\sum mc$, and Poisson's ratio $\upsilon = n/(1 + 2n)$.

Under a uniaxial load σ_1 with arbitrary magnitude

$$\begin{aligned} &\lambda_2 = (\lambda_1\lambda_2^2)^{-n}, \quad \text{since} \quad \lambda_2 = \lambda_3, \\ &\log\lambda_2 = -\upsilon\log\lambda_1. \end{aligned} \tag{5.2.24}$$

5.2.5 *Elastic constants*

Typical values of material parameters at 300 K are listed in the following tables. They relate to random aggregates of anisotropic single crystals (at least 10^{12} m^{-3}). Such aggregates are effectively isotropic and homogeneous at a macroscopic level (the volume of a test specimen being of order 10^{-5} m^3. The values are means of the best data available and have been 'normalised' so as to be theoretically consistent (within the accuracy of the decimal places given).

Common metals

	λ	μ[G]	κ[K]	E	υ
Aluminium	0.60	0.265	0.775	0.715	0.35
Copper	1.09	0.45	1.39	1.22	0.355
Nickel	1.39	0.76	1.90	2.01	0.32
Lead	0.38	0.074	0.43	0.21	0.42
Gold	1.48	0.28	1.67	0.795	0.42
Silver	0.82	0.27	1.00	0.755	0.37
Iron	1.18	0.83	1.73	2.145	0.29
Molybdenum	1.66	1.60	2.73	4.02	0.255

Material properties in the Earth

	λ	$\mu[G]$	$\kappa[K]$	E	υ
2500 km deep	4.195	2.723	6.011	7.096	0.303
1000 km deep	2.180	1.820	3.420	4.630	0.272
20 km deep	0.429	0.323	0.645	0.830	0.285

Units for λ, $\mu[G]$, $\kappa[K]$, E are $10^{11}\ \mathrm{N\,m^{-2}}$.

In metals, the elastic range of strain is terminated by plastic yielding, e.g., for Cu and Al the tensile yield stress $\approx 17.5 \times 10^6\ \mathrm{N\,m^{-2}}$ so that

$$\begin{aligned} e_{\text{yield}} &\approx 1.4 \times 10^{-4} \quad \text{for} \quad \text{Cu}, \\ e_{\text{yield}} &\approx 2.4 \times 10^{-4} \quad \text{for} \quad \text{Al}. \end{aligned} \tag{5.2.25}$$

Even for stronger constructional materials such as mild steel and aluminium alloys, the yield strains are around 10^{-3}.

For Earth materials, the linear approximation works well for strains less than 10^{-5} which is rarely exceeded except in the immediate vicinity of earthquakes and underground nuclear explosions.

5.2.6 The uniqueness theorem

Locally the stress-field in a body satisfies (3.2.4)

$$\frac{\partial}{\partial x_i}\sigma_{ij} + \rho g_j = \rho f_j, \tag{5.2.26}$$

where $\mathbf{g}$ is the body force and $\mathbf{f}$ is the acceleration.

In the linearised approximation, the acceleration $\mathbf{f} = \partial^2\mathbf{u}/\partial t^2$ in terms of the displacement $\mathbf{u}$. The advected terms $\mathbf{v} \cdot \boldsymbol{\nabla}$ can be neglected. Consider a bounded elastic region V with given boundary conditions on S:
(i) $\boldsymbol{\tau}$ specified on part of the surface,
(ii) $\mathbf{u}$ specified,
(iii) normal component of $\mathbf{u}$ and tangential component of τ (or vice versa),
and a locally positive-definite strain energy density $W = \frac{1}{2}\sigma_{ij}e_{ij}$.

Suppose we have Cauchy initial conditions

$$\mathbf{u}(\mathbf{x}, 0) = \mathbf{u}(\mathbf{x}), \quad \frac{\partial u}{\partial t}(\mathbf{x}, 0) = \mathbf{u}'(\mathbf{x}), \tag{5.2.27}$$

where $\mathbf{u}$, $\mathbf{u}'$ and the body force $\mathbf{g}$ are bounded in V. Let $\mathbf{u}$, $\mathbf{u}'$ be two distinct solutions under these conditions and denote their difference by $\delta\mathbf{u}$.

Then $\delta\mathbf{u}$ satisfies

$$\rho\frac{\partial^2}{\partial t^2}\delta u_i = \frac{\partial}{\partial x_j}\delta\sigma_{ij}, \quad \delta\mathbf{u} = \frac{\partial}{\partial t}\delta\mathbf{u} = 0 \quad \text{at} \quad t = 0, \tag{5.2.28}$$

and $\delta\mathbf{u} \cdot \delta\boldsymbol{\tau} = 0$ on the surface S.

Consider

$$0 = \int_0^t dt \int_V dV \frac{\partial}{\partial t}\delta u_i \left[\rho \frac{\partial^2}{\partial t^2}\delta u_i - \frac{\partial}{\partial x_j}\delta\sigma_{ij}\right]$$

$$= \frac{1}{2}\int_V dm \left[\frac{\partial}{\partial t}\delta u_i\right]^2 + \int_0^t dt \int_V dV \frac{\partial}{\partial t}\left[\frac{\partial}{\partial x_j}\delta u_i\right]\delta\sigma_{ij} - \int_0^t dt \int_S dS\, n_j \delta\sigma_{ij}\frac{\partial}{\partial t}\delta u_i$$

$$= \int_V dV\, K(\delta\mathbf{u}) + \int_0^t dt \int_V dV \frac{\partial}{\partial t}\delta e_{ij}\,\delta\sigma_{ij} - \int_0^t dt \int_S dS\, \delta\tau_i \frac{\partial}{\partial t}\delta u_i. \quad (5.2.29)$$

The last term vanishes identically, the first term is the positive definite kinetic energy and the second is the total elastic energy associated with $\delta\mathbf{u}$

$$W = \int_V \tfrac{1}{2}\,\delta e_{ij}\,\delta\sigma_{ij}\, dV, \quad (5.2.30)$$

which by hypothesis is positive definite.

Thus $\partial(\delta\mathbf{u})/\partial t$ and δe_{ij} are zero everywhere in V and so $\delta\mathbf{u}$ is a static displacement with $\delta\mathbf{u} = 0$ at $t = 0$. This can only be achieved if

$$\delta\mathbf{u} = 0 \text{ in } V \text{ for all } t > 0, \quad (5.2.31)$$

i.e., the solution is unique.

Kirchhoff's Theorem

Positive-definite W at every point $\mathbf{x}$ suffices for uniqueness.

For isotropic media this condition reduces to

$$\begin{array}{lll} & \kappa > 0 & \text{with} \quad \mu > 0, \\ \textbf{or} & E > 0 & \text{with} \quad -1 < \upsilon < \tfrac{1}{2}. \end{array} \quad (5.2.32)$$

EXAMPLE: STATIC ELASTIC FIELDS

Show that in an isotropic elastic body which does not include the origin and for which $\sigma_{ij} = \lambda\delta_{ij}e_{kk} + 2\mu e_{ij}$ the two displacement fields (a) $\mathbf{u}_1(\mathbf{x}) = c\mathbf{x}$, (b) $\mathbf{u}_2(\mathbf{x}) = d\mathbf{x}/r^3$, and c, d small constants, are self-equilibrating.

Consider a spherical shell of internal radius a and external radius b subject to uniform external pressure p. Neglecting body forces, find the equilibrium displacement and stress fields. Identify the principal stresses.

If the material fails as soon as the largest principal stress exceeds F and deformation can be regarded as elastic up to failure, find the shell thickness h needed to sustain a pressure p when $h \gg a$.

For a self-equilibrated field, the stress gradient $\partial\sigma_{ij}/\partial x_j = 0$.

(a) The displacement $u_i = cx_i$ and the strain

$$e_{ij} = \frac{1}{2}\left(\frac{\partial u_i}{\partial x_j} + \frac{\partial u_j}{\partial x_i}\right) = c\delta_{ij}$$

The associated stress tensor

$$\sigma_{ij} = \lambda\delta_{ij}3c + 2\mu c\delta_{ij} = (3\lambda + 2\mu)c\delta_{ij}$$

is uniform, so the stress gradient vanishes and the field is self-equilibrated.
(b) The displacement $u_i = dx_i/r^3$ and $r\partial r/\partial x_j = x_j$. The strain

$$e_{ij} = d\left[\frac{\delta_{ij}}{r^3} - 3\frac{x_i x_j}{r^5}\right]$$

and the associated stress

$$\sigma_{ij} = \lambda\delta_{ij} d\left[\frac{3}{r^3} - 3\frac{x_k x_k}{r^5}\right] + 2\mu\, d\left[\frac{\delta_{ij}}{r^3} - 3\frac{x_i x_j}{r^5}\right],$$

the first term vanishes so we are left with just the shear components. The stress gradient

$$\begin{aligned}\frac{\partial\sigma_{ij}}{\partial x_j} &= 2\mu d\left[-3\frac{\delta_{ij}x_j}{r^5} - \frac{3}{r^5}\delta_{ij}x_j - \frac{3}{r^5}3x_i + \frac{15}{r^7}x_j x_i x_j\right] \\ &= 6\frac{\mu d}{r^5}[-x - i - x_i - 3x_i + 5x_i] = 0,\end{aligned}$$

and so the field is self-equilibrated.
Both $\mathbf{u}_1$ and $\mathbf{u}_2$ are radial and we may superpose them to give a radial field to match the pressure loading.
(a) $\mathbf{u}_1$ gives a hydrostatic pressure $(3\lambda + 2\mu)c$.
(b) $\mathbf{u}_2$ in a locally radial frame has principal components $(-2, 1, 1)2\mu d/r^3$.
Set $\mathbf{u} = \mathbf{u}_1 + \mathbf{u}_2$, then the radial stress is

$$\sigma_{rr} = (3\lambda + 2\mu)c - \frac{4\mu d}{r^3}.$$

At $r = a$ we require no radial traction at the free surface, and at $r = b$ we must match the external pressure p. Thus we require

$$0 = (3\lambda + 2\mu)c - \frac{4\mu d}{a^3}, \quad -p = (3\lambda + 2\mu)c - \frac{4\mu d}{b^3}.$$

so that c and d can be determined from

$$p = 4\mu d\left[\frac{1}{b^3} - \frac{1}{a^3}\right], \quad b^3 p = (3\lambda + 2\mu)c(a^3 - b^3).$$

The displacement field is thus

$$\mathbf{u} = \frac{b^3 p}{a^3 - b^3}\left(\frac{1}{3\lambda + 2\mu} + \frac{1}{4\mu}\frac{a^3}{r^3}\right)\mathbf{x},$$

with associated stress field

$$\sigma_{ij} = \frac{b^3 p}{a^3 - b^3}\left(\delta_{ij} + \tfrac{1}{2}a^3\left[\frac{\delta_{ij}}{r^3} - \frac{3}{r^5}x_i x_j\right]\right).$$

Consider a locally radial frame, e.g., pointing along $(x, 0, 0)$, then

$$\{\sigma_{xx}, \sigma_{yy}, \sigma_{zz}\} = \frac{b^3 p}{a^3 - b^3}\left\{1 - \frac{a^3}{r^3}, 1 + \frac{a^3}{2r^3}, 1 + \frac{a^3}{2r^3}\right\}.$$

The radial principal stress has functional form $1 - a^3/r^3$ and the hoop stresses in the

orthogonal directions behave as $1 + a^3/2r^3$.
The largest principal stress is $\eta = b^3 p(1 + a^3/2r^3)/(a^3 - b^3)$. When $h = b - a$ is small, $\eta \approx a^3 \frac{1}{3}p/3ha^2 = \frac{1}{2}pa/h$.
Thus if η must be less than F to avoid failure, $\frac{1}{2}pa/h \leq F$, and so $h \geq pa/2F$ to sustain the pressure p. •

5.3 Integral representations

We can supplement the differential equations for a single elastic field with a set of integral relations that connect the properties of two different fields. We are thereby able to use the behaviour of known solutions to infer information about other situations without undertaking a full solution. We can also derive a useful representation for the displacement field in terms of the Green's tensor for the medium. This relation will prove to be very useful in the description of sources of elastic energy, e.g., faulting in an earthquake.

Consider the displacement field in an elastic body under different loadings

(i) $\mathbf{u}(\mathbf{x})$ with $\boldsymbol{\tau}$ on S, $\mathbf{g}$ in V ,

(ii) $\mathbf{u}^*(\mathbf{x})$ with $\boldsymbol{\tau}^*$ on S, $\mathbf{g}^*$ in V.

Then

$$\rho \partial_{tt} u_i = \rho g_i + \partial_j \sigma_{ij}, \tag{5.3.1}$$

$$\rho \partial_{tt} u_i^* = \rho g_i^* + \partial_j \sigma_{ij}^*. \tag{5.3.2}$$

Take the scalar product of (5.3.1) with u_i^* and of (5.3.2) with u_i and then take their difference. Now integrate throughout V and over time to get

$$\begin{aligned} &\int_{t_0}^{t} dt \int_V dm\,(\partial_{tt} u_i u_i^* - u_i \partial_{tt} u_i^*) \\ &\quad = \int_{t_0}^{t} dt \int_V dm\,(g_i u_i^* - u_i g_i^*) + \int_{t_0}^{t} dt \int_V dV\,(\partial_j \sigma_{ij} u_i^* - u_i \partial_j \sigma_{ij}^*). \end{aligned} \tag{5.3.3}$$

The volume integral over the stress

$$\int_V dV (\partial_j \sigma_{ij} u_i^* - u_i \partial_j \sigma_{ij}^*) = \int_V dV \partial_j (\sigma_{ij} u_i^* - u_i \sigma_{ij}^*) - \int_V dV (\sigma_{ij} e_{ij}^* - e_{ij} \sigma_{ij}^*), \tag{5.3.4}$$

and

$$\sigma_{ij} e_{ij}^* = c_{ijkl} e_{kl} e_{ij}^* = c_{klij} e_{ij}^* e_{kl} = \sigma_{kl}^* e_{kl}. \tag{5.3.5}$$

Thus (5.3.3) can be written as

$$\int_V dV\rho \left[\left(\frac{\partial u_i}{\partial t} u_i^* - u_i \frac{\partial u_i^*}{\partial t} \right) \right]_{t_0}^{t}$$
$$= \int_{t_0}^{t} dt \int_V dm(g_i u_i^* - u_i g_i^*) + \int_{t_0}^{t} dt \int_S dS(\tau_i u_i^* - u_i \tau_i^*). \quad (5.3.6)$$

5.3.1 The reciprocal theorem

Take the time integration up to infinity and assume that **u**, $\mathbf{u}^*$ are such that the left-hand side of (5.3.6) vanishes, then

$$\int_{-\infty}^{\infty} dt \left[\int_V dm\, g_i u_i^* + \int_S dS\, \tau_i u_i^* \right] = \int_{-\infty}^{\infty} dt \left[\int_V dm\, g_i^* u_i + \int_S dS\, \tau_i^* u_i^* \right], \quad (5.3.7)$$

and for a *static* field we can drop the time integral

$$\int_V dm\, g_i u_i^* + \int_S dS\, \tau_i u_i^* = \int_V dm\, g_i^* u_i + \int_S dS\, \tau_i^* u_i^*. \quad (5.3.8)$$

This is Betti's reciprocal theorem.

In applications of Betti's reciprocal theorem (5.3.8) we take the required field to be **u** and choose an appropriate starred field for which the solution is known and which has as close a character as we can achieve to the desired field.

EXAMPLES: APPLICATIONS OF THE RECIPROCAL THEOREM

(i) *Influence of self-gravitation*

Self gravity is switched on in a homogeneous sphere – find the decrease in radius:

At the surface $r = a$, $\tau = 0$. Let g be the acceleration due to self gravity, then the interior body force per unit mass is $\mathbf{g} = -g\mathbf{x}/a$.

Choose $\mathbf{u}^* = \epsilon \mathbf{x}$ implying uniform $\sigma_{ij}^* = 3\kappa\epsilon\delta_{ij}$, then $\mathbf{g}^* = 0$ because the field is self-equilibrated. From (5.3.8)

$$\int_V -\frac{g\mathbf{x}}{a} \cdot \epsilon \mathbf{x} \rho \, dV = 3\kappa\epsilon 4\pi a^2 u(a),$$

and hence

$$\frac{-g\rho\epsilon}{a} \frac{4\pi}{5} a^5 = 3\kappa\epsilon 4\pi a^2 u(a),$$

i.e., the decrease in radius is $-u(a) = \rho g a^2/(15\kappa)$. •

(ii) *Loaded cylinder*

A solid circular cylinder of height h and radius a is subjected to compressive loads L applied via rigid plates in contact with its ends. Radial displacements over the plates are opposed by a uniform frictional force τ (Figure 5.1). Find the reduction in height δh when L is large enough to cause sliding everywhere over the ends.

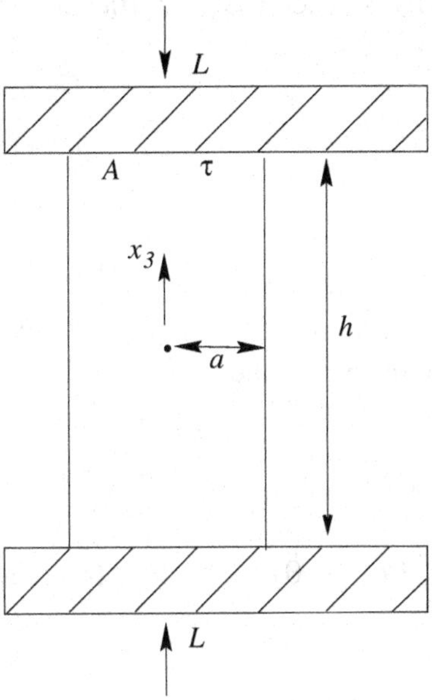

Figure 5.1. Configuration of loaded cylinder and force distribution

Let x_3 be measured axially from the centroid, and choose

$$\mathbf{u}^* = (\upsilon x_1, \upsilon x_2, -x_3)p^*/E,$$

corresponding to a uniaxial pressure p^*, so $\mathbf{g}^* = 0$. Now also $\mathbf{g} = 0$ in V, $\boldsymbol{\tau}^* = \boldsymbol{\tau} = 0$ over the curved surfaces. Thus, denoting an end by A,

$$\int \boldsymbol{\tau} \cdot \mathbf{u}^* dA = \int \boldsymbol{\tau}^* \cdot \mathbf{u}\, dA,$$

the load $L = -\int \tau^3 dA$, and so

$$\frac{p^*}{E}\left(\tfrac{1}{2}HL - \upsilon \int_0^a r\tau 2\pi r\, dr\right) = p^* \tfrac{1}{2}\pi a^2\, \delta h.$$

Thus

$$\delta h = \frac{h}{E}\left(\frac{L}{\pi a^2} - \frac{4\upsilon a \tau}{3h}\right). \bullet$$

5.3.2 The representation theorem

Consider a delta-function body force

$$g_i = \delta_{ik}\delta(\mathbf{x} - \boldsymbol{\xi})\delta(t - \theta), \tag{5.3.9}$$

acting in a volume V. The resulting displacement at $\mathbf{x}$, t is specified by the Green's tensor with components

$$G_i^k(\mathbf{x}, t; \boldsymbol{\xi}, \theta), \tag{5.3.10}$$

associated with the force in the k-direction at $\boldsymbol{\xi}, \theta$. Even when the properties of the region vary G_i^k will depend on the time difference $t - \theta$.

If we take g_i as above and set

$$g_i^* = \delta_{il}\delta(\mathbf{x} - \boldsymbol{\zeta})\delta(\sigma - t), \tag{5.3.11}$$

the displacement

$$u_i^* = G_i^l(\mathbf{x}, -t; \boldsymbol{\zeta}, -\sigma). \tag{5.3.12}$$

We assume that both G_i^k and G_i^l obey the *same* homogeneous boundary conditions on S, and then when we apply the reciprocal theorem (5.3.7) the surface integrals vanish

$$\begin{aligned}\int_{-\infty}^{\infty} dt \int_V dV\, \delta_{ik}\delta(\mathbf{x} - \boldsymbol{\xi})\delta(t - \theta)G_i^l(\mathbf{x}, -t; \boldsymbol{\zeta}, -\sigma)\\ = \int_{-\infty}^{\infty} dt \int_V dV\, \delta_{ik}\delta_{il}\delta(\mathbf{x} - \boldsymbol{\zeta})\delta(\sigma - t)G_i^k(\mathbf{x}, t; \boldsymbol{\xi}, \theta),\end{aligned} \tag{5.3.13}$$

so that when $\boldsymbol{\xi}, \boldsymbol{\zeta}$ are in V

$$G_k^l(\boldsymbol{\xi}, -\theta; \boldsymbol{\zeta}, -\sigma) = G_l^k(\boldsymbol{\zeta}, \sigma; \boldsymbol{\xi}, \theta). \tag{5.3.14}$$

Equation (5.3.14) expresses Green's tensor reciprocity.

Now set

$$g_i^* = \delta_{ik}\delta(\mathbf{x} - \boldsymbol{\xi})\delta(\theta - t) \tag{5.3.15}$$

in (5.3.6) and then

$$u_i^* = G_i^k(\mathbf{x}, -t; \boldsymbol{\xi}, -\theta), \tag{5.3.16}$$

and so

$$\begin{aligned}0 = \int_{t_0} d\theta \int_V dV_x\{g_i(\mathbf{x})G_i^k(\mathbf{x}, -t; \boldsymbol{\xi}, -\theta) - u_i(\mathbf{x})\delta(\mathbf{x} - \boldsymbol{\xi})\delta(\theta - t)\}\\ + \int_{t_0} d\theta \int_S dS_x\{\tau_i(\mathbf{x})G_i^k(\mathbf{x}, -t; \boldsymbol{\xi}, -\theta) - u_i(\mathbf{x})H_i^k(\mathbf{x}, -t; \boldsymbol{\xi}, -\theta)\}\\ + \int_V dV_x[\partial_t u_i(\mathbf{x})G_i^k - u_i\partial_t G_i^k]_{t=t_0},\end{aligned} \tag{5.3.17}$$

where H_i^k is the traction associated with G_i^k.

Now, using the reciprocity of the Green's tensor we can obtain a representation of the displacement as an integral over the boundaries, body forces and initial conditions

$$\begin{aligned}\Theta(\boldsymbol{\xi})u_k(\boldsymbol{\xi}, t) = \int_{t_0} ds \int_V dV_x\, g_i(\mathbf{x})G_k^i(\boldsymbol{\xi}, t; \mathbf{x}, s)\\ + \int_{t_0} ds \int_S dS_x\{\tau_i(\mathbf{x})G_k^i(\boldsymbol{\xi}, t; \mathbf{x}, s) - u_i(\mathbf{x})H_k^i(\boldsymbol{\xi}, t; \mathbf{x}, s)\}\\ + \int_V dV_x[\partial_t u_i(\mathbf{x})G_k^i - u_i\partial_t G_k^i]_{t=t_0}.\end{aligned} \tag{5.3.18}$$

where

$$\Theta(\boldsymbol{\xi}) = 1, \quad \boldsymbol{\xi} \text{ in } V,$$
$$= 0, \quad \text{otherwise}.$$

5.4 Elastic waves

The equation of motion and the stress-strain relation for small disturbances in an elastic medium, in the absence of body forces, can be combined to give

$$\frac{\partial}{\partial x_j}\left(c_{ijkl}\frac{\partial u_k}{\partial x_l}\right) = \rho\frac{\partial^2 u_i}{\partial t^2}, \tag{5.4.1}$$

as the governing differential equation for displacement $\mathbf{u}$. Equation (5.4.1) controls the spatial and temporal development of the displacement field and admits solutions in the form of travelling waves. Consider then a plane wave travelling in the anisotropic medium with frequency ω:

$$\begin{aligned} u_i &= U_i \exp[i\omega p\mathbf{n}\cdot\mathbf{x} - i\omega t] \\ &= U_i \exp[i\omega p n_k x_k - i\omega t]. \end{aligned} \tag{5.4.2}$$

$\mathbf{n}$ represents the direction of travel of the phase fronts and p the apparent slowness (inverse of wave velocity) in that direction. On substituting this plane wave form into (5.4.1) we obtain

$$\omega^2 p^2 c_{ijkl} n_j n_l U_k = -\omega^2 \rho U_i, \tag{5.4.3}$$

which constitutes an eigenvalue problem for the slowness p for waves travelling in the direction $\mathbf{n}$

$$[p^2 c_{ijkl} n_j n_l - \rho\delta_{ik}]U_k = 0. \tag{5.4.4}$$

Equation (5.4.4) has three roots for p^2 for each direction $\mathbf{n}$, associated with polarisations specified by the eigenvectors $\mathbf{U}^{(r)}(\mathbf{n})$. The slownesses p_r are independent of frequency.

In a general anisotropic medium, the slownesses vary with direction and can give quite complex slowness and wave surfaces as, e.g., those illustrated in Figure 5.2.

5.4.1 Isotropic media

For an isotropic medium, the eigenvalue equation for slowness takes the form

$$[p^2(\lambda+\mu)n_i n_k + p^2\mu\delta_{ik} - \rho\delta_{ik}]U_k = 0, \tag{5.4.5}$$

where we have used the isotropic form for the elastic modulus tensor c_{ijkl}.

(i) P waves

Consider a displacement field oriented along the propagation direction, i.e.,

$$\mathbf{u}_P = c\mathbf{n}, \tag{5.4.6}$$

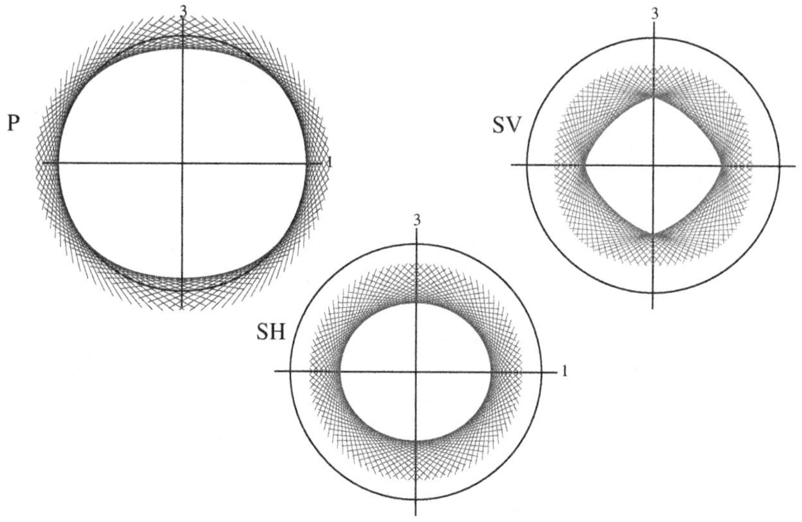

Figure 5.2. Wave surfaces for a transversely isotropic (hexagonal) medium formed from the envelope of plane wavefronts for a unit time of propagation.

then the slowness equation is

$$[(\lambda + 2\mu)p^2 - \rho]n_i = 0, \tag{5.4.7}$$

so that we have a wave disturbance with slowness

$$a^2 = \rho/(\lambda + 2\mu), \tag{5.4.8}$$

and associated phase velocity

$$\alpha = [(\lambda + 2\mu)/\rho]^{1/2}. \tag{5.4.9}$$

This longitudinal wave solution is called the *P* wave

$$\mathbf{u}_P = A_P \mathbf{n} \exp[i\omega(a\mathbf{n} \cdot \mathbf{x} - t)]. \tag{5.4.10}$$

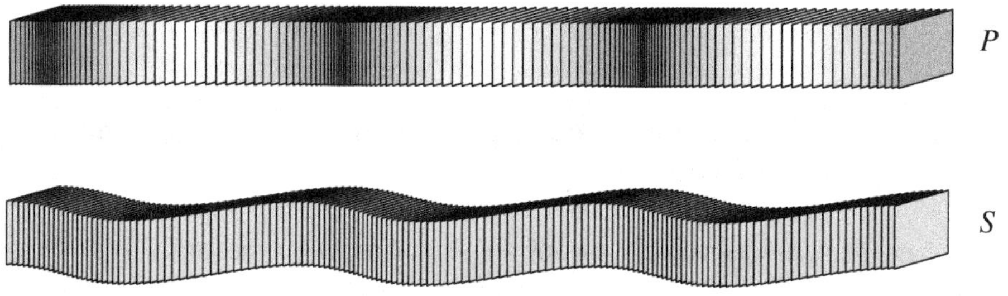

Figure 5.3. Elastic waves in a uniform isotropic medium: *P* waves with longitudinal motion and *S* waves with transverse (shear) motion.

(ii) S waves

There is an alternative type of elastic wave motion in which the displacement is transverse to the direction of motion. Consider a displacement field

$$\mathbf{u}_S = c\mathbf{s}, \quad \text{with} \quad \mathbf{s}\cdot\mathbf{n} = 0, \tag{5.4.11}$$

for which the slowness equation reduces to

$$[\mu p^2 - \rho]s_i = 0, \tag{5.4.12}$$

so that we have a wave disturbance with slowness

$$b^2 = \rho/\mu \tag{5.4.13}$$

and associated phase velocity

$$\beta = [\mu/\rho]^{1/2}. \tag{5.4.14}$$

This form of solution holds for any direction orthogonal to the direction of motion, i.e., we have a degenerate eigenvalue problem from which we can choose two orthogonal *S* wave vectors. It is conventional to choose one vector in the vertical plane (denoted *SV*) and the other purely horizontal (denoted *SH*). This choice simplifies the analysis of wave propagation in horizontally stratified media, and so we represent the *S* wave field as

$$\mathbf{u}_S = \{B_v\mathbf{s}_V + B_H\mathbf{s}_H\}\exp[i\omega(b\mathbf{n}\cdot\mathbf{x} - t)], \tag{5.4.15}$$

where $\mathbf{s}_V$ lies in the vertical plane through $\mathbf{n}$ and $\mathbf{s}_H$ in the horizontal plane $[\mathbf{s}_V\cdot\mathbf{n} = \mathbf{s}_H\cdot\mathbf{n} = 0]$.

5.4.2 Green's tensor for isotropic media

The wave disturbance at an observation point $\mathbf{r}$ produced by a delta-function point force in the direction $\mathbf{d}$ at the origin in an unbounded isotropic medium can be represented as

$$4\pi\mathbf{u}(\mathbf{r}) = (\mathbf{d}\cdot\hat{\mathbf{r}})\hat{\mathbf{r}}\frac{\delta(t - r/\alpha)}{r\alpha^2} + \hat{\mathbf{r}}\times(\mathbf{d}\times\hat{\mathbf{r}})\frac{\delta(t - r/\beta)}{r\beta^2} + (3(\mathbf{d}\cdot\hat{\mathbf{r}})\hat{\mathbf{r}} - \mathbf{d})\frac{t}{r^3}[H(t - r/\alpha) - H(t - r/\beta)], \tag{5.4.16}$$

where $\hat{\mathbf{r}}$ is a unit vector in the direction $\mathbf{r}$ and $H(t)$ is the Heaviside step function. This result was first derived by Stokes in 1851 and can be deduced by many different techniques, e.g., Fourier transform methods or the superposition of potential solutions.

The disturbance consists of two delta functions in time, one longitudinal travelling with the faster P-wave velocity α and the other with transverse displacement and the slower shear wave velocity β (this can be represented in terms of *SV* and *SH* components if required). Between these two wavefronts there

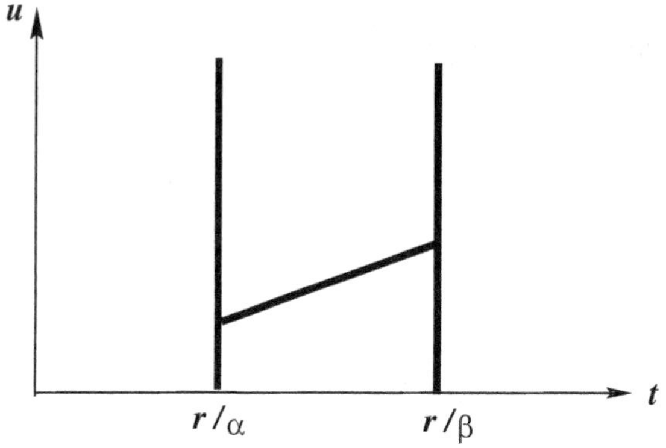

Figure 5.4. The displacement from a point force in an unbounded medium with distinct *P* and *S* arrivals linked by a 'near-field' ramp disturbance.

is a 'near-field' disturbance with intermediate character which decays more rapidly with distance r (Figure 5.4).

The specific form of the Green's tensor $G_i^k(\mathbf{x}, t; 0, 0)$ for an observation point with direction cosines γ_i is thus

$$G_i^k(\mathbf{x}, t) = (3\gamma_i\gamma_j - \delta_{ij})\frac{t}{r^3}[H(t - r/\alpha) - H(t - r/\beta)]$$
$$+\gamma_i\gamma_j\frac{1}{r\alpha^2}\delta(t - r/\alpha) + (\gamma_i\gamma_j - \delta_{ij})\frac{1}{r\beta^2}\delta(t - r/\beta). \qquad (5.4.17)$$

The radiation pattern for the far-field terms which have a r^{-1} decay with distance r away from the source takes a fairly simple form: the *P* waves depends on the cosine of the inclination to the force direction and the *S* wave pattern on the sine of the inclination (Figure 5.5).

More realistic sources can be obtained by using couples and dipoles whose associated displacements can be derived by differentiation of the Green's tensor. The sense of differentiation is that of the displacement in the couple or dipole. The double couple in Figure 5.5(c) is the equivalent force system for an infinitesimal shear displacement on a surface aligned with one of the arms of the couple, and the resulting radiation patterns are used in the interpretation of faulting mechanisms from seismological observations.

5.4.3 Interfaces

The solution presented in the previous section is for elastic waves in an unbounded medium, and more complicated behaviour results once material interfaces or a free surface is introduced.

At a free surface, the requirement is that the traction should vanish. The surface acts as a mirror for horizontally polarised *S* waves (*SH*); an incident plane *SH* wave is reflected without change of amplitude and the *SH* component of the Green's

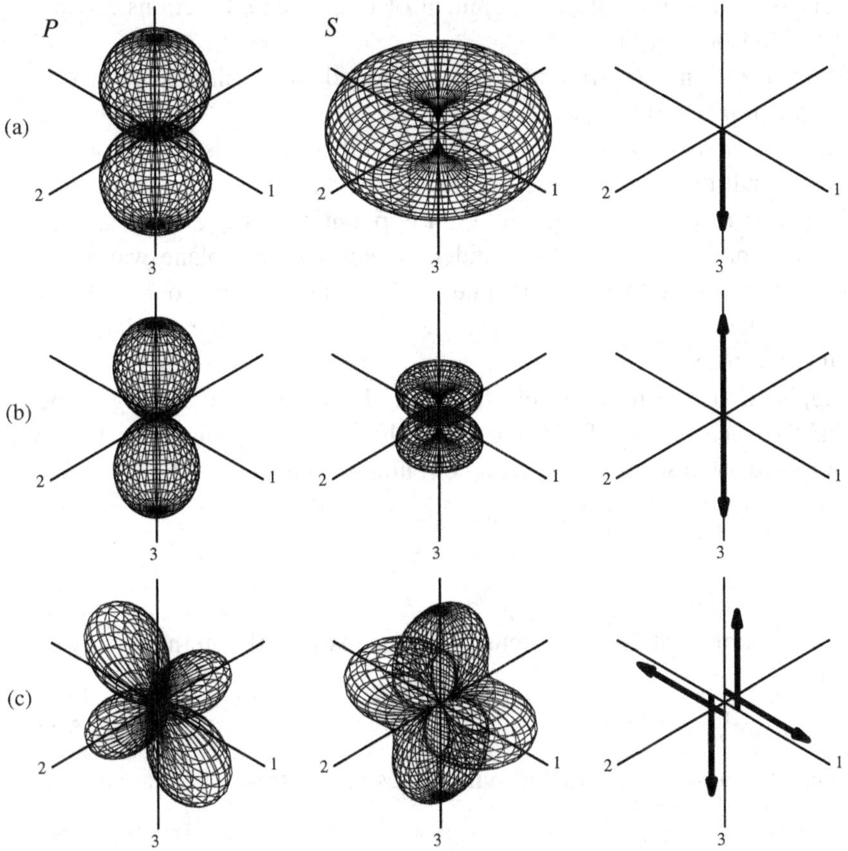

Figure 5.5. Far-field radiation patterns for *P* and *S* waves from (a) a single point force, (b) a vertical dipole, (c) a double couple.

function within the half-space can be found by introducing a fictitious mirrored source lying above the free surface. The situation is more complex for vertically polarised waves. An incident plane *P* or *SV* wave will be reflected from the free surface with an equal inclination to the normal to the surface, in addition there will be conversion between *P* and *SV* waves with the inclination of the converted wave determined by Snell's law

$$\frac{\sin \mathrm{i}}{\alpha_0} = \frac{\sin \mathrm{j}}{\beta_0}, \tag{5.4.18}$$

where i is the inclination to the normal of the *P* wave with surface wavespeed α_0, and j is the inclination of the *SV* wave with surface wavespeed β_0. In addition there is a special form of wave that satisfies the vanishing traction condition called a Rayleigh wave. This consists of coupled exponentially decaying *P* and *S* waves travelling with a horizontal velocity about 0.9 of the shear wave velocity. A surface source in a half-space excites the Rayleigh wave in addition to the *P* and *SV* body waves; an elegant treatment of the development of the Green's functions for this case is given by Hudson (1981).

At an interface between two different media the boundary conditions are that
(i) the displacement $\mathbf{u}$ is continuous,
(ii) the traction $\boldsymbol{\tau}(\mathbf{n})$ associated with the normal $\mathbf{n}$ to the interface is continuous.
These boundary conditions can only be satisfied by the coupling of *P* and *S* waves at the interface. For a plane wave incident on a horizontal interface the *P* and *SV* wave fields are coupled, but *SH* is independent. Such incident plane waves are both reflected and transmitted through the interface with inclinations to the vertical dictated by Snell's law, which may be viewed as a condition for the continuity of phase along the interface.

A major application of linearised elastic wave theory is in seismology, and a comprehensive development of seismic wave theory with applications to the interpretation of observed seismograms can be found in Kennett (2001, 2002), see also Chapter 11.

5.5 Linear viscoelasticity

As in (4.6.1) we assume that the stress tensor $\boldsymbol{\sigma}$ depends linearly on the history of strain $\mathbf{e}^t$ so that

$$\sigma_{ij}(\mathbf{x}, t) = C_{ijpq}\{e^t_{pq}\}, \tag{5.5.1}$$

with ij, pq symmetry associated with the symmetries of the stress and strain tensor.

Creep and relaxation

Consider imposing a step-function stress cycle on a wire (i.e., a one-dimensional problem)

$$\sigma(t) = \sigma_0 H(t), \tag{5.5.2}$$

then the resulting strain for $t > 0$ can be written as

$$e(t) = \frac{\sigma_0}{E_D}[1 + \psi(t)], \quad \psi(0) = 0, \tag{5.5.3}$$

where E_D is the *dynamic modulus* and σ_0/E_D is an instantaneous elastic response. The monotonically increasing function $\psi(t)$ is known as the *creep function*. If $\psi(t)$ approaches a constant as $t \to \infty$, then we have long-term solid behaviour, whereas for materials like pitch whose ultimate behaviour is that of a (viscous) fluid $\psi(t)$ increases without limit.

We can extend this concept to a more general stress history, by considering the superposition of many small steps and then taking the limit as the step interval tends to zero, to obtain

$$e(t) = E_D^{-1}\left[\sigma(t) + \int_{-\infty}^{t} ds\, \dot{\sigma}(s)\psi(t-s)\right], \tag{5.5.4}$$

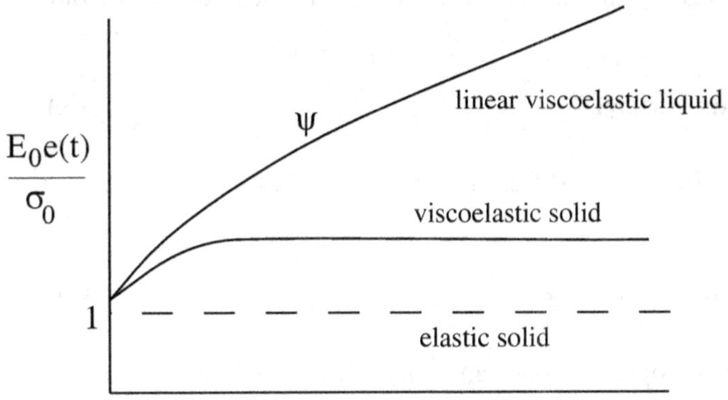

Figure 5.6. Strain evolution as a function of time.

a convolution integral over stress rate. If the creep function is differentiable for $t \geq 0$, we may use the properties of the convolution to rewrite the strain as

$$e(t) = E_D^{-1}\left[\sigma(t) + \int_{-\infty}^{t} ds\, \sigma(s)\dot{\psi}(t-s)\right]. \tag{5.5.5}$$

An alternative viewpoint is to consider a strain cycle for which

$$e(t) = e_0 H(t), \tag{5.5.6}$$

with associated stress for $t > 0$

$$\sigma(t) = E_D e_0[1 - \phi(t)], \quad \phi(0) = 0. \tag{5.5.7}$$

$\phi(t)$ is termed the relaxation function and is a monotonically increasing function with a monotonically decreasing slope. If the long-term behaviour $\phi(\infty) = 0$ we have a liquid.

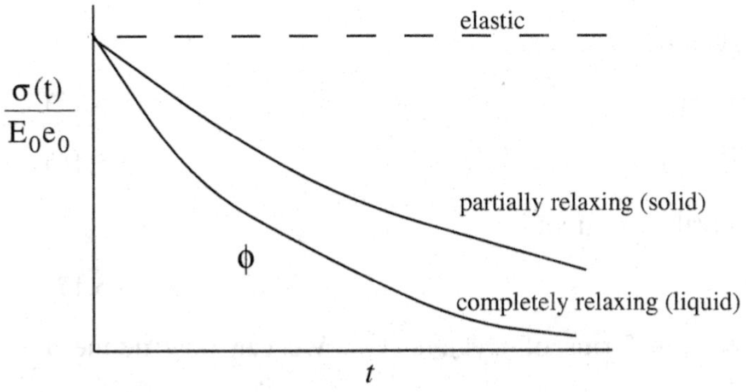

Figure 5.7. Stress evolution as a function of time.

For an arbitrary differentiable strain history we can represent the stress behaviour as

$$\sigma(t) = E_D\Big[e(t) - \int_{-\infty}^{t} ds\, \dot{e}(s)\phi(t-s)\Big]. \tag{5.5.8}$$

Once again, if the relaxation function is differentiable for $t \geq 0$, we can reorganise the convolution to give

$$\sigma(t) = E_D\Big[e(t) - \int_{-\infty}^{t} ds\, e(s)\dot{\phi}(t-s)\Big]. \tag{5.5.9}$$

Since (5.5.4), (5.5.8) have to represent the same mechanical behaviour we can establish the relation between the creep and relaxation functions. Consider the application of a constant unit stress, then

$$e(t) = E_D^{-1}[1 + \psi(t)], \tag{5.5.10}$$

and so from (5.5.8) for $t > 0$

$$1 = 1 + \psi(t) - \int_{0}^{t} ds\dot{\psi}(s)\phi(t-s). \tag{5.5.11}$$

Simplifying (5.5.11), we require

$$\psi(t) = \int_{0}^{t} ds\, \dot{\psi}(s)\phi(t-s), \quad t > 0. \tag{5.5.12}$$

Similarly, we find

$$\phi(t) = \int_{0}^{t} ds\, \dot{\phi}(s)\psi(t-s), \quad t > 0. \tag{5.5.13}$$

These two coupled integral equations have their simplest solution in the Laplace transform domain. We define, e.g.,

$$\bar{\psi}(p) = \int_{0}^{\infty} dt\, \psi(t)e^{-pt}, \tag{5.5.14}$$

and then (5.5.5) and (5.5.9) can be written as

$$\bar{e}(p) = E_D^{-1}[1 + p\bar{\psi}(p)]\bar{\sigma}(p), \tag{5.5.15}$$

$$\bar{\sigma}(p) = E_D[1 - p\bar{\phi}(p)]\bar{e}(p). \tag{5.5.16}$$

Since these relations are equivalent we require

$$[1 + p\bar{\psi}(p)][1 - p\bar{\phi}(p)] = 1, \tag{5.5.17}$$

which connects the Laplace transforms of $\psi(t)$, $\phi(t)$. We can now define a transform modulus

$$E(p) = E_D[1 - p\bar{\phi}(p)] = E_D[1 + p\bar{\psi}(p)]^{-1}, \tag{5.5.18}$$

in terms of which the one-dimensional constitutive relation for linear viscoelasticity can be written as

$$\bar{\sigma}(p) = E(p)\bar{e}(p). \qquad (5.5.19)$$

The formal analogy between the Laplace transform domain result (5.5.19) and the equivalent elastic relation provides the formal basis for solving many viscoelastic problems via the *correspondence principle*.

5.6 Viscoelastic behaviour

The relaxation function for the standard linear solid model introduced in Section 4.6 is

$$\phi(t) = [1 - (\tau_R/\tau_C)][1 - e^{-t/\tau_R}], \qquad (5.6.1)$$

and the corresponding creep function is

$$\psi(t) = [(\tau_C/\tau_R) - 1][1 - e^{-t/\tau_C}]. \qquad (5.6.2)$$

Although the standard linear solid model goes a long way towards providing a description of a viscoelastic material it does not provide an adequate description of any real material. More satisfactory representations are provided by joining standard linear solid elements in series or parallel.

In the series case, the stress in each element is the same whilst the strain is the sum of the strains in each element. The creep function is then a sum of terms of the form (5.6.2)

$$\psi(t) = \sum_n f(\tau_n)[1 - e^{-t/\tau_n}], \qquad (5.6.3)$$

where $f(\tau_n)$ may be regarded as the contribution of a creep process with a characteristic time τ_n.

For a set of parallel elements, the stress is cumulative and the relaxation function can be built from a set of terms of the form (5.6.1)

$$\phi(t) = \sum_n g(\tau'_n)[1 - e^{-t/\tau'_n}]. \qquad (5.6.4)$$

In the most general case we can consider a spectrum of relaxation and creep processes.

The creep function for the Burgers material takes the form

$$\psi(t) = \psi_U + \Delta\psi[1 - e^{-t/\tau}] + \frac{t}{\eta_M}, \qquad (5.6.5)$$

where ψ_U is the reciprocal of the relevant instantaneous modulus, $\Delta\psi$ describes the extent of viscoelastic relaxation with a relaxation time τ_C and η_M is the viscosity of the Newtonian viscous element. The equivalent Maxwell relaxation time for the influence of viscosity is $\tau_M = \eta_M\psi_U$. Once again more general behaviour can be produced by a sum or spectrum of creep and relaxation processes, and such a combination of Burgers elements can be used as a versatile representation of deformation in the mantle of the Earth.

Isotropic linear viscoelasticity

For an isotropic medium the tensor function C_{ijpq} in (5.1.5) can be represented as an isotropic tensor of the form

$$C_{ijpq} = L\delta_{ij}\delta_{pq} + M(\delta_{ip}\delta_{jq} + \delta_{iq}\delta_{jp}), \tag{5.6.6}$$

where L and M are scalar functions, and so

$$\sigma_{ij}(\mathbf{x}, t) = \delta_{ij}L\{e^t_{kk}\} + 2M\{e^t_{ij}\}, \tag{5.6.7}$$

and we can separate the dependence of the dilatation component from the deviatoric stress and strain.

By analogy with the equations of isotropic elasticity we can write the stress–strain relation as

$$\begin{aligned}\sigma_{ij}(\mathbf{x}, t) &= \lambda_0\delta_{ij}e_{kk}(\mathbf{x}, t) + 2\mu_0 e_{ij}(\mathbf{x}, t)\\ &\quad + \int_0^t ds\, \delta_{ij}e_{kk}(\mathbf{x}, s)\dot{R}_\lambda(t-s) + \int_0^t ds\, 2e_{ij}(\mathbf{x}, s)\dot{R}_\mu(t-s),\end{aligned} \tag{5.6.8}$$

in terms of relaxation functions R_λ, R_μ and a representation analogous to (5.5.9) for the one-dimensional case.

An equivalent representation to (5.6.8) expressing the strain as a function of the stress history can be made in terms of creep functions C_λ, C_μ. These are related to the relaxation functions via coupled integral equations similar to (5.5.13). The correspondence principle in the Laplace transform domain can be extended to these three-dimensional problems and provides a means of solving many viscoelastic problems.

5.7 Damping of harmonic oscillations

We consider a harmonic strain cycle of the form

$$e_{ij} = E_{ij}e^{-i\omega t}, \tag{5.7.1}$$

and then the corresponding stress is

$$\begin{aligned}\sigma_{ij}e^{i\omega t} &= \lambda_0\delta_{ij}E_{kk} + 2\mu_0 E_{ij}\\ &\quad + \int_{-T}^{t} ds\, [\delta_{ij}E_{kk}\dot{R}_\lambda(t-s) + 2E_{ij}\dot{R}_\mu(t-s)]e^{i\omega(t-s)},\end{aligned} \tag{5.7.2}$$

where $t = -T$ is the time at which the disturbance starts. Now let $T \to \infty$ and introduce the Fourier transforms of the time derivatives of the relaxation functions

$$\lambda_1(\omega) = \int_0^\infty dt\, \dot{R}_\lambda(t)e^{i\omega t}, \quad \mu_1(\omega) = \int_0^\infty dt\, \dot{R}_\mu(t)e^{i\omega t}. \tag{5.7.3}$$

We can express the stress as

$$\sigma_{ij} = \{[\lambda_0 + \lambda_1(\omega)]\delta_{ij}E_{kk} + 2[\mu_0 + \mu_1(\omega)]E_{ij}\}e^{-i\omega t}, \tag{5.7.4}$$

with a stress–strain relation in the same form as for an elastic medium but with frequency-dependent complex moduli

$$\bar{\lambda} = \lambda_0 + \lambda_1(\omega), \quad \bar{\mu} = \mu_0 + \mu_1(\omega). \tag{5.7.5}$$

We may therefore employ the descriptions of *P* and *S* waves developed in section 5.4, but have to allow for the frequency dependent velocities and consequent energy loss.

The rate of energy dissipation per unit volume due to deformation of the medium is from (4.2.11)

$$\rho \frac{D}{Dt} U = \sigma_{ij} \frac{\partial v_j}{\partial x_i} \approx \sigma_{ij} \dot{e}_{ij}, \tag{5.7.6}$$

in the linear approximation. The average work dissipated per unit time and volume is thus

$$\bar{W} = \tfrac{1}{2} \mathrm{Re}[i\omega\sigma_{ij}E^*_{ij}e^{i\omega t}] = -\tfrac{1}{2}\omega\, \mathrm{Im}[\bar{\kappa}\Delta\Delta^* + 2\bar{\mu}\hat{E}_{ij}\hat{E}^*_{ij}], \tag{5.7.7}$$

where the dilatation $\Delta = E_{kk}$, and the deviatoric strain $\hat{E}_{ij} = E_{ij} - \frac{1}{3}\Delta\delta_{ij}$. The complex bulk modulus

$$\bar{\kappa} = \bar{\lambda} + \tfrac{2}{3}\bar{\mu} = \kappa_0 + \kappa_1(\omega). \tag{5.7.8}$$

A convenient measure of the rate of energy dissipation is the loss factor $Q^{-1}(\omega)$ which may be defined as

$$Q^{-1} = \frac{\Delta\bar{W}(\omega)}{2\pi W_0(\omega)}, \tag{5.7.9}$$

where $\Delta\bar{W}(\omega)$ is the energy loss in a cycle at frequency ω and $W_0(\omega)$ is the 'elastic' energy stored in the oscillation, i.e., the sum of the strain and kinetic energy associated with just the instantaneous elastic moduli. For purely deviatoric disturbances

$$Q_\mu^{-1}(\omega) = \frac{-\frac{1}{2}\mathrm{Im}\,\bar{\mu}(\omega)\hat{E}_{ij}\hat{E}^*_{ij}}{\frac{1}{2}\hat{E}_{ij}\hat{E}^*_{ij}} = \frac{-\mathrm{Im}\,\mu_1(\omega)}{\mu_0}. \tag{5.7.10}$$

Similarly for purely dilatational disturbances

$$Q_\kappa^{-1}(\omega) = \frac{-\mathrm{Im}\,\kappa_1(\omega)}{\kappa_0}; \tag{5.7.11}$$

normally loss in pure dilatation is much less significant than loss in shear and $Q_\kappa^{-1} \ll Q_\mu^{-1}$.

In general the loss factor for dilatation or shear can be represented as

$$Q_m^{-1}(\omega) = \frac{-\mathrm{Im}\,m_1(\omega)}{m_0}, \tag{5.7.12}$$

where $\mathfrak{m}$ is the appropriate complex modulus. Thus, for the standard linear solid model,

$$\mathfrak{m}(\omega) = \mathfrak{m}_2 \frac{\tau_C^{-1} - i\omega}{\tau_R^{-1} - i\omega} \tag{5.7.13}$$

$$= \mathfrak{m}_2 \frac{\omega^2 + \tau_C^{-1}\tau_R^{-1} - i\omega(\tau_R^{-1} - \tau_C^{-1})}{\omega^2 + \tau_R^{-2}}, \tag{5.7.14}$$

and the instantaneous modulus is $\mathfrak{m}_2$, and thus the loss factor

$$Q_{\mathfrak{m}}^{-1}(\omega) = \omega \frac{\tau_R^{-1} - \tau_C^{-1}}{\omega^2 + \tau_R^{-2}}. \tag{5.7.15}$$

By superposing a spectrum of relaxation times, a wide range of frequency behaviour can be simulated.

The dissipation via $Q^{-1}(\omega)$ is accompanied by a frequency dependent correction to the real part of the complex modulus and hence to the apparent wavespeed

$$\operatorname{Re} \mathfrak{m}(\omega) = \mathfrak{m}_2 \frac{\omega^2 + \tau_C^{-1}\tau_R^{-1}}{\omega^2 + \tau_R^{-2}}. \tag{5.7.16}$$

This property is a consequence of a causal dissipation mechanism. Since the relaxation contributions depend only on the past history of the strain $\dot{R}_{\mathfrak{m}}(t)$ vanishes for $t < 0$, so that the transform $\mathfrak{m}_1(\omega)$ must be analytic in the upper half-plane. In consequence the real and imaginary parts of $\mathfrak{m}_1(\omega)$ are Hilbert transforms of each other,

$$\operatorname{Re} \mathfrak{m}_1(\omega) = \frac{1}{\pi} P \int_{-\infty}^{\infty} d\omega' \, \frac{\operatorname{Im} \mathfrak{m}_1(\omega')}{\omega' - \omega} = -\frac{2\mathfrak{m}_0}{\pi} P \int_{-\infty}^{\infty} d\omega' \, \frac{\omega' Q^{-1}(\omega')}{\omega'^2 - \omega^2}, \tag{5.7.17}$$

where P denotes the Cauchy principal value of the integral.

6

Continua under Pressure

The linearised development of elasticity in the previous chapter can be applied about a stress-free state and has therefore direct application to the shallower parts of the Earth. As we shall see, we can make an incremental treatment about the stress-state at depth, which will be dominantly hydrostatic, and derive the incremental elastic moduli for small disturbances such as those created by the passage of seismic waves.

It is difficult in experimental studies to achieve the pressures and temperatures appropriate to a direct study of the Earth. For the lower mantle, in particular, recourse needs to be made to extrapolation from available conditions. A convenient summary of the property of Earth materials at depth is provided by the equation of state, which is commonly expressed as a relation between the pressure, the specific volume and the temperature. This simplified approach ignores non-hydrostatic stresses and any non-isotropic response to the assumed hydrostatic pressure. Finite strain representations are used in the Birch–Murnaghan equations to relate the conditions at depths to the properties at zero pressure.

Long-term response to large non-hydrostatic stresses leads to irreversible (plastic) deformation, and so over geological time materials are expected to have yielded to such stresses, leaving just the hydrostatic load at depth. Non-hydrostatic stresses are, however, important at shallow depth where they approach the magnitude of the pressure and are responsible for the occurrence of earthquakes.

Although materials within the Earth are undoubtedly anisotropic on a microscopic scale, the properties of the polycrystalline aggregates averaged over a typical seismic wavelength are close to isotropic except towards the top and bottom of the mantle.

6.1 Effect of radial stratification

A convenient simplification when considering the properties of the Earth at depth is to concentrate on the variations as a function of radius and to regard any three-dimensional variation as a perturbation around this state. This enables the use of a simple equation of state in terms of the hydrostatic pressure as a function of radius, which depends directly on knowledge of the density distribution within the Earth.

6.1.1 Hydrostatic pressure

The predominant stress-field within the Earth is imposed by the hydrostatic pressure induced by gravitational effects. The gravitational force is directed radially inwards and so $\mathbf{g}(\mathbf{r}) = -g\hat{\mathbf{e}}_r$, with the acceleration due to gravity g given by

$$g = \frac{4\pi G}{r^2}\int_0^r ds\, s^2\rho(s), \tag{6.1.1}$$

in terms of the radial distribution of density $\rho(r)$. For hydrostatic balance, the pressure is determined by

$$\frac{\partial p}{\partial r} + \rho g = 0, \quad \text{i.e.,}\ p(r) = \int_r^{r_e} ds\, \rho(s)g(s), \tag{6.1.2}$$

with the integration taken from the free surface at r_e, where $p(r_e) = 0$, down to radius r.

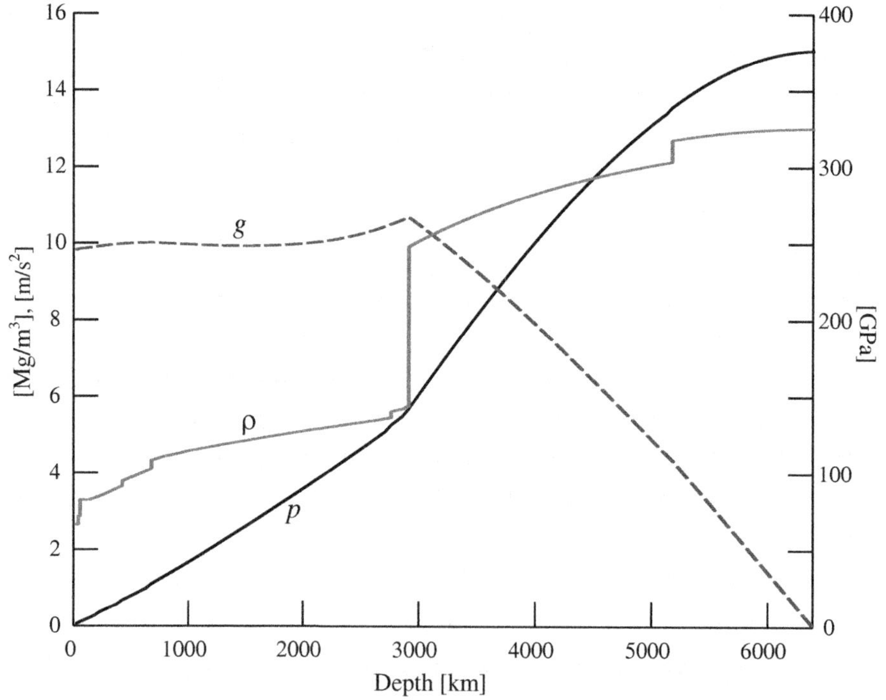

Figure 6.1. Pressure p, gravitational acceleration g for the density distribution associated with the AK135M model.

The behaviour of the pressure and the gravitational acceleration within the Earth is illustrated in Figure 6.1, using the density distribution ρ for the AK135 model (Figure 1.6) modified to remove the low density zone in the upper mantle. Both the pressure and the gravitational acceleration are continuous across the discontinuities in density associated with the phase transitions in the upper mantle, the core–mantle boundary and the boundary between the inner and outer core.

The gravitational acceleration grows through the core as we move from the centre of the Earth outwards and reaches its maximum value at the core–mantle boundary, because the mantle material is lighter and the increased volume associated with spherical shells closer to the surface of the Earth does not compensate for the lowered density. The secondary maximum at the base of the transition zone again reflects the step down to lower densities in the outer layers.

The pressure increases steadily with depth with slight changes in slope at each of the discontinuities in density. The pressure tends to a constant value at the centre of the Earth of about 370 GPa; the reduction in slope for the inner core is associated with the steady decrease of g towards zero as the centre of the Earth is approached.

6.1.2 Thermodynamic relations

Most equations of state can be regarded as statements of the conservation of energy. In terms of the total energy $\mathcal{U}$ of a system, the first law of thermodynamics can be written as

$$d\mathcal{U}(S, V) = T\, d\mathcal{S} - p\, dV, \tag{6.1.3}$$

as a function of temperature T, entropy $\mathcal{S}$, pressure p and volume V. Because we want to consider the thermal state of a material it is most convenient from a theoretical viewpoint to work in terms of temperature T and volume V, since these can be related directly to the reference state, whereas stresses are most naturally introduced in the deformed state (Chapter 3). The first law can then be written in terms of the Helmholtz free energy $\mathcal{F} = \mathcal{U} - T\mathcal{S}$:

$$d\mathcal{F}(T, V) = -\mathcal{S}\, dT - p\, dV. \tag{6.1.4}$$

In experiments on earth materials, it is usually the pressure rather than the volume which is adjusted, and we have noted in section 4.2 that propagation is adiabatic rather than isothermal. From (6.1.4)

$$p = \left(\frac{\partial \mathcal{F}}{\partial V}\right)_T. \tag{6.1.5}$$

To provide a key to the mineral physics literature we will use the notation K rather than κ for the bulk modulus and G for the shear modulus instead of μ. The bulk modulus (incompressibility) K is defined as

$$K_T = -V\left(\frac{\partial p}{\partial V}\right)_T, \quad K_S = -V\left(\frac{\partial p}{\partial V}\right)_S. \tag{6.1.6}$$

K_T applies to isothermal conditions in slow laboratory experiments and K_S to the adiabatic conditions in transient wave propagation.

A related quantity is the 'seismic parameter'

$$\Phi = \frac{K_S}{\rho} = \left(\frac{\partial p}{\partial \rho}\right)_S. \qquad (6.1.7)$$

For an isotropic material, the bulk-sound speed

$$\phi = \sqrt{\Phi} = \sqrt{(\alpha^2 - \tfrac{4}{3}\beta^2)}, \qquad (6.1.8)$$

where α is the *P* wavespeed (5.4.9) (also designated v_p) and β is the *S* wavespeed (5.4.14) (also designated v_s).

Regarding pressure p as a function of V, T, $\mathcal{S}$

$$\left(\frac{\partial p}{\partial V}\right)_T = \left(\frac{\partial p}{\partial V}\right)_S - \left(\frac{\partial p}{\partial T}\right)_V \left(\frac{\partial T}{\partial V}\right)_S, \qquad (6.1.9)$$

and so the difference between the isothermal and adiabatic moduli is

$$K_T - K_S = V \left(\frac{\partial p}{\partial T}\right)_V \left(\frac{\partial T}{\partial V}\right)_S. \qquad (6.1.10)$$

The volume expansion coefficient α_{th}

$$\alpha_{th} = \frac{1}{V}\left(\frac{\partial V}{\partial T}\right)_p = \left(\frac{\partial \ln V}{\partial T}\right)_p; \qquad (6.1.11)$$

and so we can use the chain rule for partial derivatives to write

$$\left(\frac{\partial p}{\partial T}\right)_V = -\frac{1}{V}\left(\frac{\partial V}{\partial T}\right)_p V \left(\frac{\partial p}{\partial V}\right)_T = \alpha_{th} K_T. \qquad (6.1.12)$$

The adiabatic derivative of temperature with respect to specific volume can be written as

$$\left(\frac{\partial T}{\partial V}\right)_S = -\gamma_{th}\frac{T}{V}, \qquad (6.1.13)$$

where γ_{th} is the thermodynamic Grüneisen parameter. The specific heats at constant volume C_V and constant pressure C_p are given by

$$C_V = T\left(\frac{\partial S}{\partial T}\right)_V, \qquad C_p = T\left(\frac{\partial S}{\partial T}\right)_p, \qquad (6.1.14)$$

where S is the specific entropy.

We are therefore able to express the Grüneisen parameter in the form

$$\gamma_{th} = \frac{\alpha_{th} V K_S}{C_p} = \frac{\alpha_{th} V K_T}{C_V}. \qquad (6.1.15)$$

Note also that

$$\left(\frac{\partial T}{\partial p}\right)_S = \gamma_{th}\frac{T}{K_S}. \tag{6.1.16}$$

In terms of these new variables we can express the difference between the isothermal and adiabatic moduli as

$$K_T - K_S = -\alpha_{th} K_T T \gamma_{th}, \tag{6.1.17}$$

and so we find that

$$\frac{K_S}{K_T} = 1 + \gamma_{th}\alpha_{th}T. \tag{6.1.18}$$

The behaviour of the bulk and shear moduli in the Earth with depth, inferred from the properties of seismic waves, is illustrated in Figure 6.2 for the AK135 model.

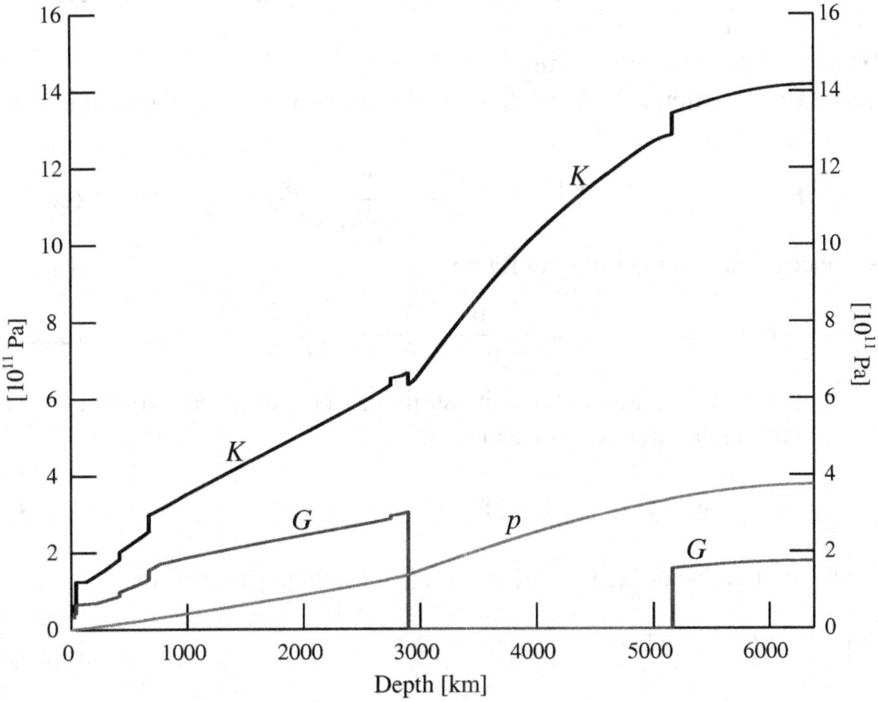

Figure 6.2. The behaviour of the bulk modulus K and shear modulus G as a function of depth for the AK135M model, compared to the variation in pressure p.

The effect of compression with depth plays a major role in determining the physical properties, but the increase in temperature with depth also plays a significant role. Seismic wave propagation will be close to isentropic conditions and hence measurements are of the adiabatic modulus K_S. For both the bulk of the lower mantle and the core, where the composition is expected to be homogeneous, the bulk modulus K is approximately proportional to pressure p.

6.2 Finite strain deformation

As we have seen in Section 4.3, the elastic properties of a material can be directly related to a local Lagrangian treatment about a suitable reference state. However, until recently, the appropriate conditions for much of the Earth's mantle were not accessible to experimental studies. In particular, even where measurements were carried out at high pressure, the temperature was close to ambient.

Estimates of the material behaviour in the Earth have therefore been made by using a systematic extrapolation from the low-pressure regime of most experiments to likely Earth conditions. A suitable equation of state for mineral systems can be derived by examining the influence of large finite strain, sufficient to change the volume of a unit cell from its zero-pressure value to that at depth. The aim of such studies is to try to understand the distribution of physical properties revealed from seismological work, as in Figure 6.2, in terms of the appropriate mineral assemblages and temperature regimes.

We recall from Section 2.2.5 the definition of the Green strain in the Lagrangian frame

$$\mathbf{E} = \tfrac{1}{2}(\mathbf{F}^T\mathbf{F} - \mathbf{I}), \qquad E_{ij} = \tfrac{1}{2}\left[\frac{\partial u_i}{\partial \xi_j} + \frac{\partial u_j}{\partial \xi_i} + \frac{\partial u_k}{\partial \xi_i}\frac{\partial u_k}{\partial \xi_j}\right], \tag{6.2.1}$$

and the Cauchy strain in the Eulerian frame

$$\mathbf{e} = \tfrac{1}{2}(\mathbf{I} - (\mathbf{F}\mathbf{F}^T)^{-1}), \qquad e_{ij} = \tfrac{1}{2}\left[\frac{\partial u_i}{\partial x_j} + \frac{\partial u_j}{\partial x_i} - \frac{\partial u_k}{\partial x_i}\frac{\partial u_k}{\partial x_j}\right]. \tag{6.2.2}$$

The volume transformation under finite strain can be expressed either in terms of the Green strain in a Lagrangian viewpoint,

$$\left(\frac{V}{V_0}\right)^2 = J^2 = \det[\mathbf{F}^T\mathbf{F}] = \det[\mathbf{I} + 2\mathbf{E}], \tag{6.2.3}$$

or alternatively in terms of the Cauchy strain in an Eulerian perspective,

$$\left(\frac{V_0}{V}\right)^2 = J^{-2} = \det[\mathbf{F}^T\mathbf{F}]^{-1} = \det[2\mathbf{e} - \mathbf{I}]. \tag{6.2.4}$$

Consider hydrostatic pressure on a cubic or isotropic material. Then in the Lagrangian formulation the Green strain tensor takes the form

$$E_{ij} = \eta\delta_{ij}, \quad \det[\mathbf{I} + 2\mathbf{E}] = (1 + 2\eta)^3, \tag{6.2.5}$$

and so, from (6.2.3),

$$\eta = \tfrac{1}{2}\left[\left(\frac{V}{V_0}\right)^{2/3} - 1\right] = \tfrac{1}{2}\left[\left(\frac{\rho_0}{\rho}\right)^{2/3} - 1\right]; \tag{6.2.6}$$

this scalar η varies from 0 to $-\frac{1}{2}$ as the pressure varies from 0 to ∞.

Alternatively, in the Eulerian formulation, the Cauchy strain is given by

$$e_{ij} = \epsilon\delta_{ij}, \quad \det[2\mathbf{e} - \mathbf{I}] = (2\epsilon - 1)^3, \tag{6.2.7}$$

and so

$$\epsilon = -\tfrac{1}{2}\left[\left(\frac{V_0}{V}\right)^{2/3} - 1\right] = -\tfrac{1}{2}\left[\left(\frac{\rho}{\rho_0}\right)^{2/3} - 1\right]. \tag{6.2.8}$$

In terms of the strain variables η and ϵ, the volume change

$$\frac{V_0}{V} = \frac{\rho}{\rho_0} = (1 + 2\eta)^{-3/2} = (1 - 2\epsilon)^{3/2}. \tag{6.2.9}$$

In an elastic material the strain energy W can be equated with the specific Helmholtz free energy $\mathcal{F}/\rho$ (we neglect thermal effects in either slow near-isothermal deformations or the nearly isentropic incremental deformation associated with the passage of seismic waves). The Helmholtz free energy can then be expressed as a function of strain, e.g., through its dependence on the Green strain $\mathbf{E}$. With a generalisation of (6.1.4) for specific quantities

$$\rho\, dW = -\rho S\, dT + \rho_0 \sigma_{ij}\, dA_{ij}, \tag{6.2.10}$$

in terms of the displacement gradient $\mathbf{A} = \mathbf{F} - \mathbf{I}$ (5.1.1). The Cauchy stress tensor σ_{ij} is thus

$$\sigma_{ij} = \frac{\rho}{\rho_0}\frac{\partial\mathcal{F}}{\partial A_{ij}}; \tag{6.2.11}$$

which is equivalent to our previous expression (4.3.10)

$$\det\mathbf{F}\,\sigma_{ij} = F_{im}\frac{\partial W}{\partial E_{mn}}F_{jn}, \tag{6.2.12}$$

The effective elastic moduli C_{ijkl} are defined by

$$C_{ijkl} = \frac{\partial\sigma_{ij}}{\partial A_{kl}} = \frac{\rho}{\rho_0}\frac{\partial^2 W}{\partial A_{ij}\partial A_{kl}} - \sigma_{ij}\delta_{kl}, \tag{6.2.13}$$

since $\partial\rho/\partial A_{kl} = -\rho\delta_{kl}$. The differentiation may be carried out either isothermally or adiabatically to give the appropriate moduli. In terms of the dependence on strain, the effective moduli in the case of hydrostatic pre-stress take the form

$$\det\mathbf{F}\,C_{ijkl} = F_{im}F_{kp}\frac{\partial^2 W}{\partial E_{mn}\partial E_{pq}}F_{jn}F_{lq} - p\delta_{ij}^{kl}, \tag{6.2.14}$$

where

$$\delta_{ij}^{kl} = \delta_{ik}\delta_{jl} + \delta_{il}\delta_{jk} - \delta_{ij}\delta_{kl}. \tag{6.2.15}$$

6.3 Expansion of Helmholtz free energy and equations of state

A commonly used approach to handling finite strain effects is to expand the Helmholtz free energy $\mathcal{F}$ as a Taylor series in the elastic strain from the zero-pressure state. This yields a power series expansion in terms of a suitable measure of finite strain.

The level of strain required to attain the required density at depth in the Earth is then employed to determine the pressure and the other desired properties of the medium in the strained, i.e. pressured, state. For the Lagrangian viewpoint we make a power-series expansion in the strain η from (6.2.6),

$$\mathcal{F} = V_0 \sum_{i=0}^{\infty} A_i \eta^i. \tag{6.3.1}$$

Whereas for the Eulerian viewpoint (Birch–Murnaghan) the expansion is in terms of the strain ϵ from (6.2.8),

$$\mathcal{F} = V_0 \sum_{i=0}^{\infty} B_i \epsilon^i. \tag{6.3.2}$$

The isothermal thermodynamic relations

$$p = -\left(\frac{\partial \mathcal{F}}{\partial V}\right)_T, \quad K_T = -V\left(\frac{\partial p}{\partial V}\right)_T, \quad K'_T = \left(\frac{\partial K}{\partial T}\right)_S \tag{6.3.3}$$

enable us to identify the coefficients in these series expansions.

For the Lagrangian expansion the coefficients in the expansion (6.3.1) for the Helmholtz free energy $\mathcal{F} = V_0 \sum_{i=0}^{\infty} A_i \eta^i$ are

$$A_0 = A_1 = 0, \tag{6.3.4}$$

$$A_2 = \tfrac{9}{2} K_0, \tag{6.3.5}$$

$$A_3 = -\tfrac{9}{2} K_0 K'_0, \tag{6.3.6}$$

$$A_4 = \tfrac{9}{2} K_0 [K_0 K''_0 + K'_0(K'_0 + 1) - \tfrac{1}{9}], \tag{6.3.7}$$

where the subscript 0 indicates evaluation at zero pressure and ambient temperature T_0. The Lagrangian equation of state is then

$$p_L = -\frac{\partial \mathcal{F}}{\partial \eta}\frac{\partial \eta}{\partial V} = -\tfrac{1}{3}\left(\frac{V}{V_0}\right)^{-1/3} \sum_{i=1}^{\infty} (i+1) A_{i+1} \eta^i, \tag{6.3.8}$$

since

$$\frac{\partial \eta}{\partial V} = \tfrac{1}{3}(1 + 2\eta)^{-1/2} = \tfrac{1}{3}\left(\frac{V}{V_0}\right)^{-1/3}. \tag{6.3.9}$$

Thus the pressure p_L is represented as

$$p_L = -3K_0 \left(\frac{V}{V_0}\right)^{-1/3} \left[\eta - \tfrac{3}{2} K'_0 \eta^2 + \cdots\right]. \tag{6.3.10}$$

and the isothermal modulus can be found in terms of η from (6.1.6).

For the Eulerian viewpoint the coefficients in the expansion (6.3.2) for the Helmholtz free energy $\mathcal{F} = V_0 \sum_{i=0}^{\infty} B_i \epsilon^i$ can be expressed as

$$B_0 = B_1 = 0, \tag{6.3.11}$$

$$B_2 = A_2 = \tfrac{9}{2}K_0, \tag{6.3.12}$$

$$B_3 = -\tfrac{9}{2}K_0(K_0' - 4), \tag{6.3.13}$$

$$B_4 = \tfrac{9}{2}K_0[K_0K_0'' + K_0'(K_0' - 7) - \tfrac{143}{9}], \tag{6.3.14}$$

and the equation of state for the pressure takes the form

$$p_E = -3K_0\left(\frac{V}{V_0}\right)^{5/3}\left[\epsilon - \tfrac{3}{2}(K_0' - 4)\epsilon^2 + \cdots\right]. \tag{6.3.15}$$

When we reinstate the full dependence on ϵ by expressing the volume change in terms of the strain, we find

$$p_E = -(1 - 2\epsilon)^{5/2}\left[C_1\epsilon - C_2\epsilon^2/2 + C_3\epsilon^3/6\right], \tag{6.3.16}$$

where

$$C_1 = \tfrac{2}{3}B_2 = 3K_0, \quad C_2 = -2B_3, \quad C_3 = 6B_4. \tag{6.3.17}$$

The corresponding expression for the isothermal bulk modulus is

$$K_T = \tfrac{1}{3}(1 - 2\epsilon)^{5/2}\left[C_1 + (C_2 - 7C_1)\epsilon + (C_3 - 9C_2)\epsilon^2 - 11C_3\epsilon^3\right] \tag{6.3.18}$$

$$= (1 - 2\epsilon)^{5/2}\left[K_0 + K_1\epsilon + K_2\epsilon^2 + \cdots\right]. \tag{6.3.19}$$

For moderate changes in volume $1.0 > V/V_0 > 0.9$ the predictions of the Lagrangian and Eulerian equations of state are essentially equivalent, but for larger strains the pressure estimate p_E increases more rapidly than p_L.

Following Birch (1952) the preferred formulation for studies of the Earth is that based on the Eulerian viewpoint because the influence of the third-order term in strain is much reduced. We note that the third-order coefficient B_3 depends on the pressure derivative of the bulk modulus through $K_0' - 4$. For most minerals the pressure derivative K_0' lies in the range 3 to 6 and is frequently close to 4 and so B_3, C_2 are small, with only a limited influence from uncertainties in measurement. The third order coefficient in the equivalent Lagrangian expression depends solely on K_0', and has a stronger influence.

The third-order truncation of the Birch–Murnaghan equation of state has the advantage that no knowledge is required of second pressure derivatives of physical properties that are more difficult to measure than the first derivatives, and hence are rather poorly known. The fourth-order representation is nevertheless more flexible because of the larger number of parameters.

The isothermal seismic parameter

$$\Phi_T = \frac{K_T}{\rho} \tag{6.3.20}$$
$$= \frac{1}{3\rho_0}(1-2\epsilon)\left[C_1(1-7\epsilon) + C_2(\epsilon - 9\epsilon^2/2) + C_3(\epsilon^2 - 11\epsilon^3/6)\right];$$

the grouping of the terms emphasises that the coefficients of the powers of ϵ are not all independent.

With the aid of an Eulerian representation comparable to (6.2.14) we can develop representations for the wavespeeds of both *P* and *S* waves in the form

$$\alpha^2 = v_P^2 = \frac{1}{\rho_0}(1-2\epsilon)\left\{L_0 + L_1\epsilon + L_2\epsilon^2 + \cdots\right\}, \tag{6.3.21}$$

$$\beta^2 = v_S^2 = \frac{1}{\rho_0}(1-2\epsilon)\left\{G_0 + G_1\epsilon + G_2\epsilon^2 + \cdots\right\}, \tag{6.3.22}$$

where

$$L_0 = K_0 + \tfrac{4}{3}G_0, \qquad L_1 = K_1 + \tfrac{4}{3}G_1, \qquad \text{etc.} \tag{6.3.23}$$

The first and second pressure derivatives of the shear and bulk moduli at zero pressure are then given by

$$\begin{aligned} G_0' &= \tfrac{1}{3}(5G_1 - G_2)/K_0, \\ K_0' &= \tfrac{1}{3}(5L_1 - L_2)/K_0 - \tfrac{4}{3}G_0', \\ G_0'' &= \tfrac{1}{9}[5G_3 + 3(K_0'-4)G_2 - 5(3K_0'-5)G_1]/K_0^2, \\ K_0'' &= \tfrac{1}{9}[5L_3 + 3(K_0'-4)L_2 - 5(3K_0'-5)L_1]/K_0^2 - \tfrac{4}{3}G_0''. \end{aligned} \tag{6.3.24}$$

The coefficients for the contributions from the bulk modulus can be related to quantities which can be measured, or estimated, from mineral physics experiments. However, the corresponding results for the shear modulus are rather poorly known, it is only recently that measurements of shear properties under conditions appropriate to the Earth's transition zone have been accomplished and the derivatives are weakly constrained. *Ab initio* quantum mechanical calculations of mechanical properties are also easier to perform for the bulk modulus, because allowance does not need to made for anelastic thermal effects (see Chapter 9).

The finite strain theory does not, by itself, provide a relation between the bulk and shear moduli. Current seismological models are consistent with a dependence of the shear modulus G on pressure p and bulk modulus K of the form

$$G = aK - bp. \tag{6.3.25}$$

This behaviour is ilustrated in Figure 6.3 for the AK135 model, by the linear segments associated with different parts of the mantle. The slopes are quite similar, except for the zone just below the 660 km discontinuity and the highly heterogeneous D″ region at the base of the mantle.

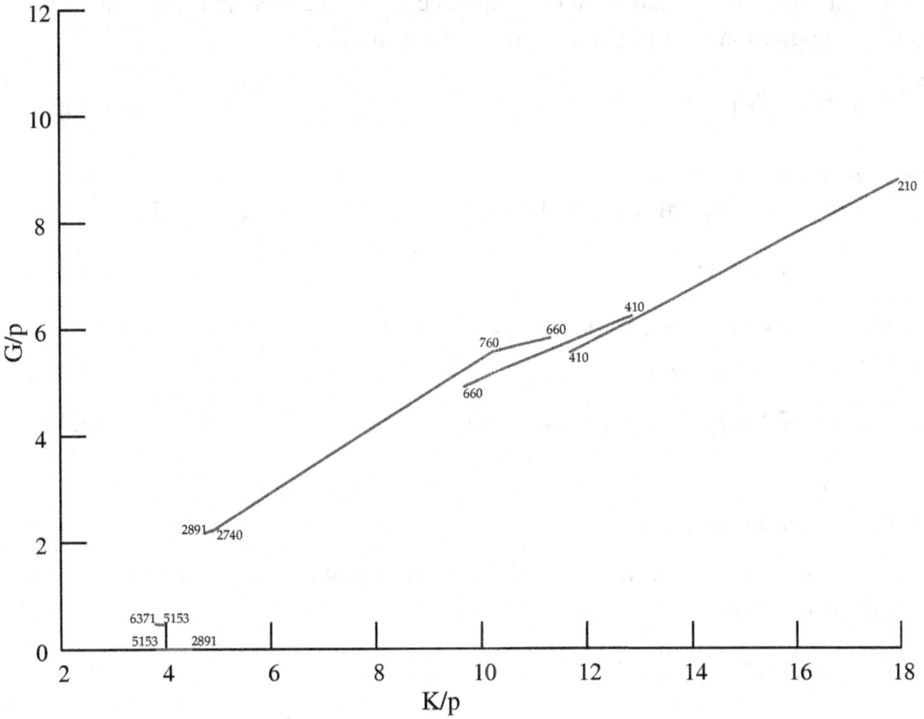

Figure 6.3. The relation of the shear modulus G and bulk modulus K scaled by the pressure p for the AK135 model. The depths corresponding to the different segments are indicated.

6.4 Incremental stress and strain

Small deviations from a pre-stressed or pre-strained state can be considered by a perturbation treatment in either the Lagrangian or the Eulerian viewpoint.

For a state variable q, the perturbation can be written as

$$q_E(\mathbf{x}, t) = q^0(\mathbf{x}) + q_E^1(\mathbf{x}, t), \tag{6.4.1}$$

$$q_L(\boldsymbol{\xi}, t) = q^0(\boldsymbol{\xi}) + q_L^1(\boldsymbol{\xi}, t), \tag{6.4.2}$$

as a result of a small change in state from $\boldsymbol{\xi}$, where q^0 is the value in the initial state. As in our treatment of linearised elasticity (Section 5.1), we can ignore the distinction between $\mathbf{x}$ and $\boldsymbol{\xi}$ in the discussion of the perturbation q^1. For an incremental displacement $\mathbf{u} = \mathbf{x} - \boldsymbol{\xi}$, the relation between the perturbations in the material and spatial frames is

$$q_L^1 = q_E^1 + \mathbf{u} \cdot \boldsymbol{\nabla} q^0, \tag{6.4.3}$$

where the advective term $\mathbf{u} \cdot \boldsymbol{\nabla} q^0$ reflects displacements through the spatial gradients in the initial state. A consequence of (6.4.3) is that the linearised conservation equation for mass transport can be written as

$$\frac{\partial}{\partial t}\rho^1 + \boldsymbol{\nabla} \cdot (\rho^0 \mathbf{v}) = 0 \tag{6.4.4}$$

where $\mathbf{v}$ is the velocity field.

In such a first-order perturbation, the volume element after deformation

$$dV^1 = (1 + \boldsymbol{\nabla} \cdot \mathbf{u})dV^0, \tag{6.4.5}$$

and similarly we can represent the changes in an areal element and its inclination, through the unit normal $\mathbf{n}$, as

$$d\sigma^1 = (1 + \boldsymbol{\nabla}^\Sigma \cdot \mathbf{u})d\sigma^0, \quad \mathbf{n}^1 = \mathbf{n}^0 - \boldsymbol{\nabla}^\Sigma \mathbf{u} \cdot \mathbf{n}^0. \tag{6.4.6}$$

6.4.1 Perturbations in stress

The Cauchy stress is defined in the spatial frame, so that we can express the effect of a small perturbation as

$$\boldsymbol{\sigma}(\mathbf{x}, t) = \boldsymbol{\sigma}^0(\mathbf{x}, t) + \boldsymbol{\sigma}^1(\mathbf{x}, t), \tag{6.4.7}$$

and if we refer this stress perturbation back to the material frame

$$\sigma^1_L = \sigma^1 + \mathbf{u} \cdot \boldsymbol{\nabla}\sigma^0. \tag{6.4.8}$$

For the Piola–Kirchhoff stress tensors in the material frame, introduced in Section 3.5, the perturbations may be obtained by linearising the deformation gradient relations and so we find

$$\sigma^{PK1} = \sigma^1_L + \sigma^0(\boldsymbol{\nabla} \cdot \mathbf{u}) - (\boldsymbol{\nabla}\mathbf{u})^T \cdot \sigma^0, \tag{6.4.9}$$

$$\sigma^{SK1} = \sigma^1_L + \sigma^0(\boldsymbol{\nabla} \cdot \mathbf{u}) - (\boldsymbol{\nabla}\mathbf{u})^T \cdot \sigma^0 - \sigma^0 \cdot \boldsymbol{\nabla}\mathbf{u}. \tag{6.4.10}$$

The linearised equation of motion for the displacement $\mathbf{u}$ takes the form

$$\rho\frac{\partial^2}{\partial t^2}\mathbf{u}_i = \frac{\partial}{\partial x_j}\sigma^1_{ij}, \tag{6.4.11}$$

since the initial stress field σ^0 is supposed to be in equilibrium. In terms of the Lagrangian stress perturbation

$$\sigma^1 = \sigma^1_L - \mathbf{u} \cdot \boldsymbol{\nabla}\sigma^0, \tag{6.4.12}$$

and thus the equation of motion can be written as

$$\rho\frac{\partial^2}{\partial t^2}\mathbf{u}_i = \frac{\partial}{\partial x_j}\left[(\sigma^1_L)_{ij} - u_k\frac{\partial}{\partial x_k}\sigma^0_{ij}\right]. \tag{6.4.13}$$

6.4.2 Perturbations in boundary conditions

The condition that there should be no slip between two solids requires that the displacement $\mathbf{u}$ in any deformation must be continuous across the boundary

$$[\mathbf{u}]_-^+ = 0. \tag{6.4.14}$$

In the case of a fluid–solid boundary, tangential slip along the boundary is allowed, and the boundary condition $[\mathbf{n}^0 \cdot \mathbf{u}]_-^+ = 0$.

For stress, the condition of continuity of traction $[\mathbf{n} \cdot \sigma]_-^+ = 0$, can be transformed into the material state as $[\mathbf{n}^0 \cdot \sigma^{PK}]_-^+ = 0$. When we take account of the distortion of the boundary in the incremental deformation the boundary condition for the linearised state, including allowance for displacement along the boundary, can be written as

$$\left[\mathbf{n}^0 \cdot \sigma^{PK1} - \nabla^{\Sigma}(\mathbf{un} \cdot \sigma^0)\right]_-^+ = 0. \tag{6.4.15}$$

The displacement term is needed in descriptions of seismic sources in terms of slip on a fault, but otherwise is not significant. For a contact between a fluid and a solid, the traction has to be normal to the distorted fluid–solid boundary

$$(\mathbf{n} \cdot \sigma) \cdot (\mathbf{I} - \mathbf{nn}) = 0, \quad \text{i.e.,} \quad n_j \sigma_{ij}(\delta_{ik} - n_i n_k) = 0, \tag{6.4.16}$$

and the equivalent linearised form is

$$\left[\mathbf{n}^0 \cdot \sigma^{PK1} + \mathbf{n}^0 \nabla^{\Sigma} \cdot (p^0 \mathbf{u}) - p^0(\nabla^{\Sigma} \cdot \mathbf{un}^0\right]_-^+ \cdot (\mathbf{I} - \mathbf{nn}) = 0. \tag{6.4.17}$$

where $p^0 = -\mathbf{n}^0 \cdot \sigma^0 \cdot \mathbf{n}^0 = n_i^0 \sigma_{ij}^0 n_j^0$ is the initial pressure on the fluid side.

6.5 Elasticity under pressure

As we have noted above we can equate the strain energy W for an elastic material to the specific Helmholtz free energy $\mathcal{F}/\rho$ because the thermal influences of deformation will be very small. In a pre-stressed state we can still use the constitutive equation (4.3.11)

$$\sigma = \frac{1}{\det \mathsf{F}} \mathsf{F} \frac{\partial W}{\partial \mathsf{E}} \mathsf{F}^T, \tag{6.5.1}$$

in terms of the derivatives of the strain energy with respect to the Lagrangian (Green) strain E. For finite deformation we need to recognise the differences in orientation between the Lagrangian and Eulerian triads reflected in the tensor transformation. We recall that the quantity $\sigma^{SK} = \partial W/\partial \mathsf{E}$ defined in the Lagrangian frame is the second Piola–Kirchhoff stress tensor (cf. Section 3.5).

Now make a quadratic approximation for $W(\mathsf{E})$ about a pre-stressed reference state

$$W(\mathsf{E}) = W_0 + \sigma_{ij}^0 E_{ij} + \tfrac{1}{2} E_{ij} C_{ijkl} E_{kl}, \tag{6.5.2}$$

The coefficient of the strain tensor itself is just the stress in the reference state σ^{o}_{ij}. C_{ijkl} has the symmetries

$$C_{ijkl} = C_{jikl} = C_{ijlk} = C_{klij}. \tag{6.5.3}$$

Thus

$$\sigma^{SK}_{ij} = \sigma^{o}_{ij} + C_{ijkl}E_{kl}, \qquad \sigma = \frac{1}{\det \mathsf{F}}\mathsf{F}\sigma^{SK}\mathsf{F}^{T}. \tag{6.5.4}$$

We now examine the behaviour for a small increment in strain superimposed on an initial stress to define the elastic moduli for the pre-stressed state. We work with the first-order incremental strain ε and rotation $\boldsymbol{\omega}$,

$$\varepsilon_{ij} = \tfrac{1}{2}\left(\frac{\partial u_i}{\partial x_j} + \frac{\partial u_j}{\partial x_i}\right), \quad \omega_{ij} = \tfrac{1}{2}\left(\frac{\partial u_i}{\partial x_j} - \frac{\partial u_j}{\partial x_i}\right), \tag{6.5.5}$$

and seek to find a representation for the incremental stress which is as close as possible to that for linearised elasticity from a zero-stress state, but with explicit representations of the influence of the hydrostatic pressure and deviatoric stress. A suitable choice is provided by setting

$$C_{ijkl} = c_{ijkl} + \tfrac{1}{2}(\sigma^{0}_{ij}\delta_{kl} + \sigma^{0}_{kl}\delta_{ij} - \sigma^{0}_{ik}\delta_{jl} - \sigma^{0}_{jk}\delta_{il} - \sigma^{0}_{il}\delta_{jk} - \sigma^{0}_{jl}\delta_{ik}), \tag{6.5.6}$$

so that c_{ijkl} has the usual $i \leftrightarrow j$, $k \leftrightarrow l$, $ij \leftrightarrow kl$ symmetries. The incremental stress $\boldsymbol{\sigma}^{SK1}$ is then related to the incremental strain ε by

$$\begin{aligned}\sigma^{SK1}_{ij} &= c_{ijkl}\varepsilon_{kl} + p^{0}(2\varepsilon_{ij} - \varepsilon_{kk}\delta_{ij}) \\ &\quad + \tfrac{1}{2}(\tau^{0}_{ij}\varepsilon_{kk} + \tau^{0}_{kl}\varepsilon_{kl}\delta_{ij}) - \varepsilon_{ik}\tau^{0}_{kj} - \varepsilon_{kj}\tau^{0}_{ik},\end{aligned} \tag{6.5.7}$$

where we have separated the terms involving the initial pressure p^{o} and the deviatoric stress $\boldsymbol{\tau}^{o}$. The corresponding incremental Lagrangian stress $\boldsymbol{\sigma}^{1}_{L}$ is given by

$$\sigma^{L1}_{ij} = c_{ijkl}\varepsilon_{kl} + \tfrac{1}{2}(-\tau^{0}_{ij}\varepsilon_{kk} + \tau^{0}_{kl}\varepsilon_{kl}\delta_{ij}) + \omega_{ik}\tau^{0}_{kj} - \omega_{kj}\tau^{0}_{ik}, \tag{6.5.8}$$

which is independent of the hydrostatic pressure.

In general c_{ijkl} will be anisotropic, but we can isolate the isotropic part by writing

$$c_{ijkl} = (\kappa - \tfrac{2}{3}\mu)\delta_{ij}\delta_{kl} + \mu(\delta_{ik}\delta_{jl} + \delta_{il}\delta_{jk}) + \gamma_{ijkl}, \tag{6.5.9}$$

where the incompressibility $K = \kappa$ and rigidity $G = \mu$ are defined by

$$\kappa = \tfrac{1}{9}c_{iijj}, \quad \mu = \tfrac{1}{10}(c_{ijij} - \tfrac{1}{3}c_{iijj}) \quad \text{so that} \quad \gamma_{iijj} = \gamma_{ijij} = 0. \tag{6.5.10}$$

Then $c_{ijkl} - \gamma_{ijkl}$ is the isotropic tensor which is the best approximation to c_{ijkl} in a least-squares sense, and γ_{ijkl} is the purely anisotropic component.

Seismic wavespeeds

As in (5.4.2) we take a plane wave representation

$$u_i = U_i \exp[i\omega p n_k x_k - i\omega t], \tag{6.5.11}$$

and insert this into the linearised equation of motion (6.4.13). For high frequencies, when we can neglect any gradients of the initial stress distribution, we find

$$\left[p^2\left\{c_{ijkl}n_j n_l - \tfrac{1}{2}(\tau^0_{kl}n_l n_i - \tau^0_{ij}n_j n_k)\right\} - \rho\delta_{ik}\right]U_k = 0. \tag{6.5.12}$$

This is an eigenvalue equation, and as in (5.4.4) there are three roots for p^2 for each direction $\mathbf{n}$, associated with polarisations specified by the eigenvectors $\mathbf{U}^r(\mathbf{n})$. The equation (6.5.12) for the wave slownesses does not depend explicitly on the initial hydrostatic pressure, but only on the deviatoric stresses τ^0. The plane wave propagation is anisotropic because the slownesses p^r and polarisations $\mathbf{U}^r$ depend on the direction $\mathbf{n}$. The anisotropy comes in part from the intrinsic anisotropy γ_{ijkl} (6.5.9), and the non-hydrostatic initial stress. In most situations in the Earth, $||\gamma|| \ll \mu$, $||\tau^0|| \ll \mu$ and then the independent wave-types will be a *quasi-P* wave with wavespeed $\alpha \approx [(\kappa + \frac{2}{3}\mu)/\rho]^{1/2}$, and two *quasi-S* waves with wavespeeds close to $\beta = [\mu/\rho]^{1/2}$.

Observations of seismic wave propagation suggest that anisotropic effects are concentrated in the lithosphere where non-hydrostatic stresses can be significant, and in the boundary layers near the 660 km discontinuity and in the complex D″ region at the base of the mantle.

7

Fluid Flow

The deformation of fluids plays an important role in geophysical phenomena, both in the long term flow of the Earth's mantle and in the more rapid motions in the core. We consider here the development of the equations describing the velocity field in a fluid and introduce the very important concept of non-dimensionalisation. In many situations with relatively slow motion we can neglect the effects of compressibility (at least to a first approximation). As a result we both simplify the equation system and eliminate the rapidly travelling acoustic waves. We illustrate the development of some simple flows, including simple considerations of the onset of thermal convection. Finally we consider the effects of rotation which are of signficance for the rapid flows in the Earth's core and, of course, for oceanic and atmospheric phenomena.

We will make the assumption of a Stokes fluid, so that the stress σ is related to the fluid velocity $\mathbf{v}$ by

$$\sigma_{ij} = -p\delta_{ij} + \eta\left(\frac{\partial v_j}{\partial x_i} + \frac{\partial v_i}{\partial x_j}\right) - \tfrac{2}{3}\eta\delta_{ij}\frac{\partial v_k}{\partial x_k}, \tag{7.1}$$

where p is the hydrostatic pressure and η the viscosity. The use of (7.1) assumes that there is no contribution from bulk viscosity.

7.1 The Navier–Stokes equation

The equation of motion for the fluid in the presence of body force $\mathbf{g}$ and mass-acceleration $\mathbf{f}$ is (3.2.4)

$$\frac{\partial \sigma_{ij}}{\partial x_i} + \rho g_j = \rho f_j, \tag{7.1.1}$$

and we can relate the acceleration $\mathbf{f}$ to the velocity field $\mathbf{v}$ using (2.4.7)

$$\mathbf{f} = \frac{D}{Dt}\mathbf{v} = \frac{\partial}{\partial t}\mathbf{v} + (\mathbf{v}\cdot\boldsymbol{\nabla})\mathbf{v}, \tag{7.1.2}$$

that can be written in component terms as,

$$f_j = \frac{D}{Dt}v_j = \frac{\partial}{\partial t}v_j + \left(v_k\frac{\partial}{\partial x_k}\right)v_j. \tag{7.1.3}$$

Now, on combining the constitutive equation (7.1) with the equation of motion (7.1.1) we require

$$\rho\frac{D}{Dt}v_j = \rho g_j - \frac{\partial}{\partial x_j}p + \frac{\partial}{\partial x_i}\left[\eta\left(\frac{\partial v_j}{\partial x_i} + \frac{\partial v_i}{\partial x_j}\right) - \tfrac{2}{3}\eta\delta_{ij}\frac{\partial v_k}{\partial x_k}\right], \tag{7.1.4}$$

which is usually called the Navier–Stokes equation for the fluid.

If the viscosity η can be taken as constant over the fluid, then we can express the Navier–Stokes equation as

$$\rho\frac{\partial}{\partial t}v_j + \rho\left(v_k\frac{\partial}{\partial x_k}\right)v_j = \rho g_j - \frac{\partial}{\partial x_j}p + \eta\frac{\partial^2}{\partial x_i\partial x_i}v_j + \tfrac{1}{3}\eta\frac{\partial^2}{\partial x_j\partial x_k}v_k, \tag{7.1.5}$$

or in vector form as

$$\rho\frac{\partial}{\partial t}\mathbf{v} + \rho(\mathbf{v}\cdot\nabla)\mathbf{v} = \rho\mathbf{g} - \nabla p + \eta\nabla^2\mathbf{v} + \tfrac{1}{3}\eta\nabla(\nabla\cdot\mathbf{v}), \tag{7.1.6}$$

which has explicit non-linearity through the convected derivative term.

If the viscosity varies with position (as may for example be associated with the effect of temperature) we have to include an additional term,

$$+\frac{\partial}{\partial x_i}\eta\left[\left(\frac{\partial v_j}{\partial x_i} + \frac{\partial v_i}{\partial x_j}\right) - \tfrac{2}{3}\delta_{ij}\frac{\partial v_k}{\partial x_k}\right], \tag{7.1.7}$$

on the right-hand side of (7.1.5). We will also need to consider the equations for the distribution of temperature or any other factor which may influence the distribution of viscosity.

7.1.1 Heat flow

We can link the equations of heat flow and fluid flow by returning to the equation for the rate of change of the internal energy U, (4.2.11),

$$\rho\frac{D}{Dt}U = h + \frac{\partial}{\partial x_k}\left[k\frac{\partial}{\partial x_k}T\right] + \sigma_{ij}\frac{\partial}{\partial x_j}v_i, \tag{7.1.8}$$

where h is the rate of internal heat production, T is the temperature and k is the thermal conductivity. From the first law of thermodynamics

$$\frac{D}{Dt}U = \frac{D}{Dt}Q + \frac{D}{Dt}W, \tag{7.1.9}$$

where Q represents the thermal contribution to the internal energy and W the external work. We isolate the contribution from $\sigma_{ij}\partial v_i/\partial x_j$ in (7.1.8) due to the pressure term

$$p\frac{\partial}{\partial x_k}v_k = -\frac{p}{\rho}\frac{D}{Dt}\rho, \tag{7.1.10}$$

using the continuity equation (conservation of mass, equation 2.5.7). We will assume that the remaining contribution to the internal energy can be represented in terms of a thermal capacity C so that

$$\rho\frac{D}{Dt}U = \rho\frac{D}{Dt}(CT) - \frac{p}{\rho}\frac{D}{Dt}\rho \tag{7.1.11}$$

– the particular form of the heat capacity C to be employed will depend on the thermodynamic state of the material.

We recast the constitutive equation for the viscous fluid (7.1) as

$$\sigma_{ij} = -p\delta_{ij} + 2\eta\hat{D}_{ij}, \tag{7.1.12}$$

in terms of the deviatoric strain rate

$$\hat{D}_{ij} = \tfrac{1}{2}\left(\frac{\partial}{\partial x_i}v_j + \frac{\partial}{\partial x_j}v_i\right) - \tfrac{1}{3}\frac{\partial}{\partial x_k}v_k\delta_{ij}, \tag{7.1.13}$$

which is constructed to have zero trace.

The stress-power contribution to the rate of change of internal energy is

$$\begin{aligned}\sigma_{ij}\frac{\partial}{\partial x_i}v_j &= -p\frac{\partial}{\partial x_k}v_k + 2\eta\hat{D}_{ij}\frac{\partial}{\partial x_i}v_j,\\ &= -p\frac{\partial}{\partial x_k}v_k + 2\eta\hat{D}_{ij}\hat{D}_{ij}\end{aligned} \tag{7.1.14}$$

(since the deviatoric term vanishes when the indices i and j coincide). We have already explicitly accounted for the pressure contribution in (7.1.11) so that we can recast (7.1.8) into an equation for the temperature field in the presence of fluid flow

$$\rho\left[\frac{\partial}{\partial t}(CT) + v_k\frac{\partial}{\partial x_k}(CT)\right] = h + \frac{\partial}{\partial x_k}\left[k\frac{\partial}{\partial x_k}T\right] + 2\eta\hat{D}_{ij}\hat{D}_{ij}, \tag{7.1.15}$$

which simplifies somewhat if C, k are constant.

7.1.2 The Prandtl number

A convenient measure of the flow properties of the fluid is provided by the *kinematic viscosity*

$$\nu = \eta/\rho, \tag{7.1.16}$$

which can be regarded as a diffusion coefficient for linear momentum [with units $m^2\,s^{-1}$].

The equivalent measure for heat transport is provided by the *thermal diffusivity*

$$\kappa_H = k/\rho C \tag{7.1.17}$$

[with units $m^2\,s^{-1}$]. The time required for thermal effects to propagate a distance l is of the order of l^2/κ_H.

The relative significance of momentum and heat transport can be measured by the ratio of ν to κ_H which is a dimensionless quantity Pr known as the Prandtl number

$$\mathrm{Pr} = \nu/\kappa_H. \tag{7.1.18}$$

A fluid with a high Prandtl number has much weaker heat diffusion than momentum transport, and under this condition we will need to take account of the influence of spatial variations in temperature on the viscosity of the fluid since the time scales of flow are very long as, e.g., in convection in the mantle of the Earth.

For the Earth's mantle:

$\rho = 4000\,\mathrm{kg\,m^{-3}}$, the average density of the upper mantle,
$\eta = 10^{21}\,\mathrm{Pa\,s}$, the dynamic viscosity of the Earth's mantle which can be estimated from studies of post-glacial rebound,
$\kappa_H = 10^{-6}\,\mathrm{m^2\,s^{-1}}$ for the thermal diffusivity,

and so the Prandtl number Pr is approximately 2.5×10^{23}.

7.2 Non-dimensional quantities

A convenient representation of many aspects of fluid behaviour is provided by working with dimensionless parameters. We have just introduced the Prandtl number to describe the ratio of viscosity and thermal diffusivity, and will shortly encounter the Reynolds number representing the relative significance of inertial and viscous forces. In the context of convection, the Rayleigh number measures the ratio of buoyancy forces to viscous forces and thermal diffusion. In the discussion of convection it is also useful to look at the Nusselt number that indicates the significance of convection in the transport of heat. In Chapter 14 we will also encounter the Péclet number providing a measure of the importance of advective processes relative to diffusive processes. In the rotating fluids the Ekman and Rossby numbers play an important role in representing the significance of interial and viscous forces relative to the Coriolis force.

These non-dimensional parameters are extremely useful. They describe the essential *nature* of the flow, independently of the particular experimental or natural setup. The Navier–Stokes equation can be put into a non-dimensional form, with some of the terms being weighted by these dimensionless parameters. By inspecting the non-dimensional form of the Navier–Stokes equation, we can estimate which terms are of leading order and which terms play only a small, perhaps irrelevant, role.

We define *dimensionless variables*, indicated by primes, through

$$x = Lx', \quad y = Ly', \quad z = Lz', \quad \mathbf{v} = U\mathbf{v}'. \tag{7.2.1}$$

Here L is a characteristic length scale, say the radius of the Earth or the depth of a laboratory tank, and is assumed to be constant. U is a characteristic speed, also assumed to be constant. This characteristic speed could be the mean velocity of a tectonic plate or the average flow velocity in a convective tank experiment. The quantities L, U provide a non-dimensional representation of time $\mathsf{t} = \mathsf{t}'\mathsf{L}/\mathsf{U}$. We also need to transform the derivative operators, so that

$$\frac{\partial}{\partial \mathsf{t}} = \frac{\mathsf{U}}{\mathsf{L}}\frac{\partial}{\partial \mathsf{t}'}, \qquad \frac{\partial}{\partial \mathbf{x}} = \frac{1}{\mathsf{L}}\frac{\partial}{\partial \mathbf{x}'}. \tag{7.2.2}$$

In terms of the non-dimensional variables the Navier–Stokes equation (7.1.6) in the absence of body forces ($\mathbf{g} = 0$) becomes

$$\rho\frac{\mathsf{U}}{\mathsf{L}}\frac{\partial}{\partial \mathsf{t}'}\mathsf{U}\mathbf{v}' + \rho\Big(\mathsf{U}\mathbf{v}' \cdot \frac{1}{\mathsf{L}}\boldsymbol{\nabla}'\Big)\mathsf{U}\mathbf{v}' =$$
$$-\frac{1}{\mathsf{L}}\boldsymbol{\nabla}'\mathsf{p} + \eta\frac{1}{\mathsf{L}^2}\boldsymbol{\nabla}'^2\mathsf{U}\mathbf{v}' + \tfrac{1}{3}\eta\frac{1}{\mathsf{L}^2}\boldsymbol{\nabla}'(\boldsymbol{\nabla}' \cdot \mathsf{U}\mathbf{v}'), \tag{7.2.3}$$

We can now simplify the equation (7.2.3) to yield

$$\frac{\partial}{\partial \mathsf{t}'}\mathbf{v}' + (\mathbf{v}' \cdot \boldsymbol{\nabla}')\mathbf{v}' = -\frac{1}{\rho}\boldsymbol{\nabla}'\mathsf{p} + \frac{\eta}{\rho\mathsf{U}\mathsf{L}}\Big(\boldsymbol{\nabla}'^2\mathbf{v}' + \tfrac{1}{3}\boldsymbol{\nabla}'(\boldsymbol{\nabla}' \cdot \mathbf{v}')\Big). \tag{7.2.4}$$

We introduce the Reynolds number

$$\mathrm{Re} = \frac{\rho\mathsf{U}\mathsf{L}}{\eta}, \tag{7.2.5}$$

and can then make a further simplification of the Navier-Stokes equation

$$\frac{\partial}{\partial \mathsf{t}'}\mathbf{v}' + (\mathbf{v}' \cdot \boldsymbol{\nabla}')\mathbf{v}' = -\frac{1}{\rho}\boldsymbol{\nabla}'\mathsf{p} + \frac{1}{\mathrm{Re}}\Big(\boldsymbol{\nabla}'^2\mathbf{v}' + \tfrac{1}{3}\boldsymbol{\nabla}'(\boldsymbol{\nabla}' \cdot \mathbf{v}')\Big). \tag{7.2.6}$$

We will normally assume that the various quantities are non-dimensionalised, and henceforth drop the indicative $'$.

The choices of the length scale L and reference speed U are guided by the nature of the problem under consideration, but different authors may well make dissimilar choices for the same problem. Thus, if we are interested in problems for the Earth's upper mantle, should we take the scale length to be just the upper part of the mantle $\mathsf{L} \sim 1\times10^3$ km or that for the whole mantle $\mathsf{L} \sim 3\times10^3$ km? We will change the non-dimensional variables by a factor of three, without changing the physical situation; there will also be consequent flow-ons to other situations.

As a guide to situations involving flow in Earth's mantle we will make a consistent set of choices throughout our development. The length scale is taken from the depth extent of the mantle, and the characteristic flow speed from plate velocities:

$\mathsf{L} = 3\times10^6$ m, both the depth of the mantle and the size of an average tectonic plate,
$\mathsf{U} = 6\times10^{-10}\,\mathrm{m\,s^{-1}}$, based on the speed of an average tectonic plate of $20\,\mathrm{mm\,yr^{-1}}$, since 1 year is approximately 3×10^7 s.

7.2.1 The Reynolds number

The Reynolds number

$$\mathrm{Re} = \frac{\rho U L}{\eta} = \frac{UL}{\nu} \tag{7.2.7}$$

measures the ratio of *inertial* forces (which have the characteristic quantity of $\rho L^2 U^2$) to *viscous* forces (which have the characteristic quantity of $\eta U L$).

With the choices of physical parameters ρ, η, scale length L and speed U appropriate to the Earth's mantle, we obtain an exceedingly small value for the Reynolds number of $\mathrm{Re} = 10^{-20}$. From this we conclude that inertial forces in the Earth's mantle are small compared with viscous forces.

7.2.2 Stokes Flow

In the limit of very small Reynolds number, the Navier–Stokes equation is often called the *Stokes* equation.

$$\mathrm{Re}\left(\frac{\partial \mathbf{v}}{\partial t} + (\mathbf{v}\cdot\nabla)\mathbf{v}\right) = -\frac{\nabla p}{\rho} + \nu\nabla^2\mathbf{v}. \tag{7.2.8}$$

As a result of the non-dimensionalisation the velocity terms can be expected to be of order unity. The very small value of the Reynolds number preceding the inertial terms allows us to discard these terms for flow in the Earth's mantle:

$$0 \approx -\frac{\nabla p}{\rho} + \nu\nabla^2\mathbf{v}. \tag{7.2.9}$$

This special class of fluid motion is termed *Stokes* flow.

7.2.3 Compressibility

The continuity equation for fluid flow (2.5.7) can be written in the form

$$\frac{1}{\rho}\frac{D}{Dt}\rho + \nabla\cdot\mathbf{v} = 0. \tag{7.2.10}$$

We can regard the medium as approximately incompressible if

$$|\nabla\cdot\mathbf{v}| \ll \frac{U}{L}, \tag{7.2.11}$$

and thus

$$\left|\frac{1}{\rho}\frac{D}{Dt}\rho\right| \ll \frac{U}{L}. \tag{7.2.12}$$

In a homogeneous fluid we can write the rate of change of the pressure in terms of the density and the entropy per unit mass S as

$$\frac{D}{Dt}p = \phi^2 \frac{D}{Dt}\rho + \left[\frac{\partial p}{\partial S}\right]_\rho \frac{D}{Dt}S, \tag{7.2.13}$$

where ϕ is the acoustic (bulk-sound) wavespeed, so that the condition (7.2.12) can be written as

$$\left| \frac{1}{\rho\phi^2}\frac{D}{Dt}p - \frac{1}{\rho\phi^2}\left[\frac{\partial p}{\partial S}\right]_\rho \frac{D}{Dt}S \right| \ll \frac{U}{L}. \tag{7.2.14}$$

The requirement (7.2.13) will normally only be satisfied if each of the two terms on the left-hand side is separately small compared with U/L. For the first term we need

$$\left|\frac{1}{K}\frac{D}{Dt}p\right| \ll \frac{U}{L}, \tag{7.2.15}$$

so that the changes in density due to pressure are negligible; we have here introduced the bulk modulus $K = \rho\phi^2$. For the second term on the left hand side of (7.2.13), we can use thermodynamic identities, as in Section 6.1.2, to write

$$\frac{1}{\rho\phi^2}\left[\frac{\partial p}{\partial S}\right]_\rho = \frac{\alpha_{th}T}{C_p}, \tag{7.2.16}$$

where α_{th} is the coefficient of thermal expansion and C_p is the specific heat at constant pressure. We can then express the requirement for this contribution to be small compared to U/L as

$$\left|\frac{\alpha_{th}}{C_p}\left\{\mathcal{V} + \frac{1}{\rho}\nabla(k\nabla T)\right\}\right| \ll \frac{U}{L}, \tag{7.2.17}$$

where $\mathcal{V}$ is the rate of dissipation of mechanical energy per unit mass of fluid due to shear viscosity and k is the thermal conductivity of the material (see Section 4.2). Variations in the density of the fluid due to internal dissipative heating or molecular conduction of heat must be small:

$$\frac{\alpha_{th}\eta U}{C_p\rho L} \ll 1, \qquad \frac{\alpha_{th}\theta k}{C_p\rho L U} \ll 1, \tag{7.2.18}$$

where θ is a measure of the magnitude of the temperature differences in the fluid. Both conditions (7.2.18) are readily satisfied in the mantle and so the primary control is provided by (7.2.15).

A useful measure of the significance of compressibility is the Mach number M, the ratio of the characteristic velocity to the bulk-sound speed

$$M = U/\phi = U\sqrt{K/\rho}. \tag{7.2.19}$$

For the mantle, the bulk-sound speed for elastic waves ϕ is around 10^4 m s^{-1} and so M is about 10^{-14}. For such small M the last term on the right-hand side of

(7.2.6) will be of no significance, and we can treat the fluid as if it is incompressible ($\nabla \cdot \mathbf{v} \approx 0$). This has the great convenience that we can concentrate on the slow time scales of the main mantle flow and do not need to include the rapid fluctuations associated with the passage of seismic waves.

7.2.4 The Péclet number

The Péclet number provides a measure of *advective* relative to *diffusive* processes in the energy equation, and is defined as

$$\mathrm{Pe} = \frac{UL}{\kappa_H}. \tag{7.2.20}$$

With our set of mantle parameters, the estimate for the Péclet number Pe is of order 10^4. This value indicates that advective processes in the mantle dominate thermal diffusion processes by four orders of magnitude, outside of thermal boundary layers.

7.3 Rectilinear shear flow

As we have seen for many fluids we can neglect the effects of compressibility (at least as a first approximation) and then the continuity equation reduces to

$$\nabla \cdot \mathbf{v} = 0. \tag{7.3.1}$$

Consider then a simple situation with flow in the z-direction induced by a pressure gradient

$$v_x = v_y = 0, \quad v_z = v_z(x, y, t), \tag{7.3.2}$$

for which $(\mathbf{v} \cdot \nabla)\mathbf{v} = 0$ and the Navier–Stokes equations reduce to

$$-\frac{\partial}{\partial x}p = -\frac{\partial}{\partial y}p = 0, \quad -\frac{\partial}{\partial z}p + \eta\left(\frac{\partial^2}{\partial x^2} + \frac{\partial^2}{\partial z^2}\right)v_z = \rho\frac{\partial^2}{\partial t^2}v_z, \tag{7.3.3}$$

so that there are no pressure variations with x or y and the pressure $p(z)$ is a function of z alone. In the case of steady flow $\partial^2 v_z/\partial t^2 \equiv 0$ and so we have

$$\eta\left[\frac{\partial^2 v_z}{\partial x^2} + \frac{\partial^2 v_z}{\partial z^2}\right] = \eta\left[\frac{1}{r}\frac{\partial}{\partial r}\left(r\frac{\partial v_z}{\partial r}\right) + \frac{1}{r^2}\frac{\partial^2 v_z}{\partial \theta^2}\right] = \frac{\partial p}{\partial z} = const, \tag{7.3.4}$$

where we have expressed the equation in both cartesian and cylindrical polar coordinates. The nature of the solution of (7.6.4) will depend on the flow field $v_z(x, y)$ in the $x-y$ plane.

We will adopt *no-slip* conditions and assume that a viscous fluid in contact with a solid boundary must have the same velocity as the boundary. For a free surface the traction must vanish and so the associated shear-stress components will be zero.

One of the simplest configurations we can envisage is a two-dimensional situation with solid boundaries at $x = \pm h$ and no dependence on y. In that case

$$\eta \frac{\partial^2 v_z}{\partial x^2} = \frac{\partial p}{\partial z}, \quad v_z(h) = v_z(-h) = 0, \tag{7.3.5}$$

with solution

$$v_z = \frac{1}{2\eta} \frac{\partial p}{\partial z}(x^2 - h^2). \tag{7.3.6}$$

which is known as Couette flow, with a velocity profile in the form of a parabola symmetric about the centre line ($y = 0$).

This simple model gives a reasonable representation of flow through a volcanic conduit in a fissure eruption, e.g., Hekla in Iceland. The flow rate per unit length of fissure

$$Q = \int_{-h}^{h} dx\, v_z = -\frac{4}{3\eta} \frac{\partial p}{\partial z} h^3, \tag{7.3.7}$$

on the assumption that the pressure gradient is constant.

Where would the pressure gradient come from in such a case? The upward flow of the magma is driven by the natural buoyancy of the lighter magma relative to the denser surrounding rock. At a depth h the *lithostatic* pressure in the rock is $\rho_s g h$ for rock density ρ_s and the corresponding *hydrostatic* pressure for a magma column is $\rho_l g h$ for magma density ρ_l. If the walls of the conduit are free to deform, the pressure gradient available to drive the magma is

$$\frac{\partial p}{\partial z} = -(\rho_s - \rho_l) g, \tag{7.3.8}$$

so that the flow rate would be $\frac{4}{3}(\rho_s - \rho_l) g h^3/\eta$.

The solution here ignores the outflow condition at the top of the pipe and assumes that we do not have to worry about magma solidification in the dyke. A more thorough treatment requires a time-dependent solution with allowance for thermal effects and a better treatment of surface conditions.

7.4 Plane two-dimensional flow

In a situation with a very slow viscous flow the acceleration terms $\rho D\mathbf{v}/Dt$ can be ignored, if we also have incompressibility $\nabla \cdot \mathbf{v} = 0$ as in (7.6.1) we are in the Stokes flow regime (7.2.9) and the Navier-Stokes equations reduce to

$$-\nabla P + \eta \nabla^2 \mathbf{v} = 0, \tag{7.4.1}$$

where P is the pressure produced by fluid flow

$$P = p - \rho g z, \tag{7.4.2}$$

allowing for a gravitational body force. Since the divergence of the velocity vanishes, the pressure P must be a harmonic function, i.e.,

$$\nabla^2 P = 0. \tag{7.4.3}$$

For a plane two-dimensional flow

$$v_x = v_x(x, z), \quad v_y = 0, \quad v_z = v_z(x, z), \tag{7.4.4}$$

we can satisfy the continuity equation by taking

$$v_x = -\frac{\partial \psi}{\partial z}, \quad v_z = \frac{\partial \psi}{\partial x}, \tag{7.4.5}$$

in terms of a stream function $\psi(x, z)$. The curves $\psi(x, z) = const$ represent streamlines since $\mathbf{v} \cdot \nabla\psi = 0$ and $\nabla\psi$ will be normal to the surfaces $\psi = const$. The x and z components of (7.4.1) take the form

$$\frac{\partial P}{\partial x} = -\eta\frac{\partial}{\partial z}(\nabla^2\psi), \quad \frac{\partial P}{\partial z} = \eta\frac{\partial}{\partial z}(\nabla^2\psi), \tag{7.4.6}$$

and we can eliminate P to give

$$\nabla^2(\nabla^2\psi) = \nabla^4\psi = 0, \tag{7.4.7}$$

which is known as the biharmonic equation.

Viscous flow model of glacial rebound

We can illustrate the application of the stream function approach for a simple model of the recovery of continental regions (e.g. Scandinavia) from the loading imposed by the ice sheets of the last Ice Age, using the response of a viscous fluid to a broadly distributed surface load.

We consider an initial vertical displacement

$$u_z = w_0 \cos(kz), \quad k = 2\pi/\lambda, \tag{7.4.8}$$

where the maximum displacement is small compared with the wavelength λ. Once we have determined the response to (7.4.8), more general deformation shapes can be generated by Fourier synthesis.

We note that the vertical velocity is to be found from $\partial_x\psi$ and so look for a stream function in the form

$$\psi = \sin(kx)\, Z(z), \tag{7.4.9}$$

and then

$$\frac{d^4}{dz^4}Z - 2k^2\frac{d^2}{dz^2}Z + k^4 Z = 0, \tag{7.4.10}$$

with solution

$$Z = Be^{-kz} + kDze^{-kz}, \tag{7.4.11}$$

where the growing exponentials have been excluded to give finite values for Z as $z \to \infty$. Thus

$$\psi = \sin(kx)\, e^{-kz}(B + kDz), \tag{7.4.12}$$

and the velocity components are

$$\begin{aligned} v_x &= \sin(kx)\, e^{-kz}[kB + (kz - 1)kD], \\ v_z &= k\cos(kx)\, e^{-kz}[B + kDz]. \end{aligned} \tag{7.4.13}$$

We will force the horizontal component of velocity to be zero at the surface, which since $u_z \ll \lambda$ can be taken as $z = 0$, and so

$$D = B. \tag{7.4.14}$$

The final constant B can be found by equating the hydrostatic pressure associated with the topography to the normal stress at the upper boundary

$$\rho g u_z = -p + 2\eta \frac{\partial v_z}{\partial z} \quad \text{at} \quad z = 0. \tag{7.4.15}$$

Now, at $z = 0$,

$$\frac{\partial p}{\partial x} = \eta \left(\frac{\partial^2 v_x}{\partial x^2} + \frac{\partial^2 v_x}{\partial z^2} \right) = -2\eta B k^3 \sin(kx), \tag{7.4.16}$$

and so

$$p = 2\eta B k^2 \cos(kx) \quad \text{on} \quad z = 0, \tag{7.4.17}$$

and

$$\frac{\partial v_z}{\partial z} = 0 \quad \text{on} \quad z = 0. \tag{7.4.18}$$

The boundary condition at $z = 0$ is thus

$$\rho g u_z = -2\eta B k^2 \cos(kx). \tag{7.4.19}$$

The surface displacement is to be found from

$$\frac{\partial u_z}{\partial t} = v_z(x, 0) = Bk\cos(kx), \tag{7.4.20}$$

and so the displacement at $z = 0$ is given by

$$\frac{\partial u_z}{\partial t} = -\frac{\rho g u_z}{2\eta k} = -\frac{g u_z}{2\nu k}, \tag{7.4.21}$$

which can be integrated to give

$$u_z = w_0 \cos(kx)\, \exp(-t/\tau_c), \tag{7.4.22}$$

where the decay time τ_c is given by

$$\tau_c = 2\nu k/g = 4\pi\nu/g\lambda, \tag{7.4.23}$$

in terms of the kinematic viscosity ν and the wavelength of the initial disturbance λ. The decay time is therefore (for long wavelength disturbances) inversely proportional to the wavelength.

7.5 Thermal convection

When a fluid is heated there is normally a decrease of density due to thermal expansion. A fluid with heating from below or internally with a cool upper surface will tend to have cool dense fluid near the upper boundary and lighter hot fluid at depth. Such a configuration is gravitationally unstable and so the hot fluid tends to rise and the cooler fluid tends to sink to give a thermal convection regime.

7.5.1 The Rayleigh and Nusselt numbers

Two important dimensionless parameters are explicitly related to convection. The Rayleigh number provides a dimensionless buoyancy ratio through the ratio of *buoyancy* forces relative to *viscous* forces and thermal diffusion:

$$\mathsf{Ra} = \frac{g\alpha_{\mathrm{th}}\,\Delta T\,L^3}{\nu\kappa_{\mathrm{H}}} = \frac{\rho g\alpha_{\mathrm{th}}\,\Delta T\,L^3}{\eta\kappa_{\mathrm{H}}}. \tag{7.5.1}$$

For the whole mantle using the scaling parameters U, L above and the additional physical parameters

$\alpha_{\mathrm{th}} = 2\times10^{-5}\,\mathrm{K}^{-1}$ for the thermal expansivity,
$g = 10\ \mathrm{m\,s}^{-2}$ for the gravitational acceleration, and
$\Delta T \approx 1000\,\mathrm{K}$ as a representative temperature difference,

the estimate for the thermal Rayleigh number Ra is around 10^6. As we shall shortly see, this value is more than 1000 times the critical value for the onset of convection and so the Earth's mantle must be in a state of convection.

The Nusselt number is a normalized heat flux, characterising the effect of convection.

$$\mathsf{Nu} = \frac{q}{q_c}, \tag{7.5.2}$$

where q is the observed heat flux across a convecting fluid, and q_c is the heat flux for the same fluid that would be conducted in the absence of convection. Estimates for the Nusselt number for the Earth are in the range 10–20.

7.5.2 The Boussinesq approximation

Because the principal driving force for convection comes from thermally induced density variations we must include these effects in the buoyancy term of the momentum equation, i.e., the gravitational force terms, but the variations will be small enough that they can be neglected elsewhere. This assumption is known as the *Boussinesq* approximation (Valentin Joseph Boussinesq 1842–1929) and leads to considerable simplification of the equations for convection.

We may also assume that density variations $\Delta\rho$ are linearly related to temperature variations ΔT. The constant of proportionality between density and temperature variations is the coefficient of thermal expansivity α_{th}.

Set

$$\rho = \rho_0 + \rho^1, \quad \rho^1 \ll \rho_0, \tag{7.5.3}$$

where ρ_0 is a reference density and ρ^1 is the thermally induced density change

$$\rho^1 = -\rho_0 \alpha_{th}(T - T_0), \tag{7.5.4}$$

where α_{th} is the volume coefficient of thermal expansion, and T_0 the temperature corresponding to the reference density ρ_0. We can then write the Navier-Stokes equation in the form

$$-\nabla P + \rho^1 g\, \hat{\mathbf{e}}_z + \eta \nabla^2 \mathbf{v} = 0, \tag{7.5.5}$$

where P accounts for the hydrostatic pressure in the reference state

$$P = p - \rho_0 g z. \tag{7.5.6}$$

The appropriate thermal equation for weak flows can be derived from (7.1.15) by neglecting the work done by the flow and treating k, ρ, and C as constants to yield

$$\frac{\partial}{\partial t} T + (\mathbf{v} \cdot \nabla) T = \kappa_H \nabla^2 T + h, \tag{7.5.7}$$

in terms of the thermal diffusivity κ_H; h represents the rate of internal heating.

7.5.3 Onset of convection

We will consider a layer of fluid of thickness L without internal heat sources (i.e. $h = 0$) with upper and lower boundaries maintained at fixed temperatures T_u, T_l respectively.

In the absence of convection when the temperature gradient is small, the temperature equation reduces to

$$\frac{\partial^2}{\partial z^2} T = 0 \quad \text{with} \quad \begin{cases} T = T_u, & z = 0, \\ T = T_l, & z = L. \end{cases} \tag{7.5.8}$$

The temperature profile is then

$$T_c(z) = T_u + (T_l - T_u) z / L = T_u + \theta z. \tag{7.5.9}$$

When the lower temperature is raised sufficiently, the system will begin to convect and we can look at the onset of convection by a perturbation analysis about the non-convecting state in which $T = T_c(z)$ and $\mathbf{v} = 0$.

Thus, we set

$$T = T_c(z) + T^1(x,z), \quad \mathbf{v} = \mathbf{v}^1(x,z), \tag{7.5.10}$$

where T^1, $\mathbf{v}^1$ are first-order perturbations. The temperature equation then takes the form

$$\frac{\partial}{\partial t}T^1 + v_z^1 = \kappa_H \left(\frac{\partial^2}{\partial x^2}T^1 + \frac{\partial^2}{\partial z^2}T^1 \right), \tag{7.5.11}$$

since the nonlinear terms $v_x^1 \partial_x T^1$, $v_z^1 \partial T^1/\partial z$ are second-order quantities and can be neglected. The two-dimensional fluid flow equations under the assumption of incompressibility are

$$\begin{aligned} &\frac{\partial}{\partial x}v_x^1 + \frac{\partial}{\partial z}v_z^1 = 0, \\ &-\frac{\partial}{\partial x}P^1 + \eta \left(\frac{\partial^2}{\partial x^2}v_x^1 + \frac{\partial^2}{\partial z^2}v_x^1 \right) = 0, \\ &-\frac{\partial}{\partial z}P^1 - \rho_0 g \alpha_{th} T^1 + \eta \left(\frac{\partial^2}{\partial x^2}v_z^1 + \frac{\partial^2}{\partial z^2}v_z^1 \right) = 0. \end{aligned} \tag{7.5.12}$$

The boundary conditions are that the surfaces $z = 0$ and $z = L$ are isothermal and that there is no vertical flow

$$T^1 = v_z^1 = 0 \quad \text{at} \quad z = 0,\ L. \tag{7.5.13}$$

We can simplify the subsequent analysis by taking free surface conditions at $z = 0,\ L$ so that the shear stress vanishes and

$$\frac{\partial v_x^1}{\partial z} = 0 \quad \text{at} \quad z = 0,\ L. \tag{7.5.14}$$

There is no contribution from $\partial v_z^1/\partial x$ since $v_z^1 = 0$ for all x on the boundaries.

As above, we introduce a stream function ψ such that

$$v_x^1 = -\frac{\partial \psi}{\partial z}, \quad v_z^1 = \frac{\partial \psi}{\partial x}, \tag{7.5.15}$$

and then we can write (7.5.11), (7.5.12) in the form of two coupled partial differential equations

$$\begin{aligned} &\frac{\partial}{\partial t}T^1 + \theta \frac{\partial}{\partial x}\psi = \kappa_H \left(\frac{\partial^2}{\partial x^2}T^1 + \frac{\partial^2}{\partial z^2}T^1 \right), \\ &\nabla^4 \psi - \rho_0 g \alpha_{th} \frac{\partial}{\partial x}T^1 = 0. \end{aligned} \tag{7.5.16}$$

The solution of these two linear equations can be found by separation of variables and to satisfy the boundary conditions (7.5.13), (7.5.14) we can take

$$\begin{aligned} \psi &= \psi_a \sin(\pi z/L) \sin(kx)\, e^{\beta t}, \\ T^1 &= T_a \sin(\pi z/L) \sin(kx)\, e^{\beta t}, \end{aligned} \tag{7.5.17}$$

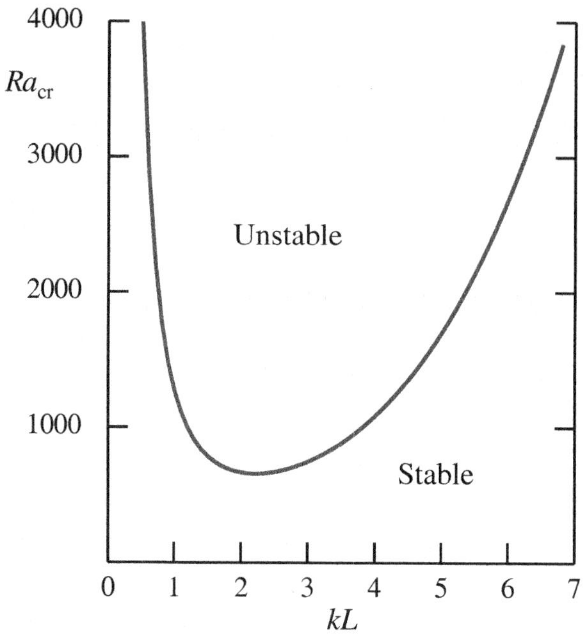

Figure 7.1. Stability condition for onset of convection in the presence of bottom heating in terms of the critical Rayleigh number.

where the horizontal wavenumber k, the growth factor β and the constants ψ_a, T_a are related by the coupled equations

$$\begin{aligned} \left(\beta^2 + \kappa_H k^2 + \kappa_H^2 \frac{\pi^2}{L^2}\right) T_a &= -k\theta\psi_a, \\ \eta\left(k^2 + \frac{\pi^2}{L^2}\right)\psi_a &= -k\rho_0 g \alpha_{th} T_a. \end{aligned} \tag{7.5.18}$$

For β positive, any perturbation away from the reference state will grow in time and the heated layer will be convectively unstable. For β negative, any perturbation will decay and so the layer will be stable against convection.

On eliminating ψ_a and T_a between the two equations (7.5.18), we can express the growth rate β in the form

$$\beta = \frac{\kappa_H}{L}\left(Ra\frac{k^2L^2}{(k^2L^2+\pi^2)^2} - (k^2L^2+\pi^2)\right), \tag{7.5.19}$$

where the dimensionless Rayleigh number

$$Ra = \frac{\rho_0 g \alpha_{th} \theta L^4}{\eta \kappa_H} = \frac{\rho_0 g \alpha_{th}(T - T_0)L^3}{\eta \kappa_H}. \tag{7.5.20}$$

The growth rate β will be positive, leading to convective instability, if

$$Ra > Ra_{cr} = \frac{(k^2L^2 + \pi^2)^3}{k^2L^2}. \tag{7.5.21}$$

If $Ra < Ra_{cr}$, $\beta < 0$ and the layer will be stable against convection. The behaviour is determined by the dimensionless wavenumber kL relating the vertical and horizontal scales of the perturbation. The critical Rayleigh number has an absolute minimum value when $kL = \pi/\sqrt{2}$ and

$$\min(Ra_{cr}) = 27\pi^4/4 \approx 657.51. \tag{7.5.22}$$

If the Rayleigh number is below this value, convection cannot occur and the layer will remain stable against bottom heating (Figure 7.1).

From the definition of the Rayleigh number (7.5.20) we can see that we can interpret the condition $Ra > Ra_{cr}$ in a variety of ways. In order to allow convection to begin we may require a sufficient temperature gradient or a temperature-dependent viscosity to drop below a certain value.

A similar calculation may be carried out for no-slip boundary conditions or for a fluid heated from within, but in these cases the critical Rayleigh number has to be found numerically.

Most studies of time-dependent convection processes have been addressed by numerical solution of the governing equations, but that does not detract from the utility of the linearised stability analysis in assessing the behaviour of different parts of the horizontal wavenumber spectrum of the fluid flow.

7.5.4 Styles of convection

There are considerable differences between the bottom heated case we have just considered and the situation with purely internal heating that are manifest through the temperature and velocity patterns. For the same vigour of convection the Rayleigh number for internal heating needs to be a factor of ten larger than for bottom heating, which already indicates that this is a less effective process for the redistribution of fluid motion.

A comparison of numerical simulations for the bottom heated and purely internally heated cases is shown in Figure 7.2 with the Rayleigh number chosen to give similar convective vigour ($Ra = 10^6$ for bottom heating). The regions shown in the figure are extracted from larger simulations to avoid any influence from the sides of the numerical domain.

The first case (Figure 7.2a) is for bottom heating with both top and bottom surfaces maintained at constant temperature with free-slip boundary conditions. Concentrated thermal boundary layers form at both the upper and lower surfaces. The flux of heat through the lower boundary for the case of bottom heating leads to concentrated upwellings of lighter hot material. The hot fluid cools at the upper boundary and becomes more dense and so descends, again in a localised fashion

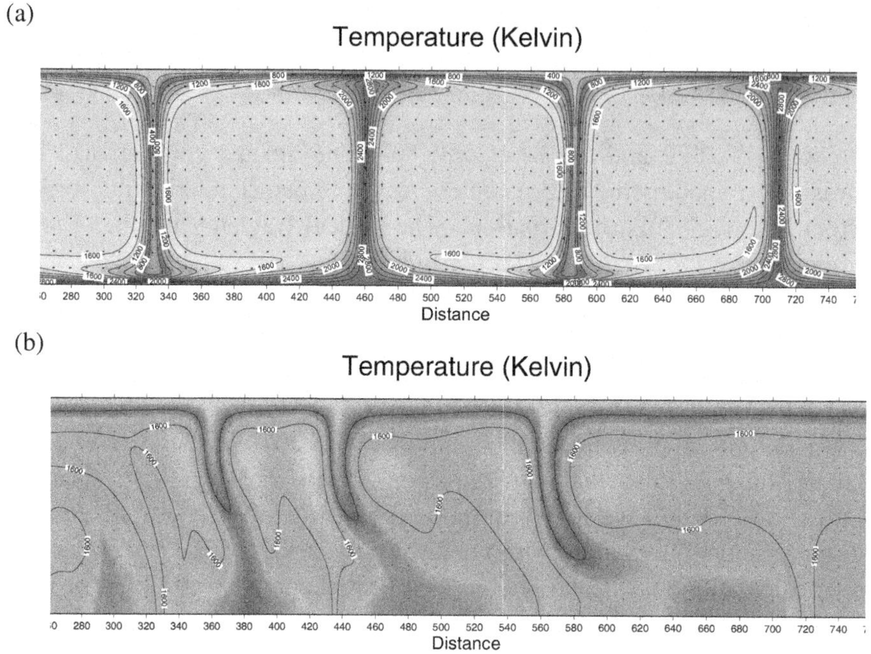

Figure 7.2. Comparison of convection from numerical simulations displayed through the temperature distribution between (a) bottom heating and (b) uniform internal heating, with the Rayleigh numbers chosen to give similar vigour of convection and a quasi-steady state. Note that the temperature contrasts are both more confined and of greater amplitude in the bottom heated case.

(Figure 7.2a). The process leads to a set of regular cells outlined by rising hot limbs and descending cold limbs with very strong temperature contrasts to the interior.

The second case (Figure 7.2b) has the same geometrical configuration but with uniform internal heating with a cold upper boundary, and no heat flux through the lower boundary. In this internally heated case the coooling effect of the upper boundary creates a thicker cool thermal boundary layer from which concentrated downwellings emerge as the dense material becomes unstable. The pattern of convection (Figure 7.2b) shows a number of irregularly spaced zones of descending cold material, but only very weak hot upwelling. Because of the relatively low temperature contrasts, the cold material warms noticeably in descent and may not even reach the lower boundary. Upwelling occurs as a pattern of diffuse flow rising between the cool downwellings; without the injection of bottom heat there is no driver for the generation of streams of light hot fluid.

7.6 The effects of rotation

In a rotating reference frame with angular velocity ω there are accelerations associated with the use of a non-inertial reference frame. These forces are commonly known as the *centrifugal* and *Coriolis* forces. The acceleration

$$\left(\frac{D\mathbf{v}}{Dt}\right)_I = \left(\frac{D\mathbf{v}}{Dt}\right)_R + \boldsymbol{\Omega} \times (\boldsymbol{\Omega} \times \mathbf{x}) + 2\boldsymbol{\Omega} \times \mathbf{v}. \tag{7.6.1}$$

Here the subscript I refers to the inertial reference frame, whereas subscript R refers to the rotating frame of reference. The term on the left-hand side, $(D\mathbf{v}/Dt)_I$, is the actual acceleration experienced by a fluid particle. The first term on the right-hand side, $(D\mathbf{v}/Dt)_R$, is the acceleration relative to the rotating reference frame. The second and third terms on the right-hand side are, respectively, the centrifugal and Coriolis forces.

We can expand the acceleration in the rotating reference frame in the usual way:

$$\left(\frac{D\mathbf{v}}{Dt}\right)_R = \frac{\partial \mathbf{v}}{\partial t} + (\mathbf{v} \cdot \boldsymbol{\nabla})\mathbf{v}. \tag{7.6.2}$$

We can drop the subscript R, because all velocities refer to the rotating reference frame for the rest of this chapter. The equation of motion for a fluid in a rotating reference frame takes the form

$$\frac{\partial \mathbf{v}}{\partial t} + (\mathbf{v} \cdot \boldsymbol{\nabla})\mathbf{v} = -\frac{\boldsymbol{\nabla} p}{\rho} - \boldsymbol{\Omega} \times (\boldsymbol{\Omega} \times \mathbf{x}) - 2\boldsymbol{\Omega} \times \mathbf{v} + \nu \boldsymbol{\nabla}^2 \mathbf{v}, \tag{7.6.3}$$

where $\nu = \eta/\rho$ is the kinematic viscosity.

7.6.1 Rapid rotation

In many cases the centrifugal force is not important. This arises because we can express it as the gradient of a scalar quantity:

$$\boldsymbol{\Omega} \times (\boldsymbol{\Omega} \times \mathbf{x}) = -\boldsymbol{\nabla}(\tfrac{1}{2}\Omega^2 d^2), \tag{7.6.4}$$

where the quantity d is the distance from the axis of rotation. Because of this relation, we can absorb the centrifugal forces into the pressure term.

We replace the pressure term by an augmented pressure $p' = (p - \frac{1}{2}\rho\Omega^2 d^2)$. This transformation reduces the problem to one that is identical in physical context, but the centrifugal force does not appear. Note that this procedure is analogous to the procedure of subtracting out the hydrostatic pressure to remove the effect of gravitational forces.

The centrifugal force is balanced by a radial pressure gradient which is present whether or not there is any flow relative to the rotating reference frame, and which does not interact with any such flow.

The limitation to this transformation is the same as for the gravitational case. First, the pressure must not appear in the boundary conditions. Second, since we have moved ρ inside the differential operator $\boldsymbol{\nabla}$, the density must be constant. Centrifugal force variations associated with density variations will give rise to body forces that can alter or even cause a flow.

The modified equation of motion for a rapidly rotating fluid is then

$$\frac{\partial \mathbf{v}}{\partial t} + (\mathbf{v} \cdot \nabla)\mathbf{v} = -\frac{\nabla p'}{\rho} - 2\Omega \times \mathbf{v} + \nu \nabla^2 \mathbf{v}. \tag{7.6.5}$$

7.6.2 The Rossby and Ekman numbers

In many geophysical flows the effects of the Coriolis force are important to the degree that they dominate the flow. Let us take this situation to the extreme and assume that the Coriolis force is large compared with the effects of inertia and viscous forces. For steady flow this means

$$|\mathbf{v} \cdot \nabla \mathbf{v}| \ll |\Omega \times \mathbf{v}|, \tag{7.6.6}$$

and

$$|\mathbf{v} \cdot \nabla^2 \mathbf{v}| \ll |\Omega \times \mathbf{v}|. \tag{7.6.7}$$

These assumptions imply that the following relations must hold for a representative velocity (U), the rotational rate (Ω) and the length (L) scale in the fluid:

$$\frac{U^2}{L} \ll \Omega U, \qquad \nu \frac{U}{L^2} \ll \Omega U, \tag{7.6.8}$$

and thus

$$\mathrm{Ro} = \frac{U}{L\Omega} \ll 1, \qquad \mathrm{Ek} = \frac{\nu}{\Omega L^2} \ll 1. \tag{7.6.9}$$

The quantities Ro and Ek are known as the Rossby number and the Ekman number. The Rossby number represents the ratio of the inertial to the Coriolis forces, and the Ekman number the ratio of the viscous and Coriolis forces.

Note: Viscous effects may become important, even when these simple scaling considerations indicate otherwise, for example in boundary layers.

7.6.3 Geostrophic flow

When both the Rossby number and the Ekman number are small the equation of motion becomes

$$2\Omega \times \mathbf{v} = -\frac{\nabla p'}{\rho}. \tag{7.6.10}$$

Here the pressure has been modified to include the centrifugal pressure.

Flows in which Coriolis and pressure forces are in balance are called *geostrophic* flows. An important property of geostrophic flow is immediately evident. The Coriolis force is always perpendicular to the flow direction, as is the pressure. This means that pressure is constant along a streamline, which is in marked contrast to flow in non-rotating systems where pressure varies along a streamline.

7.6.4 The Taylor–Proudman theorem

An interesting property of the geostrophic flow equation becomes apparent when we take the curl of the geostrophic balance. We obtain

$$\nabla \times (\boldsymbol{\Omega} \times \mathbf{v}) = 0. \tag{7.6.11}$$

Using vector identities, we can write

$$\boldsymbol{\Omega} \cdot \nabla \mathbf{v} - \mathbf{v} \cdot \nabla \boldsymbol{\Omega} + \mathbf{v}(\nabla \cdot \boldsymbol{\Omega}) - \boldsymbol{\Omega}(\nabla \cdot \mathbf{v}) = 0. \tag{7.6.12}$$

$\boldsymbol{\Omega}$ is not a function of position, so the second and third terms vanish. The fourth term is also zero by virtue of the continuity equation for an incompressible fluid $(\nabla \cdot \mathbf{v} = 0)$. We thus have

$$\boldsymbol{\Omega} \cdot \nabla \mathbf{v} = 0. \tag{7.6.13}$$

If $\boldsymbol{\Omega}$ points into the z-direction, we require

$$\Omega \frac{\partial \mathbf{v}}{\partial z} = 0, \quad \text{and so} \quad \frac{\partial \mathbf{v}}{\partial z} = 0; \tag{7.6.14}$$

there is no variation of the velocity field in the direction parallel to the rotation axis. This result is known as the *Taylor–Proudman* theorem.

In component form the Taylor–Proudman theorem is expressed as

$$\frac{\partial u}{\partial z} = \frac{\partial v}{\partial z} = \frac{\partial w}{\partial z} = 0. \tag{7.6.15}$$

If the fluid is contained by boundaries perpendicular to the rotation axis, so that $w = 0$ at the boundary, then we have

$$\frac{\partial u}{\partial z} = \frac{\partial v}{\partial z} = 0, \quad w = 0, \quad \text{everywhere in the fluid.} \tag{7.6.16}$$

The flow is entirely two-dimensional in planes perpendicular to the axis of rotation. One striking consequence of the Taylor–Proudman theorem is the formation of features known as *Taylor columns* which are expected, for example, in core flow.

7.6.5 Ekman layers

We assume a geostrophic flow is present in a fluid rotating around the z-axis bounded by a wall in the xy-plane. The flow is uniform and uni-directional with component u_0 in the x-direction:

$$2\Omega u_0 = -\frac{1}{\rho}\frac{\partial p_0}{\partial y}, \qquad 0 = -\frac{1}{\rho}\frac{\partial p_0}{\partial x}. \tag{7.6.17}$$

There is a boundary region between the geostrophic flow in the interior of the fluid and the flow just at the wall with boundary conditions $u = v = w = 0$, where viscous forces are important. This is because in the boundary region a stress exists between the boundary and the fluid. In the absence of other forces the stress can be balanced only by the Coriolis force. The Coriolis force acts at right angles to the motion. Hence we assume that flow in the boundary layer is parallel to the wall with non-zero components u, v,

$$\frac{\partial u}{\partial x} = \frac{\partial u}{\partial y} = \frac{\partial v}{\partial x} = \frac{\partial v}{\partial y} = 0. \tag{7.6.18}$$

We also require from the continuity condition that

$$w = 0. \tag{7.6.19}$$

The three components of the momentum equation are

$$\begin{aligned} -2\,\Omega v &= -\frac{1}{\rho}\frac{\partial p}{\partial x} + \nu\frac{\partial^2 u}{\partial z^2}, \\ 2\,\Omega u &= -\frac{1}{\rho}\frac{\partial p}{\partial y} + \nu\frac{\partial^2 v}{\partial z^2}, \\ 0 &= \frac{\partial p}{\partial z}, \end{aligned} \tag{7.6.20}$$

and hence

$$\frac{\partial p}{\partial x} = \frac{\partial p_0}{\partial x} = 0, \quad \frac{\partial p}{\partial y} = \frac{\partial p_0}{\partial y} = 0. \tag{7.6.21}$$

The equations to be solved for the flow are

$$\begin{aligned} -2\Omega v &= \nu\frac{\partial^2 u}{\partial z^2}, \\ -2\Omega(u_0 - u) &= \nu\frac{\partial^2 v}{\partial z^2}, \end{aligned} \tag{7.6.22}$$

with boundary conditions

$$\begin{aligned} u = v = 0, &\qquad z = 0, \\ u \to u_0, \quad v \to 0, &\qquad z \to \infty. \end{aligned} \tag{7.6.23}$$

The solution to this equation system is

$$\begin{aligned} u &= u_0\,[1 - e^{-z/\Delta}\cos(z/\Delta)], \\ v &= u_0\,e^{-z/\Delta}\sin(z/\Delta), \end{aligned} \tag{7.6.24}$$

where

$$\Delta = (\nu/\Omega)^{1/2}. \tag{7.6.25}$$

We notice that the distance over which the solution approaches the interior solution is of order Δ. In other words, Δ gives the thickness of the boundary layer; this layer is called the Ekman layer.

Comparing Δ with the expression for the Ekman number Ek confirms that the boundary layer thickness corresponds to a local Ekman number of unity.

8

Continuum Equations and Boundary Conditions

We summarise here the main equations for continuum behaviour that need to be employed in a broad range of applications, by drawing on the development in earlier chapters. We start with the conservation laws for mass, momentum and energy expressed in differential form, that need to be supplemented by the appropriate constitutive laws to express the rheological state of the medium. We then consider the boundary conditions that prevail at the surface of a continuum or an interface between two different continua. The normal components of velocity and the stress tensor are required to be continuous across a general interface. The tangential components of velocity are also continuous for solid–solid and fluid–fluid boundaries and also for a solid–viscous fluid interface under a "no-slip" condition. Heat flux is continuous across an interface, but because of variations in thermal conductivity temperature gradients can have a jump. At a phase boundary, additional thermodynamic constraints need to be applied to describe the equilibrium scenario along the interface.

Hitherto, we have concentrated on the way in which a continuum responds to deformation or imposed stress, but we also need to take into account electromagnetic phenomena. The iron-rich core of the Earth is a conducting fluid where a complex interaction of motion and electromagnetic effects leads to dynamo action and creates the Earth's internal magnetic field. We therefore provide a brief development of the topic of continuum electrodynamics and show how the continuum equations need to be modified to accommodate magnetic effects.

In many situations we have to understand the interaction of the temperature and flow fields, and so provide an introduction to the solution of diffusion problems for temperature and the corresponding heat flow.

8.1 Conservation equations

In earlier chapters we have established a number of results based on conservation properties, working from integral forms to derive the appropriate local equations. These results based on the conservation of mass, momentum and energy are brought together here and are recast into explicit differential representations of the conservation equations.

8.1.1 Conservation of mass

The velocity field $\mathbf{v}(\mathbf{x}, t)$ is related to the density field in a moving continuum by the continuity equation (2.5.6) that represents a local formulation of the conservation of mass,

$$\frac{\partial \rho}{\partial t} + \boldsymbol{\nabla} \cdot (\rho \mathbf{v}) = 0; \tag{8.1.1}$$

this relation is equivalent to the Lagrangian form (2.5.7),

$$\frac{D\rho}{Dt} + \rho \boldsymbol{\nabla} \cdot \mathbf{v} = 0; \tag{8.1.2}$$

so that incompressibility ($D\rho/Dt \equiv 0$) requires the divergence of the velocity field to vanish ($\boldsymbol{\nabla} \cdot \mathbf{v} = 0$).

8.1.2 Conservation of momentum

The acceleration of a continuum produced by the various forces acting upon it is given by (3.2.4) in conjunction with (2.4.5), in terms of the velocity $\mathbf{v}(\mathbf{x}, t)$,

$$\frac{D}{Dt}\mathbf{v} = \frac{\partial \mathbf{v}}{\partial t} + (\mathbf{v} \cdot \boldsymbol{\nabla}_{\mathbf{x}})\mathbf{v} = \boldsymbol{\nabla}_{\mathbf{x}} \sigma + \mathbf{g}, \tag{8.1.3}$$

i.e., in component terms

$$\frac{D}{Dt}v_j = \frac{\partial v_j}{\partial t} + \left(v_k \frac{\partial}{\partial x_k} \right) v_j = \frac{\partial}{\partial x_i} \sigma_{ij} + g_j. \tag{8.1.4}$$

The relation between the stress and deformation is determined by the relevant constitutive equation.

For a viscous fluid, from Section 4.5, the stress tensor is given by

$$\sigma_{ij} = -p\delta_{ij} + \eta \left(\frac{\partial v_i}{\partial x_j} + \frac{\partial v_j}{\partial x_i} - \tfrac{2}{3}\delta_{ij}\frac{\partial v_k}{\partial x_k} \right) + \zeta \delta_{ij} \frac{\partial v_k}{\partial x_k}, \tag{8.1.5}$$

where we have allowed for the possibility of bulk viscosity (ζ).

For linearised isotropic elasticity, from Section 5.2, we find that

$$\sigma_{ij} = (K - \tfrac{2}{3}G)\delta_{ij}\frac{\partial u_k}{\partial x_k} + 2G \left(\frac{\partial v_i}{\partial x_j} + \frac{\partial v_j}{\partial x_i} \right), \tag{8.1.6}$$

in terms of the displacement $\mathbf{u}$, the bulk modulus K and shear modulus $G \equiv \mu$.

For linearised viscoelasticity we also need to include terms that depend on the history of deformation.

We can write (8.1.3) in a form that more directly reflects the conservation of momentum in the current frame,

$$\frac{\partial}{\partial t}(\rho v_j) = -\frac{\partial}{\partial x_i}\Pi_{ij} + \rho g_j, \tag{8.1.7}$$

where Π_{ij} is the momentum flux density tensor

$$\Pi_{ij} = \rho v_i v_j - \sigma_{ij}; \tag{8.1.8}$$

as can be verified by direct differentiation.

8.1.3 Conservation of energy

In the discussion of energy balance in Section 4.2 we have established the relation

$$\rho \frac{D}{Dt} U = h - \nabla_x \mathbf{q} + \sigma . \nabla_x \mathbf{v}, \tag{8.1.9}$$

where h is the rate of internal heat production and the heat flux vector $\mathbf{q}$ is determined by the temperature gradient

$$\mathbf{q} = -k \nabla_x T, \tag{8.1.10}$$

where k is the thermal conductivity. When we include the kinetic energy contribution as in (4.2.9) we have

$$\rho \frac{D}{Dt}(U + \tfrac{1}{2}v^2) = h - \nabla_x \mathbf{q} + \nabla_x \cdot [\sigma \cdot \mathbf{v}] + \mathbf{g} \cdot \mathbf{v}, \tag{8.1.11}$$

where $\mathbf{g}$ is the external force field (normally gravitational). We express $\mathbf{g}$ in terms of a potential φ

$$\mathbf{g} = -\nabla_x \varphi. \tag{8.1.12}$$

When $\mathbf{g}$ is a function of position rather than time

$$\mathbf{g} \cdot \mathbf{v} = -\mathbf{v} \cdot \nabla_x \varphi = -\frac{D}{Dt}\varphi, \tag{8.1.13}$$

and then we can include the potential energy of the gravitational field φ as an additional energy contribution. For flow in the mantle and core, self-gravitational effects will be small so the time-independent approximation in (8.1.13) is well satisfied.

The material derivative of the specific energy

$$\rho \frac{D}{Dt}(\tfrac{1}{2}v^2 + U + \varphi) = \rho \frac{\partial}{\partial t}(\tfrac{1}{2}v^2 + U + \varphi) + \rho \mathbf{v} \cdot \nabla_x(\tfrac{1}{2}v^2 + U + \varphi), \tag{8.1.14}$$

and the partial derivative of the total energy

$$\frac{\partial}{\partial t}[\rho(\tfrac{1}{2}v^2 + U + \varphi)] = \rho \frac{\partial}{\partial t}(\tfrac{1}{2}v^2 + U + \varphi) + \frac{\partial \rho}{\partial t}(\tfrac{1}{2}v^2 + U + \varphi); \tag{8.1.15}$$

further, from the continuity condition (2.5.8),

$$\frac{\partial \rho}{\partial t} = -\nabla_x \cdot (\rho \mathbf{v}). \tag{8.1.16}$$

We can combine the expressions (8.1.11) and (8.1.14)–(8.1.16) to produce a representation of the rate of change of energy in the spatial domain

$$\frac{\partial}{\partial t}(\tfrac{1}{2}\rho v^2 + \rho U + \rho\varphi) = h - \nabla_{\mathbf{x}}\mathbf{Q}, \tag{8.1.17}$$

in terms of the energy flux density

$$\mathbf{Q} = (\tfrac{1}{2}\rho v^2 + \rho U)\mathbf{v} - \mathbf{v}\cdot\boldsymbol{\sigma} - k\nabla_{\mathbf{x}}T. \tag{8.1.18}$$

Equation (8.1.17) is a direct differential representation of the conservation of energy.

8.2 Interface conditions

The surface of the Earth represents a substantial density contrast between rock and air, and for many purposes the presence of the air can be disregarded and the surface treated as if it were a free surface in vacuum. Thus the surface traction $\boldsymbol{\tau}(\mathbf{n}_f)$ must vanish, where $\mathbf{n}_f$ is the normal to the free surface. If the surface can be regarded as locally horizontal, the free surface condition translates to a requirement on the 3-components of the stress tensor

$$\sigma_{13} = \sigma_{23} = \sigma_{33} = 0. \tag{8.2.1}$$

The corresponding condition for a static fluid would be the vanishing of pressure. However, the density contrasts between the ocean and atmosphere are smaller than for rock and so it is preferable to use the full fluid–fluid boundary conditions.

For a stationary rigid boundary S_B of a fluid, with normal $\mathbf{n}_B$ we require

$$\mathbf{v}\cdot\mathbf{n}_B = 0. \tag{8.2.2}$$

At an internal interface between two solids or fluids we require continuity of the traction vector acting on the surface (3.2.11)

$$[\tau_j]_-^+ = n_i[\sigma_{ij}]_-^+ = 0, \tag{8.2.3}$$

where $\mathbf{n}$ is the interface normal. For the isotropic pressure distribution in static fluids, (8.2.3) requires continuity of the pressure.

At welded contact between two solids or a fluid–fluid interface, the velocity must be continuous so that

$$[\mathbf{v}]_-^+ = 0. \tag{8.2.4}$$

At a fluid–solid boundary, the normal component of the velocity will be continuous, so we need

$$[\mathbf{v}\cdot\mathbf{n}]_-^+ = 0. \tag{8.2.5}$$

For a viscous fluid in contact with a solid boundary, the *no-slip* boundary condition requires the velocity to be continuous as in (8.2.4). For a boundary between an inviscid fluid and a solid slip can occur, and there is no requirement that the tangential velocities need to be the same.

Where sediments with slow elastic wavespeeds lie close to the surface, coupling between air and ground can take place and the fluid–solid boundary conditions need to be used.

Across a material interface we require the heat flux to be continuous and so

$$\left[k^{(1)}\nabla T^{(1)} - k^{(2)}\nabla T^{(2)}\right] \cdot \mathbf{n} = 0 \tag{8.2.6}$$

so that there will, in general, be a discontinuity in the temperature gradient $\mathbf{n} \cdot \nabla T$ across the boundary, since the thermal conductivity will be different on the two sides.

At a *phase transition* in a material, the physical properties will normally differ on the two sides of the phase boundary and so we need to impose the appropriate continuity conditions as derived above on the deformation and force fields. The different phases have to be in equilibrium at the phase interface and as a result the position of the phase boundary will be a function of the conditions to which it is subjected. Thermodynamic arguments require that there be no net change in the Gibbs free energy across the phase transition. The slope of the pressure–temperature ($p_b - T_b$) relation for the mutual phase boundary is then given by the Clausius–Clapeyron equation

$$\frac{dp_b}{dT_b} = \frac{L}{T_b \Delta v}, \tag{8.2.7}$$

where L is the latent heat of the phase transition and Δv is the change in specific volume across the phase interface. The specific entropy change is $\Delta S_b = L/T_b$.

8.3 Continuum electrodynamics

We have so far concentrated on the effects of deformation, but alongside mechanical effects we have the possibility of electromagnetic influences. The electrical conductivity of the mantle modulates how external magnetic disturbances interact with the Earth, and the character of the magnetic field of the geodynamo as seen at the surface. If a conducting fluid moves in a magnetic field, electric fields are induced with resulting flow of electric current. The magnetic field exerts forces on the electric currents that can alter the fluid flow; the currents themselves also alter the magnetic field. Such intertwining of fluid flow and electromagnetic phenomena within the Earth's core creates dynamo action to produce a sustained internal magnetic field.

8.3.1 Maxwell's equations

The electromagnetic field can be described by the electric vector $\mathbf{E}$ and the magnetic induction $\mathbf{B}$, and the interaction of the field with matter requires the specification of a further set of vectors: the electric displacement **D**, the magnetic vector **H** and the electric current density **j**. The space and time derivatives of the five vectors in the current state are related by

$$\nabla \times \mathbf{H} - \frac{\partial \mathbf{D}}{\partial t} = \mathbf{j}, \qquad \nabla \times \mathbf{E} + \frac{\partial \mathbf{B}}{\partial t} = 0,$$
$$\nabla \cdot \mathbf{D} = \breve{\rho}, \qquad \nabla \cdot \mathbf{B} = 0, \tag{8.3.1}$$

where $\breve{\rho}$ is the density of electric charge. Charge conservation requires

$$\nabla \cdot \mathbf{j} + \frac{\partial \breve{\rho}}{\partial t} = 0. \tag{8.3.2}$$

In free space

$$\mathbf{D} = \epsilon_0 \mathbf{E}, \qquad \mathbf{B} = \mu_0 \mathbf{H}, \tag{8.3.3}$$

where ϵ_0 is the permittivity and μ_0 the magnetic permeability of free space, such that $\epsilon_0 \mu_0 = 1/c^2$, where c is the speed of light.

8.3.2 Electromagnetic constitutive equations

The levels of electromagnetic fields within the Earth are such that the constitutive equations are linear relations between fields. The frequencies of interest normally lie well below those of crystal vibrations so that the dielectric constant ϵ and the magnetic permeability μ are essentially constant and show no noticeable frequency dependence.

For a conductor, the current **j** is related to the electric vector **E** by

$$\mathbf{j} = \breve{\sigma} \mathbf{E}, \tag{8.3.4}$$

where $\breve{\sigma}$ is the specific conductivity; note that we have written $\breve{\sigma}$ rather than the more conventional σ to avoid conflict with the notation we have used for stress.

Inside a material the electric displacement **D** is related to the electric vector **E** through the dielectric constant ϵ,

$$\mathbf{D} = \epsilon_0 \epsilon \mathbf{E}. \tag{8.3.5}$$

In a similar way the magnetic induction **B** is derived from the magnetic vector **H** by the action of the magnetic permeability μ,

$$\mathbf{B} = \mu_0 \mu \mathbf{H}. \tag{8.3.6}$$

Except for a few magnetic minerals, $\mu \approx 1$ for Earth materials. The temperature coefficient of the dielectric coefficient can be significant, and there will be distinct changes in physical properties associated with the influence of phase transitions.

We can therefore write the Maxwell equation for the curl of the magnetic vector **H** as

$$\nabla \times \mathbf{H} = \mathbf{j} + \frac{\partial \mathbf{D}}{\partial t} = \breve{\sigma} \mathbf{E} + \epsilon_0 \epsilon \frac{\partial \mathbf{E}}{\partial t}, \tag{8.3.7}$$

where we have assumed that the rate of change of geological materials is slow compared with the frequency of electromagnetic phenomena. For good electrical conductors such as metals, the second term $\epsilon_0\epsilon\dot{\mathbf{E}}$ on the right-hand side of (8.3.7) is negligible, and this will be the regime that prevails in the iron-rich core of the Earth. Nevertheless, for less good conductors such as the silicate minerals in the mantle this contribution can be significant at high frequencies $\omega > \breve{\sigma}/\epsilon_0\epsilon$.

For crystalline materials, the relations (8.3.4)–(8.3.6) are replaced by tensor relations

$$j_k = \breve{\sigma}_{kl}E_l, \quad D_k = \epsilon_0\epsilon_{kl}E_l, \quad H_k = \mu_0\mu_{kl}B_l. \tag{8.3.8}$$

The tensors $\breve{\sigma}$, ϵ, μ will be invariant under the symmetry group of the crystal. From thermodynamic arguments, the conductivity tensor $\breve{\sigma}$, the dielectric tensor ϵ, and the magnetic permeability tensor μ are all symmetric.

When a dielectric is deformed, the dielectric tensor has strain dependence

$$\epsilon_{ij} = \epsilon^0_{ij} + a_{ijkl}e_{kl}, \tag{8.3.9}$$

in terms of the linearised strain e_{kl}. For certain classes of crystals deformation is accompanied by the appearance of an electric field proportional to the accompanying stress σ. In this case, the electric induction takes the form

$$D_i = \epsilon_0(\epsilon_{ij}E_j + \gamma_{i,kl}\sigma_{kl}), \tag{8.3.10}$$

where the piezoelectric tensor has the symmetries

$$\gamma_{i,kl} = \gamma_{i,lk}, \tag{8.3.11}$$

because of the symmetry of the stress tensor σ. An analogous relation to (8.3.11) describes piezomagnetism. In each case the electromagnetic fields have to be determined along with the deformation. In general the ‘piezo’ coefficients are rather small and it requires very considerable stress to produce significant electromagnetic fields. Nevertheless, such contributions have been invoked to explain electromagnetic phenomena accompanying earthquakes.

8.3.3 Electromagnetic continuity conditions

At a material discontinuity where the conductivity and dielectric properties change, there is the possibility of the build up of a surface charge density $\hat{\rho}$ and a surface current $\hat{\mathbf{j}}$, and discontinuities in the electromagnetic field quantities. By an analysis similar to that in Section 3.2.2 using small volumes and surfaces crossing the interface between media (1) and (2), with local unit normal $\mathbf{n}_{12}$, we can derive the continuity conditions from the Maxwell equations (8.3.1).

The two equations for the divergence of the **B, D** fields require the application of the divergence theorem to a small volume crossing the interface. The normal component of the magnetic induction $\mathbf{B}$ is continuous across such a material discontinuity

$$\mathbf{n}_{12} \cdot (\mathbf{B}^{(2)} - \mathbf{B}^{(1)}) = 0, \tag{8.3.12}$$

but, as in Section 3.2.2, this means that the tangential components are not constrained. In the presence of a layer of surface charge density $\hat{\rho}$, the normal component of the electric displacement $\mathbf{D}$ has a discontinuity equal to $\hat{\rho}$,

$$\mathbf{n}_{12} \cdot (\mathbf{D}^{(2)} - \mathbf{D}^{(1)}) = \hat{\rho}. \tag{8.3.13}$$

For $\mathbf{E}$, $\mathbf{H}$, whose curl enters the Maxwell equations, Stokes' theorem is applied to a surface crossing the interface. The tangential component of the electric vector $\mathbf{E}$ is continuous across the material discontinuity

$$\mathbf{n}_{12} \times (\mathbf{E}^{(2)} - \mathbf{E}^{(1)}) = 0; \tag{8.3.14}$$

whereas the tangential component of the magnetic vector $\mathbf{H}$ has a discontinuity equal to the surface current density $\hat{\mathbf{j}}$,

$$\mathbf{n}_{12} \times (\mathbf{H}^{(2)} - \mathbf{H}^{(1)}) = \hat{\mathbf{j}}. \tag{8.3.15}$$

8.3.4 Energy equation for the electromagnetic field

From the Maxwell equations (8.3.1),

$$\mathbf{E} \cdot (\nabla \times \mathbf{H}) - \mathbf{H} \cdot (\nabla \times \mathbf{E}) = \mathbf{E} \cdot \mathbf{j} + \mathbf{E} \cdot \frac{\partial \mathbf{D}}{\partial t} + \mathbf{H} \cdot \frac{\partial \mathbf{B}}{\partial t}, \tag{8.3.16}$$

and from a standard vector identity, the left-hand side can be recognised as $\nabla \cdot (\mathbf{E} \times \mathbf{H})$,

$$-\nabla \cdot (\mathbf{E} \times \mathbf{H}) = \mathbf{E} \cdot (\nabla \times \mathbf{H}) - \mathbf{H} \cdot (\nabla \times \mathbf{E}). \tag{8.3.17}$$

Hence we can write (8.3.16) in the form

$$\mathbf{E} \cdot \frac{\partial \mathbf{D}}{\partial t} + \mathbf{H} \cdot \frac{\partial \mathbf{B}}{\partial t} + \mathbf{j} \cdot \mathbf{E} + \nabla \cdot (\mathbf{E} \times \mathbf{H}) = 0. \tag{8.3.18}$$

Consider then an arbitrary volume V with surface S and integrate (8.3.18) throughout the volume

$$\int_V \left\{ \mathbf{E} \cdot \frac{\partial \mathbf{D}}{\partial t} + \mathbf{H} \cdot \frac{\partial \mathbf{B}}{\partial t} \right\} dV + \int_V \mathbf{j} \cdot \mathbf{E} \, dV + \int_S (\mathbf{E} \times \mathbf{H}) \cdot \hat{\mathbf{n}} \, dS = 0. \tag{8.3.19}$$

where $\hat{\mathbf{n}}$ is the unit outward normal to S, and we have used the divergence to convert the last term into a surface integral. The relation (8.3.19) is a direct consequence of the Maxwell equations, and does not depend on the form of the material equations.

If we assume the material properties do not vary in time we can write (8.3.19) as

$$\frac{\partial}{\partial t} \int_V \frac{1}{2} (\mathbf{E} \cdot \mathbf{D} + \mathbf{B} \cdot \mathbf{H}) \, dV + \int_V \mathbf{j} \cdot \mathbf{E} \, dV + \int_S (\mathbf{E} \times \mathbf{H}) \cdot \hat{\mathbf{n}} \, dS = 0, \tag{8.3.20}$$

which we can interpret as the energy conservation equation

$$\frac{\partial \mathcal{W}}{\partial t} + \mathcal{Q} + \int_S \mathbf{S} \cdot \hat{\mathbf{n}}\, dS = 0, \tag{8.3.21}$$

where

$\mathcal{W} = \int_V \frac{1}{2}(\mathbf{E} \cdot \mathbf{D} + \mathbf{B} \cdot \mathbf{H})\, dV$ is the stored energy in the magnetic field,

$\mathcal{Q} = \int_V \mathbf{j} \cdot \mathbf{E}\, dV$ is the dissipation due to Joule heating, and

$\mathbf{S} = \mathbf{E} \times \mathbf{H}$ is the Poynting vector.

$\mathbf{S}$ can be interpreted as the energy flux per unit area, normal to $\mathbf{E}$ and $\mathbf{H}$, associated with the field.

Although flow in the mantle might modify material properties it will do so on a scale so much longer than the frequency of any electromagnetic phenomena that the time derivatives of the conductivity $\breve{\sigma}$ and dielectric constant ϵ can be ignored. For the Earth's core efficient stirring ensures a sustained radial profile of variation that does not vary significantly in time.

8.3.5 Electromagnetic disturbances

In a dielectric material described by (8.3.4)–(8.3.6), we can reduce the Maxwell equations (8.3.1) to a suitable form to describe the propagation of electromagnetic disturbances in a region without charge density. We can generate two equations for $\nabla \times \dot{\mathbf{H}}$ from the equation for the curl of the electric vector $\mathbf{E}$,

$$\nabla \times \left(\frac{1}{\mu_0\mu} \nabla \times \mathbf{E}\right) = -\nabla \times \frac{\partial \mathbf{H}}{\partial t}, \tag{8.3.22}$$

and the time derivative of the equation for the curl of the magnetic vector $\mathbf{H}$,

$$\nabla \times \frac{\partial \mathbf{H}}{\partial t} = \frac{\partial \mathbf{j}}{\partial t} + \frac{\partial^2 \mathbf{D}}{\partial t^2}. \tag{8.3.23}$$

We can eliminate $\nabla \times \partial_t \mathbf{H}$ between (8.3.22) and (8.3.23) to give

$$-\nabla \times \left(\frac{1}{\mu_0\mu} \nabla \times \mathbf{E}\right) = \frac{\partial}{\partial t}(\breve{\sigma}\mathbf{E}) + \frac{\partial^2}{\partial t^2}(\epsilon_0\epsilon\mathbf{E}). \tag{8.3.24}$$

Equation (8.3.24) can be simplified with the aid of the two vector identities

$$\nabla \times (u\mathbf{v}) = u\nabla \times \mathbf{v} + \nabla u \times \mathbf{v}, \quad \nabla \times (\nabla \times \mathbf{v}) = \nabla(\nabla \cdot \mathbf{v}) - \nabla^2 \mathbf{v}. \tag{8.3.25}$$

Thus, we find

$$-\nabla(\nabla \cdot \mathbf{E}) + \nabla^2 \mathbf{E} = \mu_0\mu\breve{\sigma}\frac{\partial \mathbf{E}}{\partial t} + \epsilon_0\mu_0\epsilon\mu\frac{\partial^2}{\partial t^2} + \mu_0\mu\nabla\left(\frac{1}{\mu_0\mu}\right) \times \mathbf{E}. \tag{8.3.26}$$

We use $\nabla \cdot \mathbf{D} = 0$ in association with the constitutive relation $\mathbf{D} = \epsilon_0\epsilon\mathbf{E}$, and then

$$\nabla(\epsilon_0\epsilon\mathbf{E}) = \epsilon_0\epsilon\nabla \cdot \mathbf{E} + \epsilon_0\nabla\epsilon \cdot \mathbf{E} = 0. \tag{8.3.27}$$

On combining (8.3.26) and (8.3.27) we have

$$\nabla^2\mathbf{E} - \mu_0\mu\breve{\sigma}\frac{\partial\mathbf{E}}{\partial t} - \epsilon_0\mu_0\epsilon\mu\frac{\partial^2\mathbf{E}}{\partial t^2} = -\nabla(\log\mu)\times(\nabla\times\mathbf{E}) - \nabla(\nabla(\log\epsilon)\cdot\mathbf{E}). \qquad (8.3.28)$$

For a homogeneous dielectric the material gradients vanish and (8.3.28) takes the form

$$\nabla^2\mathbf{E} - \mu_0\mu\breve{\sigma}\frac{\partial\mathbf{E}}{\partial t} - \epsilon_0\mu_0\epsilon\mu\frac{\partial^2\mathbf{E}}{\partial t^2} = 0, \qquad (8.3.29)$$

with a comparable form for the magnetic vector **H**.

In the absence of conductivity, $\breve{\sigma} = 0$, the equation (8.3.29) reduces to the wave equation

$$\nabla^2\mathbf{E} = \frac{\epsilon\mu}{c^2}\frac{\partial^2\mathbf{E}}{\partial t^2}, \qquad (8.3.30)$$

where we have introduced the velocity of light in free space c through $\epsilon_0\mu_0 = 1/c^2$. In the homogeneous dielectric we have

$$\nabla^2\mathbf{E} = \frac{n^2}{c^2}\frac{\partial^2\mathbf{E}}{\partial t^2}, \qquad (8.3.31)$$

with refractive index $n = \sqrt{\epsilon\mu}$, so that the velocity of electromagnetic disturbances in the material is $c/\sqrt{\epsilon\mu} = c/n$. The transverse electromagnetic waves have energy flow in the direction of the Poynting vector $\mathbf{E}\times\mathbf{H}$.

For a disturbance with frequency ω represented via an complex exponential time dependence $\exp(-i\omega t)$ we can write the equation (8.3.29) for the electric vector in a conductive medium in the form

$$\nabla^2\mathbf{E} + \left(\frac{\epsilon\mu}{c^2}\omega^2 - i\mu_0\mu\breve{\sigma}\omega\right)\mathbf{E}, \qquad (8.3.32)$$

which can be viewed as a wave equation for a complex refractive index, corresponding to attenuation as the wave propagates. The effect can be readily illustrated by considering a plane wave travelling in the x direction for which we will have a y-component of the electric field E_y and a z-component of the magnetic field H_z,

$$E_y = \exp\left(i\omega\frac{n}{c}x - i\omega t\right)\hat{E}_y, \qquad H_z = \exp\left(i\omega\frac{n}{c}x - i\omega t\right)\hat{H}_z, \qquad (8.3.33)$$

From (8.3.32) we have

$$-\frac{\omega^2 n^2}{c^2} = -\frac{\mu\epsilon\omega^2}{c^2} - i\omega\mu\mu_0\breve{\sigma}, \qquad (8.3.34)$$

and thus the complex refractive index for an isotropic medium takes the form

$$n^2 = \mu\epsilon + i\frac{\breve{\sigma}\mu}{\epsilon_0\omega}. \qquad (8.3.35)$$

We set $n = n' + in''$ and then

$$n^2 = n'^2 - n''^2 + 2in'n'' = \mu\epsilon + i\frac{\sigma\mu}{\epsilon_0\omega}. \quad (8.3.36)$$

which can be solved for n', n''. The propagating electric field takes the form

$$E_y = \exp\left(-\frac{\omega}{c}n''x\right)\exp\left(i\frac{\omega}{c}n'x - i\omega t\right)\hat{E}_y, \quad (8.3.37)$$

so that the plane wave decays as it travels in the x-direction. The amplitude falls to e^{-1} after the wave has advanced a distance

$$d = \frac{c}{\omega n''}, \quad (8.3.38)$$

known as the *penetration depth* or *skin depth*, which is often a small fraction of the wavelength in a vacuum. The relation of the electric and magnetic components is

$$\hat{E}_y = Z\hat{H}_z = \frac{c\mu\mu_0}{n}\hat{H}_z = \left(\frac{\mu\mu_0}{\epsilon\epsilon_0 + i\breve{\sigma}/\omega}\right)^{1/2}\hat{H}_z, \quad (8.3.39)$$

where Z is the intrinsic impedance of the medium. In the decaying transverse wave the ratio of the amplitudes of E_y and H_z is everywhere the same $|Z|$, but a phase difference ϕ is introduced between the electric and magnetic vectors, where

$$\tan\phi = n''/n'. \quad (8.3.40)$$

For a very good conductor such as a metal, the conduction term dominates and then

$$n' + in'' \approx \left(\frac{i\breve{\sigma}\mu}{\epsilon_0\omega}\right)^{1/2} = \frac{1+i}{\sqrt{2}}\left(\frac{\breve{\sigma}\mu}{\epsilon_0\omega}\right)^{1/2}, \quad (8.3.41)$$

with penetration depth

$$d = \left(\frac{2}{\breve{\sigma}\omega\mu\mu_0}\right)^{1/2}. \quad (8.3.42)$$

The varying depth of penetration of electromagnetic disturbances with frequency is exploited in the analysis of magnetotelluric signals to determine the conductivity structure of the Earth. Such signals may be man-made or exploit the induced electric and magnetic fields produced by the varying magnetic fields in the ionosphere. Long period waves ($\sim 10^4$ s period) can penetrate into the transition zone in the upper mantle, which appears as an increase in electrical conductivity.

8.3.6 Magnetic fluid dynamics

When a conducting fluid moves in a magnetic field, an electric field is induced and electric currents flow. The magnetic field exerts forces on the currents that can alter the flow, and the presence of the currents modifies the magnetic field. The result is a complex interaction between electromagnetic effects and fluid dynamics.

We will assume that the magnetic permeability of the conducting fluid is very close to unity ($\mu \approx 1$), and work in terms of the **E** and **B** fields. From Maxwell's equations (8.3.1)

$$\nabla \cdot \mathbf{B} = 0, \qquad \nabla \times \mathbf{E} = -\frac{\partial \mathbf{B}}{\partial t}; \tag{8.3.43}$$

the current density in a moving conductor is

$$\mathbf{j} = \breve{\sigma}(\mathbf{E} + \mathbf{v} \times \mathbf{B}). \tag{8.3.44}$$

so that

$$\frac{1}{\mu_0}\nabla \times \mathbf{B} = \breve{\sigma}\mathbf{E} + \breve{\sigma}\mathbf{v} \times \mathbf{B}. \tag{8.3.45}$$

The displacement current $\partial \mathbf{D}/\partial t$ is negligible for phenomena whose time scale is long compared with the time for passage of electromagnetic waves across the region of interest. The neglect of the displacement current acts as a filter to remove electromagnetic waves from the situation in much the same way as we used the incompressibility condition in Chapter 7 to suppress acoustic waves from fluid flow. We can use (8.3.45) to represent the electric field in terms of the magnetic field as

$$\mathbf{E} = \frac{1}{\breve{\sigma}\mu_0}\nabla \times \mathbf{B} - \mathbf{v} \times \mathbf{B}, \tag{8.3.46}$$

and then from (8.3.43) and (8.3.44),

$$\nabla \times \left(\frac{1}{\breve{\sigma}\mu_0}\nabla \times \mathbf{B}\right) = -\frac{\partial \mathbf{B}}{\partial t} + \nabla \times (\mathbf{v} \times \mathbf{B}). \tag{8.3.47}$$

In a homogeneous conductor with constant conductivity, we can simplify (8.3.47) to

$$\frac{\partial \mathbf{B}}{\partial t} = \nabla \times (\mathbf{v} \times \mathbf{B}) + \frac{1}{\breve{\sigma}\mu_0}\nabla^2 \mathbf{B}, \tag{8.3.48}$$

since $\nabla \cdot \mathbf{B} = 0$. The quantity $\lambda = (\breve{\sigma}\mu_0)^{-1}$ is known as the *magnetic diffusivity* of the fluid.

The force acting on the magnetic fluid can be expressed as

$$\mathbf{F} = \mu_0 \mathbf{j} \times \mathbf{B} = (\nabla \times \mathbf{B}) \times \mathbf{B} = (\mathbf{B} \cdot \nabla)\mathbf{B} - \tfrac{1}{2}\nabla B^2. \tag{8.3.49}$$

For a magnetic continuum the continuity equation for conservation of mass is unchanged, but in the equation of motion we need to include the electromagnetic force (8.3.49). The Navier–Stokes equation for the magnetic fluid thus has the form

$$\begin{aligned} \rho\left\{\frac{\partial \mathbf{v}}{\partial t} + (\mathbf{v} \cdot \nabla)\mathbf{v}\right\} &= \nabla\boldsymbol{\sigma} + \rho\mathbf{g} + (\nabla \times \mathbf{B}) \times \mathbf{B} \\ &= \nabla[\boldsymbol{\sigma} - \tfrac{1}{2}B^2] + \mu_0(\mathbf{B} \cdot \nabla)\mathbf{B} - \rho\nabla\varphi, \end{aligned} \tag{8.3.50}$$

where the stress tensor $\boldsymbol{\sigma}$ for the viscous fluid is related to the velocity field through (8.1.5), and ρ is the density of the fluid.

As in Section 8.1, we can recast (8.3.51) into a direct expression of the conservation of momentum

$$\frac{\partial}{\partial t}(\rho v_j) = -\frac{\partial}{\partial x_i}\Pi^m_{ij} + \rho g_j \tag{8.3.51}$$

where the momentum flux density tensor Π^m_{ij} now takes the form

$$\Pi^m_{ij} = \rho v_i v_j - \sigma_{ij} - \mu_0(B_i B_j - \tfrac{1}{2}B^2\delta_{ij}), \tag{8.3.52}$$

and the additional term relative to (8.1.8) is the Maxwell stress tensor for the magnetic field

The rate of change of internal energy for the magnetic fluid has now to include the influence of Joule heating from current dissipation ($j^2/\breve{\sigma}$). As a result (8.1.9) is replaced by

$$\rho\frac{D}{Dt}U = h + \nabla\mathbf{q} + \boldsymbol{\sigma}.\nabla\mathbf{v} + \frac{1}{\breve{\sigma}\mu_0^2}(\nabla\times\mathbf{B})^2. \tag{8.3.53}$$

The representation of the conservation of energy for the magnetic fluid has to include the magnetic energy $\frac{1}{2}B^2/\mu_0^2$ and also the electromagnetic energy flux from the Poynting vector $\mathbf{E}\times\mathbf{B}/\mu_0$. Thus we find that

$$\frac{\partial}{\partial t}\left(\tfrac{1}{2}\rho v^2 + \rho U + \rho\varphi + \tfrac{1}{2}\frac{B^2}{\mu_0^2}\right) = h - \nabla\cdot\mathbf{Q}^m. \tag{8.3.54}$$

where the energy flux density vector $\mathbf{Q}^m$ is

$$\begin{aligned}\mathbf{Q}^m &= (\tfrac{1}{2}\rho v^2 + \rho U)\mathbf{v} - \mathbf{v}\cdot\boldsymbol{\sigma} - k\nabla T \\ &\quad + \frac{1}{\mu_0^2}\mathbf{B}\times(\mathbf{v}\times\mathbf{B}) + \frac{1}{\breve{\sigma}\mu_0^2}(\nabla\times\mathbf{B})\times\mathbf{B}.\end{aligned} \tag{8.3.55}$$

As in our discussion of non-magnetic fluids in Chapter 7 we can usually make the approximation that the fluid is incompressible, and thereby both suppress acoustic waves and simplify the equation system.

Frozen flux

In the limiting case of very high conductivity, the magnetic diffusivity $\lambda \to 0$ and (8.3.48) simplifies to

$$\frac{\partial\mathbf{B}}{\partial t} = \nabla\times(\mathbf{v}\times\mathbf{B}) = (\mathbf{B}\cdot\nabla)\mathbf{v} - (\mathbf{v}\cdot\nabla)\mathbf{B} - (\nabla\cdot\mathbf{v})\mathbf{B}. \tag{8.3.56}$$

After rearrangement of the terms in (8.3.56) we see that

$$\frac{D}{Dt}\mathbf{B} = \frac{\partial\mathbf{B}}{\partial t} + (\mathbf{v}\cdot\nabla)\mathbf{B} = (\mathbf{B}\cdot\nabla)\mathbf{v} - (\nabla\cdot\mathbf{v})\mathbf{B}. \tag{8.3.57}$$

With the aid of the continuity equation (8.1.1) we obtain

$$\frac{D}{Dt}\left(\frac{\mathbf{B}}{\rho}\right) = \frac{1}{\rho}\frac{D\mathbf{B}}{Dt} - \frac{\mathbf{B}}{\rho^2}\frac{D\rho}{Dt} = \left(\frac{\mathbf{B}}{\rho}\cdot\nabla\right)\mathbf{v}. \tag{8.3.58}$$

The rate of change of the quantity $\mathbf{B}/\rho$ satisfies the same equation as that for a material line element $d\mathbf{x}$ (2.4.8). As a result, if these vectors are initially in the same direction they will remain parallel, with a constant ratio of length. Thus the line of force $\mathbf{B}$ moves with the fluid particles that lie on it. We can regard the $\mathbf{B}$-lines as 'frozen' in the fluid, so that the topological structure of the field cannot change with time.

Hence as any closed fluid contour moves over time it will not cut any $\mathbf{B}$-lines and so the flux of $\mathbf{B}$ passing through the contour remains constant. This property of *frozen flux* for a highly conducting fluid has been invoked for the Earth's core to try to resolve ambiguities in the relation of the magnetic field and the velocity field that are related to the requirement $\nabla \cdot \mathbf{B} = 0$ (see Chapter 15).

EXAMPLE: MAGNETOHYDRODYNAMIC WAVES

Consider the influence of small wave disturbances in a constant magnetic field, when dissipative effects from viscosity, thermal conductivity and electrical resistance can be neglected.

We assume high-frequency disturbances so that an adiabatic state is maintained and introduce a small perturbation to the magnetic field, density and pressure

$$\mathbf{B} = \mathbf{B}_0 + \mathbf{B}', \quad \rho = \rho_0 + \rho', \quad p = p_0 + p', \tag{8.3.e1}$$

where the subscript $_0$ indicates the constant equilibrium value. From the adiabatic condition $p' = c_0^2 \rho'$, where c_0 is the sound speed. To a first-order approximation the pertubations satisfy the magnetic equations

$$\nabla \cdot \mathbf{B}' = 0, \qquad \frac{\partial \mathbf{B}'}{\partial t} = \nabla \times (\mathbf{v} \times \mathbf{B}_0), \tag{8.3.e2}$$

the continuity condition

$$\frac{\partial \rho'}{\partial t} + \rho_0 \nabla \cdot \mathbf{v} = 0, \tag{8.3.e3}$$

and the Navier–Stokes equation

$$\rho_0 \frac{\partial \mathbf{v}}{\partial t} = -c_0^2 \nabla \cdot p' + \mathbf{B}_0 \times (\nabla \times \mathbf{B}'). \tag{8.3.e4}$$

We now seek solutions for the additional disturbances in the form of plane waves with slowness vector $\mathbf{p}$ and frequency ω so that, e.g.,

$$\mathbf{B}' \propto \exp[i\omega(\mathbf{p} \cdot \mathbf{x} - t)],$$

the corresponding wavevector $\mathbf{k} = \omega \mathbf{p}$. From (8.3.e2)–(8.3.e3) we find

$$-\mathbf{B}' = \mathbf{p} \times (\mathbf{v} \times \mathbf{B}_0) = \mathbf{v}(\mathbf{p} \cdot \mathbf{B}_0) - \mathbf{B}_0(\mathbf{p} \cdot \mathbf{v}), \tag{8.3.e5}$$

$$\rho' = \rho_0 \mathbf{p} \cdot \mathbf{v}, \tag{8.3.e6}$$

$$-\rho_0 \mathbf{v} + c_0^2 \rho' \mathbf{p} = -\mathbf{B}_0 \times (\mathbf{p} \times \mathbf{B}'). \tag{8.3.e7}$$

The first equation (8.3.e5) requires that the vector $\mathbf{B}'$ is perpendicular to the slowness vector $\mathbf{p}$, i.e., $\mathbf{p} \cdot \mathbf{B}' = 0$. We can eliminate ρ' from (8.3.e7) with the aid of (8.3.e6) to yield

$$\rho_0\{c_0^2(\mathbf{p}\cdot\mathbf{v})\mathbf{p} - \mathbf{v}\} = -\mathbf{B}_0 \times (\mathbf{p}\times\mathbf{B}') = -\mathbf{p}(\mathbf{B}_0\cdot\mathbf{B}') - \mathbf{B}'(\mathbf{p}\cdot\mathbf{B}_0). \tag{8.3.e8}$$

The properties of the waves are governed by the two coupled equations (8.3.e5), (8.3.e8) for $\mathbf{B}'$ and $\mathbf{v}$.
Consider a disturbance transverse to the plane containing the slowness vector $\mathbf{p}$ and the ambient magnetic field $\mathbf{B}_0$, such that $\mathbf{B}'\cdot\mathbf{B}_0 = 0$, $\mathbf{p}\cdot\mathbf{v} = 0$, then

$$\mathbf{B}' = -\mathbf{v}(\mathbf{p}\cdot\mathbf{B}_0), \qquad -\rho\mathbf{v} = \mathbf{v}(\mathbf{p}\cdot\mathbf{B}_0)^2; \tag{8.3.e9}$$

so that $\mathbf{p}\cdot\mathbf{B}_0 = \sqrt{\rho_0}$ and the wavespeed $u = 1/|p| = |\mathbf{B}_{\|}|/\sqrt{\rho_0}$, where $\mathbf{B}_{\|}$ is the component of $\mathbf{B}$ parallel to $\mathbf{p}$. The velocity

$$\mathbf{v} = -\mathbf{B}'/\sqrt{\rho_0}, \tag{8.3.e10}$$

and the dispersion relation for these transverse disturbances is

$$\omega = -\mathbf{B}_0\cdot\mathbf{k}/\sqrt{\rho_0}, \tag{8.3.e11}$$

with group velocity

$$\frac{\partial\omega}{\partial\mathbf{k}} = -\mathbf{B}_0/\sqrt{\rho_0}, \tag{8.3.e12}$$

which does not depend on the wavevector $\mathbf{k}$.
A disturbance with $\mathbf{v}$ and $\mathbf{B}'$ lying in the plane of $\mathbf{p}$ and $\mathbf{B}_0$ will have associated density perturbations $\rho' = \rho(\mathbf{p}\cdot\mathbf{v})$. By eliminating $\mathbf{B}'$ between (8.3.e5) and (8.3.e8) we find

$$\rho_0\{c_0^2p^2 - 1\}(\mathbf{p}\cdot\mathbf{v}) = p^2\left[(\mathbf{p}\cdot\mathbf{v})\mathbf{B}_0^2 - (\mathbf{p}\cdot\mathbf{B}_0)(\mathbf{B}_0\cdot\mathbf{v})\right], \tag{8.3.e13}$$

which leads to a quartic in the slowness p.
For very weak ambient magnetic fields we have one class of waves for which $c_0^2p^2 \approx 1$ so that the magnetohydrodynamic waves reduce to ordinary sound waves. In the second class,

$$\mathbf{v} \approx -\mathbf{B}'/\sqrt{\rho_0}, \tag{8.3.e14}$$

so that the situation is similar to (8.3.e10), but now with $\mathbf{v}$, $\mathbf{B}'$ in the plane of $\mathbf{p}$ and $\mathbf{B}_0$.
In an *incompressible fluid*, for which $\rho' = 0$, we have a single kind of wave propagation with two independent directions of polarisation; both $\mathbf{v}$ and $\mathbf{B}'$ are perpendicular to $\mathbf{p}$ with

$$\mathbf{v} = -\mathbf{B}'/\sqrt{\rho_0}, \tag{8.3.e15}$$

such Alfvén waves are exact solutions of the equations and do not depend on the assumption of small perturbations. •

8.4 Diffusion and heat flow

The mechanical forms of deformation we have introduced include material flow and the transmission of wave phenomena, but in many circumstances diffusive phenomena are important, as in the way that temperature effects spread through a medium.

We have encountered the influence of heat effects in our discussion of energy balance in Section 4.2, where we demonstrated that the local form of the conservation of energy can be written as

$$\rho \frac{D}{Dt} U = h - \nabla_{\mathbf{x}} \cdot \mathbf{q} + \sigma \cdot \nabla_{\mathbf{x}} \mathbf{v}, \tag{8.4.1}$$

for internal energy density U, heat production h, heat flow vector $\mathbf{q}$, velocity field $\mathbf{v}$ and stress tensor σ. The heat flux $\mathbf{q}$ is related to the temperature gradient by

$$\mathbf{q} = -k \nabla_{\mathbf{x}} T, \tag{8.4.2}$$

where k is the thermal conductivity.

The internal energy density U can be related to temperature through a thermal capacity C and from (7.1.15), in the absence of deviatoric strain, the equation for the temperature field can be written as

$$\rho \frac{D}{Dt}(CT) = \rho \left[\frac{\partial}{\partial t}(CT) + \mathbf{v} \cdot \nabla_{\mathbf{x}}(CT) \right] = h + \nabla_{\mathbf{x}} \cdot [k \nabla_{\mathbf{x}} T]. \tag{8.4.3}$$

The particular thermal capacity C depends on the thermodynamic state; normally we would use C_p the thermal capacity at constant pressure. In equilibrium, $\partial T/\partial t = 0$ and so

$$\rho \mathbf{v} \cdot \nabla_{\mathbf{x}}(C_p T) - \nabla_{\mathbf{x}} \cdot [k \nabla_{\mathbf{x}} T] = h. \tag{8.4.4}$$

In a material at rest with constant properties C_p, k the equation for temperature simplifies substantially to

$$\frac{\partial}{\partial t} T = h + \kappa_H \nabla_{\mathbf{x}}^2 T. \tag{8.4.5}$$

A comprehensive set of analytical solutions to heat flow problems based on (8.4.5) are to be found in Carslaw & Jaeger (1959).

8.4.1 Equilibrium heat flow

Consider the simple case where the temperature T and the thermal conductivity k depend only on depth z. Then the conduction of heat in steady state is governed by

$$\frac{d}{dz}\left(k(z) \frac{d}{dz} T(z) \right) = 0, \tag{8.4.6}$$

which can be readily integrated to yield

$$T(z) = a + b \int^z \frac{d\zeta}{k(\zeta)}. \tag{8.4.7}$$

The constants a, b have to be determined by the boundary conditions in the specific situation. Suppose we have a layer whose boundaries are at fixed temperatures $T(z_u) = T_u$, $T(z_l) = T_l$, then

$$T(z) = T_u + (T_l - T_u) \frac{\int_{z_u}^{z} d\zeta/k(\zeta)}{\int_{z_u}^{z_l} d\zeta/k(\zeta)}. \tag{8.4.8}$$

In the special case where k is constant, (8.4.8) reduces to a linear gradient in temperature, independent of k:

$$T(z) = T_u + (T_l - T_u)\left(\frac{z - z_u}{z_l - z_u}\right). \tag{8.4.9}$$

When heat sources are evenly distributed through a region so that

$$\frac{d}{dz}\left(k(z)\frac{d}{dz}T(z)\right) = -h_o, \tag{8.4.10}$$

the solution (8.4.7) is augmented by the influence of the heat generation

$$T(z) = a + b\int^z \frac{d\zeta}{k(\zeta)} - h_o\int^z d\zeta\,\frac{\zeta}{k(\zeta)}; \tag{8.4.11}$$

once again a, b are to be found from the imposed boundary conditions. For a homogeneous layer the additional temperature contribution is quadratic in depth.

The exponentially decaying heat production $h = h_r e^{-z/L}$ represents an approximation to the likely radiogenic heat production in the Earth. For this situation the temperature distribution in $z > 0$ satisfies

$$\frac{d}{dz}\left(k(z)\frac{d}{dz}T(z)\right) = h_r e^{-z/L}. \tag{8.4.12}$$

The heat flow

$$q(z) = -k(z)\frac{d}{dz}T(z) = h_r L e^{-z/L} + q_m, \tag{8.4.13}$$

if the heat flow at great depth is q_m. Thus the surface heat flow

$$q(0) = q_m + h_r L, \tag{8.4.14}$$

and is linear in the surface heat production rate.

EXAMPLE: STEADY-STATE PROBLEMS FOR A CYLINDER OR SPHERE

For a homogeneous material the equilibrium temperature satisfies

$$k\nabla^2 T = -h.$$

Cylindrical and spherical problems can be tacked with the appropriate form for the Laplacian operator ∇^2.

Cylindrical symmetry: for steady radial flow in the presence of heat production h_c, the temperature distribution is controlled by

$$k\frac{1}{\varpi}\frac{\partial}{\partial\varpi}\left(\varpi\frac{\partial T}{\partial\varpi}\right) = -h_c,$$

where ϖ is the cylindrical radial variable. The general solution of this differential equation is

$$T(\varpi) = A + B\ln\varpi - h_c\frac{\varpi^2}{2k}.$$

For a solid cylinder the term in $\ln \varpi$ is not allowed, but needs to be included for problems with hollow cylinders.
Spherical symmetry: the radial temperature distribution is controlled by

$$k\frac{1}{r^2}\frac{\partial}{\partial r}\left(r^2\frac{\partial T}{\partial r}\right) = -h_s,$$

with solution

$$T(r) = A + \frac{B}{r} + \frac{h_s}{6k}r^2;$$

once again the singular term has to be excluded if the domain includes the origin $r = 0$.•

A simple model of the motion of a fluid through a porous rock is provided by the scenario where there are no heat sources, but the advective term is retained. For a homogeneous region with porosity g and a fluid with vertical transport speed v_z, the temperature has to satisfy

$$\gamma\frac{dT}{dz} = \frac{d^2T}{dz^2}, \tag{8.4.15}$$

where $\gamma = \rho C_p v_z g/k = v_z g/\kappa_H$. The temperature solution is

$$T(z) = c - d\left(\frac{1}{\gamma k}\right)e^{\gamma z}, \tag{8.4.16}$$

with c, d determined by the boundary conditions. The departure of the temperature distribution (8.4.16) from the linear solution for conductive heat transport in a uniform medium can be used to infer the value of v_z.

8.4.2 Time-varying problems

We continue to consider problems that depend just on the depth coordinate z, but now allow for the evolution of the temperature field with time.

Consider the response to an impulsive source of heat $h(z,t) = \delta(z)\delta(t)$ in a uniform medium, the Green's function for the temperature has to satisfy

$$\frac{\partial T_G}{\partial t} - \kappa\frac{\partial^2 T_G}{\partial t^2} = \delta(z)\delta(t), \tag{8.4.17}$$

where for notational convenience we have written κ for the thermal diffusivity κ_H. We can find a solution by means of a Fourier transform. We express $T(z,t)$ as a combination of frequency and wavenumber components as

$$T_G(z,t) = \frac{1}{2\pi}\int_{-\infty}^{\infty} dq \int_{-\infty}^{\infty} d\omega\, \bar{T}(q,\omega)e^{i(qz-\omega t)}, \tag{8.4.18}$$

and then

$$-i\omega\bar{T}(q,\omega) + \kappa q^2\bar{T}(q,\omega) = 1/2\pi. \tag{8.4.19}$$

We now have to evaluate the multiple integrals over vertical wavenumber q and frequency ω

$$T_G(z,t) = \frac{1}{(2\pi)^2}\int_{-\infty}^{\infty} dq \int_{-\infty}^{\infty} d\omega \, \frac{e^{i(qz-\omega t)}}{\kappa q^2 - i\omega}. \tag{8.4.20}$$

Performing the frequency integral first using contour integration we can extract the contribution from the polar residue at $\omega = -i\kappa q^2$, when $t > 0$, to give

$$T_G(z,t) = \frac{1}{2\pi}\int_{-\infty}^{\infty} dq \, e^{i(qz-\kappa q^2 t)} H(t), \tag{8.4.21}$$

where $H(t)$ is the Heaviside step function. We can rewrite (8.4.21) as

$$\begin{aligned} T_G(z,t) &= \frac{1}{2\pi} e^{-z^2/4\kappa t} H(t) \int_{-\infty}^{\infty} dq \, e^{-\kappa t(q+iz/2\kappa t)^2}, \\ &= \left(\frac{1}{4\pi\kappa t}\right)^{1/2} e^{-z^2/4\kappa t} H(t), \end{aligned} \tag{8.4.22}$$

with the aid of a change of variables. Subsequent to the impulse of heat the temperature spreads out with a Gaussian profile whose scale length depends on the inverse square root of the time $t^{-1/2}$.

We can find a comparable solution for the situation where the initial temperature profile $T_0(z)$ is specified at $t = 0$,

$$T(z,t) = \int_{-\infty}^{\infty} d\zeta \, T_0(\zeta) T_G(z-\zeta, t). \tag{8.4.23}$$

An important special case is provided by

$$T_0(z) = T_0 H(z), \tag{8.4.24}$$

which is equivalent to the sudden juxtaposition of two uniform half-spaces with different temperatures. The resulting temperature distribution

$$\begin{aligned} T(z,t) &= \frac{T_0}{\sqrt{4\pi\kappa t}} \int_0^{\infty} d\zeta \, e^{-(z-\zeta)^2/4\kappa t} = \tfrac{1}{2} T_0 \frac{2}{\sqrt{\pi}} \int_0^{z/\sqrt{4\kappa t}} d\xi \, e^{-\xi^2}, \\ &= \tfrac{1}{2} T_0 \left[1 + \operatorname{erf}\left(\frac{z}{\sqrt{4\kappa t}}\right)\right], \end{aligned} \tag{8.4.25}$$

where the error function $\operatorname{erf}(x)$ is defined as

$$\operatorname{erf}(x) = \frac{2}{\sqrt{\pi}} \int_0^{x} d\xi \, e^{-\xi^2}, \tag{8.4.26}$$

and $\operatorname{erf}(-x) = -\operatorname{erf}(x)$.

The situation of the cooling of a half-space initially at temperature T_0 with a requirement that the surface temperature vanishes is equivalent in $z > 0$ to the choice

$$T_0(z) = T_0 \operatorname{sgn}(z), \tag{8.4.27}$$

and then

$$T(z,t) = \tfrac{1}{2}T_0\left[1 + \text{erf}\left(\frac{z}{\sqrt{4\kappa t}}\right)\right] - \tfrac{1}{2}T_0\left[1 + \text{erf}\left(\frac{-z}{\sqrt{4\kappa t}}\right)\right]$$
$$= T_0\,\text{erf}\left(\frac{z}{\sqrt{4\kappa t}}\right). \tag{8.4.28}$$

EXAMPLE: DIURNAL AND ANNUAL TEMPERATURE FLUCTUATIONS

The relative penetration of temperature fluctuations imposed at the surface can be assessed by the analysis of a harmonic field.

Consider a temperature field with a frequency dependence $\exp(-i\omega t)$, then

$$\kappa\frac{d^2T}{dz^2} + i\omega T = 0$$

with solutions of the form

$$T(z,t) = \exp\left(-\left[\frac{\omega}{2\kappa}\right]^{\frac{1}{2}} z\right)\exp\left(i\left[\frac{\omega}{2\kappa}\right]^{\frac{1}{2}} z - i\omega t\right).$$

The temperature behaviour is in the form of a diffusive wave, decaying as it penetrates into the medium (cf. (8.3.39)). The rate of decay with depth is proportional to the square root of frequency ($\omega^{1/2}$). This means that for a comparable reduction of the amplitude of temperature variation relative to the surface, the penetration of the yearly cycle will be $\sqrt{365}$ greater than the diurnal variation. The diurnal cycle can be perceptible through a masonry wall about 20 cm thick, whereas in a 4 m deep cellar the yearly variations are nearly out of phase with the surface variations and serve to minimise the overall temperature fluctuations. •

For a situation where boundary conditions are imposed on a layer the solution may be found in terms of a Fourier series in z rather than a Fourier transform. Thus, for example, for the situation where a layer of thickness L has an initial temperature T_a and boundary conditions $T = 0$ at $z = 0$, $T = T_a$ at $z = L$, a solution for the subsequent temperature distribution can be found as

$$T(z,t) = T_a\left[\frac{z}{L} + \sum_{n=1}^{\infty}\frac{2}{n\pi}\sin\left(\frac{n\pi z}{L}\right)\exp\left(-\frac{n^2\pi^2\kappa t}{L^2}\right)\right], \tag{8.4.29}$$

with heat flow $Q_0(t)$ at $z = 0$ of

$$Q_0(t) = \frac{kT_a}{L}\left[1 + 2\sum_{n=1}^{\infty}\exp\left(-\frac{n^2\pi^2\kappa t}{L^2}\right)\right]. \tag{8.4.30}$$

Part II

Earth Deformation

9

From the Atomic Scale to the Continuum

The continuum equations we have encountered in previous chapters make use of the device of a constitutive equation representing the relation between stress and strain at the macroscopic level. Such constitutive equations are therefore the intermediaries by which properties at the atomic level are translated to their implications at much larger length scales. Behaviour at the atomic scale depends on the properties of the constituent crystals, e.g., elastic moduli, and on departures of the material from ideality. Grain boundaries, vacancies, dislocations and other forms of defects play their role in controlling transport properties in the crystals. Such transport phenomena in turn control the way in which composite materials respond to externally imposed forces.

A wide variety of semi-empirical constitutive relation have been devised to represent specific aspects of material behaviour. The variety of functional forms represents the balance between different classes of deformation mechanisms at the atomic level. The net result is a non-linear relation between stress and strain (or strain-rate).

9.1 Transport properties and material defects

9.1.1 Grains and crystal defects

The polycrystalline aggregates that make up the Earth's silicate mantle are composed of multiple minerals, each with finite grain size within which there may be further classes of interruption of the regular crystal lattices.

The distribution of grain size with depth is therefore a very important, but imperfectly known, property of the mantle. It is tempting to assume that the millimetre to centimetre size crystals of mafic materials brought to the surface in xenoliths are representative of the depths from which they have been detached. Kimberlite and lamprophyre pipes show every evidence of very rapid eruption so that material has been projected to the surface very rapidly, but interaction with the eruptive process cannot be excluded.

Some exposures of former upper mantle rocks in mountain belts suggest that the original crystal sizes were rather large. Nevertheless, the common view is that grain sizes in the continental lithosphere are unlikely to be larger than a few centimetres and would be expected to decrease in size in the asthenospheric regime. Faul & Jackson (2005) have included a grain size dependence in a model of the behaviour of shear moduli and attenuation with temperature for olivine systems; fits to velocity profiles for both continental and oceanic domains suggest the need for a significant increase in grain size in the lower part of the asthenosphere. In the transition zone and below there are no direct indicators of grain size. However, there are suggestions that there should be a grain size reduction at the phase transitions in silicate minerals that lead to closer packed structures, particularly just below the 660 km discontinuity due to the spinel to perovskite phase change. Increasing evidence suggests the presence of significant amounts of water at depth, and variations in water content could well affect grain size.

Within the crystal grains, the lattice structures are not perfect; there may be *point, line, surface* or *volume* defects that disturb locally the regular arrangements of the atoms. Minor departures from the theoretical composition (non-stoichiometry) will lead to point defects, either *vacancies* due to the absence of one atomic species, or *interstitials* where there too many atoms of one variety. A common feature of the silicate minerals is the substitution of small amounts of different atoms. The presence of some iron atoms in a magnesio-silicate will lead to some lattice strain around the iron sites, that may need to be accommodated by other defects. Even more significant is the case when elements with different valency such as aluminium are introduced into the lattice since the charge distribution will also be affected (see, e.g., Li et al., 2005, for a discussion of substitution in perovskite and the effect on material properties).

The most significant class of line defects is the presence of *dislocations* that offset parts of the regular crystal structure. Such dislocations can be described by two vectors: the line direction **l**, which is the tangent to the defect at any point, and the Burgers vector **b**, which describes the closure failure in a circuit made around the line defect. The line defect moves in the glide plane with normal $\mathbf{n} = \mathbf{l} \times \mathbf{b}$ (Figure 9.1a). In an edge dislocation (Figure 9.1b) an extra half plane of atoms is inserted into the lattice, so that the Burgers vector **b** is perpendicular to the line direction **l**. For a screw dislocation (Figure 9.1c), which may be thought of as a shear of part of the lattice, **b** and **l** are parallel so that glide can occur on any plane including **l**. The most general dislocation will have both both edge and screw components, as in Figure 9.1a.

Dislocation lines can end at the surface of a crystal and at grain boundaries, but never inside a crystal. Dislocations must therefore form closed loops or branch into other dislocations (for which the net Burgers vector must vanish). The result is a complex net of dislocations permeating the crystal that will be affected by crystal deformation.

The importance of dislocations is that they provide an energetically favourable way for a crystal lattice to take up deformation. Rather than the high energy

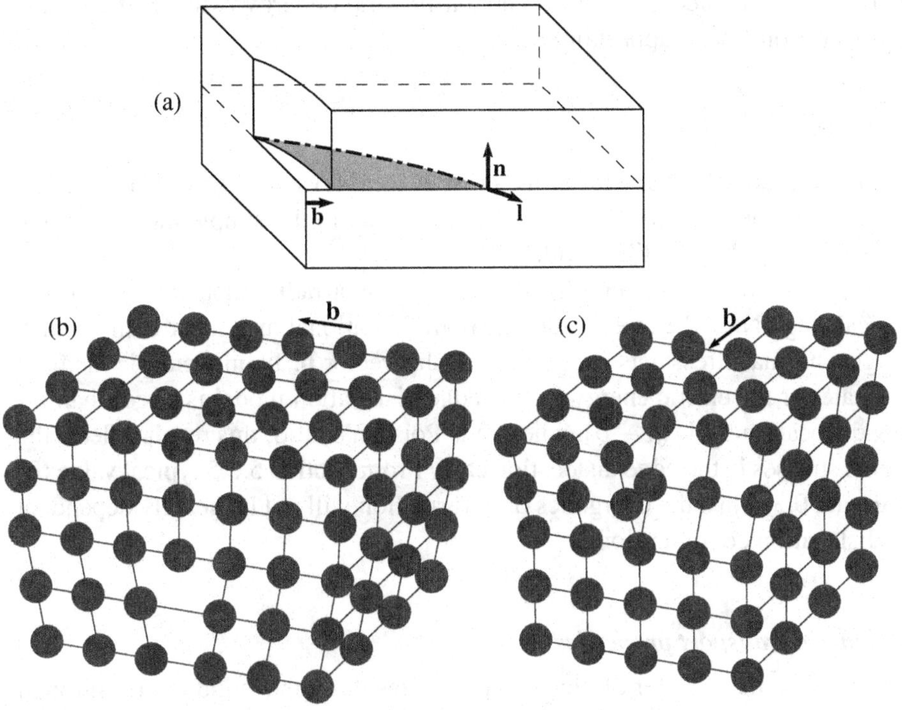

Figure 9.1. (a) The geometry of a dislocation line defect is described by the Burgers vector **b**, line direction **l** and the glide plane with normal **n**. The Burgers vector describes the displacement produced by the passage of the dislocation. (b) Configuration of an edge dislocation, where the Burgers vector is perpendicular to the line direction. (c) Configuration of a screw dislocation, where the Burgers vector is parallel to the line direction.

requirement to force one atom past another, a dislocation can progress by the breaking and re-formation of atomic bonds. Thus, e.g., an edge dislocation can move in the slip direction defined by the Burgers vector **b** by the transfer of the additional half plane of atoms under relatively small applied stress.

Estimates of the stress and strain field around a dislocation can be obtained by using continuum approximations. A simple model of a screw dislocation in an isotropic medium is provided by the distortion of a cylindrical ring surrounding the line vector **l**. For a uniform shear strain along **l**, taken as the z-axis, the associated stress field for displacement b at radius r from the dislocation line is

$$\sigma_{z\theta} = \sigma_{\theta z} = \frac{Gb}{2\pi r}, \tag{9.1.1}$$

where G is the shear modulus. All other stress tensor components vanish. The stress field thus has shear components $\sigma_{\theta z}$ in radial planes parallel to the line direction z, and $\sigma_{z\theta}$ in planes normal to the z-axis perpendicular to the radius vector from the dislocation line. The immediate neighbourhood of the dislocation line has to be excluded; within the core radius r_0 the atomic displacements and forces have to be treated explicitly. The elastic strain energy associated with the screw dislocation is thus approximately

$$E_{el} = \frac{Gb^2}{4\pi} \ln\left(\frac{r_1}{r_0}\right), \tag{9.1.2}$$

where r_1 is the external diameter of the crystal. Real crystals are not isotropic and in consequence the stress field is more complex than in the simple model, but the energy estimate (9.1.2) is still useful.

A comparable model for an edge dislocation has a radial displacement b in a cylindrical ring. Now the stress field has both dilatational and shear components. The largest normal stress acts parallel to the slip vector $\mathbf{b}$. In an isotropic medium the estimate for the elastic energy for the edge dislocation model is larger than for a screw dislocation by $1/(1-\upsilon)$, where υ is Poisson's ratio; this result reflects the additional stresses in the edge dislocation case. From Section 5.2.5 typical values of υ are around 0.3. Thus the energetics of deformation will not be heavily dependent on the dislocation configuration.

9.1.2 General transport properties

We can describe the transfer of physical properties through a solid matrix through the action of 'carriers' that are associated with specific properties. Electrical conductivity depends on the transfer of charge by electrons and holes and the conduction of heat depends on lattice vibrations (quantised as phonons). We therefore have to take account of carrier motion (diffusion), the influence of external fields and scattering of the carriers from the atomic lattice or its imperfections.

We introduce the local concentration of carriers $f_{\mathbf{k}}(\mathbf{x})$ for a state $\mathbf{k}$, with velocity $\mathbf{v}_{\mathbf{k}}$. The total rate of change of the carrier concentration, which determines the local transport properties, is given by

$$\frac{\partial f_{\mathbf{k}}}{\partial t} = \left.\frac{\partial f_{\mathbf{k}}}{\partial t}\right|_{\text{diff}} + \left.\frac{\partial f_{\mathbf{k}}}{\partial t}\right|_{\text{field}} + \left.\frac{\partial f_{\mathbf{k}}}{\partial t}\right|_{\text{scatt}}, \tag{9.1.3}$$

and will vanish in a steady-state situation.

The first component on the right hand side of (9.1.3), $\partial f_{\mathbf{k}}/\partial t|_{\text{diff}}$, describes the migration of the carriers by diffusion. In the presence of a temperature gradient the diffusive term

$$\left.\frac{\partial f_{\mathbf{k}}}{\partial t}\right|_{\text{diff}} = -\mathbf{v}_{\mathbf{k}} \cdot \nabla f_{\mathbf{k}} = -\mathbf{v}_{\mathbf{k}} \frac{\partial f_{\mathbf{k}}}{\partial T} \cdot \nabla T\,. \tag{9.1.4}$$

The second contribution $\partial f_{\mathbf{k}}/\partial t|_{\text{field}}$ arises mainly from the interaction of the electromagnetic field with charged carriers, such as electrons, holes or ions, and is of particular significance for electrical conductivity.

The final contribution from scattering, $\partial f_{\mathbf{k}}/\partial t|_{\text{scatt}}$, can be described through a relaxation time that is inversely proportional to a weighted integral of scattering probability over all directions. The deviations from the reference state $f^0_{\mathbf{k}}$ are determined by

$$\left.\frac{\partial f_{\mathbf{k}}}{\partial t}\right|_{\text{scatt}} = \frac{1}{\tau} g_{\mathbf{k}} = \frac{1}{\tau}\left(f_{\mathbf{k}} - f^0_{\mathbf{k}}\right), \tag{9.1.5}$$

with an exponentially decaying solution

$$g_{\mathbf{k}}(t) = g_{\mathbf{k}}(0)e^{-t/\tau}. \tag{9.1.6}$$

The mean free path for the carriers Λ associated with the scattering is given by $\Lambda = \tau \mathbf{v}_{\mathbf{k}}$. Where several different classes of scattering phenomena are present, such as the various classes of defects, the effects can be described approximately as due to the addition of contributions with different relaxation times.

In a perfect lattice the dominant scattering contribution comes from thermal vibrations of the lattice. This effect is very important in metals at low temperatures with a low concentration of defects. However, for the semi-conducting materials of the silicate mantle at relatively high temperatures the influence of defects is dominant.

9.1.3 Atomic diffusion

The driving force for the diffusion of an atomic species A comes from gradients in the local concentration c_A, i.e., the number of atoms per unit volume. The local flux of the species A,

$$\mathbf{J}_A = -D_A \boldsymbol{\nabla} c_A, \tag{9.1.7}$$

where D_A is the *diffusivity* of species A, with units $m^2\,s^{-1}$. Mass conservation requires

$$\frac{\partial c_A}{\partial t} = -\boldsymbol{\nabla}\cdot\mathbf{J}_A = \boldsymbol{\nabla}\cdot(D_A \boldsymbol{\nabla} c_A). \tag{9.1.8}$$

For a dilute concentration of species A, the diffusivity is only weakly dependent on c_A and then we obtain a diffusion equation (Fick's equation),

$$\frac{\partial c_A}{\partial t} = D_A \boldsymbol{\nabla}^2 c_A. \tag{9.1.9}$$

The concentration c_A can be related to the chemical potential μ_A through the atomic fraction of the species $N_A = c_A/n$, where n is the total number of atoms per unit volume. The chemical potential

$$\mu_A = \mu^0_A + RT\ln[\gamma_A N_A] = \mu^0_A + RT\ln\left[\frac{\gamma_A c_A}{n}\right], \tag{9.1.10}$$

where μ^0_A is the chemical potential in the standard state and γ_A is the activity coefficient for species A.

Direct diffusion through the lattice carries a high energy penalty. It is much more efficient for diffusion to occur by the migration of vacancies. Even so an energy barrier has to be overcome to shift an atom from one site to another that requires thermal activation to accomplish. Similar considerations apply for other classes of interaction, including the migration of charged species where electrical neutrality has to be maintained. In general, the diffusivity has a temperature dependence that takes the Arrhenius form

$$D = D_0 \exp[-H^*/RT], \tag{9.1.11}$$

in terms of an activation enthalpy H^*, for absolute temperature T and gas constant R. The activation enthalpy $H^* = E^* + pV^*$ is a function of pressure, since the effect of pressure is to reduce the number of vacancies and to increase the potential barrier between lattice sites.

An alternative expression for the diffusivity is in terms of the *homologous temperature* (T/T_M), the ratio of the temperature T to the melting temperature T_M,

$$D = D_0 \exp[-aT_M/T], \tag{9.1.12}$$

where the pressure dependence of diffusivity is included through the variation of the melting temperature with pressure.

9.2 Lattice vibrations

Atoms in a crystal vibrate about their equilibrium positions, and analysis of such vibrations allows a link to be made between thermal and elastic properties. The low-frequency parts of the vibrational spectrum are equivalent to elastic waves; the higher-frequency parts correspond to thermal vibrations.

We define the lattice displacement $\mathbf{u}_{sl}$ as the displacement of the sth atom from its equilibrium position in the unit cell with origin vector $\mathbf{l}$. The kinetic energy associated with such vibrations is

$$\mathcal{K} = \sum_{sl} \tfrac{1}{2} M_s |\partial_t \mathbf{u}_{sl}|^2, \tag{9.2.1}$$

where M_s is the mass of the sth atom. The potential energy of the distorted crystal can be expanded about the equilibrium position, where there are no atomic displacements,

$$\mathcal{V} = \mathcal{V}_0 + \sum_{ss',\mathbf{ll}'} \mathbf{u}_{sl}^{T} \left[\frac{\partial^2 \mathcal{V}}{\partial \mathbf{u}_{sl} \partial \mathbf{u}_{s'l'}} \right] \mathbf{u}_{s'l'}. \tag{9.2.2}$$

The equation of motion of the atoms can therefore be written as

$$M_s \partial_{tt} \mathbf{u}_{sl} = -\sum_{s'l'} \left[\frac{\partial^2 \mathcal{V}}{\partial \mathbf{u}_{sl} \partial \mathbf{u}_{s'l'}} \right] \mathbf{u}_{s'l'} = -\sum_{s'l'} \mathbf{G}_{ss';\mathbf{ll}'} \cdot \mathbf{u}_{sl}, \tag{9.2.3}$$

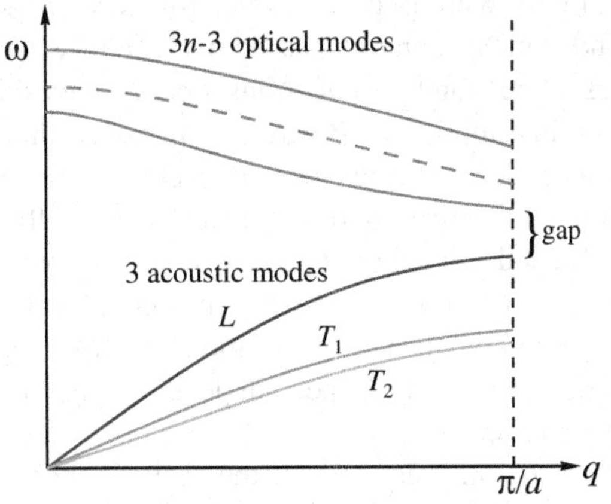

Figure 9.2. Dispersion curve of an infinite lattice with n atoms per unit cell, mapped into half the first Brillouin zone. There are 3 acoustic modes (a longitudinal *P* mode and two transverse *S* modes polarized at right angles), and $3n-3$ optical modes with a gap of excluded frequencies at the edge of the Brillouin zone.

where we have introduced the tensor **G** summarising the atomic forces. Such forces between atoms depend only on the relative position of $\mathbf{l}'$ and $\mathbf{l}$ so that

$$\mathbf{G}_{ss';\mathbf{ll}'} = \mathbf{G}_{ss'}(\mathbf{h}) \quad \text{where} \quad \mathbf{h} = \mathbf{l}' - \mathbf{l}. \tag{9.2.4}$$

With this force representation the equations of motion can be cast as

$$M_s \partial_{tt} \mathbf{u}_{s\mathbf{l}}^{T} = -\sum_{s'\mathbf{h}} \mathbf{G}_{ss'}(\mathbf{h}) \cdot \mathbf{u}_{s,\mathbf{l}+\mathbf{h}}, \tag{9.2.5}$$

and we can seek a solution in terms of functions that preserve the translational invariance between different unit cells,

$$\mathbf{u}_{s\mathbf{l}} = e^{i\mathbf{q}\cdot\mathbf{l} - i\omega t}\mathbf{U}_{s',\mathbf{q}}. \tag{9.2.6}$$

The possible values of the wave vector **q** are determined by the boundary conditions on the crystal; the periodic nature of the crystal means that the wave vectors **q** can be restricted to a Brillouin zone in wavenumber space (see, e.g., Poirier, 1991). The frequency ω for the wavevector **q** must satisfy

$$\omega^2 M_s \mathbf{U}_{s,\mathbf{q}} = \sum_{s'} \left\{ \sum_{\mathbf{h}} \mathbf{G}_{ss'}(\mathbf{h}) e^{i\mathbf{q}\cdot\mathbf{h}} \right\} \mathbf{U}_{s',\mathbf{q}} = \sum_{s'} \bar{\mathbf{G}}_{ss'}(\mathbf{q}) \mathbf{U}_{s',\mathbf{q}}, \tag{9.2.7}$$

where $\bar{\mathbf{G}}_{ss'}$ is the Fourier transform of the force tensor **G**.

For N unit cells with n atoms there are Nn vectorial equations (9.2.3), but the periodic form (9.2.6) for $\mathbf{u}_{s\mathbf{l}}$ reduces the system to just n vector equations. The 3n frequencies ω and the eigenvectors $\mathbf{U}_{s,\mathbf{q}}$ are to be found from the equation system

$$\sum_{s'} \left\{ \bar{\mathbf{G}}_{ss'}(\mathbf{q}) - \omega M_s \delta_{ss'} \right\} \mathbf{U}_{s',\mathbf{q}} = 0. \tag{9.2.8}$$

The lowest frequency modes, with longest wavelength, correspond to movements of the entire lattice and can be regarded as equivalent to elastic waves. The three *acoustic* branches correspond to two transverse *S* waves with orthogonal polarisation, and longitudinal motion in *P* waves. The remaining $3n - 3$ *optical* branches involve more localised disturbances with motion of neighbouring planes out of phase for small $|\mathbf{q}|$. The optical mode frequencies generally fall in the range of infrared or visible light, and can lead to optical absorption. Near the edge of the Brillouin zone, there is a gap in frequency between the optical and acoustical modes (Figure 9.2). At high temperatures, the optical modes carry a significant part of the thermal energy and their propagation speed depends on frequency; interactions between the modes are also important.

For small wavenumbers $|\mathbf{q}|$, the frequency of the acoustic modes is proportional to the wavenumber and the modes are non-dispersive: the group velocity $\partial\omega/\partial q$ is equal to the phase velocity ω/q, which is the macroscopic phase velocity. Because crystals are generally anisotropic, the phase velocity will be a function of the direction of the wave vector $\mathbf{q}$. As the boundary of the Brillouin zone is approached there will be dispersion for the acoustic modes for any given direction as indicated in Figure 9.2.

Lattice specific heat

The thermal excitation of lattice waves makes a distinct contribution to the specific heat of the solid. We have shown in (9.2.7) that we are able to reduce the lattice vibration equations to a set of $3n$ independent harmonic oscillators for each wave vector $\mathbf{q}$. Using a Bose–Einstein quantum representation for each oscillator, the average total energy of the system for temperature T is given by

$$\langle E\rangle = \frac{NV_L}{(2\pi)^3}\sum_{\mathcal{P}}\int d^3\mathbf{q}\,\frac{\hbar\omega_{\mathbf{q}}e^{\hbar\omega_{\mathbf{q}}/k_B T}}{e^{\hbar\omega_{\mathbf{q}}/k_B T}-1}, \tag{9.2.9}$$

where k_B is the Boltzmann constant and V_L is the volume of a unit cell. The summation over $\mathcal{P}$ includes all modes. We introduce the density of states $g(\omega)$ as the number of modes per unit frequency range, and can then write

$$\langle E\rangle = NV_L\int d\omega\, g(\omega)\frac{\hbar\omega\, e^{\hbar\omega/k_B T}}{e^{\hbar\omega/k_B T}-1}, \tag{9.2.10}$$

The specific heat is found by differentiating the energy with respect to temperature,

$$C_V = \frac{1}{NV_L}\frac{\partial\langle E\rangle}{\partial T} = \int d\omega\, g(\omega)k_B\frac{(\hbar\omega/k_B T)^2 e^{\hbar\omega/k_B T}}{\left(e^{\hbar\omega/k_B T}-1\right)^2}, \tag{9.2.11}$$

A useful, but restricted, approximation can be made with a highly simplified representation of the density of states $g(\omega)$ under the assumptions:

(i) only the acoustic modes need to be considered, and all are assigned the same constant sound velocity v_m;

(ii) the Brillouin zone, which limits the range of allowed values of **q**, is replaced by a sphere of the same volume in wavenumber space.
There is therefore a maximum wavenumber for the lattice waves, the *Debye wavenumber* q_D such that

$$\tfrac{4}{3}\pi q_D^3 = \frac{(2\pi)^3}{V_L}, \quad \text{i.e.,} \quad q_D = \left(\frac{6\pi^2}{V_L}\right)^{1/3}. \tag{9.2.12}$$

The associated cut-off frequency

$$\omega_D = q_D v_m, \quad \text{with} \quad v_m = 3^{1/3}\left(\frac{1}{v_P^3} + \frac{2}{v_S^3}\right)^{-1/3}, \tag{9.2.13}$$

where v_P and v_S are the *P* and *S* wavespeeds. The Debye lattice spectrum is therefore assumed to be quadratic in frequency

$$g_D(\omega)d\omega = \frac{3nNV_L}{2\pi^2 v_m^3}\omega^2 d\omega, \qquad 0 \le \omega \le \omega_D. \tag{9.2.14}$$

With the substitution $z = \hbar\omega/k_BT$ and the approximation (9.2.14) for the spectrum $g(\omega)$ we can express the specific heat as

$$C_v = 3nNk_B\left(\frac{T}{\Theta_D}\right)^3 3\int_0^{\Theta_D/T} dz\,\frac{z^4 e^z}{(e^z - 1)^2}, \tag{9.2.15}$$

where the *Debye temperature* $\Theta_D = \hbar\omega_D/k_B$. When the temperature is large, i.e., $T \gg \Theta_D$, the upper limit on the integral is small and the integrand can be expanded in powers of z to give

$$\int_0^{\Theta_D/T} dz\,\frac{z^4}{z^2} = \frac{1}{3}\left(\frac{\Theta_D}{T}\right)^3, \tag{9.2.16}$$

so that we obtain the classical value for the specific heat

$$C_v = 3nNk_B, \quad T \gg \Theta_D, \tag{9.2.17}$$

the Dulong and Petit Law. For very low temperatures, the upper limit of the integral can be regarded as tending to infinity, and so the integral tends to a constant $(4\pi^4/15)$. The specific heat is then

$$C_v \sim \frac{12\pi^4}{15}nNk_B\left(\frac{T}{\Theta_D}\right)^3. \tag{9.2.18}$$

Despite the drastic simplifications the Debye formula (9.2.15) works well for most solids. The relation $g(\omega) \propto \omega^2$ should hold near $\omega = 0$, when the material behaves like an elastic continuum, but the sharp cut-off at ω_D is not justified. Exact calculations show that the lattice spectrum has a large spread with several peaks corresponding to the very different propagation velocities of modes with different polarisations. A slightly better physical approximation is to use a sum of Debye-type contributions for each of the acoustic branches, but the simplicity of the single Debye temperature is then lost.

The vibrational motion of the atoms leads to a thermal contribution to the pressure, that can be estimated from the Debye treatment as

$$P_{th}(V,T) = \left(\frac{\gamma}{V}\right) \langle E \rangle_D = \left(\frac{\gamma}{V}\right) 9nNk_BT \left(\frac{T}{\Theta_D}\right)^3 \int_0^{\Theta_D/T} dz \frac{z^3}{e^z - 1}, \quad (9.2.19)$$

where γ is the Grüneisen constant, which under the Debye assumptions is equivalent to γ_{th} introduced in (6.1.13).

The optical modes can be significant at high temperatures. The frequency variation across the optical modes is limited. With the simple approximation of a constant frequency ω_E we can get a rough approximation to their contribution to the specific heat,

$$C_v^{opt} = 3nNk_B \frac{(\Theta_E/T)^2 e^{\Theta_E/T}}{\left(e^{\Theta_E/T} - 1\right)^2}, \quad (9.2.20)$$

where $\Theta_E = \hbar\omega_E/k_B$.

9.3 Creep and rheology

The ultimate control on the behaviour of Earth materials under deformation comes from processes at the atomic level, and the cumulative effects of the atomic level interactions manifest themselves through the nature of the constitutive relation. Many different mathematical forms for constitutive relations have been devised to provide a macroscopic description of behaviour as different classes of atomic processes become dominant (see, e.g., Evans & Kohlstedt, 1995). Such constitutive relations have been deduced from experimental tests that, by necessity, are of limited duration. In consequence the strain rates in the laboratory are rarely smaller than $10^{-5}\,s^{-1}$, which is still many orders of magnitude larger than the fastest natural strain rates in rocks (around $10^{-9}\,s^{-1}$).

9.3.1 Crystal elasticity

The elastic properties of single crystals can be related to the statistical properties of multi-atom assemblages through the same concept of strain energy as we have used in the continuum description in Chapter 4, and in the treatment of the potential energy of lattice vibrations in Section 9.2. The natural coordinates are the positions of the atoms rather than their reference positions. Thus an Eulerian description is appropriate and the Cauchy strain (2.2.27) needs to be used for large deformation. This semi-classical treatment provides a good description of behaviour in conjunction with the periodic nature of the atomic lattice. The fluctuations in the quantum phase vector in the periodic lattice depend on the inverse elastic moduli for the material; these fluctuations are manifest through reduction of the intensity in the X-ray diffraction peaks associated with different lattice parameters (see, e.g., Chaikan & Lubensky, 1995). Careful X-ray intensity measurements can therefore provide some constraints on elastic properties of the lattice.

The influence of crystal defects can be incorporated into the description of the elastic state through the way that the strain energy is modified. The Eulerian stress tensor in a material with defects can be expressed in terms of the Helmholtz free energy $\mathcal{F}$ as

$$\sigma_{ij} = \rho \left[\mathcal{F} - n \left.\frac{\partial \mathcal{F}}{\partial n}\right)_{\mathbf{A}} \right] \delta_{ij} + \left.\frac{\partial \mathcal{F}}{\partial \mathbf{A}}\right)_n, \tag{9.3.1}$$

where n is the number of particles per unit volume and $\mathbf{A}$ is the displacement gradient (5.1.1). The contribution $h_{ij} = \partial \mathcal{F}/\partial \mathbf{A})_n$ can be recognised as the normal stress term, thermodynamically conjugate to strain, as in (6.2.11). The first term on the right hand side of (9.3.1) allows for the presence of vacancies and interstitials:

$$\left.\frac{\partial \mathcal{F}}{\partial n}\right)_{\mathbf{A}} = \mu, \tag{9.3.2}$$

where μ is the chemical potential. The stress tensor can therefore change in an isothermal state as a consequence of changes in either h_{ij} or μ. In the periodic atomic lattice h_{ij} couples to the separation between layers, but μ is linked to changes in density (Chaikan & Lubensky, 1995) and merely changes the isotropic (pressure) component of stress.

9.3.2 Deformation behaviour

The deformation of materials becomes more pronounced as the melting temperature T_M is approached, and this dependence of the creep of a solid can be represented in the general form (see, e.g., Evans & Kohlstedt, 1995)

$$\dot{\epsilon} = B \left[\frac{\sigma}{G}\right]^n \left[\frac{d_0}{d}\right]^m \exp\left(-g\frac{T_M}{T}\right), \tag{9.3.3}$$

with a dependence of the strain rate $\dot{\epsilon}$ on the grain size d, through the granularity index m. g is a dimensionless quantity, and the stress σ is normalised by the shear modulus G. If $n = 1$ we have a Newtonian relation, but generally $n > 1$ and we have *power-law creep*; a typical value for n would be 3–5.

The apparent viscosity at constant stress for the general relation (9.3.3) is therefore

$$\eta_\sigma = \frac{\sigma}{\dot{\epsilon}(\sigma)} = \frac{\sigma}{B}\left[\frac{G}{\sigma}\right]^n \left[\frac{d}{d_0}\right]^m \exp\left(g\frac{T_M}{T}\right). \tag{9.3.4}$$

Alternatively the viscosity at constant strain rate is

$$\eta_{\dot{\epsilon}} = \frac{\sigma(\dot{\epsilon})}{\dot{\epsilon}} = \frac{G}{\dot{\epsilon}}\left[\frac{B}{\dot{\epsilon}}\right]^{1/n} \left[\frac{d}{d_0}\right]^{m/n} \exp\left(g\frac{T_M}{nT}\right), \tag{9.3.5}$$

which is proportional to $\sigma^{(1-n)/n}$. The sensitivity to temperature will be lower at constant strain rate for $n > 1$.

Creep deformation occurs through the transport of matter by the movement of defects in the lattice. Diffusion controlled creep involves the motion of atoms and vacancies in opposite directions and is Newtonian ($n = 1$). In contrast the movement of dislocations through the lattice leads to non-Newtonian behaviour.

Diffusion creep

Under weak stress diffusion creep is a favoured mechanism. Different styles of diffusion can be recognised depending on whether diffusion is faster through the matrix or along the grain boundaries. Diffusion of the slowest ion along the fastest path determines the nature of the diffusion and thus the granularity index m. For matrix diffusion (Nabarro–Herring creep) $m = 2$, and for grain-boundary creep (Coble creep) $m = 3$. Under compression the formation of vacancies in a lattice will be suppressed, but under tension such vacancy formation will be enhanced. There will therefore be a gradient in the concentration of vacancies in a solid under stress leading to a flow of vacancies from (9.1.7), with a consequent reverse flow of atoms. The strain rate is approximately inversely proportional to temperature, so that the viscosity from (9.3.4) will be close to a linear function of T. The activation energy for this class of diffusion is just that for self-diffusion in the lattice (Poirier, 1991).

When the diffusion through the lattice is very slow, the dominant mode of diffusion will be the migration of atomic species along the grain boundaries. In such Coble creep the viscosity is again approximately linear in T, and in order for grains to remain in contact there needs to be some grain-boundary sliding.

Dislocation creep

For higher stresses, the deformation of a crystal is controlled by the movement and interactions of dislocations and the thermal effects become more pronounced.

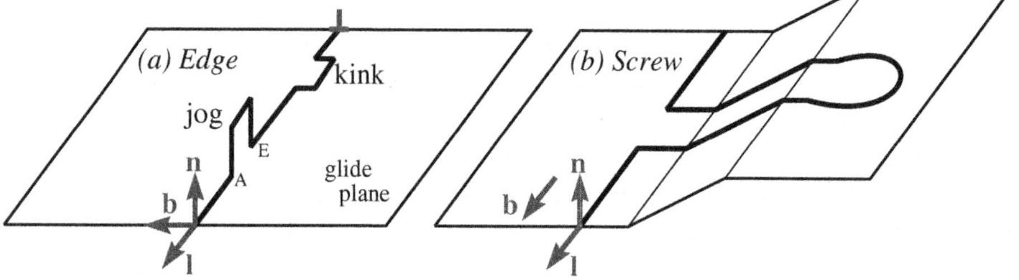

Figure 9.3. The migration of dislocations is controlled by kinks and jogs. (a) Migration of an edge dislocation in the glide plane by the migration of kinks does not involve diffusion processes. Jog motion out of the glide plane requires a diffusive component, with the absorption (A) or emission (E) of vacancies. (b) Screw dislocations can slip on any plane containing the line direction.

Dislocations move by the energy-efficient process of the progressive breaking and reforming of atomic bonds, and are the carriers of plastic deformation through a solid.

The effect of deformation on a crystal is to force migration of existing dislocations, that then interact with possible annihilation of the defects. For larger strains, dislocations will be created and migrate leading to complex dislocation tangles. Each of the processes of creation, migration and annihilation requires work from the applied stresses.

The net local stress acting on a dislocation is the superposition of the externally applied stresses and internal stresses due to the influence of other dislocations, defects and interfaces. The internal stress will normally resist the migration of the dislocation and may increase with time, the process of hardening, or decrease over time (recovery) depending on the way that the microstructure is modified.

There are two basic types of dislocation movement, *glide* in which the dislocation moves in the surface defined by its line and the Burgers vector, and *climb* in which the dislocation moves out of the glide surface. For a density of mobile dislocations ρ_d of a similar type with average dislocation velocity $\bar{v}$ the creep rate is

$$\dot{\epsilon} = \rho_d b \bar{v}, \tag{9.3.6}$$

where b is the length of the Burgers vector. The dislocation velocity is related linearly to the deviatoric stress $\Delta\sigma$

$$\bar{v} = k \, \Delta\sigma \exp[-Q/RT], \tag{9.3.7}$$

with strong temperature dependence because of thermal activation barriers. The hardening of materials is reduced at higher temperatures because dislocation pile-ups can be avoided.

In silicate minerals, only small portions of the dislocation, called kinks and jogs, are mobile at any moment. Both represent abrupt changes in the line direction; kinks lie within the glide plane, and jogs take the dislocation locally out of the glide plane. Thus dislocation migration in the glide plane occurs by kink migration, and does not require atomic diffusion. Whereas climb out of the glide plane through the movement of jogs needs atomic diffusion into the extra half plane, either along the dislocation core or through the lattice (Figure 9.3a). The screw component can slip along any glide plane containing the line direction (Figure 9.3b). The migration of a kink has to overcome the resistance of the crystal structure, since there is an energy barrier to be overcome to transfer the dislocation position from one location to another. However, this energy barrier is much less than that required to force atoms past each other so that the shear resistance is orders of magnitude less than it would be in a perfect crystal. The mobility and the number of jogs and kinks will be influenced by the concentration of point defects in the crystal.

The initial application of significant stress can produce significant transient effects through the migration of dislocations to grain boundaries. Steady-state rheology is governed by the rate-limiting step in the sequence of generation, movement and elimination of the dislocations that have a strong dependence on the local crystal environment. Generation of dislocations may arise at boundaries, intersections of other dislocations or localised defects. Dislocations may be annihilated by reaction with dislocations of opposite Burgers vector or by interaction with a boundary in the process of recrystallisation. In the steady state the dislocation density ρ_d is proportional to the square of the deviatoric stress

$$\rho_d = C(\Delta\sigma)^2, \tag{9.3.8}$$

where C is nearly independent of temperature. Thus for a single style of dislocations we can combine (9.3.6), (9.3.7) and (9.3.9) to obtain a simple power-law creep model

$$\dot{\varepsilon} = Ckb(\Delta\sigma)^3 \exp[-Q/RT], \tag{9.3.9}$$

with an exponent $n = 3$. Evans & Kohlstedt (1995) present a representative sub-set of constitutive equations for dislocation-controlled creep under a variety of other scenarios.

9.4 Material properties at high temperatures and pressures

The conditions prevailing in the deep interior of the Earth lie far from standard laboratory conditions. Although very significant progress has been made in achieving very high pressures, it is particularly difficult to achieve high pressure and calibrated uniform temperatures across a specimen.

The pressure and temperature regime that needs to be explored to study the Earth's interior is illustrated in figure 9.4. The pressure variation, here for model AK135 (Kennett et al., 1995), depends on the radial density distribution as shown in (6.1.2), but is not very sensitive to the choice of Earth model. However, the temperature distribution is much less certain and estimates of temperature at great depth have considerable uncertainty (1000 K or more), particularly in the core. The challenge is therefore to achieve pressures of 150 GPa and temperatures approaching 4000 K so that the full range of conditions in Earth's mantle can be investigated. Pressure represents the force applied per unit area, and thus ultra-high pressures require either or both of very large force and a very small area of application.

9.4.1 Shock-wave techniques

The first technique to achieve pressures comparable to those at the centre of the Earth (370 GPa) was the use of shock waves induced by explosions or the impact of high velocity projectiles. The impact or explosion produces a compressional shock front that passes through the material faster than the local sound wavespeed, until it is reflected by the far side of the target to return as a rarefaction. Shock waves heat as well as compress the target material. Temperatures can reach higher than 2000

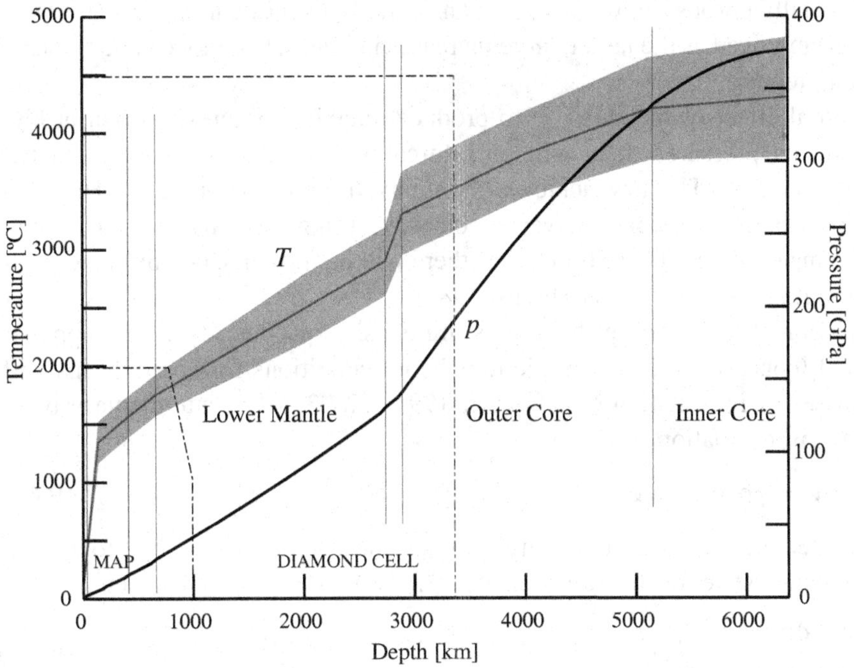

Figure 9.4. The pressure variation from the AK135 model (Kennett et al., 1995) and the likely temperature band for the interior of the Earth as a function of depth. The regimes accessible to multi-anvil (MAP) and diamond cells with laser heating are also indicated.

K for silicate minerals, and melting can sometimes occur under shock conditions. The experiments measure the shock-wave velocity U and the particle velocity u, imparted to the particles of the material, and the temperature. A common means of achieving high incident projectile velocities is the use of a two-stage light-gas gun in which a large primary projectile imparts its momentum to a small flyer plate that impinges on the target. The experimental conditions can be varied by adjusting the projectile velocity, but only one shock state can be investigated at a time. As a result a very substantial investment is required to map out the relation between shock-induced pressure and density known as *Hugoniot* curves.

The passage of a shock-front changes the state of the material and the properties in front of the shock (a) and after its passing (b) are related by the conservation of mass, momentum, internal energy (or enthalpy) across the pressure discontinuity of the shock-front. The density ρ, pressure p and specific internal energy e and enthalpy h are thus related by the Rankine–Hugoniot conservation equations:

$$\rho_b = \rho_a(U - u_a)/(U - u_b), \tag{9.4.1}$$

$$p_b = p_a + \rho_a(U - u_a)(u_b - u_a), \tag{9.4.2}$$

$$e_a = e_b + \tfrac{1}{2}(p_a + p_b)/(1/\rho_a - 1/\rho_b), \tag{9.4.3}$$

$$h_a = h_b + \tfrac{1}{2}(p_a - p_b)(1/\rho_a + 1/\rho_b). \tag{9.4.4}$$

If an artificially porous sample with ρ_a considerably lower than the normal crystal density is employed as the target, large increases in internal energy can be achieved at a given pressure.

The initial effect of the shock is to produce uniaxial compression, but at high shock-induced pressures there will be bifurcation of the shock wave with two successive shocks. The first achieves the uniaxial compression to the Hugoniot elastic limit and the second slower shock wave brings the material to its final state. For material with phase transitions there will be further shock bifurcations to produce multiple shock-fronts (Ahrens, 1987).

The Hugoniot trajectory in shock pressure-density space needs to be mapped to isothermal (constant T) or isentropic (adiabatic) conditions for comparisons with other classes of information (see Poirier, 1991, §4.7.3). The intermediary is the Mie–Grüneisen equation

$$p_H - p_K = \gamma\rho(e_H - e_K), \tag{9.4.5}$$

which relates the pressures and energies along the Hugoniot (p_H, e_H) to another thermodynamic state for the same density. The Grüneisen ratio γ is given by

$$\gamma = \frac{1}{\rho}\left[\frac{\partial p}{\partial E}\right]_\rho = \frac{\alpha_{th} K_S}{\rho C_p}, \tag{9.4.6}$$

where, as in (6.1.15), α_{th} is the coefficient of thermal expansion, K_S is the adiabatic bulk modulus and C_p is the specific heat at constant pressure. The ratio γ is frequently assumed to depend only on density.

Ahrens & Johnson (1995a,b) provide a summary of shock-wave data for many classes of rocks and minerals. The properties of silicate minerals at high pressure can be estimated quite well from those of mixtures of the relevant oxides. Shock-wave experiments provide direct calibration of pressure and such results are therefore important for pressure measurements in more indirect methods.

9.4.2 Pressure concentration by reduction of area

The simplest device to subject samples to pressure is a piston cylinder apparatus in which uniaxial load is applied directly to relatively large specimens. In this way pressures of the order of 7 GPa can be achieved, comparable to conditions at 100 km depth.

A more uniform loading and much higher pressures can be achieved in a multi-anvil press. An externally applied uniaxial load is transmitted to a small octahedral sample assembly through a secondary set of anvils. Six anvil wedges press on the faces of eight cubes, typically made of tungsten carbide for hardness. These cubes squeeze the octahedral sample assembly nestling in the inner space created by truncating the cube corners. A ceramic octahedron acts as the pressure-transmitting medium. With large presses (nominal ratings up to 5000 tons) pressures approaching 30 GPa can be reached for sample volumes of several cubic millimetres. This means that the pressure conditions throughout the mantle transition zone to about 800 km are accessible; mantle temperatures can also be matched. There is a trade-off between the size of the sample and the pressure that can be achieved. The largest presses tend to be used for synthesising significant quantities of high-pressure materials or in-situ measurements such as thermal and electrical conductivity.

Studies with the multi-anvil press have played a major role in elucidating the phase changes in the mineral assemblage through the upper mantle and the mantle transition zone (Figure 9.5). Recent developments have included establishing multi-anvil presses at synchrotrons so that the high-intensity X-ray beams can be used to examine the properties of the material during deformation. Thus, e.g., the density can be monitored across a phase transition from the change in unit cell size. The elastic properties of unquenchable high-pressure phases that cannot be brought back to normal conditions have been measured with a combination of synchrotron X-ray methods and ultrasonic interferometry. With a suitably cut transducer both P and S waves can be injected into the specimen through the sample holder. The change of length of the specimen can be monitored during the course of the experiments and hence the timing of the multiply reflected waves can be unravelled to provide wavespeed information.

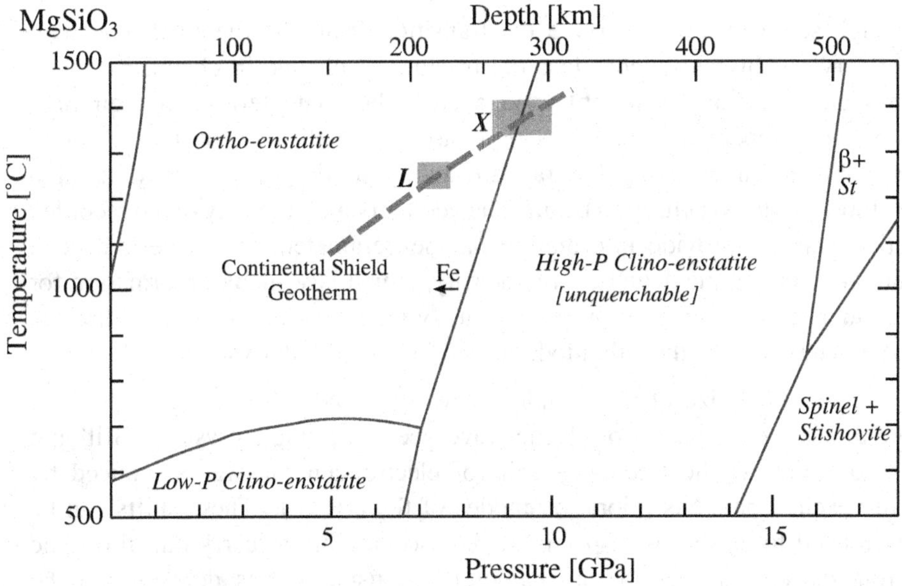

Figure 9.5. The pressure and temperature (P,T) phase diagram for the $MgSiO_3$ composition. In-situ measurements have revealed the elastic properties of the unquenchable phase high-pressure clino-enstatite and the small discontinuity in elastic properties from ortho-enstatite is compatible with the seismic X discontinuity beneath continental shields, but not the shallower Lehmann discontinuity (L). The movement of the phase boundary with increasing Fe content is indicated. [Courtesy of J. Kung.]

Higher pressures can be achieved in diamond anvil cells in which two anvils made from superhard materials squeeze a very small sample (10–500 μm in diameter). In this way, with relatively modest forces, it is possible to attain pressures up to several hundred GPa. Heating can be provided externally (up to 1000 K) or internally by laser heating. The diamond anvils are transparent to a wide range of the electromagnetic spectrum (X-rays, visible and infrared radiation) and so samples can be monitored during an experiment. The laser heating can be induced by directing beams from high-intensity infrared lasers through the anvils with a focus at the specimen. Temperatures can be determined from the local Planck emission spectrum with a 1 μm resolution. It is difficult to avoid temperature gradients across the specimen away from the directly heated region, but now uniformity can be achieved for samples as large as 100 μm. Fortunately the diamonds themselves stay cool because of their high thermal conductivity.

The pressure in the diamond cell is determined indirectly from the shift in the fluorescence line emitted by tiny ruby chips embedded in the sample. The pressure scale is calibrated by making simultaneous measurements of the density of metals in the cell, using in-situ X-ray diffraction measurements, and the shift of the ruby fluorescence line. The pressure is derived from the density through the equation of state for the metal derived from shock-wave experiments, with corrections for thermal pressure (typically a few GPa).

Although some materials survive the transition from the diamond anvil cell to laboratory conditions without loss of structure or change in chemistry, many high-pressure states are metastable and need to be characterised in their high pressure and temperature state. The relatively small size of the diamond cell assembly means that rapid X-ray diffraction studies can be made at high temperatures using a synchrotron beam. The geometry and intensity of the recorded diffraction patterns provide information on the parameters of the periodic cell structure and also on the actual charge density, which is valuable information for computational modelling. The pressure–density relationships can then be used to infer parameters such as the bulk modulus and its pressure derivative.

Despite the small size of the samples in the diamond anvil cell, it is possible to make direct measurements of elastic wavespeeds for single crystals. Brillouin spectroscopy detects the frequency shift of electromagnetic waves induced by thermally excited phonons (vibration modes of the crystal). These shifts can be directly related to elastic wavespeeds. It is also possible to carry out ultrasonic interferometry on samples in the diamond cell using waves delivered to the specimen through one of the anvils (Jacobsen et al., 2002). From the interference pattern of reflected waves the travel times of P and S waves can be determined, and then converted to wavespeed with knowledge of the sample size under pressure (Figure 9.6). In either case the full set of anisotropic elastic moduli can be determined from multiple experiments for different crystal orientations. Pressure derivatives can also be found using different experimental conditions.

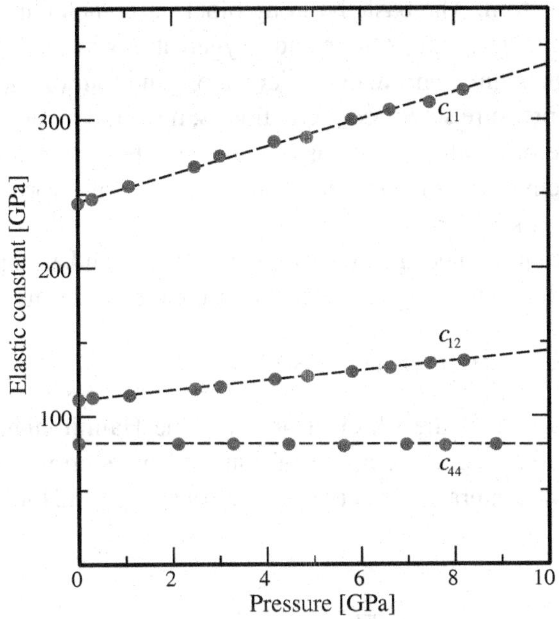

Figure 9.6. The variation of the three independent elastic moduli of cubic ferropericlase (Mg,Fe)O with pressure to 10 GPa determined by ultrasonic interferometry. Ferropericlase coexists with silicate perovskite $(Mg,Fe)SiO_3$ in the lower mantle.

Pressures beyond the range of the conventional multi-anvil press (80 GPa) have been achieved with a three stage system based on the use of a superhard nanodiamond aggregate (T. Irifune - personal communication). The same material offers the possibility of enlarged diamond anvils. In each case this means that larger samples can be investigated at high pressures.

9.5 Computational methods

Since the conditions deep in the Earth are difficult to reach with experimental methods, a useful alternative source of information comes from computational methods applied at the atomic level. There are two broad classes of method depending on the treatment of electronic structure. In *electronic structure* calculations the Schrödinger equations for the quantum behaviour of a set of atoms are solved at some degree of approximations to provide direct access to material properties. On the other hand the explicit solution of the Schrödinger equations can be avoided by using *interatomic potentials* that represent the energy of the system through nuclear coordinates.

9.5.1 Electronic structure calculations

Such first-principles methods make no initial assumptions regarding the physical nature of a mineral system, instead they rely on basic quantum mechanics and the atomic number of an atom. The basic building blocks are nuclei and electrons. In principle, if we place magnesium, silicon and oxygen atoms in a suitable box in the proportions two magnesium atoms to one silicon atom and four oxygen atoms, with the temperature and pressure set to room conditions, it should be possible to obtain the magnesio-silicate mineral forsterite as a final state. However, this is currently too complex a problem to be solved directly so some level of information needs to be supplied about the likely structure.

The energy of a quantum mechanical system of electrons and atoms can be found from the solution of the Schrödinger equations for the electron states

$$H\psi(\mathbf{r}_1, \mathbf{r}_2, \cdots, \mathbf{t}_N) = E\psi(\mathbf{r}_1, \mathbf{r}_2, \cdots, \mathbf{t}_N), \tag{9.5.1}$$

where $\mathbf{r}$ is the coordinate of the ith electron, H is the Hamiltonian and E is the energy. The Hamiltonian operator consists of a sum of three terms, the influence of the atomic nuclei (V_{en}), interactions between electrons (V_{ee}) and the kinetic energy of the system, so that

$$H = V_{en} + \sum_{i<j}^{N} \frac{1}{|\mathbf{r}_i - \mathbf{r}_j|} - \frac{1}{2}\sum_{i=1}^{N} \nabla_i^2. \tag{9.5.2}$$

The external potential V_{en} arises from the interaction of electrons with the M atomic nuclei,

$$V_{en} = -\sum_{p}^{M}\sum_{i}^{N} \frac{Z_p}{|\mathbf{r}_i - \mathbf{R}_p|}, \tag{9.5.3}$$

where the charge on the nucleus at $\mathbf{R}_p$ is Z_p. In addition the influence of electronic spin needs to be considered explicitly in some circumstances, such as the presence of a magnetic field.

The coupled Schrödinger equations for the N electrons (9.5.1) cannot be solved directly because of the complexity and large dimension of the equation. The representation of the $3N$-dimensional wavefunction ψ scales as g^N, where g is the number of degrees of freedom for a single electron. Nevertheless, there are a number of approximations based on the combination of self-consistent solutions for a single-electron Schrödinger equation with summation over all electrons in the system. The best approximations using correlated methods scale as N^7 and can be used to determine chemical properties of small molecules (Harrison, 2003) but are not feasible for larger systems.

Fortunately it is not necessary to solve for the $3N$-dimensional wavefunction ψ in order to extract information on the energetics of the system. The electron density $\rho_e(\mathbf{r})$ uniquely determines the positions and charges of the nuclei and thus the Hamiltonian operator H (9.5.2). The energy functional contains three elements, the kinetic energy T, the electrostatic potential W (from both nuclei and electronic charge density) and a residual contribution E_{xc} associated with correlations between electrons and exchange effects associated with electron spin:

$$E[\rho_e] = T[\rho_e] + W[\rho_e] + E_{xc}[\rho_e]. \tag{9.5.4}$$

The *density functional* approach of Kohn and Sham exploits this property to break the N-body Schrödinger equation into a set of N coupled equations that closely resemble those for a single electron. The Kohn–Sham equations for a set of *orbitals* $\{\phi_i\}$ are

$$\left(\tfrac{1}{2}\nabla^2 + V_{en} + V_{ee}[\rho_e(\mathbf{r})] + V_{xc}[\rho_e(\mathbf{r})]\right)\phi_i = \epsilon_i\phi_i, \tag{9.5.5}$$

where the electron density $\rho_e(\mathbf{r})$ and energy E are given by

$$\rho_e(\mathbf{r}) = \sum_i^N |\phi_i(\mathbf{r})|^2, \qquad E[\rho_e] = \sum_i^N \epsilon_i. \tag{9.5.6}$$

The exchange correlation potential V_{xc} is the functional derivative of E_{xc} with respect to ρ_e, i.e., $V_{xc} = \partial E_{xc}/\partial\rho_e$.

The set of non-linear equations (9.5.6) describes the behaviour of non-interacting 'electrons' in an effective local potential that actually has an unknown component V_{xc}. If the exact functional were known, then there would be a complete correspondence between the many-body situations and the uncoupled equations. For any imperfect representation of the density functional the energy estimate $E[\rho_e]$ will be an upper bound on the true ground state energy E_0. The development of excellent approximations to the density functional allows unbiased and predictive studies of a wide range of materials, and they are often referred to as *ab initio* methods.

The exact form of the exchange-correlation potential is not known, but two styles of approximations have proved to be quite effective (see, e.g., Harrison, 2003). The local density approximation (LDA) defines the exchange correlation as a function of the electron density at a point, and works well for most insulating systems. However, the computation of energy differences between rather different structures can have significant errors; thus the binding energy tends to be overestimated and energy barriers in diffusion are likely to be too small. A more sophisticated approximation, the generalised gradient approximation (GGA) includes information from the gradient of the electron density as well, which is very effective for metals but not for strongly correlated oxides such as FeO or Fe_3O_4.

In our sketch of the theory we have not included the influence of electron spin, but both spin up and spin down states need to be considered separately for understanding magnetic effects that are important for minerals containing transition metals. The density functional theory as described above is a ground state approach, based on an athermal state at absolute zero Kelvin without any zero-point energy. The absence of thermal excitations means that perfect periodicity can be assumed for the lattice of a crystal, and hence an infinite medium can be simulated with a single primitive unit cell subject to periodic boundary conditions. For the orthorhombic symmetry of forsterite we need to employ four formula units and thus 28 atoms. Experimental work cannot attain direct correspondence with this athermal state, but in many cases room-temperature thermal excitation of the electrons is small and electronic band structure and material properties can be compared between experiment and the calculations from first principles. Temperature effects tend to be more pronounced for magnetic and anisotropic systems.

The efficient description of the ϕ_i is an important part of the implementation of density functional theory, and these are usually assembled from a superposition of basis functions. Many approaches use plane waves that would correspond to free electron states; since solutions can be improved through the addition of further plane waves.

Only a fraction of the full set of electrons is involved in chemical bonding and hence it is possible to develop a scheme of *pseudopotentials* in which the interactions of valence electrons and nuclei are represented by a weak effective potential. The pseudopotential is much smoother than the bare core potential around a nucleus and so not only are fewer electrons considered, but also fewer basis functions are needed. Eliminating the tightly bound electrons in the system can make significant reductions in the number of effective ϕ_i, e.g., a drop from 280 to 48 electrons for forsterite. Careful use of pseudopotentials cuts down on the computational effort, so that large mineral systems such as clays can be studied, or alternatively supercells with hundreds of atoms can be used to describe scenarios without full periodicity, as in thermal phenomena or crystals containing defects.

9.5.2 Atomistic simulations

The potential energy of an assembly of N atoms $U(\mathbf{r}_1, \mathbf{r}_2, \ldots, \mathbf{t}_N)$ is expressed as a function of the position vectors $\{\mathbf{r}_j\}$ from the various nuclei. In this approximation the electronic structure of the system is subsumed into the potential function. U is commonly expanded into a series of different levels of interatomic interaction

$$U = \frac{1}{2}\sum_{i=1}^{N}\sum_{j=1}^{N}{}' \phi_{ij}(\mathbf{r}_i, \mathbf{r}_j) + \frac{1}{3}\sum_{i=1}^{N}\sum_{j=1}^{N}\sum_{k=1}^{N}{}' \phi_{ijk}(\mathbf{r}_i, \mathbf{r}_j, \mathbf{r}_k) + \cdots, \tag{9.5.7}$$

where the $'$ symbol indicates that multiple counting of terms is avoided. The *two-body* functions ϕ_{ij} depend only on the positions of pairs of atoms (i, j), whereas the *three-body* terms depend on the atomic triplet (i, j, k). The two-body term ϕ_{ij} can be represented in terms of an electrostatic component and a residual potential V_{ij} that has both attractive and repulsive components,

$$\phi_{ij}(\mathbf{r}_i, \mathbf{r}_j) = \frac{q_i q_j}{|\mathbf{r}_i - \mathbf{r}_j|} + V_{ij}(|\mathbf{r}_i - \mathbf{r}_j|), \tag{9.5.8}$$

where q_i is the charge on the ith atom (or ion). Analytic functions are used to represent V_{ij} such as the Buckingham potential, suitable for ionic solids,

$$V(r) = Ae^{-r/\rho} - Cr^{-6}; \tag{9.5.9}$$

the repulsive component $A\exp(-r/\rho)$ represents the overlap of atomic energy shells, and the longer range attractive term Cr^{-6} includes induced dipole effects. The parameters A and C are empirical and need to be adjusted to observed crystal properties. Many-body effects can be included by incorporating, e.g., angle-dependent forces or triple dipole terms (Catlow, 2003).

Methods which rely on parametric fits are valid for the range of physical parameters for which the fit is made and extrapolation outside this range has to be undertaken with caution. Thus, for example, the presence of a high-pressure phase transition could only be predicted with confidence if the physical parameters for both the phases were included in the fitting procedure.

9.5.3 Simulation of crystal structures

The basic tool kit of *ab initio* methods allow the construction of lattice energy and electronic charge density for a given set of atomic positions and species. If spin-polarisation effects are included then magnetic moments can also be found. Such calculations have demonstrated that the high-temperature polymorph of iron, with a face-centred cubic structure, has a complex magnetic ground state that gives rise to anomalous thermal expansion.

For understanding Earth materials further properties are desirable, including the equilibrium volume, which require the forces acting on the atoms and the

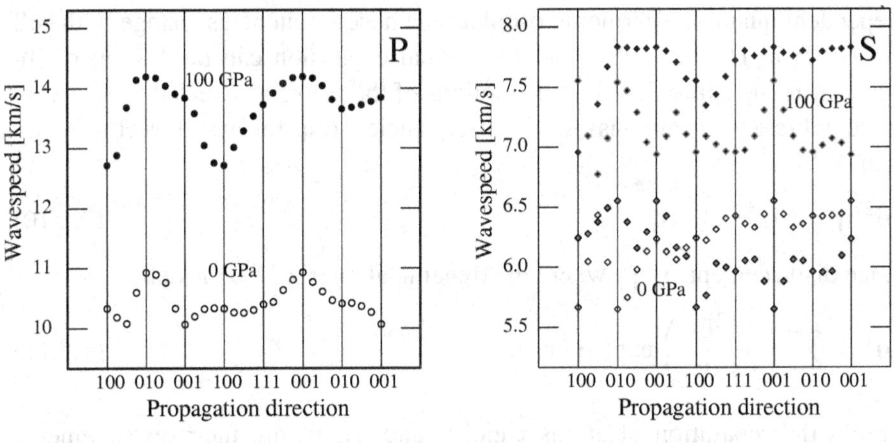

Figure 9.7. Variation of (a) longitudinal and (b) shear wave speeds of $(Mg,Fe)(Si,Al)O_3$ perovskite as a function of propagation direction and pressure calculated using *ab initio* methods [after Li et al. (2005)].

stress on the cell. These properties can be extracted from the first derivatives of the energy. The first derivative with respect to atomic positions provides the forces, and the tensor derivative with respect to small strain yields the stress. The equilibrium volume is best determined by finding the minimum of the energy–volume relationship for variable cell volume.

The computation of the stress for a strained configuration enables the direct evaluation of the corresponding elastic constants (Kiefer et al., 2001). Alternatively the elastic constants can be derived from the variation of strain-energy density with lattice strains (Steinle-Neumann et al., 1999).

An example of the use of density functional theory with a plane wave basis set, using the generalised gradient approximation, is provided by the work of Li et al. (2005) who have examined the elasticity of ferro-magnesium perovskite with allowance for aluminium substitution for silicon. This is the most abundant mineral on Earth through the lower mantle, but there is little experimental constraint on properties. The elastic properties are illustrated in Figure 9.7 as a function of propagation direction and pressure for one configuration of aluminium substitution. An extensive set of calculations suggests that the spin state of the iron atoms has a smaller influence on the elastic properties than the inclusion of aluminium instead of iron.

9.5.4 Finite temperature

An important aspect of material properties is the influence of temperature and in particular the temperature derivatives of material properties such as elastic moduli and wavespeeds. The methods for finite temperature are based on thermodynamic properties in terms of the Helmholtz free energy derived from statistical mechanics.

(a) Lattice dynamics

This is a semi-classical approach that uses a representation of the unit cell in terms of independent quantized harmonic oscillators whose frequencies change with cell volume (Born & Huang, 1954) and so thermal expansion can be described. In this quasiharmonic approximation the motions of the atoms are treated collectively as lattice vibrations (phonons) whose frequencies $\omega\mathbf{q}$ for wave vector $\mathbf{q}$ are determined by solving

$$m\omega^2(\mathbf{q})s_k(\mathbf{q}) = D_{kl}(\mathbf{q})s_l(\mathbf{q}), \tag{9.5.10}$$

for lattice displacements $\mathbf{s}(\mathbf{q})$, where the dynamical matrix $D(\mathbf{q})$ is defined as

$$D(\mathbf{q}) = \sum_{ij} \left(\frac{\partial^2 E}{\partial s_i \partial s_j} \right) \exp(i\mathbf{q} \cdot \mathbf{r}_{ij}). \tag{9.5.11}$$

Here $\mathbf{r}_{ij}$ is the separation of atoms i and j, and u_i, u_j are their displacements from the equilibrium position. For a unit cell with n atoms, there are $3n$ sets of eigenvalues $\omega^2(\mathbf{q})$ for a given wave vector $\mathbf{q}$ associated with $3n$ sets of eigenvectors $\mathbf{s}(\mathbf{q})$ that describe the patterns of atomic displacement for the mode.

The vibrational frequencies can be calculated from first principles by looking at the influence of small displacements (Kresse et al., 1995).

Once the eigenfrequencies have been determined, the thermodynamical properties can be found by using standard methods from statistical mechanics for Bose–Einstein oscillators. The Helmholtz free energy, as a function of volume, takes the form

$$\mathcal{F} = \frac{1}{2K}\left(\frac{\Delta V}{V}\right)^2 + k_B T \sum_{\mathbf{q}} \ln\left[2 \sinh \frac{\hbar\omega_{\mathbf{q}}}{2k_B T}\right], \tag{9.5.12}$$

where k_B is the Boltzmann constant. The first term on the right-hand side of (9.5.12) is the potential energy associated with the compressibility K of the solid as an elastic continuum. The second term is the sum of the free energies in the lattice modes.

If we make the simplifying assumption that a change of volume ΔV produces the same relative change of frequency for each mode

$$\frac{\Delta\omega}{\omega} = \gamma \frac{\Delta V}{V}, \tag{9.5.13}$$

then the condition for minimum free energy from (9.5.12) is

$$\frac{1}{K}\left(\frac{\Delta V}{V}\right) = \sum_{\mathbf{q}} \gamma \hbar\omega_{\mathbf{q}} \tfrac{1}{2} \coth\left(\frac{\hbar\omega_{\mathbf{q}}}{2k_B T}\right) = \gamma \bar{\mathcal{E}}(T). \tag{9.5.14}$$

Here $\bar{\mathcal{E}}(T)$ is the energy in the lattice modes at temperature T and the Grüneisen constant γ relates the dilatation to the mean thermal energy density

$$\frac{\Delta V}{V} = K\gamma\bar{\mathcal{E}}(T). \tag{9.5.15}$$

The quasiharmonic approximation assumes that the modes can be treated as independent. However, as the vibrational amplitudes increase at higher temperatures phonon–phonon interactions become important and the independence of the different lattice modes is destroyed. Up to the Debye temperature Θ_D at which all vibrational modes are excited, the assumption that the frequency of the lattice modes is just a function of volume is reasonable. At ambient pressure Θ_D is in the range of a few hundred K for metals and 600–1000 K for typical silicate minerals.

(b) Molecular dynamics

An alternative approach to the inclusion of the effects of temperature is provided by the use of molecular dynamics techniques. This approach explores the time evolution of one representation of a crystal system. The ergodic hypothesis of statistical thermodynamics is then used to equate the time average of a physical property $\bar{X}$ to the ensemble average over configurations $\langle X \rangle$.

For the n atoms in the configuration, the time variation is found by integrating the set of n coupled equations

$$m_i \frac{\partial^2 \mathbf{r}_i}{\partial t^2} = \mathbf{f}_i(\mathbf{r}_1, \mathbf{r}_2, \ldots, \mathbf{r}_n), \quad i = 1, \ldots, n, \tag{9.5.16}$$

for the positions of the atoms $\{\mathbf{r}_i\}$ controlled by the atomic forces $\mathbf{f}$ and how these are specified. The numerical solution is normally achieved with periodic boundary conditions on a supercell; $2 \times 2 \times 2$ unit cells is usually sufficient for the computation of thermodynamic variables. The fastest thermal vibration of the lattice needs to be covered in a few time steps, so typical steps are of the order of femto-seconds (10^{-15} s) with a total simulation duration of a few pico-seconds (10^{-12} s). The initial conditions are the positions of the atoms and their velocities drawn randomly from a Gaussian distribution corresponding to the temperature of interest. To encourage the equation system towards convergence during the interval of numerical integration, the temperature can be temporarily increased to overcome energy barriers.

The equations of motion (9.5.16) conserve energy, and so the natural thermodynamic ensemble is the microcanonical where the internal energy E is fixed. However, the normal experimental state has temperature controlled. To undertake molecular dynamics simulations in the canonical ensemble, with n, V, T fixed, a thermostat has to be applied to the system, e.g., via friction terms applied to the equations of motion (Allen & Tildesley, 1989). Free energies cannot be obtained directly from molecular dynamics simulations, but the difference between the free energy and that of a reference system ($\Delta\mathcal{F}$) can be found by thermodynamic integration. $\Delta\mathcal{F}$ is the work done in a reversible and isothermal change from the total energy of the reference system to that with the actual total energy; the calculations need to be performed with care (de Wijs et al., 1998).

9.5.5 *Influence of defects*

As we have seen in Sections 9.1, 9.3 the presence of defects has a profound influence on the properties of materials and so there is need for computational methods to include the presence of such defects. Two styles of approach are currently employed: (a) the construction of a periodic array of defects, (b) the defect is embedded in an infinite representation of the surrounding crystal.

(a) Supercell methods

The simplest procedure for calculating the energies of defects is to build a periodic construct with a defect at the centre of a large 'supercell'. The influence of the defect can then be determined by comparing the results of calculations for the perfect lattice with those when the defect is present. The difference in energy between the two states is the energy of formation for the defect, and calculations may be performed for either constant volume or constant pressure.

Rather large supercells are needed if complex defects are being considered. Even

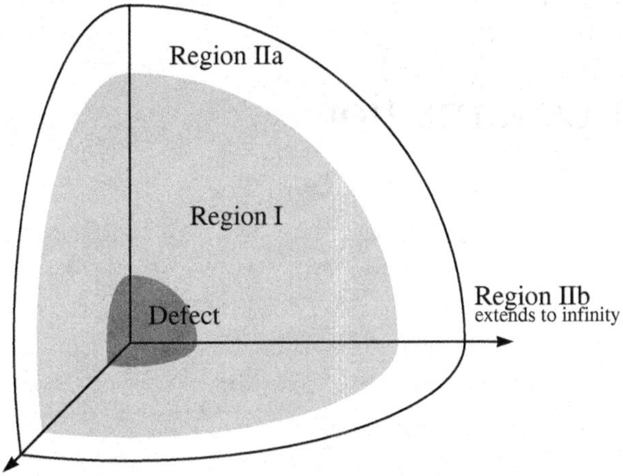

Figure 9.8. Schematic representation of the Mott-Littleton technique for treating the influence of defects, with an inner zone (Region 1) surrounding the defect, a transition zone (Region IIa) and a perfect crystal extending to infinity (Region IIb). Only one octant is shown.

with such large supercells there will be a contribution from defect interaction as well as formation. However, this contribution can be turned to advantage by calculating defect energy for different sizes of supercells to determine defect interactions as a function of defect spacing.

A further complication that can arise when investigating different possibilities for the presence of a class of defect in a crystal structure is the number of different configurations that have to be tested to find the structure with the smallest energy of formation and hence highest likelihood.

(b) Embedded defects

The alternative approach is based on ideas introduced by Mott and Littleton, in which an isolated defect or defect cluster is placed at the centre of an idealised crystal structure extending to infinity (Figure 9.8).

In region I, immediately surrounding the defect, with 100–300 atoms a minimisation procedure is used to bring the atomic forces to zero. In the interface region IIa, the displacements are evaluated as the sum of the response to individual defects including the polarisation effects of defect charge, together with short-range interactions between the ions in region I and those in this interface zone. In the outermost region IIb only the polarisation due to the net charge of the defect configuration is considered.

10
Geological Deformation

In Part I we have concentrated on materials which in the aggregate, at least, have relatively simple properties and focussed on the impact of the latest phase of deformation. These concepts are of considerable utility in understanding the nature of evolution of structure in rocks, but we have to take into consideration the nature of the complex mineralogical composites whose properties depend strongly on physical conditions.

We have abundant evidence of past deformation in rocks on many length scales. The structures that are seen both in the fine scale microfabric and on much larger scales in field exposures frequently reflect the influence of multiple phases of deformation and spatially varying strain.

Different classes of rocks have a wide range of chemical and mineralogical compositions, which have experienced physical and chemical environments that themselves may have shaped the actual fabric. Rock masses contain materials with a large variation in strength, determined by local mineralogy, which respond in very different ways to the same stress field, leading to differences in local strain. Further, the specific temperature and pressure conditions due, e.g., to depth of burial will affect the way in which the rock units behave.

The action of the stress system will lead to a pattern of deformation that will vary from point to point as parts of the body move relative to each other, in part because of the differences in mechanical properties of the different parts of the rock mass.

One of the difficulties that has to be faced in developing descriptions of deformation is that most of the flow laws which have been developed for geological materials are based on low strain experiments, yet normally with a much higher strain rate than in nature. It is now possible to examine high-strain conditions in torsion, but still at high strain rate. The link from the laboratory to the conditions prevailing in a major orogeny involves substantial extrapolation, in particular with regard to rates of deformation.

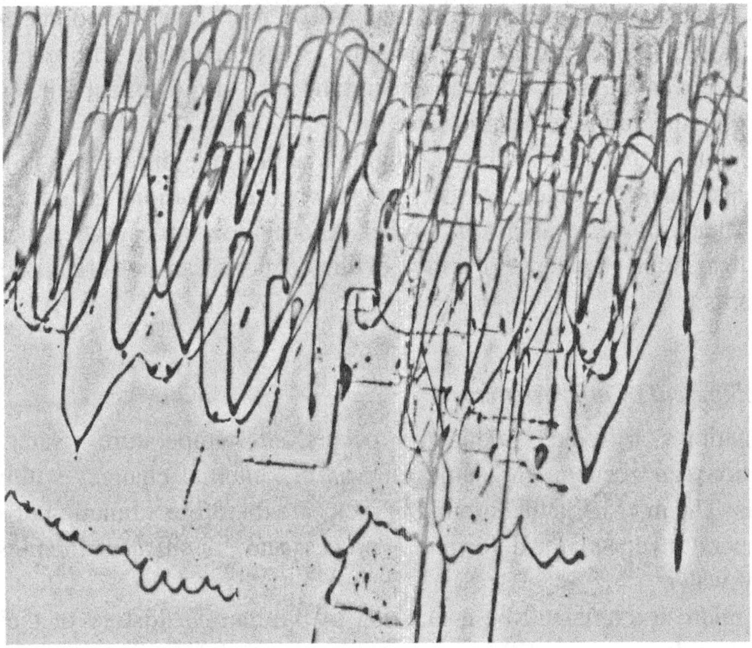

Figure 10.1. Example of dislocation structures responsible for plastic deformation in a quartz foil. The width of the transmission electron microscope image is approximately 5 μm. Clear screw dislocations are seen at the bottom of the image in contrast to the dominantly edge dislocations above. [Courtesy of J. Fitz Gerald.]

10.1 Microfabrics

10.1.1 Crystal defects

The crystal defect systems that we introduced in the previous chapter play an important role in the microstructure of rocks through the way in which their arrangement is influenced by the thermal history.

For silicate minerals a particularly important class of point defects comes from the diffusion of (OH) ions through the structure from the presence of water. The Si–O–Si bridge bonds are replaced by Si–OH–HO–Si, so that the silicon–oxygen Si–O bonding is replaced by the weaker OH–HO bond. The result is a significant decrease in the mechanical strength of silicates, a greater ease of recrystallisation to form new grains, and an enhanced rate of chemical reactivity.

The dislocation line defects have the most direct influence on the mechanical behaviour of crystals. Figure 10.1 shows part of a transmission electron micrograph through a foil prepared from a quartz crystal deformed at 800°C, and a strain rate near $10^{-5}\,s^{-1}$ in creep deformation under a differential stress of 435 MPa. The dislocations show points of rapid growth that are dragging the edge dislocations behind them. Slip on crystallographic planes allows one part of the crystal to undergo shear with respect to a neighbouring part, so that the shape of a grain can change. Quite high densities of dislocations can occur in silicate minerals with deformed materials reaching densities as high as 10^{16} m^{-2} at which points tangles of dislocations largely neutralise slip. The slip systems operating within a crystal are determined by energetic considerations and the ease of movement on a system will depend on temperature and strain rate.

Within a crystal grain there are often subgrains with subtle differences in lattice orientation, less than 5°, which can be formed when dislocations group to form planar walls that can move as a unit. After deformation the crystal will try to reduce the dislocation density and arrays of dislocations can get organised into planar networks.

10.1.2 Development of microstructure

The deformation of rocks frequently occurs at temperatures such that recrystallisation can occur at the same time as the internal changes within the crystal grains. The microstructures found in rocks are therefore a function of strain rate, the temperature during the deformation process and, for silicates, the presence or absence of water.

The final microstructure will be a function of the thermal history of the rock mass. For example, we expect that rocks deformed at high strain rates or lower temperatures will show strong evidence of the deformation with grains flattened by the deformation, and internal structure within the grains. However, if the rock is subsequently maintained at moderate temperatures there will be a process of recovery as the internal energy of the crystal grains is reduced by the reduction of dislocation density, with diffusion allowing dislocation climb to take place. In addition, recrystallisation will occur whereby new crystal grains are developed with internal strains removed. The orientation of these new grains will be significantly different from that of the strain, so that high-angle boundaries will tend to form and migrate through the material, removing most of the dislocations remaining from the recovery stage. There may then be a process of grain growth as internal energy is again reduced by the elimination of grain boundaries, as large grains absorb smaller ones.

When deformation occurs at higher temperatures or lower strain rates, solid-state diffusion can become important at the grain scale and dislocations can then climb under the influence of the stress regime causing the deformation. The result is the migration of grain boundaries and possible growth of recrystallised zones and the final microstructure will depend on the relative importance of a variety of competing processes.

The majority of experimental apparatus for studying rock-deformation relies on some class of compression through, e.g., a piston cylinder system. Such equipment is of considerable value for many scenarios, particularly the influence of pressure and temperature on a rock, or the influence of fluids. However, many natural features are associated with very large shear strains that cannot be produced in such a configuration. Experimental techniques have therefore been developed in

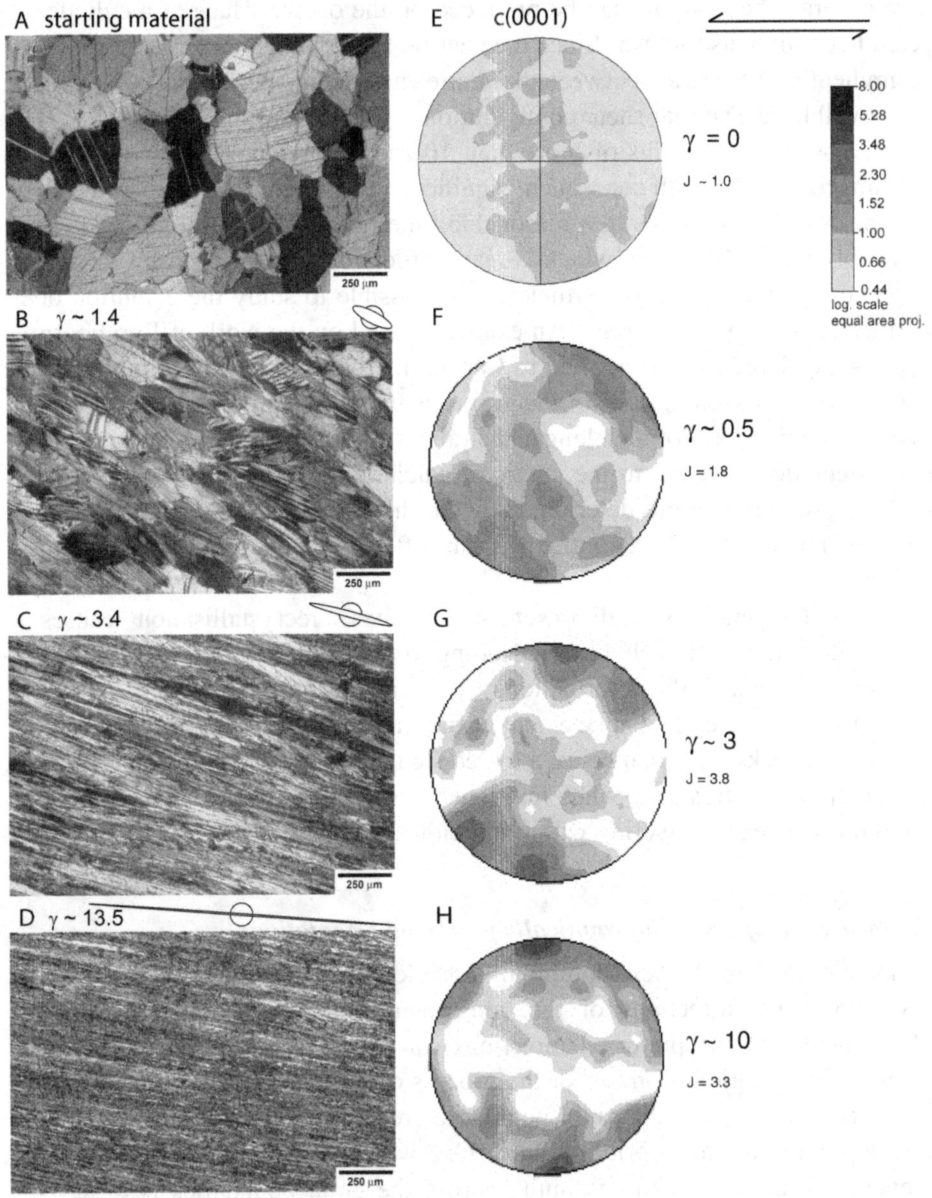

Figure 10.2. Development of shear fabric in calcite (Carrara marble) subjected to large finite torque at 500°C, 300 MPa confining pressure and shear strain rates of $1\times10^{-3}\ s^{-1}$ to various amounts of finite shear strain: (A–D) Cross-polarised transmitted light images of the calcite microstructures; the amount of final strain is indicated for each sample and represented by a non-deformed circle and the corresponding simple shear ellipse. (E–H) Contoured c-axis pole figures of calcite orientations measured by electron backscatter diffraction analyses; the strength of the crystallographic preferred orientations is indicated by the texture index γ. Data in (E) are taken from Pieri et al. (2001), and the data in (B–D) and (F–H) from Barnhoorn et al. (2004). [Courtesy of A. Barnhoorn.]

which very large shear strains can be produced on the outer surface of a cylinder subjected to continuous torques. Since the centre of the cylinder is fixed there is a strong gradient in shear strain between the centre and the outside, but the outermost 5 per cent will have a similar shear configuration.

In such a way shear strains of more than 10 can be produced under elevated temperature conditions, with consequent significant changes in microstructure. To achieve the large shear strains in a reasonable time frame the strain rate needs to be reasonably high (10^{-3} s^{-1}), which is many orders of magnitude faster than in actual geological processes. Nevertheless it is possible to study the evolution of microstructure under severe strain. An example based on the work of Barnhoorn et al. (2004) for large torques applied to Carrara marble is shown in Figure 10.2. There is a progressive elongation of the grains as the torque progresses and many approach the simple form of the elongated strain ellipsoid. The initial broad range of grain orientations relative to the c-axis of calcite is progressively replaced by coordinated grain orientations determined by the shear process. The sense of shear and the amount of strain in natural rocks can often be deduced from the grain elongation.

At higher temperatures and lower strain rates, recrystallisation causes progressive destruction of such a fabric leading to the formation of a fine-grained ultra-mylonite, for which the deformation history can generally not be recognised. Other features such as dyke displacement at the transition from the shear zone to the surrounding rocks may then be used to deduce the shear sense and strain.

More complex textures arise when multiphase materials are subjected to severe shear and this can lead to layering within mylonites.

10.1.3 Formation of crystallographically preferred orientations

As rocks are deformed they commonly develop a preferred orientation of the crystallographic directions of the constituent minerals (cf. Figure 10.2). The development of the preferred directions can be associated with two main mechanisms: the rotation of grains or the process of recrystallisation. A rock has to deform coherently with the grain boundaries remaining in contact, or cracks will develop with ultimately brittle failure. One way in which a rock can submit to a general shear strain with continuity across the grain boundaries is to have homogeneous strain across the constituent grains. In order for this to be achieved by slip within the grains, five independent slip systems have to act within each grain. The condition for preservation of volume means that the trace of the incremental strain must vanish, i.e., $\sum_i e_{ii} = 0$. In consequence only five of the elements of the strain tensor e_{ij} can be independent. To produce this condition five independent simple shears are needed, and hence the requirement of five independent slip systems.

When solid-state diffusion is sufficient to allow climb, fewer slip systems are needed. At such high temperatures some grain boundary sliding could also

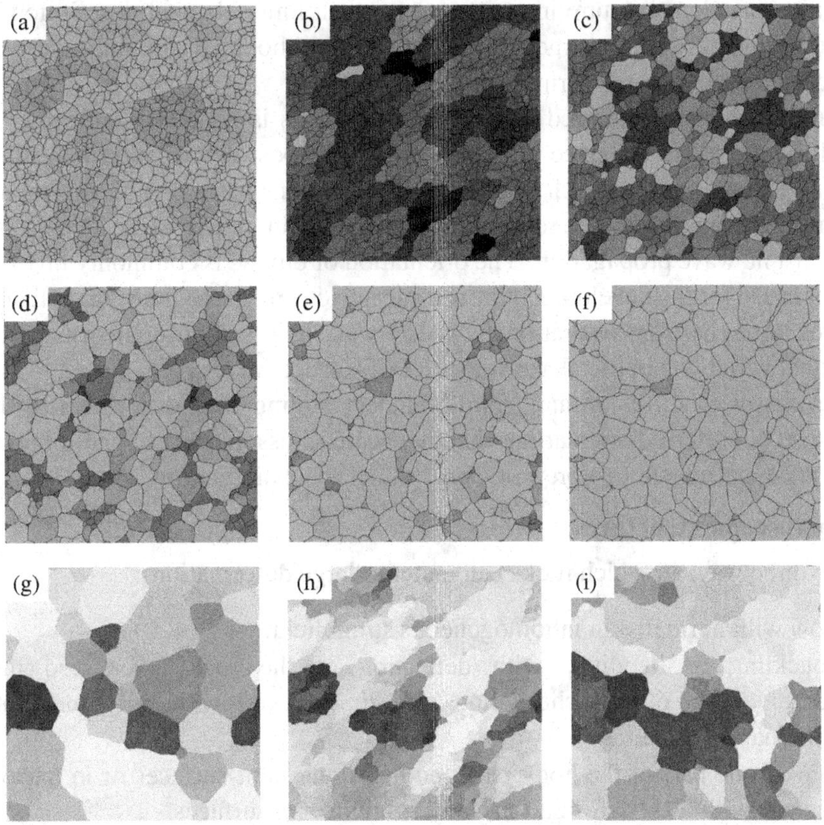

Figure 10.3. This sequence of frames simulates the process of dynamic recrystallisation in quartz (Jessell & Bons, 2002). (a) Starting grain boundary geometry. The shading reflects the internal energy of grains and subgrains, darker shading indicates higher energy. (b) Situation after the imposition of a dextral shear of 0.65, modelled using 13 steps. (c–f) Post deformation recovery and grain growth in intervals of 26 steps. (g) Representation of the microstructure in the original state (a) with shading indicating lattice orientation. (h),(i) Comparable lattice orientations for states (b),(f) showing the retention of the orientations. [Courtesy of M. Jessell.]

contribute to crystal orientation. Other processes such as dissolution–precipitation creep with solution of, e.g., calcite or quartz under pressure, can also move material within a polycrystalline framework in the presence of some hydrous fluids to develop fabric. Material is moved from the crystal faces under most compression to those that are less compressed so that grains change shape without changing mass.

Numerical simulations of microstructures have proved a useful tool to guide understanding of the way in which natural materials acquire their characteristics (see, e.g., Jessell & Bons, 2002). Because of the different physical character of the various processes, the topology and geometry of the polycrystalline aggregate has to be carefully described and transferred to modules that represent the effect of specific grain-scale deformation processes. The simulation shown in Figure 10.3

for evolution of microstructure in quartz combines a finite-element description of deformation linked to a front-tracking model for grain boundary migration driven by surface and defect energy terms.

Crystallographically preferred orientation of crystals is generally indicative of prior deformation history, since it can often survive post-deformation annealing processes. The presence of such preferential orientations has a significant impact on the physical properties of the rock mass, especially with regard to the anisotropy, e.g., for seismic wave propagation. The orientation of olivine is commonly invoked as a major cause for the development of seismic anisotropy in the lithosphere that is manifested through a time separation between *S* wave pulses of different orientation or azimuthal variations in the phase velocities of surface waves.

The processes of diffusion and grain-boundary sliding should not lead to the development of preferred orientations, but there can be dissolution and precipitation processes that are related to a preferred crystallographic orientation

10.2 Macroscopic structures

There are three ways in which rocks can undergo large deformations:

(1) by flow with normally an inhomogeneous strain field,

(2) by buckling or bending, with deflection of the rock layers, so that considerable overall shortening can be achieved with only moderate internal deformation

(3) by slip of one part of the body past another along fault surfaces or in narrow zones, with little deformation away from the slip surfaces.

The deformation observed in rock outcrops is generally complex, reflecting multiple phases of inhomogeneous strain that have interacted with the relative strengths of different components of the rock. The outcome can be complicated folding such as is illustrated in Figure 10.4 with an outcrop from the Proterozoic

Figure 10.4. Example of complex folding structures in the Proterozoic fold belt of the King Leopold Range, Western Australia

rocks of the King Leopold fold belt in northwestern Australia. Such convoluted structures with tight folds and rapid changes of orientation imply substantial finite compression.

10.2.1 Multiple phases of deformation

Although we have developed the theoretical formulation of inhomogeneous strain in Chapter 2, most of our prior examples have been based on the simpler condition of homogeneous strain through a body. Once the strain field becomes both finite and inhomogeneous a much wider range of behaviour is possible. The mathematical development can be described through the use of a spatially varying deformation gradient F and, as in (2.1.5), the influence of multiple phases of deformation is described by the ordered product of the deformation gradients $\mathsf{F}_1\mathsf{F}_2$.

A useful way to follow the nature of such complex deformation is to look at the development of the local Eulerian ellipsoid generated from a sphere in the original reference state (Section 2.2.2). This process is illustrated for multiple two-dimensional deformation in Figure 10.5. A set of small circles is drawn on the surface of the undeformed material, and the action of progressive deformation can

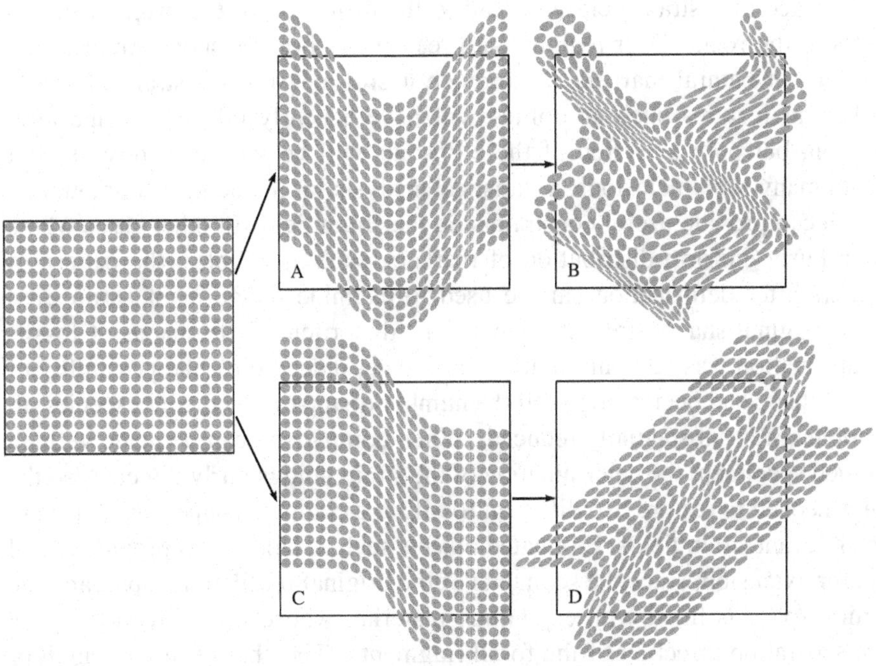

Figure 10.5. Illustrations of multistage inhomogeneous deformation leading to concentrated belts of strong deformation indicated by the distortion of the initially circular markers. (A),(B) Superimposition of oblique tighter folding (at a 60°) angle on a simple fold. (C),(D) Simple shear superimposed on an inhomogeneous shear profile.

be seen through the spatial behaviour of the Eulerian ellipses in the inhomogeneous strain field.

Two simple cases are shown that lead to complex patterns of behaviour with only two phases of deformation. In the upper row of Figure 10.5 A-B we consider the application of a sinusoidal variation of strain, followed by a further sinusoidal variation with shorter wavelength at an angle of 60° to the original. In A the strain is concentrated on the flanks of the fold, although the centre is strongly displaced, there is only modest distortion of the circles near the axis. However, with the superimposition of the second finite strain in B, the pattern is much more complex. Narrow bands of severe deformation develop and also the deformation is sufficient that the circles have been distorted to non-elliptical shapes.

In the lower row, Figure 10.5 C–D, we consider the imposition of shear strain. At first a concentrated band of shear deformation is applied with a hyperbolic tangent (tanh) profile in C, so that the strain fades away towards the edges. This is followed by a simple shear at right angles in D. The combination significantly modifies the direction of the Eulerian triad in D compared to C, and some care is needed to infer the original deformation field after removal of the homogeneous simple shear. The situation becomes rapidly more difficult when the second phase of deformation is itself inhomogeneous.

The presence of a strain field will lead to the distortion of the original shapes of objects in the rock. Over regions with reasonably homogeneous strain fields, the reactions of natural markers can provide useful measures of strain. Initially near spherical features such as oolites will have a nearly ellipsoid shape after deformation, but only a section of the ellipsoids will be seen in exposure. For rocks with many such inclusions, a statistical analysis for the dominant orientation and stretch can provide useful constraints on the strain field. In other cases where there is a fairly uniform distribution of markers, such as pebbles, the pattern of the markers after deformation can be used to examine the form of strain even where the original shapes are not known. The distortion of the shapes of fossils under deformation was recognised at an early date, indeed once the possibility of changes of shape had been recognised the number of apparently distinct species of, e.g., trilobites was substantially reduced. When comparison can be made with an undeformed specimen, the determination of the strain field is fairly direct. Also the mappings needed to restore two different fossils to a common shape can be used to infer the relevant strain fields. Elongated objects such as belemnites provide useful markers for extension, since the segments of the original fossil break apart and the intervening space is filled with crystalline material, whose fibre directions align with the separation direction of the fossil fragments. The change in the length of the fossil and the fibre directions provide direct measures of the properties of the strain ellipsoid.

Many of the features in rocks are associated with progressive deformation during geological processes, occurring over a substantial time frame, so that the strain field can change in time. The influence of the strain rate can then be felt through the rheology of the material, particularly the effective viscosity of different rock components. The differences in resistance to deformation can have very significant effects on the way that the different components of the rock behave under the same deformation field.

In a shear zone a variety of mechanisms can operate to provide weakening of the material. Geometric softening can arise from the development of crystallographically preferred orientations for minerals associated with shear deviations. Softening can also occur due to grain size reduction (e.g., by dynamic recrystallisation); this allows a switch from dislocation creep to diffusion creep and grain-boundary sliding creep with a consequent modification of the constitutive equation. In the presence of reactive fluids, there can be chemical effects that induce softening. Many chemical reactions produce a weaker mineral assemblage, e.g., the breakdown of alkali feldspars to produce micas, or the production of chlorite from amphiboles. There is a widespread belief that material softening leads to strain localisation.

Even during plastic deformation, the permeability of rocks can be enhanced during straining due to the presence of microcracks. This permeability facilitates the localisation of fluid flow and reaction in shear zones. The presence of fluid commonly leads to reaction softening, but very intense hydrothermal alteration can result in micas being replaced by assemblages rich in quartz and feldspar with consequent reaction hardening. The presence of elevated pore fluid pressures relative to hydrostatic pressure can lead to modification of the normal behaviour. Thus, e.g., with high pore fluid pressures p_f material may move from the ductile to the brittle field even deep within the crust.

10.2.2 Folding and boudinage

Folding in rocks represents a common response to deformation and can be observed over a very wide range of scales from the microscopic in fine grained metamorphic rocks to scales of tens to hundreds of kilometres (Figure 10.6). Frequently multiple scales of folding can coexist representing different phases of deformation. The diversity of observed fold features come from the interaction of a range of different factors:

(a) the composition of the layers and the primary rheologies of each rock type,

(b) the way in which the rheological parameters change with pressure and temperature during the period of fold formation,

(c) the development of oriented mineral grains during the deformation leading to anisotropy in material properties,

(d) the mechanical properties of the interfaces between layers

(e) the thickness of the different rock layers, and the spacing between the most competent layers,

(f) the nature of the boundary constraints on the rock units undergoing folding, e.g., an upper free boundary but a lower boundary defined by sliding on a decollement

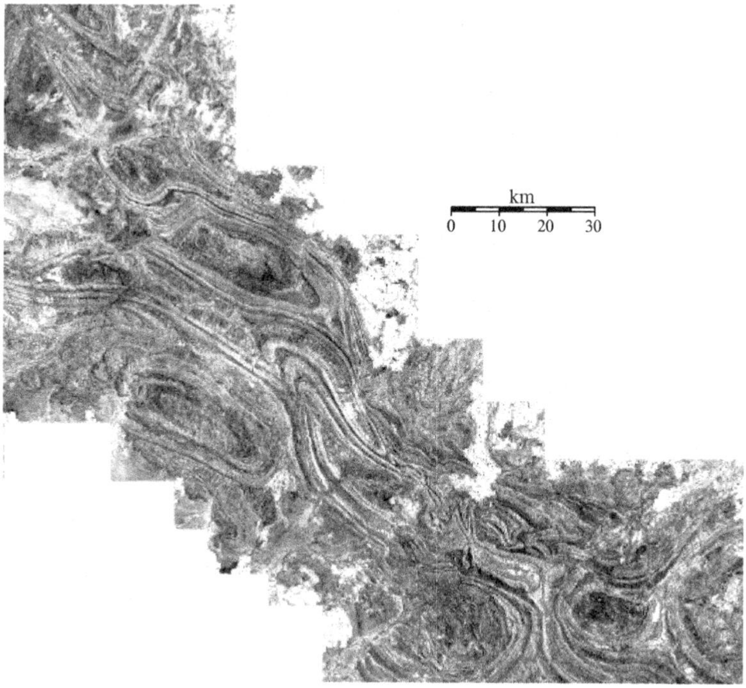

Figure 10.6. Large-scale and smaller-scale folds exposed by differential weathering of the Proterozoic meta-sediments in the Davenport and Murchison Ranges, Northern Territory, Australia.

surface (see Section 10.2.4), and
(g) the overall scale of the rock mass being folded.

One way in which fold structures can be developed that does not depend on the actual rheology is by differential flow with well-defined flow lines (passive or shear folds – Figure 10.7). The deformation of each layer is then the same; this mechanism can be important in providing an amplification and distortion of existing folds; in which case the folded layers provide geometric markers for the imposed deformation.

Nevertheless, most of the observed deformation in rocks is related to the layering and differential strength of a geological sequence. Under compression the stronger layers tend to bend and then buckle, while weaker layers flow (Figure 10.8), and the nature of the folded structures will depend on the direction of the prevailing stress relative to the layering.

When a compression is applied parallel to rock layers the initial effect is homogeneous shortening along their length, but as the process continues there will be a tendency for buckling with transverse deflection. The physical effects will vary depending on the environment of the stronger layers. If they are widely spaced in a softer matrix then the layers will buckle independently; whereas if they are close

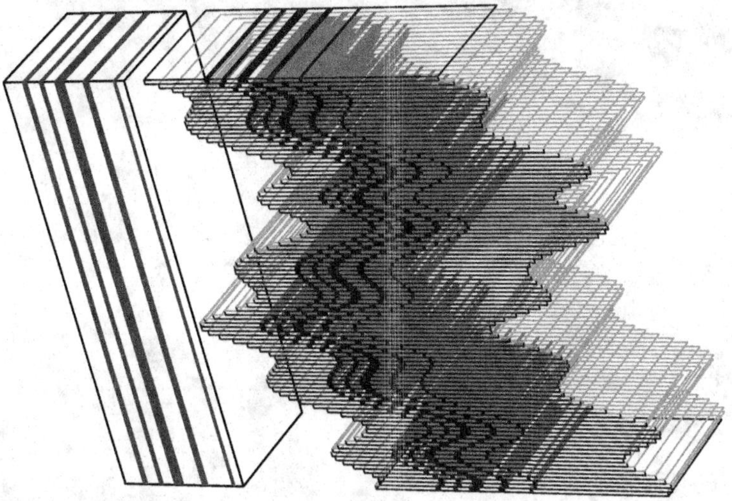

Figure 10.7. Development of folds by spatial varying but coherent flow.

together the configuration is generally imposed by the most competent material and the effect is that of a composite, thicker layer.

A simple model for the development of folds can be found for a linear viscous rheology by examining the dependence on the viscosity ratio ζ between an isolated, competent layer with viscosity η_1 and its less viscous surroundings (viscosity η_2). Consider the initiation of a sinusoidal fold of long wavelength L_b relative to the layer thickness d of the competent layer, in a two-dimensional situation with plane strain and volume conservation. There will be two parts to the resistance to fold formation. The first will come from the more competent layer through the force required to stretch the outer arc of each fold and compress the inner arc; the force can be represented as

$$F_{int} = \eta_1 \frac{2\pi^2}{3} \frac{d^3}{L_b^2} \frac{\partial}{\partial t} \ln e_x, \qquad (10.2.1)$$

where e_x is the shortening strain along the orientation of the original layer; the force will be least for long wavelengths L_b. The second resistive component comes from the matrix through the transverse deflection,

$$F_{ext} = \eta_2 \frac{L_b}{\pi} \frac{\partial}{\partial t} \ln e_x, \qquad (10.2.2)$$

and here the least energy would be associated with the shortest wavelength. The balance of the two forces leads to a minimum when

$$\frac{\partial}{\partial L_b}(F_{int} + F_{ext}) = -\eta_1 \frac{4\pi^2}{3} \frac{d^3}{L_b^3} \frac{\partial}{\partial t} \ln e_x + \eta_2 \frac{1}{\pi} \frac{\partial}{\partial t} \ln e_x = 0, \qquad (10.2.3)$$

Figure 10.8. Influence of compression on a multi-layered rock, showing the differential effects of buckling depending on the resistance of the layers. [Courtesy of S. Cox.]

and thus when the wavelength L_b is given by

$$L_b^3 = \frac{4\pi^3}{3}\frac{\eta_1}{\eta_2}d^3. \qquad (10.2.4)$$

For such an isolated layer, the ratio of the wavelength of buckling L_b to the layer thickness d thus depends on the relative strength of the layer to its surroundings ζ

$$L_b = 2\pi h[\zeta/6]^{1/3}, \qquad (10.2.5)$$

with a similar result for elastic buckling where ζ is the ratio of the elastic moduli. Buckling of layers with a rheology governed by a power-law dependence of strain rate on stress also displays a proportionality between wavelength and layer thickness, but the dependence on relative strength can be reduced.

The effect of extension on a multi-layered material is somewhat different. The stronger layers resist extension along their length and as a result tend to become segmented (Figure 10.9). The material necks and may even separate to give a cross-section reminiscent of sausages, from whence the name *boudin* from a French sausage. There is a flow of the softer material material into the necks and gaps that imposes a distinct local fabric.

When both the stronger and weaker materials are in the ductile regime, the boudins can form with wavelike regularity in a similar way to folds.

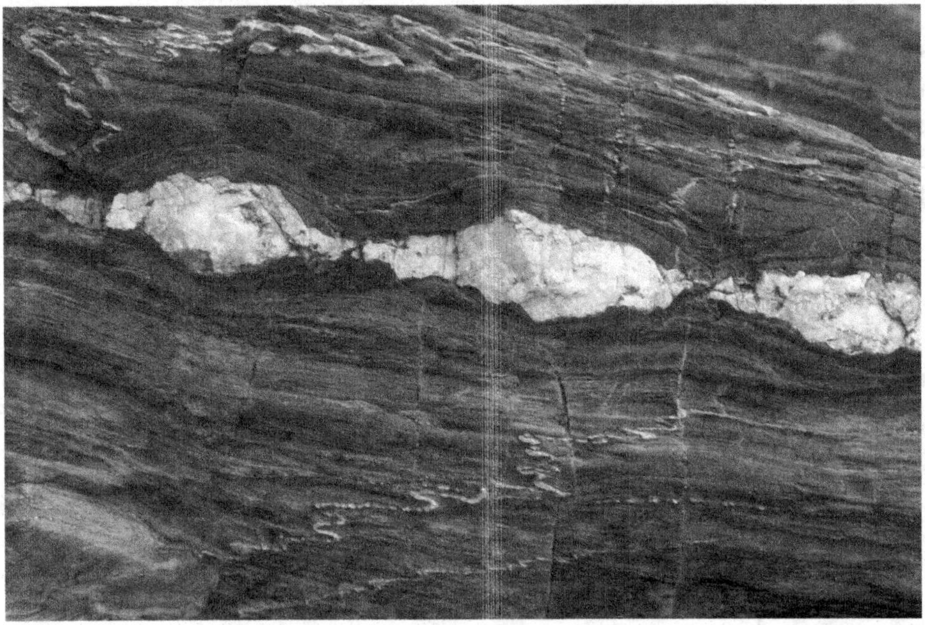

Figure 10.9. Influence of extension on a multi-layered rock, showing the development of boudins in the most resistant layers. [Courtesy of S. Cox.]

10.2.3 Fractures and faulting

In the crust at shallow depth the rheology is brittle and as a result the response to large deformation is for rocks to fail along narrow zones of frictional slip. At greater depth the brittle to ductile transition reflects the change from localised to distributed failure, with more plastic flow. The interaction of plastic flow mechanisms and brittle cracking can occur in a number of ways. Cracks may be nucleated at pile-ups of dislocations, from more rigid secondary phases or boundaries. The nature of the growth of such cracks depends strongly on the ambient conditions particularly with regard to mineral properties, stress conditions including the pressure from pore fluids and temperature.

The consequence of brittle failure is that coherent bodies of rock slip past each other to create a *fault* in which equivalent strata on the two sides of the slip surface are displaced. Repeated slippage in the same place leads to a zone of localised deformation with broken material between the fault surfaces.

Material failure

A convenient representation of the stress conditions associated with material failure is provided by working with the normal stress $\sigma_N = \mathbf{n} \cdot \boldsymbol{\tau}$ and the shear stress $\sigma_S = \mathbf{m} \cdot \boldsymbol{\tau}$ on a plane where, as in Section 3.3, the vector $\mathbf{m}$ is orthogonal to the current normal vector $\mathbf{n}$ to the plane. We have already seen that we can represent the stress state on a plane by the stress-circle construction as in Figure 3.9. We can also represent the failure surface in the same domain, as a limit on acceptable shear stresses (Figure 10.10).

In the presence of fluid pore pressure p_f the entire effective stress state is shifted, so that $\sigma_N \rightarrow \sigma'_N = \sigma_N - p_f$. This shift is indicated by the movement between the solid stress circle in Figure 10.10 and the chain-dashed variant in the presence of fluid pressure. The stress state for a normal at an angle θ to the σ_1 principal stress axis is still found at the point on the stress circle with inclination 2θ.

If the state of stress in the material plots on or above the fracture envelope the material will fracture. The onset of fracture will occur when the stress circle touches the fracture envelope. The final macroscopic fracture will generally be oriented close to the plane that is a tangent to the fracture envelope (orientation θ to the σ_1 axis in Figure 10.10). However, the presence of flaws in the rock or

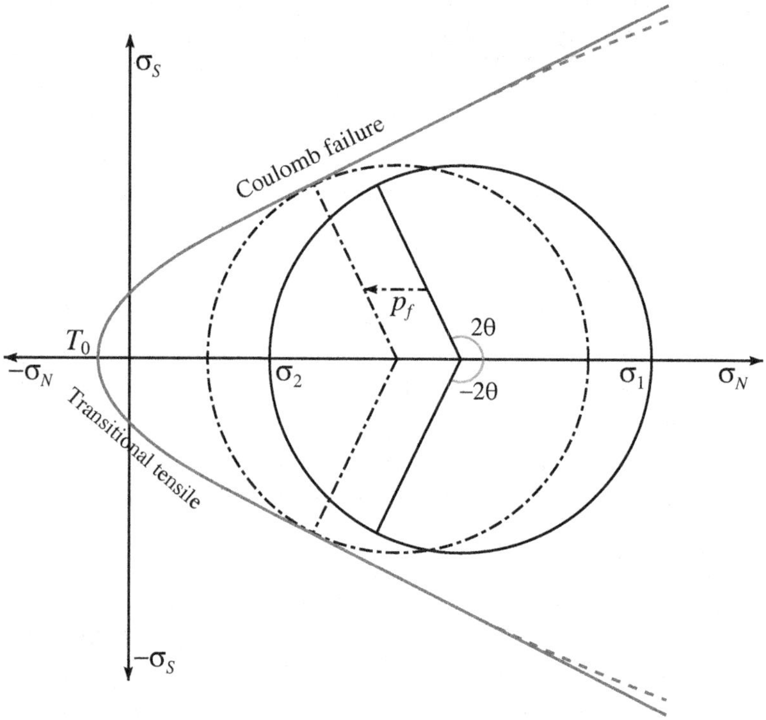

Figure 10.10. Three fields of the Mohr fracture envelope in the normal and shear stress domain. The tensile fracture field shows a fixed strength T_0. Coulomb failure shows a linear increase of shear strength σ_S with normal stress σ_N associated with frictional sliding. Transitional tensile behaviour occurs for normal stresses between the tensile and Coulomb states. The stress circle is shifted to lower normal stresses by the presence of pore fluid pressure p_f. The onset of fracture occurs when the relevant stress circle touches the fracture envelope, in this case for a stress state on the planes inclined at angle θ to the σ_1 axis. For high normal stresses the shear strength tends to drop below the straight line predicted by the Coulomb relation.

existing zones of weakness can modify the actual fracture development. There is only one point at which the stress circle is a tangent to the fracture envelope and only a single direction for potential failure.

For material under tension, or effective tensile stress, $\sigma_N < 0$, *tensile fractures* can form perpendicular to the direction of maximum tensile stress. The tensile strength T_0 is independent of the other principal stresses and is typically in the range −5 to 20 MPa for dry rocks.

For intermediate confining pressures $\sigma_1 \gtrsim |5T_0|$, and $\sigma_1 - \sigma_3 \geq 4T_0$, the fracture strength increases approximately linearly with increasing normal stress. This regime is characterised by a shear stress at failure on a fault plane

$$\sigma_S = c + \mu\sigma_N, \tag{10.2.6}$$

where c is known as the cohesion and μ is the coefficient of internal friction. This form was suggested by Coulomb in 1773 on the basis of an oversimplified physical model, but similar behaviour can be derived for a model with small flaws (*Griffiths cracks*) which can enlarge and propagate under the influence of an applied stress. The cracks will tend to close at high pressure and the equivalent form to (10.2.6) is

$$\sigma_S = 2T_0 + \mu\sigma_N, \tag{10.2.7}$$

with c identified as $2T_0$ in terms of the tensile strength. As illustrated in Figure 10.10 the stress circle touches the fracture envelope twice so that there are two possible directions for failure. For many rocks the coefficient μ is close to 0.8, which also appears appropriate for rock friction. A linear relation of the form (10.2.6) also applies approximately to many less consolidated materials such as sand and concrete, but with a reduced value for μ.

At high confining pressures the shear stress for fracture tends to drop below the straight line predicted from (10.2.6) as flow begins, and the strength increases more slowly with pressure.

The transitional region between tensile failure and the Coulomb regime has two potential directions of fracture since the stress circle will touch the fracture envelope for both positive and negative shear stresses. This transitional region is characterised by a rapid increase in the strength with increasing normal stress, which for the Griffith crack model takes the form

$$\sigma_S^2 = 4T_0(\sigma_N - T_0). \tag{10.2.8}$$

The fracture orientation is typically oriented at about 30° to the maximum compression. Such transitional tensile behaviour is likely to be important in the development of joints in rocks.

Faults

Faults represent discontinuities across which there is a relative displacement of equivalent geological features. If the corresponding strata can be recognised on the two sides of the fault then slip across the discontinuity can measured. By their nature as zones of concentrated slip, faults are commonly zones of crushing, with a weak infill and often hot-water circulation. Fault zones are therefore rather easily eroded and hence can be less visible than the surrounding rocks. As a result, many faults are inferred from offsets in geological patterns across an unexposed zone that is presumed to be the surface expression of the zone of slip. In some circumstances faults can be directly observed in natural exposures, as, e.g., coastal cliffs (Figure 10.11) and canyons, roadcuts, quarries and mines.

The grinding of the two fault surfaces against each other during the process of slip can produce polished and grooved surfaces called *slickenslides*. The scratches and polish are produced by abrasion by the wall rock and the fault zone fragments during the frictional sliding. The grooves record the orientation of the fault slip.

Multiple episodes of slip along a fault at low temperatures tend to produce a weak infill material, *fault gouge*, with frequently a significant clay component. For major faults the gouge zone can be quite wide, e.g., on the North Anatolian fault in Turkey fault gouge zones about 50 m wide are commercially quarried for sands for building materials. The configuration within the gouge can be rather complex without a simple relation between materials of different ages.

Slip on a fault zone at higher temperatures is accompanied by crystal plasticity

Figure 10.11. Faulting in Palaeozoic rocks exposed on the south coast of New South Wales, Australia, which juxtaposes grey shales on the left against red sandstone on the right. The near horizontal quartz veins imply that σ_3 is close to vertical, and thus this is a reverse fault.

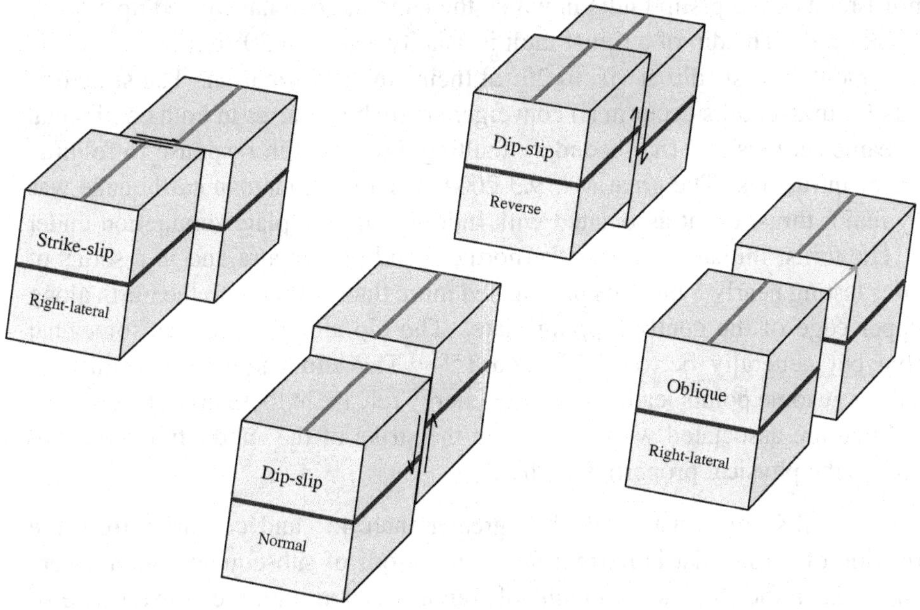

Figure 10.12. Examples of faulting displacements in dip-slip, strike-slip and oblique faults.

rather than brittle failure. The plastic fault zones are generally composed of *mylonite*, which is a fine-grained laminate with common streaking of contrasting materials. The mylonite is often very hard because of the fine grains and possible work-hardening of the crystals. Mylonites do not always retain the fabric and texture of the faulting event since they can be subject to recrystallisation, and possibly annealing, after deformation. Despite the recrystallisation, the crystallographic preferred orientation is commonly preserved, as are the mesoscopic fabrics that indicate shear strain.

Types of faults

The main classes of faults are illustrated in Figure 10.12. In a normal fault the hanging wall drops relative to the foot wall and such faults are common in extensional environments. The dip of a normal fault is commonly quite steep, usually in the range 55° to 70°, and can sometimes approach near vertical in the near-surface. Frequently normal faults occur in conjugate sets with parallel strikes, but opposite dip; although the slip displacements are frequently significantly different between the two sets. Good examples of normal faults are provided by the bounding faults of the East African rift, and other graben structures. Collapse structures such those above an emptied magma chamber are marked by a ring of normal faults. Deep seated normal faults associated with crustal extension tend to show lower dips at depth as they pass into the zone of plastic flow in the lower crust (listric faults).

Thrust faults are dip-slip faults in which the hanging wall has moved up relative to the foot wall. The dip of a thrust fault is usually less than 30° during active slip and commonly thrusts dip at 10° to 20° at their time of formation. The structural settings for thrust faults occur in (i) convergent plate boundaries in both continental and oceanic settings and (ii) secondary faulting developed in response to folding, flexure or intrusions. The great Mw 9.3 2004 Sumatra–Andaman earthquake was a very major thrust event associated with Indo–Australian plate subduction under Asia. The thrust initiated towards the northern end of Sumatra and in a series of episodes lasting nearly 5 minutes propagated more than 1300 km to the north along the upper edge of the gently dipping plate. The dip along strike was somewhat variable but generally between 10° and 15°. The slip pattern shows distinct reductions at some points leading to lower energy release in these areas (see Figure 11.26), that are associated with changes in the strike of the subduction zone and changes in the physical properties of the slab.

Reverse faults are thrusts with dips greater than 45° and can arise from the reactivation of former listric normal faults in periods of subsequent compression, such as, e.g., in the Zagros mountains of Iran associated with the convergence of Africa and Asia.

Strike-slip faults have nearly horizontal slip parallel to the strike of vertical or very steeply dipping faults. Such prominent strike-strip faults such as the San Andreas fault in California or the North Anatolian fault in Turkey are often regarded as vertical faults, but can have some variations in dip. For example the segment of the San Andreas fault near Santa Cruz has an inclination to the vertical as indicated by earthquake locations; this dip is needed to compensate for the bend in the fault line. Other prominent continental strike-slip systems are the Dead Sea system and the Denali fault in Alaska with lengths of hundreds to thousands of kilometres.

A common feature of strike-slip faults is that the fault surface is not a continuous plane, but is instead composed of en-echelon segments that accommodate slip by compressive or extensional deformation, producing local uplift or depressions ('pull-apart' basins). Figure 10.13(a) shows how the offset will require the movement on the strike-slip fault to lead to uplift. The effect is illustrated in Figure 10.13(b) by a photograph taken immediately after the 1891 Neo Valley earthquake in Central Japan that produced a 16 m vertical scarp in a strike-slip event. The horizontal displacement away from the uplift is clearly indicated by the deformation of the pre-1891 fields, which are still outlined by tea-bushes (Figure 10.13b). A large-scale example of a similar type is the zone of thrusting and folding in the Transverse Ranges of southern California associated with an en-echelon step to the left of the San Andreas fault (viewed from the north or south).

Many faults have some oblique component superimposed on the main motion that can be deduced from the displacement of the strata on the two sides of the fault. Thus, the thrust associated with the Mw 7.6 1999 Chi-Chi earthquake in

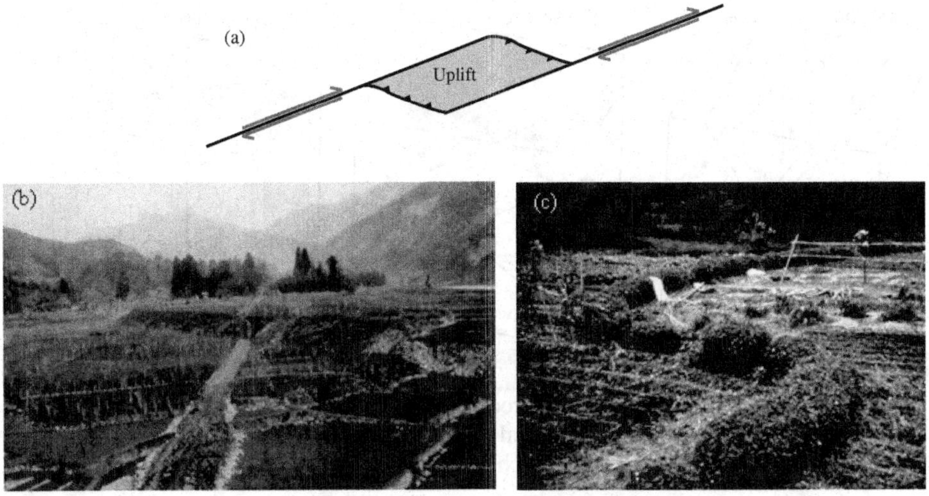

Figure 10.13. En-echelon segments of strike-slip faults have the effect of producing local compression in the overlap region between the segments: (a) illustration of the development of an uplift block, (b) the 16 m fault scarp in Central Japan formed during the 1891 Neo Valley strike-slip earthquake [1891 photograph courtesy of the Seismological Society of Japan], (c) the strike-slip component illustrated by the offset of tea-bushes marking the edge of fields before the earthquake [1991 photograph, B. Kennett].

Taiwan has significant oblique motion, and the dominantly strike-slip Alpine fault in New Zealand has associated vertical uplift.

A simple mechanical theory of faulting was developed by Anderson (1951) on the basis of the assumption that shallow-level faults are Coulomb fractures (Figure 10.10) in an isotropic medium. The line of the intersection of the two fracture orientations is parallel to the intermediate principal stress σ_2. The direction of maximum compression, σ_1 bisects the acute angle between the two fracture directions and the axis of least principal compression σ_3 will bisect the obtuse angle. The material will shorten parallel to σ_1 and expand parallel to σ_3 as a result of slip along the fractures. Thus, given the orientation of the principal stresses and the slope of the fracture envelope μ, the orientation and sense of the slip in the Coulomb fractures can be determined.

The Earth's surface is a plane without shear stress, as a fluid–solid boundary, and as a result one of the principal stress directions must be perpendicular to the surface. For shallow faulting it can reasonably be assumed that this principal stress direction is maintained and the other two principal stress directions will then be horizontal. Different configurations will be achieved depending on which of σ_1, σ_2 or σ_3 is vertical (Figure 10.14). Normal faults will be associated with σ_1 vertical with fractures dipping more than 45° and slip will be downdip. When σ_3 is vertical the fractures will be inclined at less than 45° and slip will be updip in thrust faults.

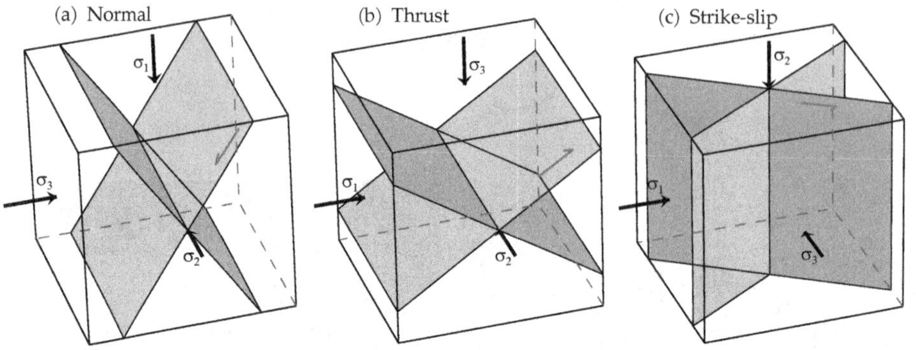

Figure 10.14. Relation of principal stress directions and fault planes (indicated by shading) predicted from Coulomb failure: (a) Normal faulting, (b) Thrust faulting, (c) Strike-slip faulting. The sense of slip is indicated in each case.

For the intermediate case, σ_2 vertical, the fractures will be vertical and the slip horizontal. With the failure criterion (10.2.7) and the typical value of μ of 0.85, the predicted angle of the failure plane to the σ_3 principal axis will be about 25 degrees, in general agreement with the patterns in natural fault systems.

This simple theory captures much of the behaviour of near-surface faulting, but fails to predict, e.g., high-angle reverse faults. Thrust faults often dip at smaller angles than predicted (less than 20°) because their properties are dominated by bedding slip. Rocks are anisotropic and when the maximum principal stress is close to foliation, the anisotropy can have profound effects on the orientation of faulting. The stress field is also somewhat variable in space, and models that incorporate the stress field due to tectonic forces indicate modification of the Coulomb fracture directions in ways that are closer to field observations. The stress state considered in the Anderson theory is definitely a considerable oversimplification, because the presence of faults changes the stress state in their vicinity that can give rise to secondary faulting.

The process of slip in an earthquake has to be accommodated by the surrounding material, and complex residual stress and strain patterns can be left around the edge of the region in which there has been major slip. The discomfort is relieved by subsequent smaller earthquakes and so the initial *aftershock* pattern tends to map out the edges of the slipped zone or regions within that have proved resistive to slip.

The nature of the termination of fault segments depends on the level of displacements relative to the fault length. If displacement is modest, the space and continuity issues associated with fault termination can be taken up by gradual reduction in displacement towards the end of the fault with distributed elastic strain (Figure 10.15). Blind thrusts as in Figure 10.15a have proved to be important in recent earthquakes in California, such as the 1994 Northridge event in the San Fernando valley, north of Los Angeles, where significant surface disruption and

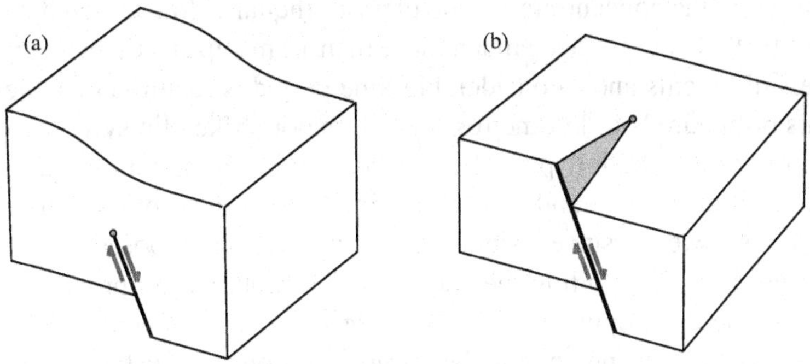

Figure 10.15. Accommodation of thrust faulting in three dimensions: (a) blind thrust with surface uplift, (b) termination at the surface. Note the similarities to the structures in screw and edge dislocations.

very strong ground acceleration occurred without faulting actually reaching the surface.

Where displacement is larger relative to the fault dimensions, accommodation can still be by elastic strains provided that they are distributed through a large volume of rock with, e.g., a set of splay faults branching from near the end of the main fault. Such splays may occur both vertically and horizontally. A very fine example is provided by the set of dominantly strike-slip faults branching from the northern end of the Alpine fault in New Zealand (Figure 10.16).

A somewhat different scenario applies at the depth for major strike-slip faults such as the San Andreas fault. The lower part of the fault lies within the ductile regime so that part of the motion has to be taken up by plastic flow that will become more widespread, but less intense, as the depth increases.

Motion on a fault

The observed displacements on faults are typically built up from many episodes of slip. Although there are some examples of slow progressive slip, the most common

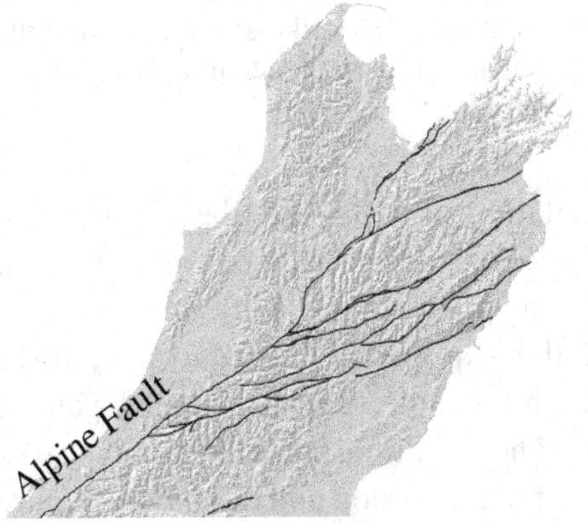

Figure 10.16. The configuration of faults at the northern end of South Island, New Zealand, showing the well-developed set of splay faults branching from the northern end of the Alpine Fault. [Active faults from the QMAP: Geological Map of New Zealand and Active Faults databases courtesy of GNS Science, New Zealand.]

mode of failure is in the concentrated mode of an earthquake. Since even the largest earthquakes, Mw 9, do not produce much more than 12 m slip at a time, a very large number of seismic events and a considerable time period is required to achieve the displacements of hundreds of kilometres seen on major strike-slip systems. On the thrusts at the top of the shallow part of the subducting slab the faulting process is part of the subduction system, and recurrence times are of the order of hundreds of years for the largest events, since a significant strain must be imposed.

Early progress in understanding the way in which ruptures can propagate came from models of the development of cracks. However, one of the difficulties in understanding the faulting process is the complex nature of the fault zone. The surfaces that move are not completely planar and en-echelon features can occur at many scales. Further, the fault may contain gouge material and be subjected to fluid pressures. The specific conditions have therefore a strong influence on the apparent friction between the two sides of the fault, and it is therefore very difficult to specify a constitutive equation for the gouge material. Such information is needed for dynamically consistent simulations of fault motion that take into consideration the interaction of the fault with its environment. Considerable development has been made for such computational models that are needed for understanding strong ground motion in the immediate vicinity of the fault.

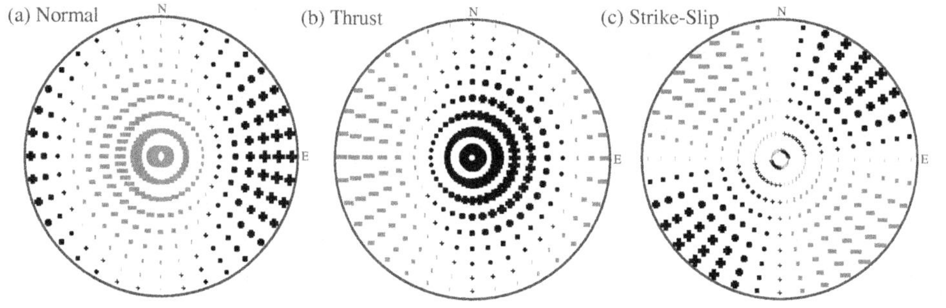

Figure 10.17. P wave radiation patterns, lower hemisphere equal-area projection, for the three basic fault types for a north–south strike with compression indicated by pluses and dilatation by minuses, the size of the symbol is scaled by relative amplitude: (a) normal fault with a dip of 60°, (b) thrust fault with a dip of 30°, and (c) strike-slip fault. Note the ambiguity between the fault plane and the perpendicular auxiliary plane, and the relatively broad zone of diminished amplitude around the nodal planes.

For most purposes a rather simpler formulation of a slip distribution across a fault surface is adequate to explain seismological observations (see Section 11.2). Small events can be adequately described by a 'point source' with an associated moment that allows for the actual size of the slip area. The equivalent force system to localised slip on a plane is a double-couple, i.e., a matched pair of couples with no net moment (Figure 11.4). The resulting radiation for P waves is a four-lobed pattern with alternating zones of compression and dilatation (Figure

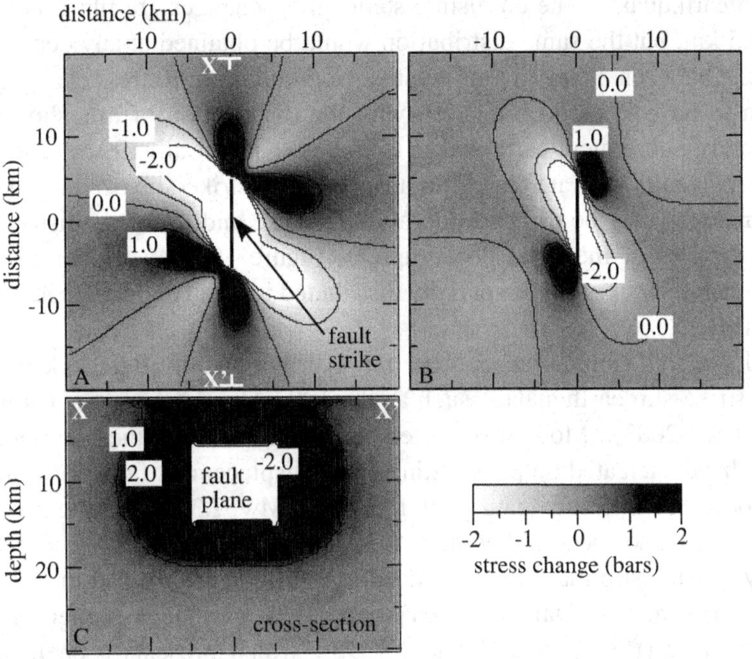

Figure 10.18. Map and cross-section views of the change in Coulomb failure stress generated by a right-lateral slip event on a vertical fault. The fault is 10 km long, extending from 5–15 km in depth. (A) Map-view calculation of the stress change that is resolved on optimally oriented faults in the wall-rock around the mainshock. In all areas other than very close to the mainshock fault, optimally oriented faults have the same orientation and slip rake as the mainshock fault. The results show that two clusters of positive stress changes are generated around both tips of the mainshock fault. (B) Map-view calculation of the stress change that is resolved on thrusts, which generate spatially restricted single clusters of positive stress change around the tips of the fault. (C) Cross-section view of the strike-slip fault along line X–X' shown in A. This figure shows a long-section projection of the rupture plane and associated stress changes with depth. Positive stress changes form a doughnut-like distribution around the rupture plane in three dimensions; such positive Coulomb failure stress changes indicate where the crust is brought closer to failure after slip on the mainshock fault. [Courtesy of S. Micklethwaite.]

10.17), with the nodal planes accompanied by zero radiation along the fault plane and perpendicular to it (see Figure 5.5). The patterns of observed ground motion at distant stations can be projected back to the immediate neighbourhood of the source and used to constrain the fault mechanism of the event.

Figure 10.18 shows the stress distribution associated with right-lateral slip on a vertical fault, calculated in three dimensions as a dislocation in an elastic, isotropic, half-space with a free surface. The imposed far-field stresses have Andersonian orientations relative to the strike-slip fault, that is the maximum principal stress is horizontal with an azimuth of 30°. The fault extends from 5–15 km in depth and the magnitude of slip is 0.5 m, which would be approximately equivalent to a magnitude 6 earthquake. The coseismic static stress changes are illustrated for a depth of 14.5 km, but the same distribution would be obtained at any depth along the fault plane between 5 and 15 km. Positive Coulomb failure stress changes have been shown to have a good correlation with the distribution of aftershocks and the triggering of subsequent earthquakes (see, e.g., King et al., 1994). The stress transfer model results indicate that, after a right-lateral strike-slip event, strike-slip related aftershocks may be triggered both along-strike and in lateral distributions away from the tips of the mainshock rupture (Figure 10.18A). In contrast, any thrust-related aftershocks will be spatially distributed in tightly constrained clusters (Figure 10.18B).

The results of this simulation replicate the distributions of aftershock reported from many strike-slip earthquakes, such as the 1992 Mw 7.3 Landers, California event (Liu et al., 2003). Most of the aftershocks, triggered by these earthquakes are likely to have nucleated on pre-existing small-displacement faults surrounding the mainshock. A striking feature of the 1992 Mw 7.3 Landers example is that thrust aftershocks occurred within the aftershock sequence, despite being triggered by a strike-slip mainshock. Indeed, the full gamut of focal mechanisms (strike-slip, thrust and normal) has been observed in aftershock sequences from recent earthquakes (Beroza and Zoback, 1993), which presents a challenge to conventional structural geology and the construction of deformation histories.

Cox & Ruming (2004) and Micklethwaite & Cox (2004) have identified a relationship between the distribution of aftershocks and fault-related mineral deposits by applying the stress transfer modelling technique to fossil fault systems. Aftershocks can occur over months to decades after a large earthquake, and can also display a large range of orientations and kinematics. Where fluids access these aftershock networks, aftershock clusters associated with strike-slip ruptures will produce near vertical pipe-like flow. Permeability is potentially enhanced in aftershock domains for months to years after a mainshock rupture, long after the mainshock fault plane would have lost permeability due to hydrothermal sealing. For a mineral deposit to form, such a process would have to have operated multiple times, with repeated mainshock arrest at certain geometrical barriers along a fault system.

10.2.4 Development of thrust complexes

Some of the most important examples of thrust faults occur in the complex belts in continents along convergent plate boundaries. In many environments there is evidence for displacement of material over long distances as in the nappe structures of the Alps. The deformed, and frequently folded, structure is detached from its base along a *decollement* surface with a narrow zone of intense shear.

Individual thrust faults within the compressional complexes tend to cut upwards through the stratigraphic layering in a sequence of ramps that impose fault-bend folding as displacement continues to occur along the decollement surface. The

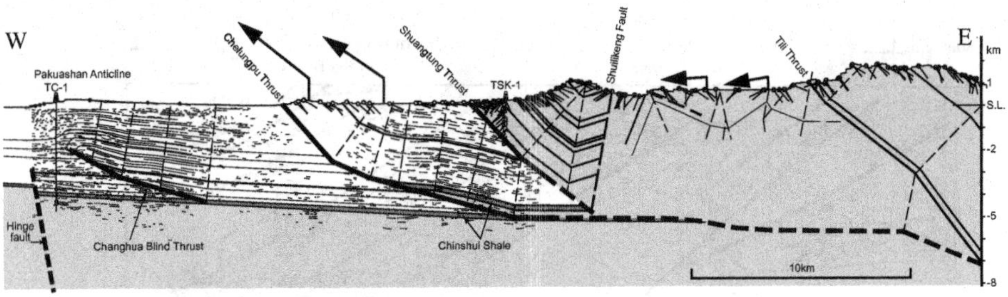

Figure 10.19. Configuration of the thrust fault associated with the 1999 Chi-Chi earthquake in Taiwan. The black arrows indicate projected GPS vectors that are consistent with lower dip thrusting at depth. [Courtesy of John Suppe.]

ramps may cut through earlier parts of the thrust sheet to give a complicated sequence of fault splays.

A good example of such thrust ramping associated with recent faulting is provided by the 1999 Chi-Chi earthquake in Taiwan (Yue, Suppe & Hong, 2005) as illustrated in Figure 10.19. The earthquake produced several metres of surface displacement along an 85 km surface break. Seismic reflection control is available on the structures away from the mountain belt, and the Chelungpu Thrust on which the main ground displacement occurred shows a distinct set of ramps. The earthquake in fact initiated on a deeper fault below the Chelungpu thrust, at about 10 km depth approximately vertically below the TSK-1 borehole, but the main motion in the event was taken up on the thrust with a very gentle dip at depth. A number of shallow seismic lines and drill holes confirm that the near-surface thrust is parallel to bedding in the upper 500–2000 m, and there therefore have to be distinct fault-bend folds to accommodate the motion along the ramped thrust. These fault-bend folds are commonly associated with localised kink bands with a rapid change in direction reflecting the changes in dip on the underlying thrust. Shallow fault-bend folds produced some distinct surface scarps of a few metres height displaced about 3–4 km to the east of the main fault trace.

In a few cases a major thrust can be followed from the surface to depth with the aid of deep seismic reflection profiling. A good example is the Red Bank zone in central Australia (Figure 10.20), a mylonite zone that separates two Proterozoic terrains. The thrust shows clearly in the surface exposure through the juxtaposition of rocks with very different colour and texture. This distinctive surface feature can be correlated with a major structure penetrating through the whole crust revealed by deep seismic reflections (Goleby et al., 1989). The quality of reflection data in this area is very high and relatively high-frequency energy is returned from depths below 40 km (more than 14 s two-way time). The reflection section (Figure 10.20) shows a distinct change in deep reflection character across the thrust structure associated with the Red Bank zone (RDZ). A splay thrust links to the Ormiston Nappe and Thrust Zone (ONTZ) to the south. To the north the

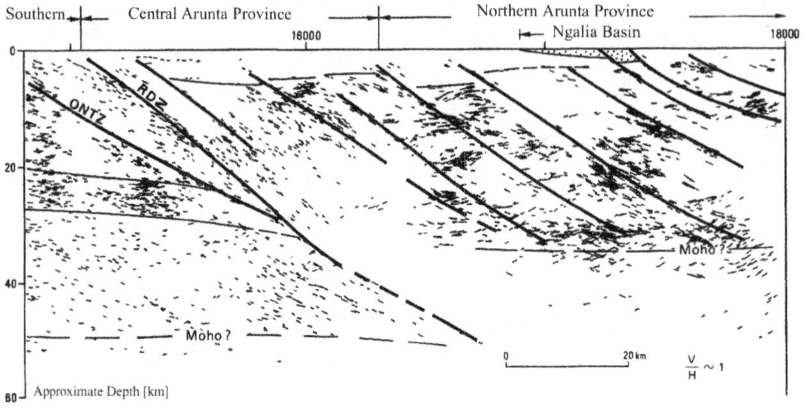

Figure 10.20. Deep seismic reflection profiling across the region around the Red Bank thrust complex in Central Australia showing the major displacement of the crust-mantle boundary associated with the thrust and a sequence of related features to the north [courtesy of B. Goleby].

pattern of reflections can be interpreted in terms of a stack of secondary thrust slices, indicating thick-skinned tectonics. The crust–mantle boundary appears to be displaced upwards on the north side of the thrust zone, bringing lower crustal material to the surface. The Red Bank zone is accompanied by a large gravity signature that can be well matched by the upward displacement of the crust–mantle boundary. The last deformation in this region is associated with the intraplate Alice Springs orogeny (~300 Ma) and the structure has remained out of isostatic equilibrium since that time.

11

Seismology and Earth Structure

The theory of linearised elasticity introduced in Chapter 5 provides the basis for understanding the behaviour of seismic waves except in the immediate vicinity of an earthquake fault or explosion. Once the incremental strains associated with wave disturbances drop below 10^{-5}, the linearised treatment provides an effective description of the situation. The dominant variation in the properties of the Earth is with radius and the treatment of wave propagation in a spherically stratified Earth gives a useful reference. The complexities of the pre-stressed, three-dimensional Earth can then generally be addressed by a perturbation treatment about the reference state. Although most minerals are anisotropic, the incremental properties of the mineral aggregates in the Earth are close to isotropic. The complexities of anisotropic propagation are therefore needed only for limited regions, mostly in boundary or transition zones.

11.1 Seismic waves

In Chapter 5 we demonstrated the properties of plane elastic waves in a uniform medium, and illustrated the radiation from simple sources. For application to the Earth we need to take account of the variations in elastic parameters with depth, the presence of material interfaces, and energy loss through intrinsic anelasticity and scattering.

The waves exploited in seismology span a considerable range of frequencies. The longest period free oscillations of the Earth have a period close to an hour, whereas frequencies above 100 Hz can be employed in seismic exploration for shallow structure. The underlying principles are the same, but the styles of analysis vary significantly between reflection seismic techniques employed for exploration and studies of the deep Earth.

In exploration work the source is under control and closely spaced seismic recorders can be deployed to achieve multiple spatial coverage of the relevant part of the wave field. The weak signals returned from depth can then be tracked by making use of the redundancy in the recording through stacking to suppress incoherent noise but enhance coherent arrivals.

For most deep Earth studies we are dependent on where earthquakes occur and have to make use of the global network of high quality seismic stations. The coverage and signal quality is best for continental regions. Major efforts are being made to install relatively dense networks of recording stations in some localities, e.g., the USarray, so that signal coherence can be exploited in appropriate frequency bands.

11.1.1 Reflection and refraction

In a spherical Earth model the influence of the wavespeed variation with depth is supplemented by the spherical geometry. When the wavespeed depends only on radius r the path of a ray through the structure is determined by a continuous version of Snell's law. The ray parameter $\wp$ is constant along the path,

$$\wp = r \sin i / v(r), \tag{11.1.1}$$

where i is the inclination to the radial vector and $v(r)$ is the current wavespeed (either *P* or *S*).

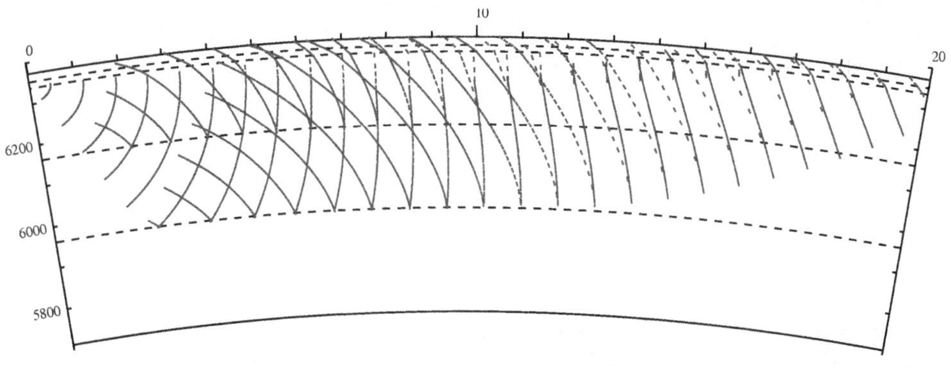

Figure 11.1. Wavefront pattern for *P* waves in the upper mantle of model AK135 with increasing velocity with depth and velocity discontinuities. The wavefronts of refracted and reflected waves are indicated at 10 s intervals.

When $v(r)/r$ grows with depth, the angle i must also increase with depth for a ray descending into the Earth, and when $\wp = r/v(r)$ the ray will travel perpendicular to the radius vector, $i = \pi/2$. Subsequently i continues to grow, but the ray climbs back to the surface. The effect can be seen in the pattern of wavefronts at 10 s intervals shown for upper mantle propagation in Figure 11.1; the wavefronts turn and sweep back to the surface with a shape determined by the variation of *P* wavespeed with depth.

At an interface, there is partition of energy into reflected and transmitted waves, with the possibility of conversion between *P* waves and vertically polarised *SV* waves. All the reflected and transmitted waves at the interface with radius r_i share the same ray parameter $\wp$ as the incident wave; so that the inclinations to the radial vector of waves with wavespeeds v_1, v_2 are determined by Snell's law,

$$\wp = r_i \sin i_1/v_1 = r_i \sin i_2/v_2. \tag{11.1.2}$$

Horizontally polarised *SH* waves do not undergo conversion at an interface that is a surface of constant radius. The amplitudes of the reflected and transmitted waves are determined by the conditions of continuity of displacement and traction for plane wave components with the slowness $\wp$; this leads to a system of four coupled equations for *P-SV* waves and two coupled equations for *SH* waves. The reflection and transmission coefficients depend on the contrasts in elastic wavespeeds and density as well as the angle of incidence (represented through the slowness $\wp$).

The wavefronts of reflected *P* waves from the upper mantle discontinuities are displayed in Figure 11.1 together with the refracted phases. At moderate inclinations the reflected wavefronts link to the incident refracted waves at the interface. For wide angle reflections the sharp junction of the reflected and refracted wavefronts separates from the interface and moves progressively towards the surface.

A detailed presentation of the properties of refracted and reflected waves, and the treatment of the interface conditions can be found in, e.g., Kennett (2001).

11.1.2 Attenuation effects

The passage of a seismic wave through the Earth is accompanied by energy loss, due to conversion of energy to heat (intrinsic anelasticity) or from the cumulative effect of interaction with small-scale heterogeneity (scattering attenuation). We can use the theory of linear viscoelasticity introduced in Chapter 5 to include the effects of intrinsic anelasticity through the use of complex elastic moduli in the frequency domain.

One useful model for intrinsic attenuation is the standard linear solid, as in (5.7.14)–(5.7.16), but now with a spectrum of relaxation times to match the observations of little frequency dependence of Q over the seismic frequency band. The form suggested by Anderson & Minster (1979) for the distribution of relaxation times τ is

$$D(\tau) = \frac{\alpha\tau^{\alpha-1}}{\tau_H^{\alpha} - \tau_L^{\alpha}}, \quad \tau_L < \tau < \tau_H \tag{11.1.3}$$

with $D(\tau) = 0$ outside the band between the lower and upper cut-off times. The exponent α characterises the weak frequency dependence.

An alternative form based on a Burgers model has been used by Faul & Jackson (2005) to interpret the results of laboratory experiments on the wavespeed and attenuation of synthetic olivine aggregates, without melt, as a function of both temperature and grain size (Figure 11.2). The influence of temperature becomes appreciable well before the solidus is reached with a very significant increase in dissipation Q^{-1} as temperature increases at fixed grain size. The experiments are carried out in torsion over the seismic frequency band from 0.001 - 1.0 Hz and the use of synthetic aggregates provides control on grain size.

The Burgers material includes elastic and viscoelastic effects with the further possibility of long term viscosity (5.6.5). For a crystalline material, the viscous relaxation time will depend on the temperature and grain size sensitivity of the viscosity and can be approximated as

$$\tau_M = A d^m \exp[E_M/RT], \tag{11.1.4}$$

where A is a constant, R is the gas constant and T is the absolute temperature. The behaviour is controlled by an activation energy E_M and the dependence on the grain size d carries exponent m. The anelastic relaxation times are expected to have a similar dependence on temperature and size.

The various relaxation times can then be written in a way that emphasises deviations from a reference state with temperature T_R and grain size d_R

$$\tau_i = \tau_{iR}\left(\frac{d}{d_R}\right)^m \exp\left[\frac{E_i}{R}\left(\frac{1}{T}-\frac{1}{T_R}\right)\right]\exp\left[\frac{pV_i^*}{R}\left(\frac{1}{T}-\frac{1}{T_R}\right)\right], \tag{11.1.5}$$

where τ_{iR} is the relaxation time in the reference state, p is the pressure, and V_i^* is an activation volume.

With a distribution of anelastic relaxation times, the full representation of the creep function for the Burgers solid, including pressure dependence, then takes the form

$$\psi(t, d, T, p) = \psi_U(p) + \delta\psi_U(T) + \Delta\psi\int_{\tau_L}^{\tau_H} d\tau\, D(\tau)[1-e^{-t/\tau}] + \frac{\psi_U(p)t}{\tau_M}, \tag{11.1.6}$$

where ψ_U is the reciprocal of the relevant instantaneous modulus and $\delta\psi_U(T)$ is an adjustment for the effect of temperature. $\Delta\psi$ describes the extent of viscoelastic relaxation with a relaxation time distribution and the Maxwell relaxation time for the influence of viscosity is $\tau_M = \eta\psi_U$, where η is the Newtonian viscosity.

The single expression (11.1.6) is able to provide a satisfactory representation of the variation of both shear modulus and dissipation for the synthetic olivines as a function of period (1–1020 s), temperature (1000–1300°C) and grain size (3-165 μm). The experiments do not cover the full range of the absorption band; the temperature correction $\delta\psi_U(T)$ compensates for the cumulative effect of the relaxation times shorter than 1 s, and can be approximated as

$$\delta\psi_U(T) = \psi_U(p)\left(\frac{\partial \ln \psi_U}{\partial T}\right)_R \left(\frac{d}{d_R}\right)^{-m_J}(T - T_R). \tag{11.1.7}$$

For a harmonic oscillation with frequency ω, the complex compliance (the reciprocal of the shear modulus)

$$\psi^*(\omega, d, t, p) = \psi_1(\omega, d, t, p) + i\psi_2(\omega, d, t, p), \tag{11.1.8}$$

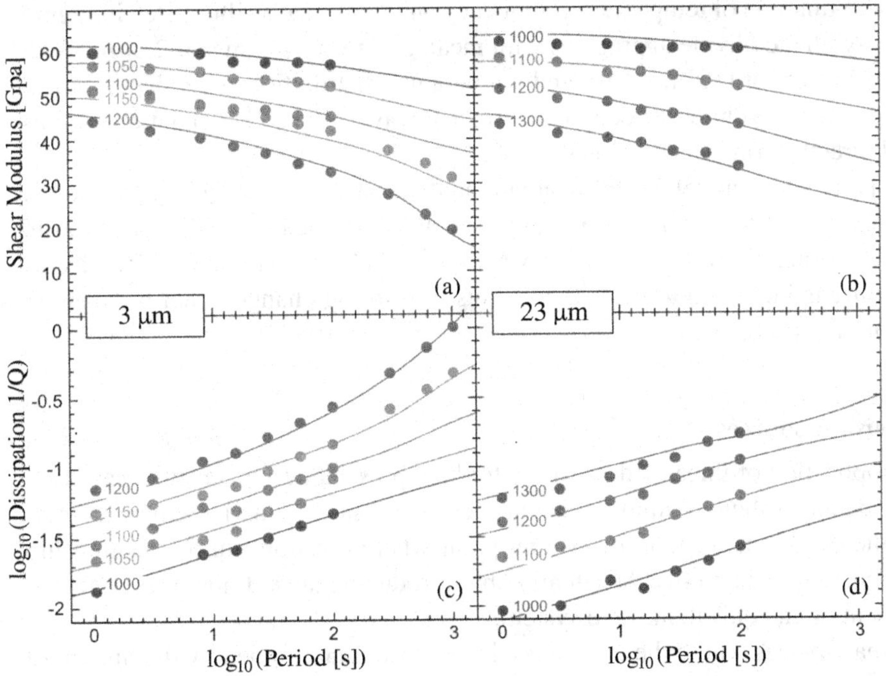

Figure 11.2. Shear modulus and strain energy dissipation as a function of temperature (in °C) and oscillation period for two samples of synthetic olivine aggregate with grain sizes of 3 µm and 23 µm. The dissipation shows a mild period dependence with constant exponent α, except for the highest temperatures and longest periods for the fine-grained sample when the transition to viscous behaviour is evident. The dots represent the experimental data and the lines the fit to the modulus from (11.1.10) and the attenuation from (11.1.11). [Courtesy of U. Faul & I. Jackson.]

with real and imaginary parts

$$\psi_1(\omega, d, t, p) = \psi_U(p)\left\{1 + \left(\frac{\partial \ln \psi_U}{\partial T}\right)_R \left(\frac{d}{d_R}\right)^{-m_J}(T - T_R) + \frac{\alpha \Delta\psi}{\tau_H^\alpha - \tau_L^\alpha}\int_{\tau_L}^{\tau_H} d\tau \frac{\tau^{\alpha-1}}{1+\omega^2\tau^2}\right\},$$

$$\psi_2(\omega, d, t, p) = \psi_U(p)\left\{\frac{\omega\alpha\Delta\psi}{\tau_H^\alpha - \tau_L^\alpha}\int_{\tau_L}^{\tau_H} d\tau \frac{\tau^{\alpha-1}}{1+\omega^2\tau^2} + \frac{1}{\omega\tau_M}\right\}, \tag{11.1.9}$$

The frequency-dependent shear modulus is then to be found from

$$m(\omega) = [\psi_1^2(\omega) + \psi_2^2(\omega)]^{-1/2}, \tag{11.1.10}$$

and the energy dissipation

$$Q^{-1}(\omega) = \psi_2(\omega)/\psi_1(\omega). \tag{11.1.11}$$

The extended Burgers model provides a good representation of the experimental results as a function of temperature, frequency and grain size as illustrated in Figure 11.2 for synthetic olivine aggregates with mean grain sizes 3 μm and 23 μm. With suitable choices for the grain size and temperature distributions with depth, it also proves possible to achieve a good match to shear wavespeed and attenuation profiles derived directly from seismological models.

This parametric model for intrinsic attenuation based on detailed experimental observations captures many significant aspects of attenuation within the Earth. The temperature derivatives of both wavespeed and attenuation vary rapidly as temperature is increased towards the solidus, but strong change is not by itself an indicator of partial melt.

11.2 Seismic sources

In the application of linearised elasticity to the representation of seismic waves we have made an explicit assumption about the relationship of the incremental stress σ and the displacement field $\mathbf{u}$. At any point where this constitutive relationship is inadequate we can rectify the situation by introducing an additional stress tensor component or an equivalent force system.

The main places where the true stress differs from that implied by the linearised relations will be in the immediate vicinity of the generation of seismic waves where strains exceed 10^{-5}. For an explosive source this zone will approximate a sphere around the point of initiation. Whereas for an earthquake, the difference $\boldsymbol{\Gamma}$ between the actual and linearised estimate of the stress tensor will be concentrated in a narrow zone surrounding the portion of a fault that has slipped, including any gouge zone.

The stress difference Γ_{ij}, which is often referred to as the 'stress glut', can be viewed as a local density of source excitation with an equivalent body force system $\mathbf{e} = -\boldsymbol{\nabla} \cdot \boldsymbol{\Gamma}$. This stress glut is an example of a moment tensor density m_{ij}. For an internal source there can be no net change in linear and angular momentum of the whole Earth, so that m_{ij} is symmetric, $m_{ij} = m_{ji}$.

We can examine the radiation from a dislocation source, representing the action on an earthquake, by exploiting the representation for displacement introduced in Section 5.3.2. We consider a region V containing some form of dynamic discontinuity across a fault surface Σ (Figure 11.3).

We apply the integral representation (5.3.18) for the displacement field $\mathbf{u}$ to a surface S consisting of the two surfaces Σ^+, Σ^- lying on either side of the fault surface Σ and joined at the termination of the dislocation (Figure 11.3). We work in the frequency domain so that the convolution integrals in time in (5.3.18) reduce to products of Fourier components, and we can include anelastic effects. When the Green's tensor corresponds to the configuration in V before the action of the discontinuity, $\mathbf{G}_k(\mathbf{x}, \xi, \omega)$ and its associated traction $\mathbf{H}_k(\mathbf{x}, \xi, \omega)$ will be continuous across Σ. The two surface integrals over Σ^+, Σ^- can therefore be

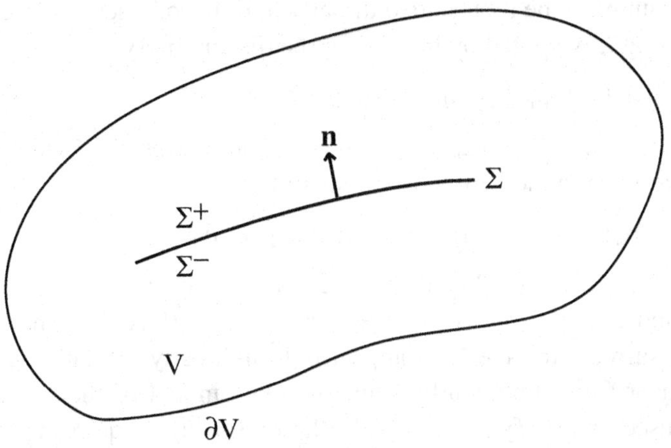

Figure 11.3. Surface of discontinuity in displacement and tractions Σ representing a source of seismic energy.

combined into a single integral

$$u_k(\mathbf{x}, \omega) = -\int_\Sigma d^2\xi \{G_{kq}(\mathbf{x}, \xi, \omega)[t_q(\xi, \omega)]_-^+ - [u_q(\xi, \omega)]_-^+ H_{kq}(\mathbf{x}, \xi, \omega)\}, \tag{11.2.1}$$

where $[t_q(\xi, \omega)]_-^+$, $[u_q(\xi, \omega)]_-^+$ represent the jumps in traction and displacement in passing from Σ^- to Σ^+. The normal $\mathbf{n}$ to Σ is taken to be directed from Σ^- to Σ^+. The equivalent source distribution will lie along the fault surface Σ. A general dislocation with both displacement and traction jumps is allowed to provide a flexible parameterisation of possible sources associated with a fault surface. The traction associated with the Green's tensor is given by

$$H_{kq}(\mathbf{x}, \xi, \omega) = n_p(\xi) c_{pqrs}(\xi, \omega) \partial_r G_{ks}(\mathbf{x}, \xi, \omega). \tag{11.2.2}$$

If the surface Σ intersects the surface or internal surfaces of material discontinuity additional surface tractions are introduced.

In (11.2.1) we are only concerned with the values of G_{kq} and h_{kq} on the fault surface Σ, so we introduce the Dirac delta function $\delta_\Sigma(\xi, \eta)$ and its derivative localised on the surface Σ. The expression for the radiated seismic displacement (11.2.1) can then be cast as a surface integral over Σ of the form

$$\begin{aligned} u_k(\mathbf{x}, \omega) = -\int_V d^3\eta\, G_{kq}(\mathbf{x}, \eta, \omega) \{&[t_q(\xi, \omega)]_-^+ \delta_\Sigma(\xi, \eta) \\ &+ [u_s(\xi, \omega)]_-^+ n_r(\xi) c_{pqrs}(\xi, \omega) \partial_p \delta_\Sigma(\xi, \eta)\}, \end{aligned} \tag{11.2.3}$$

where $-\partial_p\delta$ extracts the derivative of the function it acts upon and we have exploited the symmetry of the moduli $c_{rspq} = c_{pqrs}$.

The traction jump can be represented directly in terms of a set of force elements ϵ distributed along Σ weighted by the size of the discontinuity,

$$\epsilon_q(\boldsymbol{\eta}, \omega) = -n_r(\boldsymbol{\xi})[\tau_{qr}(\boldsymbol{\xi}, \omega)]_-^+ \delta_\Sigma(\boldsymbol{\xi}, \boldsymbol{\eta}). \tag{11.2.4}$$

The displacement jump $[u]$ leads to force doublets along Σ which are best represented in terms of a moment tensor density m_{pq},

$$m_{pq}(\boldsymbol{\eta}, \omega) = n_r(\boldsymbol{\xi}) c_{pqrs}(\boldsymbol{\xi}, \omega)[u_s(\boldsymbol{\xi}, \omega)]_-^+ \delta_\Sigma(\boldsymbol{\xi}, \boldsymbol{\eta}). \tag{11.2.5}$$

The seismic radiation predicted by (11.2.1) is determined solely by the displacement and traction jumps imposed across the surface Σ. The properties of the material surrounding the fault appear only indirectly through the nature of the Green's tensor G_{kq}. Frequently, some assumed model of the slip behaviour on the fault is used to specify $[\mathbf{u}]_-^+$. A full solution for $[\mathbf{u}]_-^+$ requires a much more complex calculation in which the propagating fault interacts with its surroundings; such calculations are just becoming feasible for simple situations.

Earthquake models normally prescribe only tangential displacement jumps and then $n_r[\mathbf{u}]_-^+ = 0$. A normal displacement will occur for an opening crack that occurs in, e.g., rock bursts in mines.

In general we can represent the effect of a distributed dislocation source (as in a major earthquake) through an integral over the source volume

$$u_k(\mathbf{x}, \omega) = \int_V d^3\eta\, \partial_p G_{kq}(\mathbf{x}, \boldsymbol{\eta}, \omega) m_{pq}(\boldsymbol{\eta}, \omega). \tag{11.2.6}$$

Frequently, we are interested in the best equivalent *point* source for which

$$u_k(\mathbf{x}, \omega) \approx \partial_p G_{kq}(\mathbf{x}, \mathbf{x}_s, \omega) M_{pq}(\omega), \tag{11.2.7}$$

where the point moment tensor

$$M_{pq}(\omega) = \int_V d^3\eta\, m_{pq}(\boldsymbol{\eta}, \omega). \tag{11.2.8}$$

The moment tensor M_{pq} is the most commonly used description of a seismic source, and as we see from (11.2.7) represents the integral of the moment tensor density across the source region. The centroid of the disturbance may therefore be displaced from the point of initiation of the event. We note that the moment tensor acts on $\partial_p G_{kq}$ or, equivalently, the strain associated with the Green's tensor $\mathbf{G}$.

The moment tensor $\mathbf{M}$, represents the weights to be applied to the set of 9 force doublets produced by displacements along the coordinate axes, as illustrated in Figure 11.4. The diagonal elements of $\mathbf{M}$ correspond to dipoles and the off-diagonal elements to pure couples. For an internal source $\mathbf{M}$ is required to be symmetric, e.g., $M_{12} = M_{21}$ and so each couple is accompanied by another that neutralises the induced moment, giving an effective double-couple source.

If we represent an earthquake by displacement $[u]$ in the slip direction $\boldsymbol{\nu}$, over an area A, on a planar fault surface with normal $\mathbf{n}$, the point moment tensor is

$$M_{ij}(\omega) = A[\bar{u}(\omega)]\{(\kappa - \tfrac{2}{3}\mu) n_k \nu_k + \mu(n_i \nu_j + n_j \nu_i)\}, \tag{11.2.9}$$

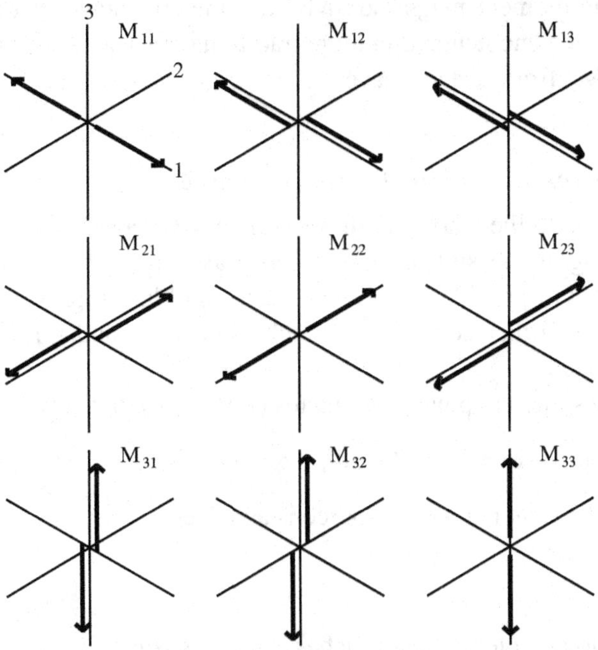

Figure 11.4. The representation of the elements of the moment tensor as weights for a set of dipoles and couples.

in terms of the bulk modulus κ and shear modulus μ. This is equivalent to a double-couple mechanism with axes along the orthogonal directions $\mathbf{n}$, $\boldsymbol{\nu}$. The symmetry of $\mathbf{n}$, $\boldsymbol{\nu}$ in (11.2.9) means that the radiation patterns carry no distinction between the fault plane and the orthogonal auxiliary plane. For *P* wave radiation both will be nodal planes; hence additional evidence is needed to resolve the ambiguity.

The trace of the moment tensor for this displacement source vanishes, $\mathrm{tr}\, M = \sum_i M_{ii} = 0$, since there is no volumetric component. Estimates of earthquake mechanisms from waveform fitting often suggest the presence of a minor isotropic component. Such effects can arise from a non-planar rupture surface, directivity in rupture inadequately captured in a point estimate, and the influence of three-dimensional structure in the Earth.

The scalar moment spectrum

$$M_o = M_o(\omega) = A[\bar{u}(\omega)], \tag{11.2.10}$$

characterises the behaviour of the event. The extrapolation to zero frequency $M_o(0)$ provides a convenient measure of the size of the event, and forms the basis of the earthquake magnitude scale M_w introduced by Kanamori:

$$M_w = (\log M_o/1.5) - 10.7, \tag{11.2.11}$$

when M_o is measured in dyn cm [10^{-7} N m]; the constant is introduced to tie to earlier scales. The moment magnitude M_w has the advantage that it avoids the problems of saturation encountered in more traditional measures such as m_b, using body waves, and M_s from surface waves.

11.3 Building the response of the Earth to a source

As an approximation to the whole Earth we will consider a spherical, non-rotating model incorporating the possible effects of self-gravitation associated with slower deformation. This will provide the framework for building representations of the behaviour to which further complexities such as rotation and three-dimensional structure can be added.

We will work in spherical polar coordinates (r, θ, ϕ) so that

$$x_1 = r\sin\theta\cos\phi, \quad x_2 = r\sin\theta\sin\phi, \quad x_3 = r\cos\theta. \tag{11.3.1}$$

In the undisturbed state before the source is activated

$$\frac{\partial}{\partial x_i}\sigma^0_{ij} = \rho_0\frac{\partial}{\partial x_i}\varphi_0, \tag{11.3.2}$$

where the gravitational potential φ_0 satisfies Poisson's equation

$$\nabla^2\varphi_0(r) = 4\pi G\varphi_0(r). \tag{11.3.3}$$

We will assume that the initial state is in hydrostatic equilibrium and then as in Section 6.1.1 we have

$$-\frac{\partial p_0}{\partial r} = \rho_0 g_0, \quad \frac{\partial g_0}{\partial r} + \frac{2}{r}g_0 = 4\pi G\varphi_0(r), \quad \text{where } g_0(r) = \frac{\partial\varphi_0}{\partial r}, \tag{11.3.4}$$

and the acceleration due to gravity is $\mathbf{g}_0 = -g_0\hat{\mathbf{e}}_r$, where $\hat{\mathbf{e}}_r$ is the unit vector in the radial direction. The equations (11.3.4) are subject to the boundary conditions that φ_0, g_0, p_0 are continuous across all surfaces of discontinuity in physical properties

$$[\varphi_0]_-^+ = 0, \quad [g_0]_-^+ = 0, \quad [p_0]_-^+ = 0, \tag{11.3.5}$$

and p_0 must vanish at the free surface. The gravitational acceleration and potential must diminish to zero at infinity and so in terms of the density ρ_0:

$$g_0(r) = \frac{4\pi G}{r^2}\int_0^r ds\, s^2\rho_0(s),$$

$$\varphi_0(r) = -\int_r^\infty ds\, g_0(s) = -\frac{GM}{r_e} - \int_r^{r_e} ds\, g_0 s, \tag{11.3.6}$$

$$p_0(r) = \int_r^{r_e} ds\, \rho_0(r)g_0(r), \tag{11.3.7}$$

where r_e is the radius of the Earth and M is the total mass

$$M = 4\pi\int_0^{r_e} ds\, s^2\rho_0(s). \tag{11.3.8}$$

With a time-dependent displacement $\mathbf{u}(\mathbf{x}, t)$ induced by a force system $\mathbf{f}$, the linearised equation of motion takes the form

$$\rho_0 \frac{\partial^2 \mathbf{u}}{\partial t^2} = \mathbf{f} + \boldsymbol{\nabla} \cdot \boldsymbol{\sigma} + \boldsymbol{\nabla}(\mathbf{u} \cdot \boldsymbol{\nabla} p_0) - \rho_0 \boldsymbol{\nabla} \varphi_1 - \rho_1 \boldsymbol{\nabla} \varphi_0; \tag{11.3.9}$$

the second term on the right hand side of (11.3.9) is the familiar gradient of the incremental stress $\boldsymbol{\sigma}$ associated with $\mathbf{u}$, and the remaining contributions come from the perturbations in the density distribution induced by $\mathbf{u}$. From the conservation of mass

$$\rho_1 = -\boldsymbol{\nabla}(\rho_0 \mathbf{u}), \tag{11.3.10}$$

so the incremental gravitational potential satisfies the Poisson equation

$$\nabla^2 \varphi_1 = 4\pi G \rho_1 = -4\pi G\, \boldsymbol{\nabla}(\rho_0 \mathbf{u}). \tag{11.3.11}$$

With the aid of (11.3.3) and (11.3.11) we can write the momentum equation (11.3.9) in component form

$$\rho_0 \frac{\partial^2 u_i}{\partial t^2} = f_i + \frac{\partial \sigma_{ij}}{\partial x_j} + \rho_0 \left[\frac{\partial u_k}{\partial x_k} \frac{\partial \varphi_0}{\partial x_i} - \frac{\partial \varphi_1}{\partial x_i} - \frac{\partial}{\partial x_i} \left(u_k \frac{\partial \varphi_0}{\partial x_k} \right) \right] + \left[\frac{\partial \rho_0}{\partial x_k} \frac{\partial \varphi_0}{\partial x_i} - \frac{\partial \rho_0}{\partial x_i} \frac{\partial \varphi_0}{\partial x_k} \right] u_k. \tag{11.3.12}$$

In the hydrostatic state both ρ_0 and φ_0 depend only on r and so the final term in (11.3.12) vanishes.

The system of partial differential equations for the displacement $\mathbf{u}$ and incremental gravitational potential φ_1 (which depends on $\mathbf{u}$) thus takes the form

$$\rho_0 \frac{\partial^2 u_i}{\partial t^2} = f_i + \frac{\partial}{\partial x_j} \left\{ c_{ijkl} \frac{\partial u_k}{\partial x_k} \right\} + \rho_0 \left[\frac{\partial u_k}{\partial x_k} \frac{\partial \varphi_0}{\partial x_i} - \frac{\partial \varphi_1}{\partial x_i} - \frac{\partial}{\partial x_i} \left(u_k \frac{\partial \varphi_0}{\partial x_k} \right) \right],$$

$$\nabla^2 \varphi_1 = 4\pi G \frac{\partial}{\partial x_k} (\rho_0 u_k), \tag{11.3.13}$$

where we have used the generalised Hooke's law (5.2.7) to express the incremental stress in terms of the displacement gradient.

The boundary conditions on the incremental gravitational potential φ_1 are that it vanishes at infinity and the continuity conditions

$$\left[\varphi_1 \right]_-^+ = 0, \quad \left[\frac{\partial \varphi_1}{\partial r} + 4\pi G \rho_0 u_r \right]_-^+ = 0 \tag{11.3.14}$$

apply at all surfaces of material discontinuity, including the free surface. Here $u_r = \mathbf{u} \cdot \hat{\mathbf{e}}_r$ is the radial component of displacement.

The boundary conditions on the displacement depend on the nature of the discontinuity surface: at the free surface $r = r_e$ the traction must vanish

$$\sigma_{ij} n_j = 0, \quad \text{i.e., } c_{ijkl} \frac{\partial u_k}{\partial x_l} n_j = 0; \tag{11.3.15}$$

at fluid–solid boundaries, we require continuous traction

$$[\sigma_{ij}n_j]_-^+ = 0, \tag{11.3.16}$$

the traction must be normal to the boundary

$$(\delta_{ik} - n_i n_k)\sigma_{kj}n_j = 0; \tag{11.3.17}$$

and the normal displacement must be continuous

$$[u_k n_k]_-^+ = 0; \tag{11.3.18}$$

whereas at solid–solid (welded) boundaries both traction and displacement must be continuous

$$[\sigma_{ij}n_j]_-^+ = 0, \quad [u_i]_-^+ = 0. \tag{11.3.19}$$

The displacement must also be finite at the centre of the Earth ($r = 0$).

In the frequency domain, (11.3.13) becomes

$$\rho_0\omega^2 u_i + \frac{\partial}{\partial x_j}\left\{c_{ijkl}\frac{\partial u_k}{\partial x_k}\right\} + \rho_0\left[\frac{\partial u_k}{\partial x_k}\frac{\partial \varphi_0}{\partial x_i} - \frac{\partial \varphi_1}{\partial x_i} - \frac{\partial}{\partial x_i}\left(u_k\frac{\partial \varphi_0}{\partial x_k}\right)\right] = -f_i, \tag{11.3.20}$$

There is a set of frequencies ω_I for which a displacement solution $\mathbf{u}_I^e$ exists that satisfies (11.3.20) and the full set of boundary conditions in the absence of external forcing (i.e., $\mathbf{f} = 0$). These normal modes with index I form a basis from which we can build the response to excitation.

11.3.1 Displacements as a normal mode sum

The eigendisplacements corresponding to different modes of oscillation are orthogonal and can be normalised so that

$$\int_{V_e} d^3\mathbf{x}\, \rho_0(r)\mathbf{u}_I^e(\mathbf{x}) \cdot [\mathbf{u}_J^e(\mathbf{x})]^* = \delta_{IJ}, \tag{11.3.21}$$

where the integral includes the entire volume of the Earth and the star denotes a complex conjugate. We expand the displacement field in the presence of a source as a sum of the normal modes

$$\mathbf{u}(\mathbf{x}, \omega) = \sum_{I=0}^{\infty} c_I \mathbf{u}_I^e(\mathbf{x}), \tag{11.3.22}$$

and from the orthonormality of the eigenfunctions

$$c_I = \int_{V_e} d^3\mathbf{x}\, \rho_0(r)\mathbf{u}(\mathbf{x}) \cdot [\mathbf{u}_I^e(\mathbf{x})]^*. \tag{11.3.23}$$

We can write the equation of motion (11.3.20) in the operator form

$$\mathfrak{H}(\mathbf{u}) + \rho_0\omega^2\mathbf{u} = -\mathbf{f}, \tag{11.3.24}$$

and then

$$\sum_J c_J \mathfrak{H}(\mathbf{u}_J^e) + \rho_0 \omega^2 \sum_J c_J \mathbf{u}_J^e = -\mathbf{f}, \tag{11.3.25}$$

and thus, taking the scalar product of (11.3.25) with $[\mathbf{u}_K^e]^*$,

$$\int_{V_e} d^3\mathbf{x} \sum_J c_J \mathfrak{H}(\mathbf{u}_J^e) \cdot [\mathbf{u}_K^e]^* + \omega^2 c_K = -\int_{V_e} d^3\mathbf{x}\, \mathbf{f} \cdot [\mathbf{u}_K^e]^*. \tag{11.3.26}$$

Each eigenvector $\mathbf{u}_I^e$ satisfies the equation

$$\mathfrak{H}(\mathbf{u}_I^e) + \rho_0 \omega^2 \mathbf{u}_I^e = 0, \tag{11.3.27}$$

and hence

$$\int_{V_e} d^3\mathbf{x}\, \mathfrak{H}(\mathbf{u}_J^e) \cdot [\mathbf{u}_K^e]^* + \omega^2 \delta_{JK} = 0. \tag{11.3.28}$$

With this substitution in (11.3.26) the coefficients in the modal expansion can be found as

$$c_K = \frac{1}{\omega_K^2 - \omega^2} \int_{V_e} d^3\mathbf{x}\, \mathbf{f} \cdot [\mathbf{u}_K^e]^*. \tag{11.3.29}$$

We have therefore a frequency-domain representation of the displacement induced by the force system **f**:

$$\mathbf{u}(\mathbf{x}, \omega) = \sum_K \frac{\mathbf{u}_K^e}{\omega_K^2 - \omega^2} \int_{V_e} d^3\mathbf{x}\, \mathbf{f} \cdot [\mathbf{u}_K^e]^*. \tag{11.3.30}$$

For a step function time history $H(t)$ for **f** we can perform the inverse Fourier transform and recover the time domain response, extracting the residue at each pole ω_K using Cauchy's theorem, to produce

$$\mathbf{u}(\mathbf{x}, t) = \sum_K \mathbf{u}_K^e \frac{1 - \cos(\omega_K t)}{\omega_K^2} \int_{V_e} d^3\mathbf{x}\, \mathbf{f} \cdot [\mathbf{u}_K^e]^*. \tag{11.3.31}$$

When we allow for the decay of the eigenmodes produced by slight anelasticity

$$\mathbf{u}(\mathbf{x}, t) = \sum_K \mathbf{u}_K^e \frac{1 - \cos(\omega_K t) e^{-i\omega_K t/2Q_K}}{\omega_K^2} \int_{V_e} d^3\mathbf{x}\, \mathbf{f} \cdot [\mathbf{u}_K^e]^*. \tag{11.3.32}$$

The static response as $t \to \infty$ is also given by a normal mode sum

$$\mathbf{u}(\mathbf{x}, t) = \sum_K \mathbf{u}_K^e \frac{1}{\omega_K^2} \int_{V_e} d^3\mathbf{x}\, \mathbf{f} \cdot [\mathbf{u}_K^e]^*. \tag{11.3.33}$$

So far we have not specified the nature of the force system **f**, but if we specialise to the point moment tensor formulation introduced in Section 11.2 above then

$$\int_{V_e} d^3\mathbf{x}\, \mathbf{f} \cdot [\mathbf{u}_K^e]^* = [\mathbf{e}_K^e]^* : \mathbf{M}(\mathbf{x}_s) = [e_K^e]^*_{pq} M_{pq}(\mathbf{x}_s) \tag{11.3.34}$$

where $\mathbf{e}_K$ is the strain tensor associated with the Kth mode.

We can extend the treatment to a source that depends on time by introducing the moment rate tensor

$$\mathfrak{M}(t) = \frac{\partial}{\partial t}\mathbf{M}(t). \tag{11.3.35}$$

The moment rate tensor has the advantage that it differs from zero only during the occurrence of an earthquake, e.g., if the moment tensor $\mathbf{M}$ behaves as a Heaviside step function, the moment rate tensor $\mathfrak{M}$ is a delta function $\delta(t)$.

We also introduce the modal time function

$$C_K(t) = \left[1 - \cos(\omega_K t)e^{-i\omega_K t/2Q_K}\right]H(t), \tag{11.3.36}$$

and then the complete response can be written as a time convolution of the modal response with the moment rate tensor,

$$\mathbf{u}(\mathbf{x}, t) = \sum_K \int_{-\infty}^{\infty} dt' \frac{1}{\omega_K^2} \left\{C_K(t - t')\mathbf{u}_K^e [\mathbf{e}_K^e]^* : \mathfrak{M}(\mathbf{x}_s, t)\right\}. \tag{11.3.37}$$

11.3.2 Free oscillations of the Earth

We have seen that we can build the response of the Earth to a source through the superposition of contributions from the normal modes of free oscillation of the Earth. For all except the low frequency modes the influence of the rotation of the Earth is small, and it is possible to analyse the modes in terms of two distinct class of modes in a spherical harmonic representation: the toroidal modes with purely torsional behaviour and the spheroidal modes which include the influence of self-gravitation. A comprehensive treatment of the development of the low frequency part of the seismic wave field in terms of these normal modes is presented by Dahlen & Tromp (1998).

The presence of the fluid core means that for the toroidal modes the boundary conditions are the vanishing of radial traction at both the surface and the core–mantle boundary. The spheroidal modes penetrate right to the centre of the Earth and can have significant energy in both the outer and inner core. For each spherical harmonic, the displacement $\mathbf{u}$ and the radial traction $\mathbf{t}_r$ are expressed in terms of vector surface harmonics to yield quantities which describe the displacement and radial traction fields and which are linked through first-order ordinary differential equations in radius. For spheroidal modes the influence of self-gravitation is important for low frequencies. For each angular order l there will be a set of modes described by a radial order n whose frequencies are determined by satisfying the interior and exterior boundary conditions.

The calculation of the free oscillations of the Earth depends on the numerical integration of the coupled sets of ordinary differential equations. These systems

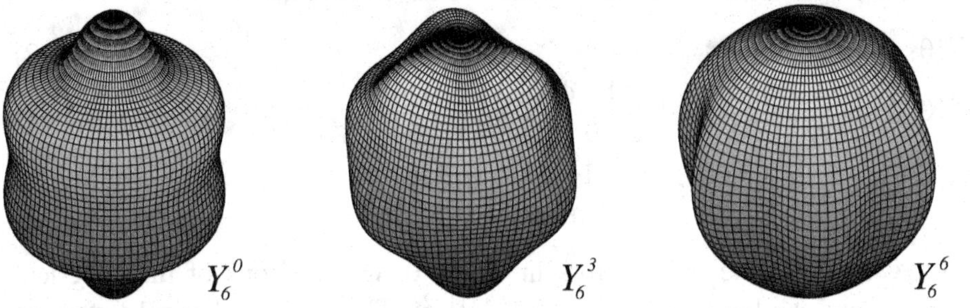

Figure 11.5. Behaviour of the spherical harmonics of degree 6. The real parts of Y_6^m are plotted for $m = 0, 3, 6$ as deviations from a sphere of radius 5.

govern the radial behaviour of the normal mode eigenfunctions with an assumed angular dependence through the spherical harmonics of order l,

$$Y_l^m(\theta, \phi) = (-1)^m \left[\frac{2l+1}{4\pi}\frac{(l-m)!}{(l+m)!}\right] P_l^m(\cos\theta) e^{im\phi}, \tag{11.3.38}$$

where P_l^m is an associated Legendre polynomial and $-l \leq m \leq l$. The behaviour of Y_l^m for $l = 6$ is illustrated in Figure 11.5.

For short wavelengths the standing waves of the spherical harmonic can be decomposed into two travelling waves circling the Earth in opposite directions with phase velocities

$$c = \omega r_e / (l + \tfrac{1}{2}). \tag{11.3.39}$$

For simplicity we will discuss the calculation of free oscillation periods for a spherical non-rotating Earth model. With these assumptions the normal mode periods are independent of the azimuthal order m, i.e., each period has $(2l+1)$-fold degeneracy. This degeneracy is broken by the Earth's rotation, the ellipticity of the figure of the Earth, and the presence of three-dimensional structure within the Earth.

We project the displacement $\mathbf{u}$ and radial traction $\mathbf{t}_r$ on a vector spherical harmonic field

$$\mathbf{u} = \sum_l \sum_{m=-l}^{l} \{U(r)\mathbf{P}_l^m(\theta, \phi) + V(r)\mathbf{B}_l^m(\theta, \phi) + W(r)\mathbf{C}_l^m(\theta, \phi)\}, \tag{11.3.40}$$

$$\mathbf{t}_r = \sum_l \sum_{m=-l}^{l} \{P(r)\mathbf{P}_l^m(\theta, \phi) + S(r)\mathbf{B}_l^m(\theta, \phi) + T(r)\mathbf{C}_l^m(\theta, \phi)\}, \tag{11.3.41}$$

where the orthogonal vector surface harmonics have the properties

$$\mathbf{P}_l^m(\theta,\phi) = Y_l^m \hat{\mathbf{e}}_r, \tag{11.3.42}$$

$$\mathbf{B}_l^m(\theta,\phi) = \frac{1}{\mathcal{L}}\left(\frac{\partial Y_l^m}{\partial\theta}\hat{\mathbf{e}}_\theta + \frac{1}{\sin\theta}\frac{\partial Y_l^m}{\partial\phi}\hat{\mathbf{e}}_\phi\right), \tag{11.3.43}$$

$$\mathbf{T}_l^m(\theta,\phi) = -\frac{1}{\mathcal{L}}\left(\frac{\partial Y_l^m}{\partial\theta}\hat{\mathbf{e}}_\theta - \frac{1}{\sin\theta}\frac{\partial Y_l^m}{\partial\phi}\hat{\mathbf{e}}_\phi\right). \tag{11.3.44}$$

Here $\hat{\mathbf{e}}_r$, $\hat{\mathbf{e}}_\theta$, $\hat{\mathbf{e}}_\phi$ are the unit vectors in the r, θ, ϕ directions of the spherical coordinates, and we have set $\mathcal{L} = [l(l+1)]^{1/2}$. The vector spherical functions are orthogonal so that, e.g.,

$$\int_0^\pi \sin\theta d\theta \int_0^{2\pi} d\phi\, [\mathbf{R}_l^m]^* \mathbf{R}_{l'}^{m'} = \delta_{mm'}\delta_{ll'},$$

$$\int_0^\pi \sin\theta d\theta \int_0^{2\pi} d\phi\, [\mathbf{R}_l^m]^* \mathbf{S}_l^m = 0. \tag{11.3.45}$$

With this representation the governing equations (11.3.20) separate into two groups of coupled differential equations in terms of radius r. The toroidal oscillations associated with the $\mathbf{C}_l^m(\theta,\phi)$ separate from the spheroidal oscillations that link the coefficients of $\mathbf{P}_l^m(\theta,\phi)$, $\mathbf{B}_l^m(\theta,\phi)$.

The first few overtone sequences of the normal modes link through the travelling wave representations at short periods to Love waves for the toroidal oscillations, and to Rayleigh waves for the spheroidal oscillations.

Toroidal oscillations

The toroidal oscillations involve no radial displacement and can be described entirely through the coefficients of the $\mathbf{C}_l^m(\theta,\phi)$ vector harmonics, i.e., the displacement coefficient W, and traction coefficient T. The toroidal modes do not penetrate into the fluid core so disturbances are confined to the mantle.

In the mantle, the quantities W and T are related by

$$\frac{d}{dr}\begin{pmatrix} W \\ T \end{pmatrix} = \begin{pmatrix} r^{-1} & \mu^{-1} \\ -\rho\omega^2 + (l+1)(l+2)\mu r^{-2} & -3r^{-1} \end{pmatrix} \begin{pmatrix} W \\ T \end{pmatrix}. \tag{11.3.46}$$

The boundary conditions that have to be applied for these equations for a free oscillation are

$$\begin{aligned} T(r_e) &= 0 \quad \text{at the surface } r = r_e, \text{ and} \\ T(r_c) &= 0 \quad \text{at the core–mantle boundary } r = r_c. \end{aligned} \tag{11.3.47}$$

The toroidal equations may thus be integrated from the core–mantle boundary ($r = r_c$) with a starting solution,

$$W(r_c) = 1, \quad T(r_c) = 0, \tag{11.3.48}$$

and the secular equation for the torsional modes that determines the allowed frequency ω to satisfy the boundary conditions is

$$F_T(\omega, l) = T(r_e, \omega, l) = 0. \quad (11.3.49)$$

For higher mode overtones of the torsional modes at large l, integration from the core–mantle boundary can lead to overflow in the computations. This is because the main variation in the eigenfunction is concentrated near the surface with only an exponential tail decaying into the lower mantle. In this case it is worth starting the integration from a higher level in the mantle. We assume that for a surface of radius r_1 the material is uniform, then a suitable starting solution is

$$\begin{aligned} W(r_1) &= 1, \\ T(r_1) &= (\mu/r_1)[(l-1) - z_l(\omega r/\beta)], \end{aligned} \quad (11.3.50)$$

where z_l is defined in terms of the spherical Bessel functions by

$$z_l(x) = x j_{l+1}(x)/j_l(x), \quad (11.3.51)$$

Takeuchi & Saito (1972) have given a recursion relation for z_l

$$z_{l-1}(x) = x^2/[(2l+1) - z_l(x)]. \quad (11.3.52)$$

For a large wavenumber this recursion formula may be applied in order of decreasing l from an initial value $z_l = x^2/(2l+3)$.

We work at fixed angular order l (i.e., fixed wavenumber) and integrate (11.3.46) up from the initial values with a trial value of frequency ω, and the surface value of $T(r_e)$ is then the required secular function. If the trial frequency does not lead to the vanishing of the secular function then the eigenfrequency estimate can be improved by using variational results. For a trial frequency ω_t an improved estimate ω_n may be found from

$$\omega_n^2 = \omega_t^2 + \mathcal{B}/\mathcal{T}, \quad (11.3.53)$$

where for this torsional mode case

$$\mathcal{T} = \int_0^{r_e} dr\, r^2 \rho(r) W^2(r), \quad \mathcal{B} = r_e^2 W(r_e) T(r_e), \quad (11.3.54)$$

with W and T calculated for the trial frequency ω_t. This scheme may be applied iteratively until convergence to the eigenfrequency of the desired normal mode is achieved. Once the eigenfrequency has been found the last set of values for $W(r)$ and $T(r)$ define the corresponding eigenfunction, which may need to be normalised.

For a given angular order l the toroidal modes are designated ${}_nT_l$ in order of increasing frequency $\omega_n, n = 0, 1, 2, \ldots$. The index n is the radial order and for the toroidal modes corresponds to the number of zero crossings in the radial eigenfunctions. The $(2l+1)$-fold degeneracy in the eigenfrequency for azimuthal order m is broken in the presence of lateral heterogeneity, and can be accounted for using perturbation theory.

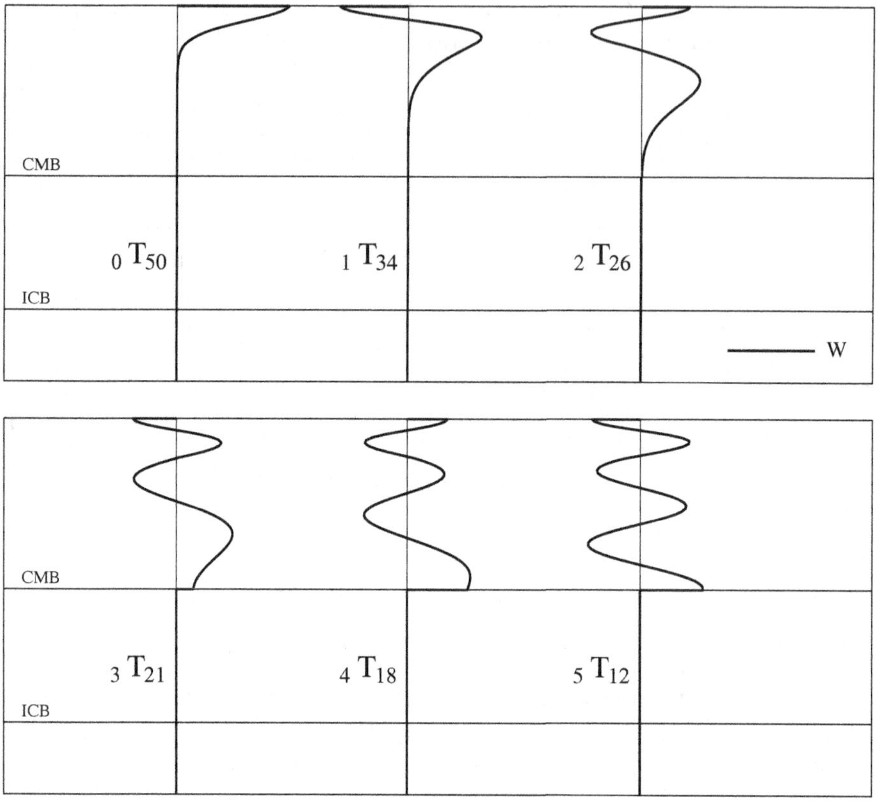

Figure 11.6. Radial variation of the eigendisplacement W associated with toroidal modes for model AK135, with frequencies close to 6 mHz. The upper row corresponds to modes that are equivalent to turning *S* waves in the mantle and the lower row to *ScS* equivalent modes.

The radial dependence of the eigenfunctions for the set of toroidal modes with eigenfrequency close to 6 mHz is shown in Figure 11.6. The displacement W is confined to the mantle. The fundamental mode $_0T_{50}$ is equivalent to a Love wave with decay of the displacement W with depth. As the radial order n increases the depth of penetration increases and the modes can be regarded as equivalent to mantle *S* waves. For the higher overtones in the lower row of Figure 11.6 the displacement fills the whole mantle; this behaviour represents *ScS* equivalent modes.

Spheroidal oscillations

The spheroidal oscillations of the Earth give rise to a perturbation in the gravitational potential; we therefore need to consider coupled elastic–gravitational disturbances and satisfy the elastic equations of motion and Poisson's equation simultaneously. Unlike the toroidal modes, the spheroidal mode involves the whole Earth with motion in both the solid and liquid parts of the core and the mantle.

The displacement and traction fields link the $\mathbf{P}_l^m(\theta,\phi)$ and $\mathbf{B}_l^m(\theta,\phi)$ vector harmonics through the quantities U, V, P, S. We express the perturbation in gravitational potential as

$$\varphi(\mathbf{x},\omega) = -\Phi(r)Y_l^m(\theta,\phi), \tag{11.3.55}$$

and introduce the additional variable

$$\Psi = \frac{d\Psi}{dr} + 4\pi G\rho U + \frac{l+1}{r}\Phi. \tag{11.3.56}$$

This scalar Ψ has the desirable properties of being continuous, zero at the Earth's surface $r = r_e$ and identically zero for $r > r_e$.

The displacement coefficient U, V, traction coefficients P, S and the gravitational terms Φ, Ψ satisfy the coupled set of ordinary differential equations with respect to radius r:

$$\begin{aligned}
\frac{dU}{dr} &= \frac{\eta}{r}(\mathcal{L}V - 2U) + \frac{1}{\lambda+2\mu}P,\\
\frac{dV}{dr} &= -\frac{1}{r}(V - \mathcal{L}U) + \frac{1}{\mu}S,\\
\frac{dP}{dr} &= \left\{-\rho\omega^2 + 4\left[\frac{\chi}{r^2} - \frac{\rho g}{r}\right]\right\}U + \mathcal{L}\left\{\frac{\rho g}{r} - 2\frac{\chi}{r^2}\right\}V\\
&\quad + \frac{2(\eta-1)}{r}P + \frac{\mathcal{L}}{r}S - \frac{(l+1)\rho}{r}\Phi + \rho\Psi,\\
\frac{dS}{dr} &= \mathcal{L}\left\{\frac{\rho g}{r} - 2\frac{\chi}{r^2}\right\}U\left\{-\rho\omega^2 + \frac{\mathcal{L}^2 - 2\mu}{r^2}\right\}V - \mathcal{L}\frac{\eta}{r}P - \frac{3}{r}S + \mathcal{L}\frac{\rho}{r}\Psi,\\
\frac{d\Phi}{dr} &= -4\pi G\rho U - \frac{l+1}{r}\Psi + \Psi,\\
\frac{d\Psi}{dr} &= \frac{4\pi G\rho}{r}[\mathcal{L}V - (l+1)U] + \frac{l-1}{r}\Psi;
\end{aligned} \tag{11.3.57}$$

with

$$\eta = \frac{\lambda}{\lambda+2\mu}, \quad \chi = \frac{\mu(3\lambda+2\mu)}{\lambda+2\mu}, \quad \nu = \frac{4\mu(\lambda+\mu)}{\lambda+2\mu}. \tag{11.3.58}$$

In this form the set of elements of the coupled differential equations reduce asymptotically to the forms for plane stratification for large ω and l.

The boundary conditions that have to be satisfied at the Earth's surface are that the surface traction and gravitational scalar Ψ should vanish, i.e.,

$$P(r_e) = S(r_e) = 0, \quad \Psi(r_e) = 0. \tag{11.3.59}$$

The set of six coupled equations (11.3.58) will have six independent solutions, but the solutions of interest are the three that are regular at the origin. We therefore start the integrations for these three solutions in the solid inner core and integrate up to the Earth's surface. The secular equation for the spheroidal oscillations is then given by the vanishing of the determinant

$$F_S(\omega, l) = \begin{vmatrix} P_1(r_e) & P_2(r_e) & P_3(r_e) \\ S_1(r_e) & S_2(r_e) & S_3(r_e) \\ \Psi_1(r_e) & \Psi_2(r_e) & \Psi_3(r_e) \end{vmatrix} = 0. \tag{11.3.60}$$

We model the outer core of the Earth as a liquid with vanishing rigidity and, in consequence, the system of differential equations is simplified since no tangential stress can be supported. In the equation system (11.3.57) $\eta = 1, \chi = 0, \nu = 0$ and

$$V = \frac{\mathcal{L}}{\rho\omega^2 r}\left[P + \rho(gU - \Psi)\right], \tag{11.3.61}$$

so that in the fluid

$$\begin{aligned}
\frac{dU}{dr} &= -\frac{2}{r}U + \frac{\mathcal{L}}{r}V + \frac{1}{\lambda}P, \\
\frac{dP}{dr} &= -\left\{\rho\omega^2 + 4\frac{\rho g}{r}\right\}U + \mathcal{L}\frac{\rho g}{r}V - \frac{(l+1)\rho}{r}\Phi + \rho\Psi, \\
\frac{d\Phi}{dr} &= -4\pi G\rho U - \frac{l+1}{r}\Psi + \Psi, \\
\frac{d\Psi}{dr} &= \frac{4\pi G\rho}{r}[\mathcal{L}V - (l+1)U] + \frac{l-1}{r}\Psi;
\end{aligned} \tag{11.3.62}$$

V becomes indeterminate at zero frequency ($\omega = 0$).

We can summarise the set of displacement, traction and gravitational variables through the column vector

$$\mathfrak{b} = [U, V, P, S, \Phi, \Psi]^T. \tag{11.3.63}$$

The solution of (11.3.57) in the inner core can be written as a superposition of the three regular solutions

$$\mathfrak{b}_i^I = C_1^I \mathfrak{b}_{1i}^I + C_2^I \mathfrak{b}_{2i}^I + C_3^I \mathfrak{b}_{3i}^I, \quad i = 1, 2, ..., 6; \tag{11.3.64}$$

whereas in the liquid outer core we have

$$\mathfrak{b}_i^K = C_1^I \mathfrak{b}_{1i}^K + C_2^I \mathfrak{b}_{2i}^K + C_3^I \mathfrak{b}_{3i}^K, \quad i = 1, 3, 5, 6; \tag{11.3.65}$$

as we are now only dealing with four coupled equations. In the solid mantle we have again

$$\mathfrak{b}_i^M = C_1^I \mathfrak{b}_{1i}^M + C_2^I \mathfrak{b}_{2i}^M + C_3^I \mathfrak{b}_{3i}^M, \quad i = 1, 2, ..., 6; \tag{11.3.66}$$

When we integrate the coupled equations outward from the inner core we must be able to connect the solutions in the different regions. We can, e.g., use the requirement that at the inner core boundary $r = r_i$ the tangential stress S^I must vanish and so one of the constants of integration can be eliminated in the inner core,

$$C_3^I = -\frac{S_1^I}{S_3^I} C_1^I - \frac{S_2^I}{S_3^I} C_2^I. \qquad (11.3.67)$$

The continuity relations at the inner core interface are

$$\mathfrak{b}_{1i}^K(r_i) = \mathfrak{b}_{1i}^I - \frac{S_1^I}{S_3^I}\mathfrak{b}_{3i}^I, \quad \mathfrak{b}_{2i}^K(r_i) = \mathfrak{b}_{2i}^I - \frac{S_2^I}{S_3^I}\mathfrak{b}_{3i}^I, \quad i = 1,3,5,6, \qquad (11.3.68)$$

with

$$C_1^K = C_1^I, \quad C_2^K = C_2^I. \qquad (11.3.69)$$

Once we reach the core mantle boundary we may continue two solutions directly into the mantle, i.e., at the interface $r = r_c$,

$$\mathfrak{b}_{1i}^K(r_c) = \mathfrak{b}_{1i}^M(r_c), \quad \mathfrak{b}_{2i}^K(r_c) = \mathfrak{b}_{2i}^M(r_c), \quad i = 1,3,5,6 \qquad (11.3.70)$$

supplemented by the requirements

$$V_1^M(r_c) = S_1^M(r_c) = V_2^M(r_c) = S_2^M(r_c) = 0, \; C_1^M = C_1^K, \; C_2^M = C_2^K. \qquad (11.3.71)$$

The third set of mantle–core boundary values can be determined by recalling that the tangential displacement V^M need not be continuous across the interface. We may thus take

$$V_3^M = 1, \quad U_3^M = P_3^M = S_3^M = \Phi_3^M = \Psi_3^M = 0. \qquad (11.3.72)$$

The constants of integration C_1^M, C_2^M, C_3^M are to be determined by satisfying the free surface boundary conditions for an eigenfrequency.

The starting conditions at the centre of the Earth are based either on a locally uniform medium or on power series. For a given trial frequency, the spheroidal equations have to be integrated to the surface using the continuation scheme we have just described. The determinantal secular function (11.3.60) has then to be constructed from the surface values of P_i, S_i, Ψ_i from the separate integrations. In this direct approach very high order accuracy is needed in the integration scheme to overcome the "stiff" nature of the differential equation system. In consequence modern programs favour working directly with the minors of the secular determinant, with a much more complex set of differential equations but the advantage of numerical stability.

Iterative improvement of the estimated frequency ω_t can be achieved with the variational result for an improved frequency ω_n

$$\omega_n^2 = \omega_t^2 + \mathcal{B}/\mathcal{T}, \qquad (11.3.73)$$

with

$$\mathcal{T} = \int_0^{r_e} dr\, r^2 \rho(r)[U^2(r) + V^2(r)],$$
$$\mathcal{B} = r_e^2[U(r_e)P(r_e) + V(r_e)S(r_e) + (4\pi G)^{-1}\Phi(r_e)\Psi(r_e)]. \qquad (11.3.74)$$

The particular set of functions $\mathbf{b}$ to be used in these equations must represent a good approximation to the eigenfunction at the true eigenfrequency. The combination of the three independent solutions must therefore be made as if ω_t were the eigenfrequency.

Spheroidal modes ${}_nS_l$ have a similar notation to that used for the toroidal modes. The index n again indicates the frequency order at fixed l, but unlike the toroidal case does not correspond directly to the number of nodal surfaces in radius.

The eigenfunctions for the spheroidal modes are illustrated in Figure 11.7, once again for eigenfrequencies close to 6 mHz. The range of behaviour is much more complex than for the toroidal modes of similar frequency shown in Figure 11.6, because the spheroidal modes include *P* wave behaviour and are no longer confined to the mantle. Whereas U, Φ are continuous, the eigendisplacement V displays jumps at the fluid–solid interfaces at the top and bottom of the inner core. The progression of the higher overtones with eigenfunctions extending through more of the Earth is interrupted by two modes that correspond to trapped Stoneley waves concentrated at these fluid–solid interfaces. The mode ${}_3S_{28}$ is trapped at the core–mantle boundary, and mode ${}_7S_{15}$ at the inner-core boundary. The variations in the gravitational scalar are concentrated in the core and have very little expression at the surface.

The fundamental mode ${}_0S_{54}$ is equivalent to a Rayleigh wave. The modes ${}_1S_{34}$ and ${}_4S_{24}$ correspond to *S* waves turning in the mantle, and ${}_5S_{19}$, ${}_6S_{17}$ are equivalent to *ScS* waves reflected from the core–mantle boundary which is no longer a specular reflector since transmitted *P* waves can make their way into the fluid core. The bottom row in Figure 11.7 shows modes with much of their energy in the inner core.

Radial oscillations

If we consider spheroidal free oscillations for angular order zero, the deformation is purely in the radial direction so we only need to consider the solution of two coupled equations

$$\frac{d}{dr}\begin{pmatrix} U \\ P \end{pmatrix} = \begin{pmatrix} -2\eta r^{-1} & (\lambda + 2\mu)^{-1} \\ -\rho\omega^2 + 4(\chi r^{-2} - \rho g r^{-1}) & (2\eta - 2)r^{-1} \end{pmatrix} \begin{pmatrix} U \\ P \end{pmatrix}. \qquad (11.3.75)$$

where η, χ are the composite moduli (11.3.58). The gravitational potential in this case can be found from

$$\frac{d\Phi}{dr} = -4\pi G \rho U. \qquad (11.3.76)$$

The surface boundary condition is as before $P(r_e) = 0$.

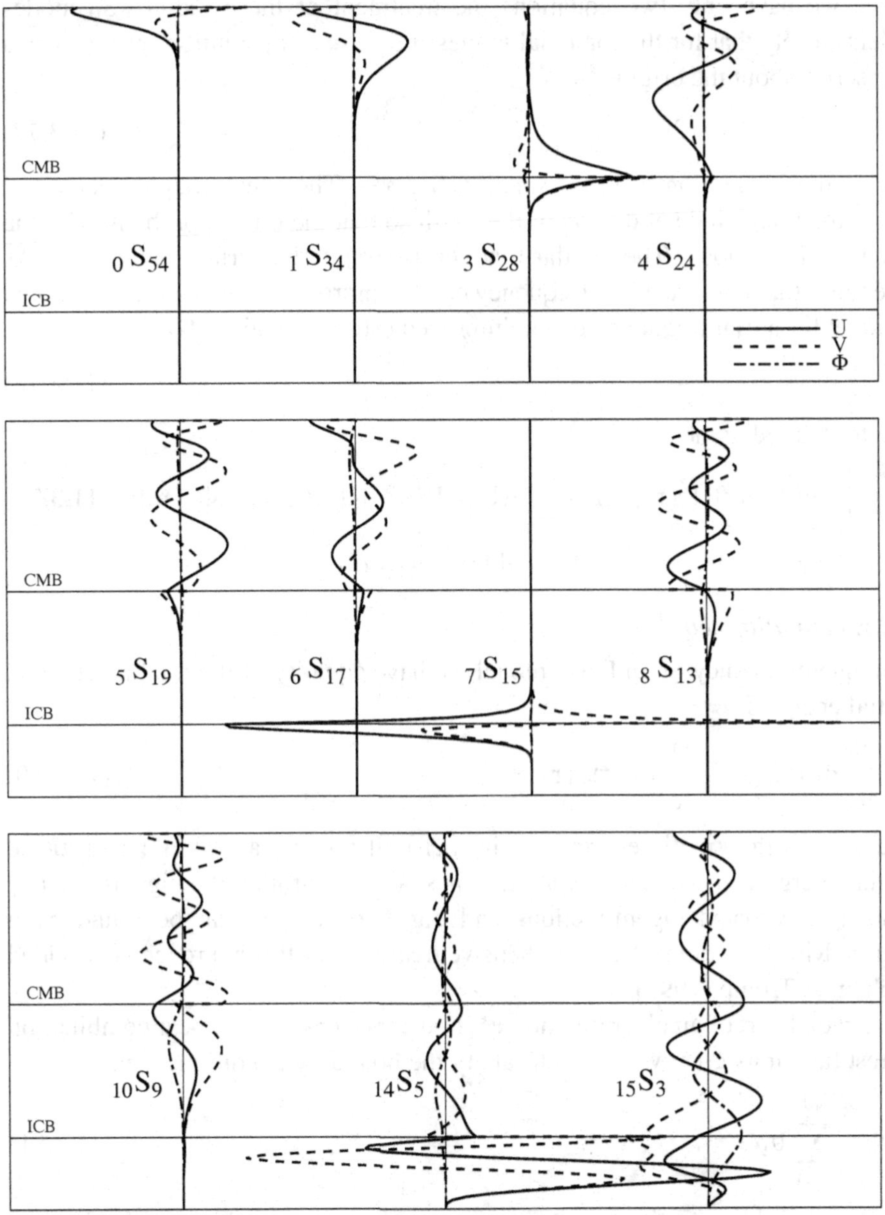

Figure 11.7. Radial variation of the displacements U, V and gravitational term Φ associated with spheroidal modes for model AK135, with frequencies close to 6 mHz. A more complex behaviour is displayed than for the toroidal modes of similar frequency shown in Figure 11.6. The low radial order modes are largely confined to the mantle, but the spheroidal modes can penetrate into the core. There are now modes that are equivalent to Stoneley waves concentrated at the core–mantle boundary $_3S_{28}$ and the boundary between the inner and outer cores $_7S_{15}$. Some modes such as $_{14}S_5$ have almost all their energy concentrated in the inner core.

Since we have only two equations the treatment of these radial ${}_nS_0$ modes parallels closely that for the torsional modes. For a starting solution we may use a power series about the origin

$$U = r, \quad P = 3\lambda + 2\mu, \tag{11.3.77}$$

and start the integration at some small radius r_1. The same coupled equations apply in the liquid and solid parts of the Earth so that there is no problem with the continuity of solutions. The secular function is simply the surface value of P. As before the estimate of the eigenfrequency can be improved by using the variational technique; for a trial frequency ω_t an improved estimate is given by

$$\omega_n^2 = \omega_t^2 + \mathcal{B}/\mathcal{T}, \tag{11.3.78}$$

where for the radial modes

$$\mathcal{T} = \int_0^{r_e} dr\, r^2 \rho(r) U^2(r), \quad \mathcal{B} = r_e^2 [U(r_e)P(r_e) + (4\pi G r_e)^{-1} \Phi^2(r_e)], \tag{11.3.79}$$

with U, P and Φ calculated for the trial frequency ω_t.

The Rayleigh–Ritz method

For an eigenfrequency of an Earth model we have equality of the total kinetic and potential energies

$$\omega^2 \int_0^{r_e} dr\, r^2 \mathfrak{T}(r) - \int_0^{r_e} dr\, r^2 \mathfrak{V}(r) = 0, \tag{11.3.80}$$

where $\omega^2 \mathfrak{T}$ is the kinetic energy density per unit volume at radius r and $\mathfrak{V}$ the potential energy density. This expression is stationary around the eigenfrequency and so approximated eigenfunctions and eigenfrequencies can be found by a Rayleigh–Ritz procedure. A comprehensive treatment of this approach is provided by Dahlen & Tromp (1998).

If the radial part of an eigenfunction $\mathbf{u}^e(r)$ is represented as a linear combination of N test functions $\zeta_i(r)$ which each satisfy the boundary conditions then

$$\mathbf{u}(r) \approx \sum_{i=1}^{N} b_i \zeta_i(r). \tag{11.3.81}$$

The energy balance equation results in a matrix eigenvalue problem of the form

$$(\omega^2 \mathsf{T} - \mathsf{V})\mathbf{b} = 0. \tag{11.3.82}$$

Each eigenvector $\mathbf{b}$ represents a projection of the modal eigenfunction $\mathbf{u}(r)$ into the space spanned by the test functions ζ_i and the corresponding value of ω^2 will be an upper bound to the eigenfrequency associated with $\mathbf{u}$. Since T and V depend on the angular order l of a normal mode, we have a different set of matrix equations for each angular order. The successive eigenvectors $\mathbf{b}$ obtained for a given angular order represent higher radial order (overtones). With a suitable set of test functions such as piecewise cubic Hermitian splines, the model-dependent parts of T and V can be evaluated by Gauss–Legendre integration with not less than six grid points per wavelength. The computation time per eigenfrequency is independent of the node, and no advantage can be taken of the properties of the mode. However, difficulties such as problems with nearly coincident eigenfrequencies are avoided.

11.4 Probing the Earth

In this section we look at various aspects of the Earth as seen by seismic waves. We start by considering relatively high-frequency wave propagation for which the pattern of arrivals can be understood through ray tracing and wavefronts. We introduce the terminology for seismic phases and show the way in which *P* and *S* waves interact with the structure in the Earth's mantle and core. Because the outer core behaves as a fluid only *P* waves propagate through the core; the *P* wavespeed at the top of the core is lower than the *P* wavespeed at the base of the mantle, but higher than the corresponding *S* wavespeed. This leads to very different behaviour for the transmitted *P* waves in the core, depending on whether they are derived from *P* or *S* waves in the mantle. Horizontally polarised *SH* waves are totally reflected from the core–mantle boundary and so are confined to the mantle.

The different classes of seismic propagation are also useful in understanding the way in which the array of frequencies of the normal modes of the Earth relate to structure. Toroidal modes correspond to *SH* waves and so have displacements only in the mantle. In contrast, the spheroidal modes penetrate the whole Earth and the pattern of modes is influenced by the structure of the core and the mantle.

The two complementary viewpoints on seismic wave propagation provide convenient ways to look at the higher and lower frequency characteristics of the seismic wave field. A spherical Earth model provides a good basis for understanding the timing and general characteristics of seismic arrivals. However, the presence of three-dimensional structure in the Earth leads to variations in the properties of seismic phases that depend on the particular path between source and receiver, rather than just their angular separation. This information can be exploited in the techniques of seismic tomography to provide images of internal structure exploiting both ray-based and normal-mode-based methods.

11.4.1 Seismic phases

The body wave portion of a seismogram is marked by the arrival of distinct bursts of energy that can be associated with different classes of propagation path through the Earth. This is illustrated in Figure 11.8, for three-component seismograms at 100° from an intermediate event in Vanuatu that displays a rich set of seismic phases that are marked with their phase code. The individual seismic arrivals sample different parts of the Earth in their passage between the source and the receiver and their properties are dictated by the structure they encounter.

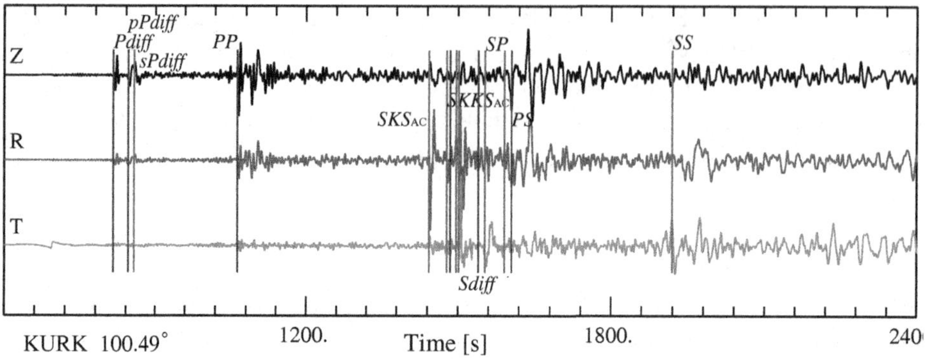

Figure 11.8. Three-component seismogram from an intermediate depth earthquake in Vanuatu recorded in Kazakhstan at approximately 100° from the source, showing a rich set of seismic phase arrivals.

The phase code describing each propagation path is built up by combining the different elements of the path through the major zones in the structure of the Earth introduced in Chapter 1:

- a leg in the mantle is denoted by *P* or *S* depending on wave type;
- a compressional leg in the outer core is labelled as *K*;
- a compressional leg in the inner core is indicated with *I*, the corresponding shear wave leg would be denoted *J*;
- waves leaving upward from the source are indicated by lowercase letters, thus *pP* represents a surface reflection for *P* near the source, and *sP* a reflection with conversion above the source; and
- reflected waves from major interfaces are indicated by lowercase letters: *m* for the Mohorovičić discontinuity (Moho), *c* at the core–mantle boundary and *i* at the boundary between the inner and outer cores.

A *P* wave which returns to the surface after propagation through the mantle and is then reflected at the surface to produce a further mantle leg will be represented by *PP*. *PKIKP* is a *P* wave which has travelled through the mantle and both the inner and outer cores, whilst *PKiKP* is reflected back from the surface of the inner core. Similarly an *S* wave that is reflected at the core–mantle boundary is indicated by *ScS*, and if conversion occurs on reflection we have *ScP*.

The times of passage of the different classes of seismic waves through the Earth are sensitive to different aspects of Earth structure. For example, the time of arrival of the *PcP* phase, which is reflected from the core–mantle boundary, is a strong function of the radius of the core. The primary control on the wavespeed structure in the AK135 model illustrated in Figure 1.6 comes from the times of *P*, *S* and the major core phases, but in all the observed times of 19 different phases were used to assess the properties of different models (Kennett et al., 1995).

PKJKP would correspond to a wave that traversed the inner core as a shear wave,

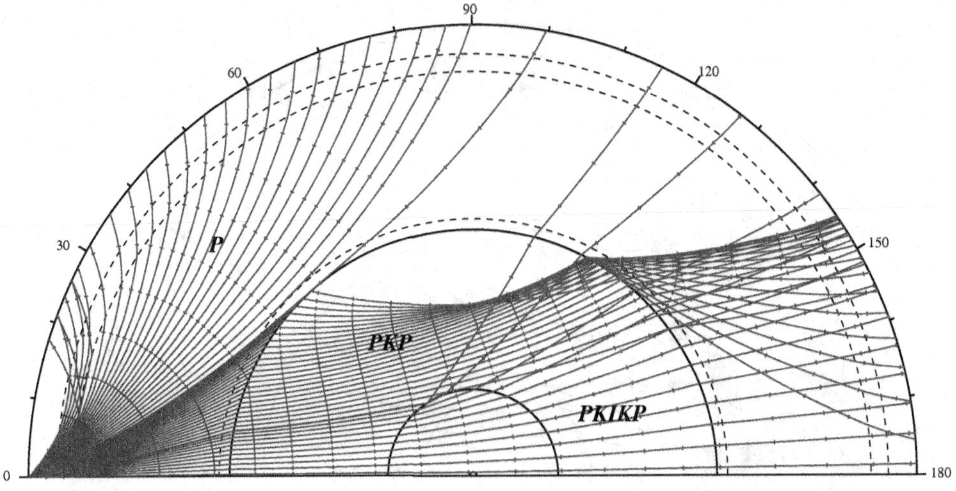

Figure 11.9. Ray paths for major *P* phases. Wavefronts are indicated by ticks on the ray paths at 1 minute intervals.

a number of claims have been made for the observation of this phase but none have been confirmed. However, a very careful analysis by Duess et al. (2000) suggests that the arrival *SKJKP* has been detected using recordings from many stations for a very deep event in the Flores Sea, Indonesia.

The 'depth phases' (*pP*, *sP* and *pS*, *sS*) can be very distinct for deep events. The time difference between the arrival of such phases reflected at the surface near the source and the main phases *P*, *S* provides a useful estimate of the source depth.

The way in which seismic energy travels through the Earth to emerge at the surface can be illustrated by the nature of the ray paths and wavefronts associated with different classes of arrivals. Figures 11.9–11.12 show the behaviour for major *P* and *S* phases for surface source in the model AK135 of Kennett et al. (1995). *P* wave legs are shown in black and *S* wave legs are indicated by using grey tone. The time progression of the waves through the Earth is shown by marking the wavefronts with ticks along the ray paths at one minute intervals; the spreading and concentration of the ray paths provides a simple visual indication of the local amplitude associated with each phase.

Kennett (2002) provides an extended treatment of the nature of seismic body wave propagation through the Earth accompanied by illustrations of seismograms for the major phases. Shearer (1999) shows stacks of seismograms recorded around the globe that reveal the complexity of the range of propagation paths through the Earth, including a variety of minor phases that are rarely recognised on single seismograms.

The direct *P* waves refracted back from the velocity gradients in the mantle extend to about 100° away from the source, but then reach the core–mantle boundary at grazing incidence. *P* waves with steeper take-off angles at the source

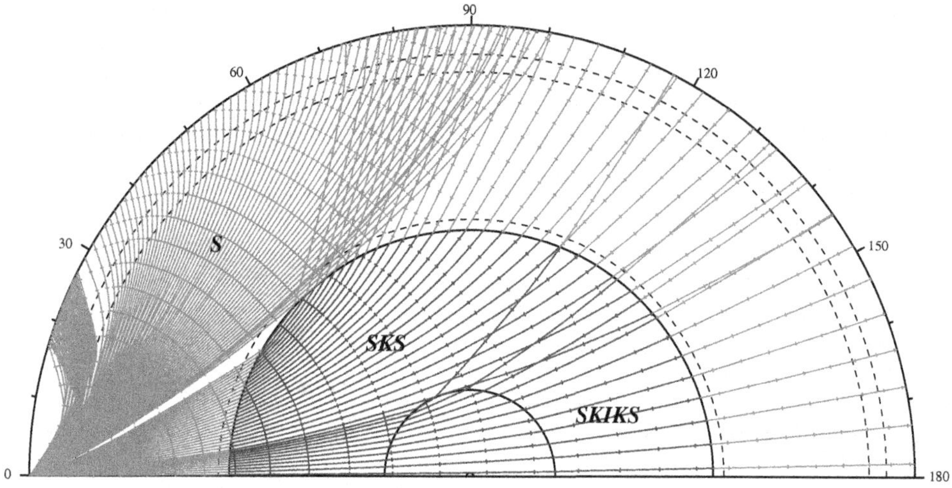

Figure 11.10. Ray paths for major *S* phases for the AK135 model of seismic wavespeeds.

are either reflected from the core–mantle boundary (*PcP*) or refracted into the core (*PKP*). The refracted *PKP* waves have a relatively complex pattern of propagation in the core (see Figure 11.14). The absence of shear strength in the fluid outer core means that the *P* wavespeed at the top of the core is markedly less than at the base of the mantle. The result of this wavespeed reduction is that *P* waves entering the core have a much steeper inclination than in the mantle as a result of Snell's law. The refraction into the lower velocity medium combined with the wavespeed gradients in the outer core produce a pronounced *PKP* caustic with a concentration of arrivals near 145° from the source. The apparent 'shadow zone' with no *P* arrivals from 100 to 145° led to the discovery of the core by Oldham, at the end of the 19th century. Reflection from the boundary between the inner and outer cores (*PKiKP*) and refraction through the outermost part of the inner core (*PKIKP*) produce small-amplitude arrivals which help to fill in the gap between direct *P* out to 100° and the *PKP* caustic. The identification of these arrivals in the 'shadow zone' led to the discovery of the inner core by Inge Lehmann. The reduction in *P* wavespeed is so large that there are no turning points for *P* waves in the upper part of the core. Eventually the effect of sphericity and the increase in wavespeed with depth is sufficient to produce turning points that combine to produce the *PKP* caustic that emerges at the surface near 144°.

For *S* waves (Figure 11.10) the pattern of propagation in the mantle is similar to that for *P*. However, since the *P* wavespeed in the outer core is slightly higher than the *S* wavespeed at the base of the core, it is possible for an *SKS* wave, with a *P* wave leg in the core, to travel faster in the core and overtake the direct *S* wave travelling just in the mantle. For distances beyond 82°, the *SKS* phase arrives before *S*. Unlike the *PKP* waves, the *SKS* family of waves (including the set of underside

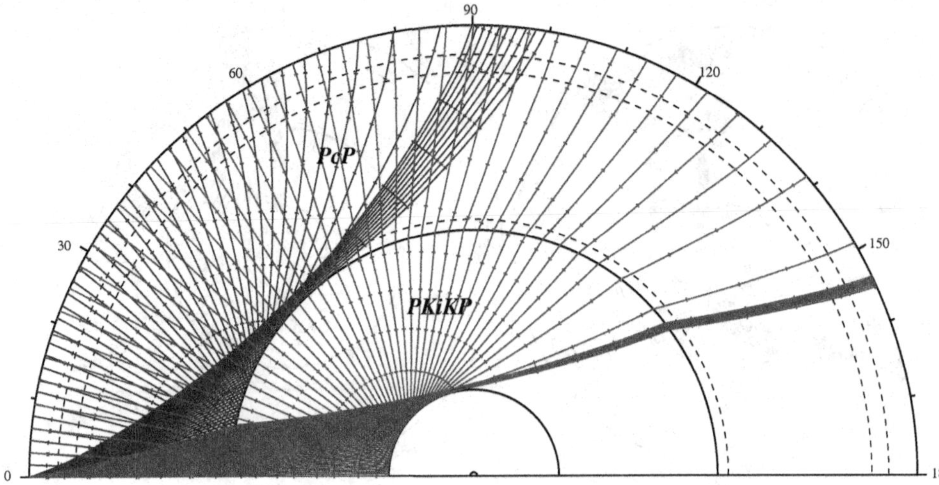

Figure 11.11. Ray paths for the reflected phases *PcP* and *PKiKP*.

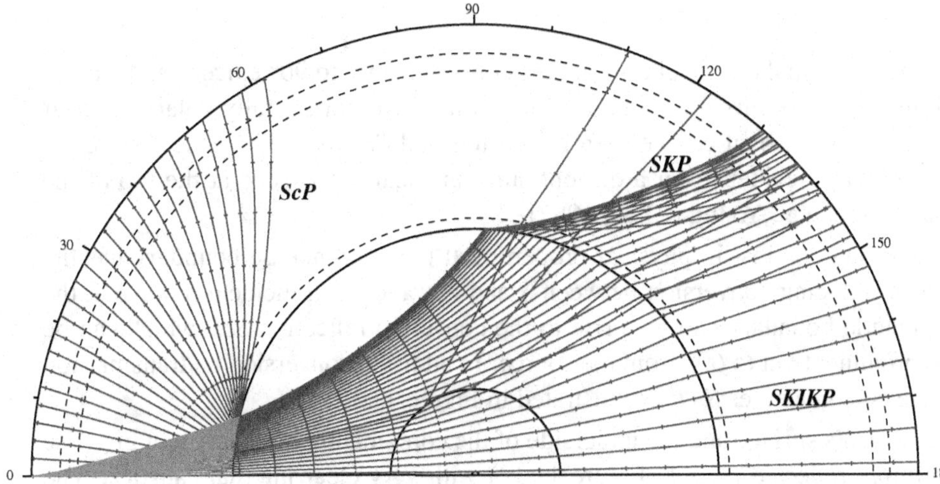

Figure 11.12. Ray paths for the converted phases *ScP* and *SKiKP*.

reflections at the core–mantle boundary as in *SKKS*) sample the whole of the core (see also Figure 11.13).

The *P* reflections from the major internal boundaries of the Earth can often be seen as distinct phases. In Figure 11.11 we show the ray paths for *PcP* reflected from the core–mantle boundary and *PKiKP* reflected from the inner-core boundary. Such reflections can sometimes be seen relatively close to the source, but their time trajectories cut across a number of other phases and so they can often be obscured by other energy. The *P* wave refracted in the mantle and the *PcP* wave reflected from the core–mantle boundary have very similar ray paths for near grazing

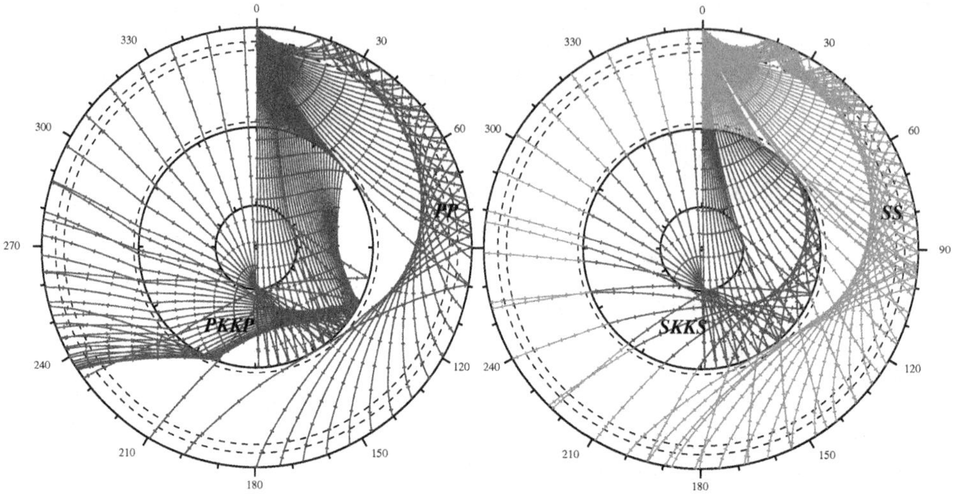

Figure 11.13. Ray paths for multiply reflected phases (a) *PP*, *PKKP*, (b) *SS*, *SKKS*.

incidence, and so their travel time curves converge close to 90° epicentral distance. A similar effect occurs for *S* and *ScS*. Sometimes, for horizontally polarised shear waves, a minor arrival can be seen between *S* and *ScS* as the branches converge, which provides evidence for a discontinuity in shear wavespeed at the top of the D″ layer (see, e.g., Young & Lay, 1987).

The strong contrast in physical properties at the core–mantle boundary has the effect of inducing conversion between *S* and *P* waves. An incident *S* wave at the core–mantle boundary can give rise to the converted reflection *ScP* which can be quite prominent out to 60° from the source. In addition conversion on transmission into the core generates *SKP* as illustrated in Figure 11.12.

The Earth's surface and the underside of the core–mantle boundary can give rise to multiply reflected phases (Figure 11.13) with very clear internal caustics. The surface multiples can be tracked for *P* and *S* to great distances and can often be recognised as distinct phases for the third or higher multiples (*PPP*, *SSS*, etc.). The internal core reflections also carry energy to substantial distance, and retain the character of the original wave system. Thus *PKKP* only has strong sensitivity to *P* wave structure in the upper part of the core near the bounce point at the core–mantle boundary, whereas *SKKS* samples the *P* wavespeed distribution through the whole core. The higher-order multiples *SKKKS* etc. provide the closest probing of the structure in the outermost parts of the core.

Multiply reflected waves from the core–mantle boundary are of particular significance for horizontally polarised shear waves (*SH*), since the solid–fluid interface is close to a perfect reflector. With efficient reflection at the Earth's surface, a long train of multiple $(ScS)_H$ can be established that carry with them information on the internal structure of the Earth, in terms of the influence of internal discontinuities.

Branches of core phases

The propagation of *P* waves into the Earth's core in the *PKP* wave group includes refraction back from the velocity gradients in the outer core and reflection (*PKiKP*) and refraction (*PKIKP*) from the inner core. The components of the wave group are also commonly designated by a notation based on the different branches of the travel-time curve (Figure 11.14).

The patterns of the travel times and the associated ray paths can be followed in Figure 11.14, where the critical slowness points corresponding to the transition between branches are clearly labelled. The details of the positions of these critical slownesses vary slightly between different models for the Earth's core, but the general pattern is maintained.

The PKP_{AB} branch corresponds to the waves entering the core at the shallowest angles and PKP_{BC} to waves refracted back from the lower part of the outer core. The CD branch corresponds to wide-angle reflection from the inner boundary (the extension of the *PKiKP* phase). The PKP_{DF} phase, which is equivalent to

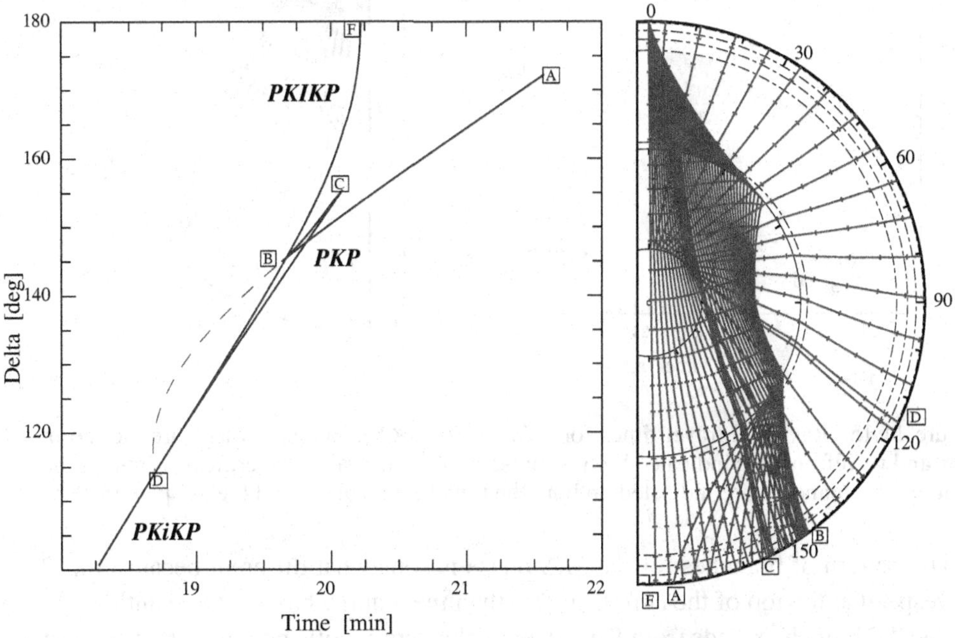

Figure 11.14. Rays and travel times for *PKP* for the AK135 model, wavefronts are indicated by tick marks at 60 s intervals. The critical points for the various *PKP* branches are indicated on both the travel-time curve and the ray pattern. The dashed segment indicates the locus of precursors to *PKIKP* from scattering at the core–mantle boundary.

PKIKP is refracted through the inner core. Rays penetrating into the upper part are strongly bent and emerge near 110° at the D point, steeper entry leads to more direct propagation through the inner core with the F point at 180° corresponding to transmission without deflection.

The concentration of rays near the *PKP* caustic at B is reflected in localised large-amplitude arrivals for *PKP* near 145°. The observations of the BC branch tend to extend beyond the ray-theoretical predictions. For epicentral distances beyond the C point, there is the possibility of diffraction around the inner core. At the B caustic the real branch does not just stop and there will be frequency-dependent decay into the shadow side of the caustic. In addition scattering in the mantle from *PKP* produces short-period arrivals as precursors to *PKIKP* that can be seen because they arrive in a quiet portion of the seismic record. The envelope of possible precursors is indicated in Figure 11.14 by a dashed line.

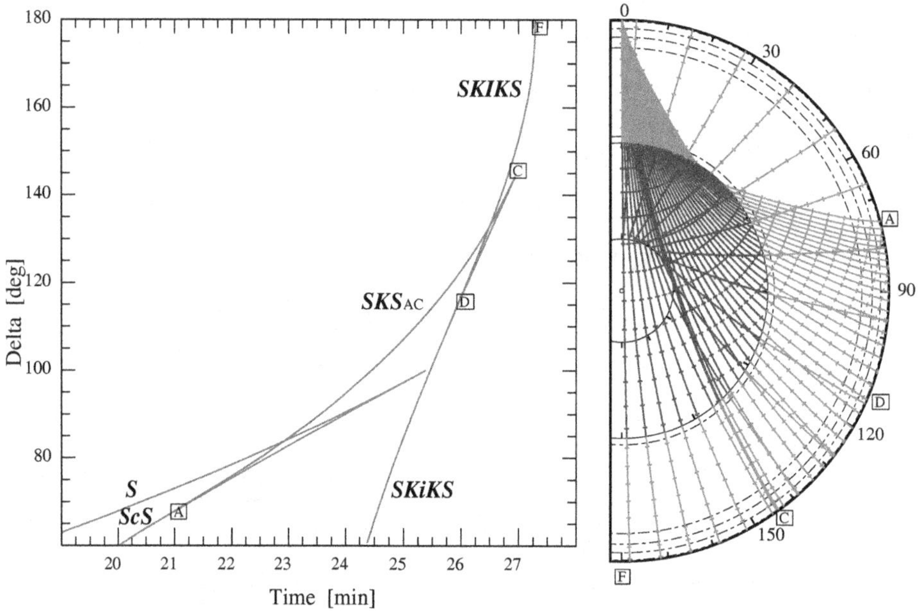

Figure 11.15. Rays and travel times for *SKS* for the AK135 model. *S* legs are plotted in grey and wavefronts are indicated by tick marks at 60 s intervals. The critical points for the various *SKS* branches are indicated on both the travel-time curves and the ray pattern.

The pattern of branches for the *SKS* phase is somewhat different because the *P* wavespeed at the top of the core is higher than the *S* at the base of the mantle.

The AC branch extends from *S* incident on the core-mantle boundary that can just propagate as *P* at the top of the outer core and emerge at A – 63°, through to grazing incidence at the inner core boundary at C. Diffracted waves around the inner core can extend the branch beyond the formal C point. The DF branch (*SKIKS*) again corresponds to refracted waves in the inner core. The post-critical reflections from the inner-core boundary (*SKiKS*) form the CD branch and connect directly into the pre-critical reflection at shorter distances than the D point at 104° (Figure 11.15).

When the refraction just begins at A, the *S* wave path to the same epicentral distance is shorter and *SKS* is about 75 s behind *S*. However, as the proportion of faster *P* wave path in the core increases, the discrepancies in *S* and *SKS* travel time are reduced. Eventually, the travel time of *SKS* becomes less than that for *S* at the same epicentral distance. Beyond 83° *SKS* becomes the onset of the shear wave group and vertically polarised *S*, *Sdiff* have to be sought in the *SKS* coda. The transversely polarised *S* wave is very distinct and small precursory *SKS* contributions (as in Figure 11.8) can arise from either anisotropy or heterogeneity in passage through the mantle.

11.4.2 Normal mode frequencies

The frequencies of the normal modes in the frequency/angular order domain are displayed in figure 11.16 for the toroidal modes and in figure 11.17 for the more complex spheroidal mode pattern.

For the toroidal modes the patterns of the modal frequencies is relatively simple. Radial lines from the origin represent lines of constant phase velocity $\mathrm{a}l/\omega$ and can be used to examine the physical character of the modes. Modes lying to the left of the *ScS* line in figure 11.16 have eigenfunctions that sample the whole mantle and can be identified with multiple reflections between the core–mantle boundary and the surface. The modes to the right of this line have their energy concentrated above the core–mantle boundary and correspond to multiple free-surface reflections of *S* body waves. As the fundamental mode is approached the crowding of the multiple reflection processes fuses into a representation of Love waves. The progression of the behaviour of the eigendisplacement W of toroidal modes for frequencies near 6 mHz has been illustrated in Figure 11.6. The modes equivalent to *ScS* have displacement throughout the mantle, whereas the modes corresponding to multiple *S* decay below the tuning depth for a body wave of the same angular slowness.

The modal pattern for spheroidal modes is much more complicated but can be understood in terms of the major features of the structure of the Earth. The critical sets of phase velocities are those corresponding to the existence ranges for different types of propagation processes as indicated in Figure 11.17. As in the toroidal case the *ScS* line separates processes which involve multiple *S* reflections in the mantle from phase velocities for which an *S* wave can be reflected back from the core–mantle boundary. For the vertically polarised *S* waves in the spheroidal case there is also the possibility of transmission into the core as a *P* wave for phases of the type *SKS*. Modes with this type of character lie to the left of the *SKS* line in figure 11.17. At even higher phase velocities *P* wave propagation in the outer core is possible, as indicated by the *PKP* phase velocity line, and finally *P* waves penetrate into the inner core. The presence of a range of different propagation patterns for

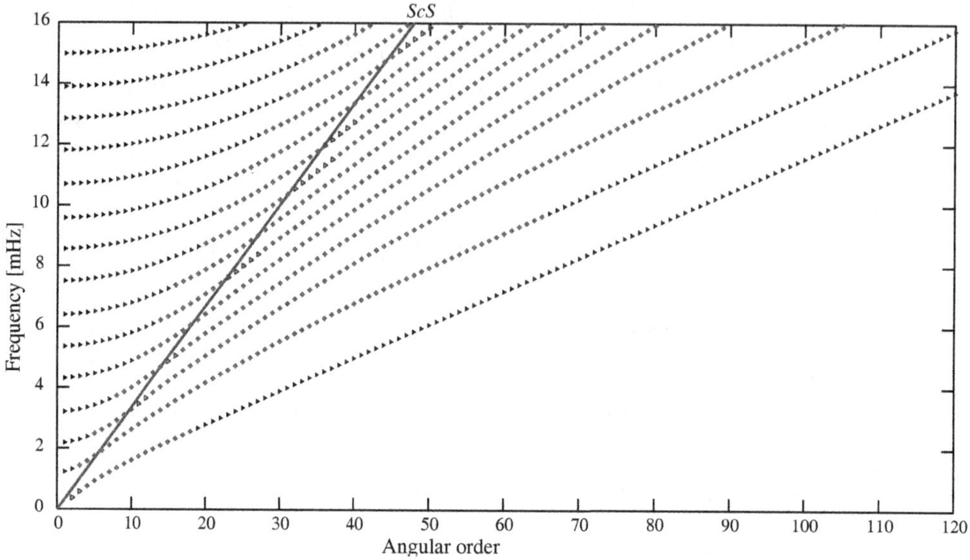

Figure 11.16. Frequencies of toroidal modes as a function of angular order l for model PREM.

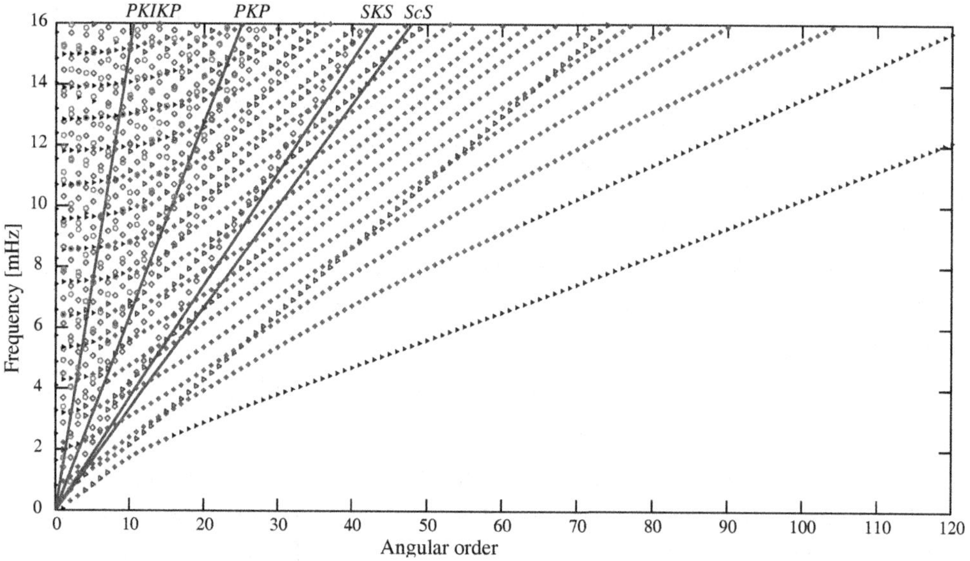

Figure 11.17. Frequencies of spheroidal modes as a function of angular order l for model PREM.

modes with high phase velocity leads to the complex observed dispersion patterns in Figure 11.17.

In both Figures 11.16 and 11.17 an attempt has been made to indicate the physical character of the propagation processes associated with each part of the mode branches, by distinguishing the individual modes in terms of their associated group velocity. Those modes with low group velocity are indicated by solid triangles, and as the group velocity increases the number of sides of the polygon is increased and for the spheroidal modes with higher group velocities we move to open symbols. This representation enhances the visibility of the mantle *S* and *ScS* equivalent modes for the spheroidal modes.

At larger angular orders, the fundamental mode branch and part of the first radial overtone branch are indicated by solid triangles; these parts are direct analogues of Love waves for toroidal modes and Rayleigh waves for spheroidal modes. The next prominent pattern (denoted by the solid diamonds) represents the trapping of *S* wave energy in the mantle and can be seen clearly in both the toroidal and spheroidal modes. For spheroidal modes some modulation of these branches is introduced by the transfer of energy into *SKS* type propagation.

Two separate classes of contribution can be recognised arising from the structure of the Earth's core (Figure 11.17). Mode segments with shallower slopes are associated with propagation in the outer core [*PKP*, *PKKP*, *SKS*, *SKKS* etc.]. The segments with steeper slopes involve energy concentrated in the inner core and so are rarely observable at the Earth's surface. The modes can be separated into different groups associated with the dominant character of the physical processes as discussed in some detail in Chapter 8 of Dahlen & Tromp (1998). The different classes of reverberation within the Earth are characterised by a spacing in frequency between the different branches for the appropriate modes that is inversely proportional to the size of the zone.

The *ScS* equivalent spheroidal modes for smaller angular orders have a similar pattern to those for the toroidal modes, but modified slightly by the changed boundary conditions at the core–mantle boundary. The tightest mode spacing comes from *PKIKP* equivalent modes that sample the whole Earth. Reverberations dominantly in the outer core and inner core provide other sequences that contribute to the complex patterns of spheroidal modes for smaller slownesses.

In addition to the main types of propagation phenomena it is also possible to get energy trapped at the core–mantle boundary or inner core–outer core boundary with exponential decay of amplitude away from the boundary. Such modes form a line of constant phase velocity cutting across the major branches and can be recognised from their differing properties in figure 11.17. The set of open triangles cutting across the mantle *S* branches arises from the presence of a Stoneley wave trapped at the fluid–solid boundary at the top of the core. There is an equivalent set of modes for the inner core boundary which lies just to the left of the *SKS* propagation line.

The illustrations of the eigenfunctions for the spheroidal modes near 6 mHz in Figure 11.7 show the concentration of displacement in the Stoneley modes at the top and bottom of the fluid outer core. The upper row of modes in Figure 11.7 represent propagation largely confined to the mantle. The middle row include *ScS* equivalent modes and progressively deeper penetration into the outer core as the radial order increases. The bottom row in Figure 11.7 shows the behaviour of modes where displacement extends through the whole Earth, even though for an inner core mode such as ${}_{14}S_5$ comparatively little displacement is actually at the surface.

Observations of modal eigenfrequencies

A single strain record for the 1952 Kamchatka earthquake recorded in Pasadena suggested that the fundamental mode of the Earth had been observed. This result stimulated work on the calculation of the frequencies of the normal modes of the Earth, so that by the time of the great Chilean earthquake in May 1960 instrumental and theoretical seismology had converged to identify a wide range of normal modes. Subsequent studies used the 1964 Alaska event and a large deep earthquake near the Peru–Chile border in 1965 to develop an extensive mode catalogue. Many additional data of very high quality were collected from the very deep magnitude 8 Bolivian event in 1994. A long observation period (several days) without significant interference from other events is desirable for high precision estimates of the normal mode frequencies. The great Sumatra–Andaman earthquake (Mw 9.3), in late 2004, generated large amounts of low-frequency energy and significantly improved measurements of the lower-frequency spheroidal modes. Indeed the fundamental radial mode ${}_0S_0$ with a period close to 20 minutes, and little attenuation, could still be detected on sensitive instruments some months after the event.

Heavy dots in Figure 11.18 show those spheroidal modes for which high precision frequency estimates have been determined. The limited coverage of seismic stations around the globe means that such frequency observations require the combination of all available records for several days after a very large earthquake.

The spectra of seismic records following a great earthquake show a sequence of frequency peaks that can be enhanced by simple summation of the spectra from many stations. These peaks are normally the fundamental modes ($n = 0$) and the angular order can be identified by comparison with theoretical calculations. More sophisticated methods can be used to extract the spectral peak associated with a particular target mode; in particular the set of available records can be combined in such a way as to enhance the target mode and reduce noise. Such 'stacking' methods are quite powerful, particularly with the enhanced coverage of the globe with high-fidelity instruments in recent years since they depend on a good knowledge of the source parameters including the moment tensor. A stack aimed at a specific mode may well contain contributions from overtones of the same type, and a further step known as 'stripping' can help to separate the different modes. For low-frequency modes, biases associated with mode coupling due to the rotation of the Earth need to be removed. Masters & Widmer (1995) present a comprehensive catalogue of modal frequencies and related information.

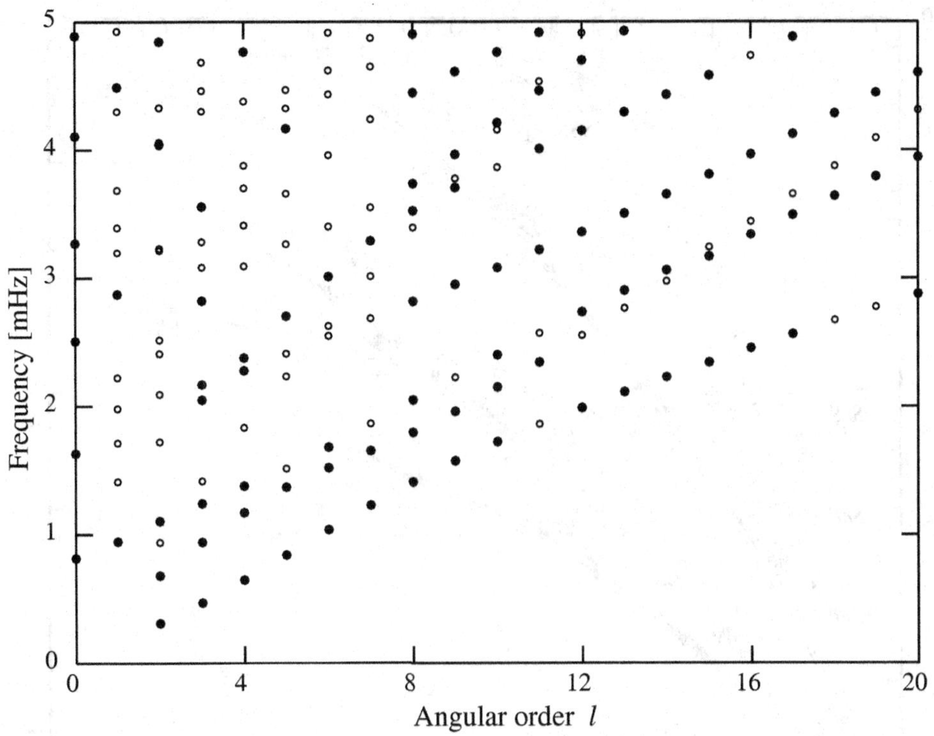

Figure 11.18. Low frequency spheroidal modes for which high precision frequency information is available are indicated by solid symbols, superimposed on the theoretical values for the PREM model.

Results from stations in low-noise environments, such as Antarctica, indicate the presence of normal mode peaks in the absence of significant earthquakes. Such observations have been made in many different parts of the world and suggest that there is continual low-level excitation of the Earth's normal modes, possibly due to coupling between the solid Earth and the atmosphere.

11.4.3 Comparison with observations

We have considered seismic wave phenomena in a spherical Earth, both the free oscillations and the travel times for higher-frequency seismic phases. Even though the dominant variation in physical properties is with depth, there are significant three-dimensional effects. So how far do the simplified models represent the observed behaviour?

In Figure 11.19 we show the travel times for the full set of picked phases for a set of 104 events (83 earthquakes, 21 explosions) that have a well controlled hypocentre. The travel times for the rich set of later phase readings (57 655 phases) are corrected to surface focus and are compared with the predictions from the AK135 model.

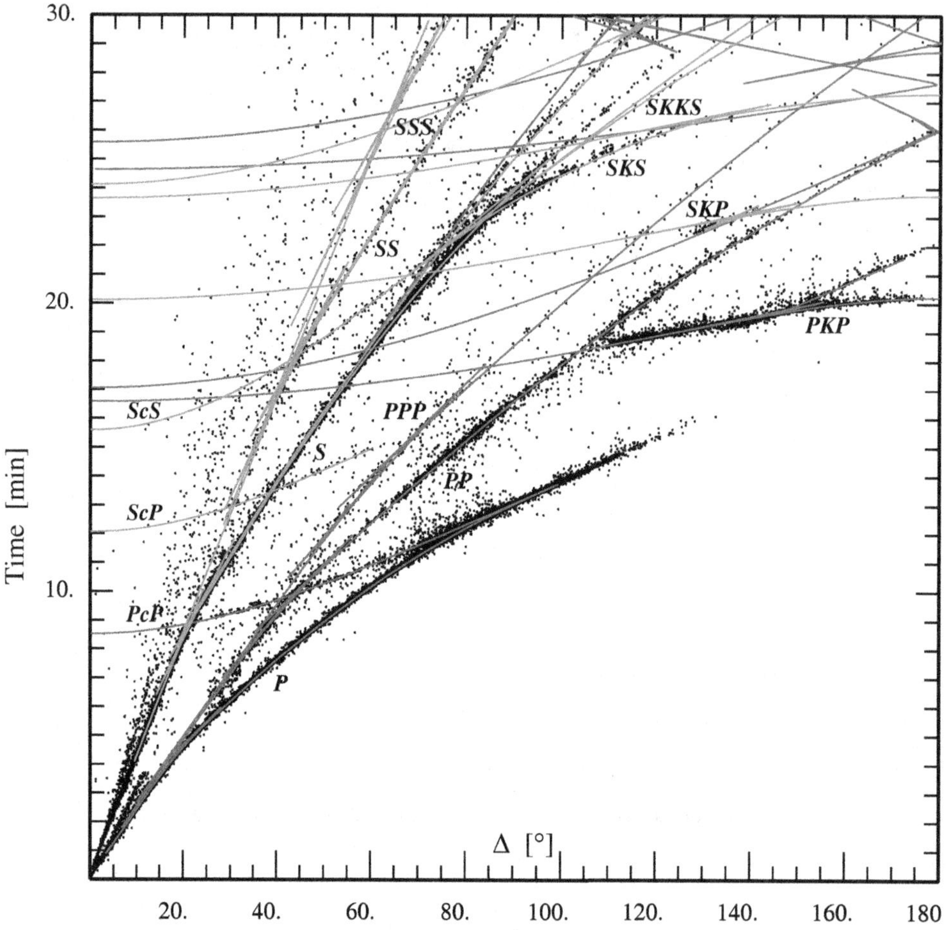

Figure 11.19. Travel times for the AK135 model for surface focus superimposed on the times of phases for a set of test events.

There is a good correspondence between the observed and calculated times, but we note some spread about the theoretical curves. The differences arise in part because of the limitations in picking the actual onset of phases on seismograms, but a significant component comes from the departure of the real Earth from the spherically symmetric model AK135. The differences in timing contain information about the three-dimensional structure encountered in passage between the source and receiver that can be exploited to delineate such structure (see Section 11.4.4).

The long-period disturbances propagating from the 2001 Mw 8.4 event in Peru across the globe are shown in Figure 11.20 for both the radial (R) and transverse (T) components of motion at the stations of the GEOSCOPE network. The radial component is taken along the great-circle between the hypocentre and the station and the transverse component is orthogonal to the radial in the horizontal plane.

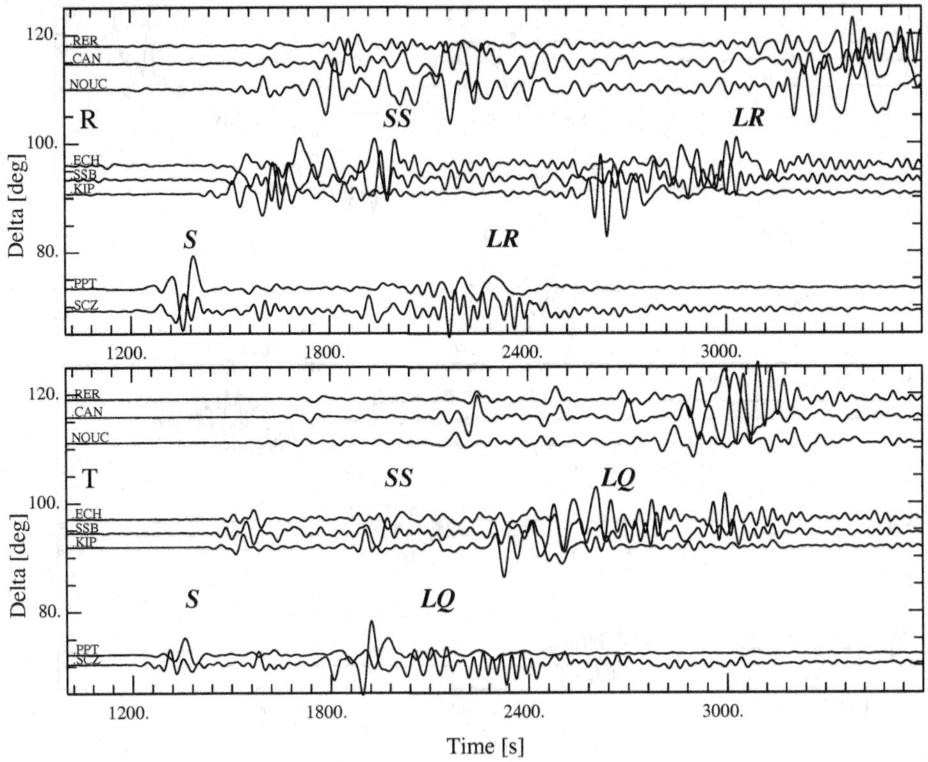

Figure 11.20. Radial (R) and transverse (T) component seismograms for the Mw 8.4 shallow focus event in Peru in June 2001 at GEOSCOPE stations, with a low pass filter to 0.025 Hz. The main seismic phases are indicated.

The main seismic phases are indicated in Figure 11.20; even for this very large event with a complex source waveform, there is a good correspondence to the predicted phase times from the AK135 model. The large late surface waves on both R and T components are the travelling wave equivalents of the free oscillations. The fundamental mode Love waves, *LQ*, on the transverse component are equivalent to the ${}_0T_l$ modes and arrive somewhat earlier in the seismograms than the Rayleigh waves, *LR*, on the radial component. The fundamental mode Rayleigh waves correspond to the ${}_0S_l$ modes, and are preceded by overtones with faster propagation speeds.

The equivalence of the travelling wave and normal mode representations for seismograms is illustrated by the synthetic seismograms presented in Figure 11.21 for the same stations. These seismograms are calculated by normal mode summation from (11.3.37) for the PREM model, using the source mechanism derived from analysis of waveforms for the Peru event (particularly the surface waves). A summation over spheroidal modes is used for the radial component

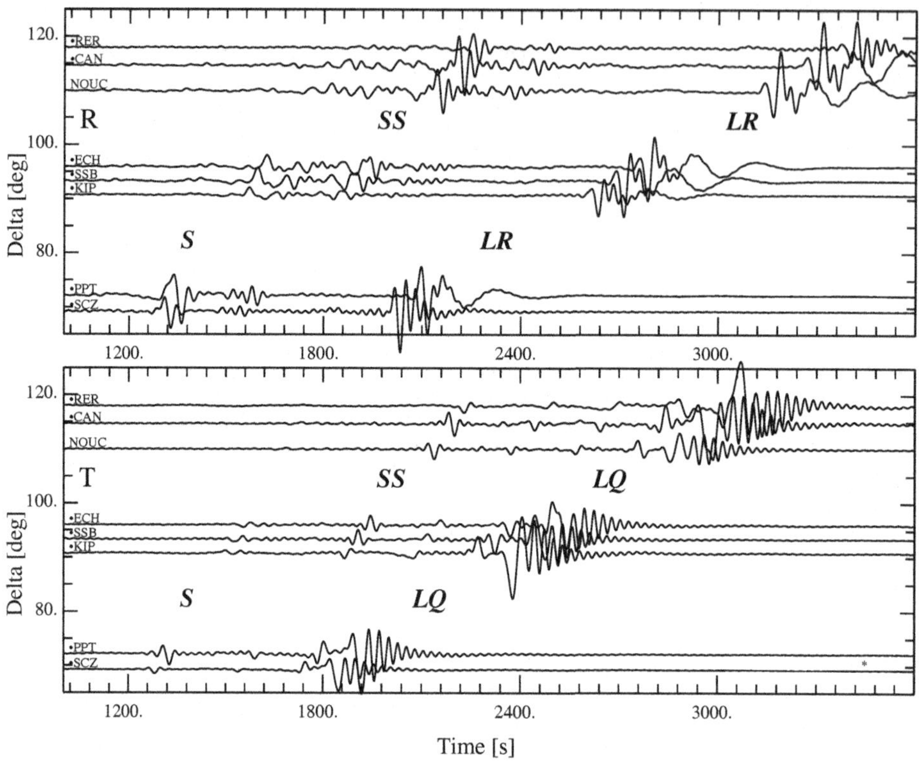

Figure 11.21. Simulation of the Peru event shown in Figure 11.20 using the sum of the first 15 normal mode branches for the PREM model, with again a low pass filter to 0.025 Hz. [Courtesy of K. Yoshizawa].

and over toroidal modes for the transverse component, fifteen mode branches are included which is sufficient to provide a synthesis of the *S* body waves.

The representation of the source of the Peru event used in Figure 11.21 is not exact, since it is derived from analysis of the observations; the point moment tensor model does not include the full effects of the source in this very large event and undoubtedly oversimplifies the seismic radiation characteristics. Even though the synthetic seismograms in Figure 11.21 have been calculated with this simple point source representation, there is a good general correspondence with the observed seismograms in Figure 11.20. However, the energy in the body waves is somewhat underestimated compared with the surface waves, and the balance between the *S* and *SS* energy is not quite correct. The spherical PREM model provides a good general prediction of the character and timing of both the Rayleigh and Love wave trains. We see from Figure 11.20 that there are noticeable differences in the nature of the observed seismograms for stations at similar distances. This arises because of the cumulative effect of three-dimensional structure on the passage of the surface waves, so that long oceanic paths impose a very different dispersion than an equivalent continental path. The differences in the long-period seismograms between the observations and the modal predictions form the basis of methods for determining the large-scale variations in three-dimensional Earth structure.

11.4.4 Imaging three-dimensional structure

As we have seen, the use of a model with purely radial variations in properties provides an excellent representation of the major behaviour of seismic phases. The presence of three-dimensional structure manifests itself in a variety of ways, which have been exploited to provide images of such structure; Kennett (2002) provides a detailed discussion of the styles of tomographic inversion and their contrasting properties.

The splitting of the frequencies of the free oscillations of the Earth with respect to the angular order m is an important source of information for the lowest frequency modes that have a sensitivity to density and seismic wavespeed. At intermediate frequencies the commonest procedure comes through the fitting of portions of seismograms using a perturbation development based on the normal mode theory outlined above. Surface waves can be described through the summation of simple mode branch contributions, but body waves need multiple branch contributions with coupling between the coefficients.

The arrivals of seismic phases are exploited in a variety of ways. High frequency information can be derived from the compilations of readings from seismic stations across the globe. Careful reprocessing of such catalogues, including relocation of events and association of arrivals, provides a major source of information for *P* and *S* waves and many later phases (see, e.g., Engdahl, Buland & van der Hilst, 1998). At somewhat lower frequencies a substantial data set has been built up for both absolute and differential times through the use of cross-correlation methods, including the use of synthetic seismograms. The way in which such lower frequency waves interact with structure is not fully described by ray theory and a number of schemes have been developed to represent the zone of interaction around the ray path.

Many recent studies use a wide range of different styles of information to try to achieve the maximum level of sampling of the Earth's mantle (see, e.g., Masters et al., 2000). However, such studies face the complication of assigning relative weighting to different classes of information and also of combining information in different frequency bands.

Three-dimensional models of the variation in seismic wavespeeds are normally displayed as deviations from a radially-stratified reference model. Many studies of seismic tomography have concentrated attention on a single wavetype (particularly *S*). Current *S* images derived from long-period seismic data (such as free oscillations, waveforms of multiple *S* phases) provide good definition of heterogeneity with horizontal scales larger than 1000 km across the whole of the mantle. The model of Megnin & Romanowicz (2000) based on just the use of *SH* waves (Figure 11.22) illustrates the capabilities of waveform inversion on a global scale. This model was derived using coupled normal modes to represent long-period body wave phenomena, with an expansion of the perturbations of wavespeed structure in terms of spherical harmonics to order 24 and a spline representation in radius. We see in Figure 11.22 the strong variations in shear wavespeed near the surface and the core–mantle boundary compared with the more modest variations in the mid-mantle.

Somewhat higher resolution of three-dimensional structure can be achieved using arrival times extracted from seismograms for both P and S waves, but at the cost of less coverage of mantle structure. A prerequisite for good quality imaging in tomographic inversion is multi-directional sampling through any

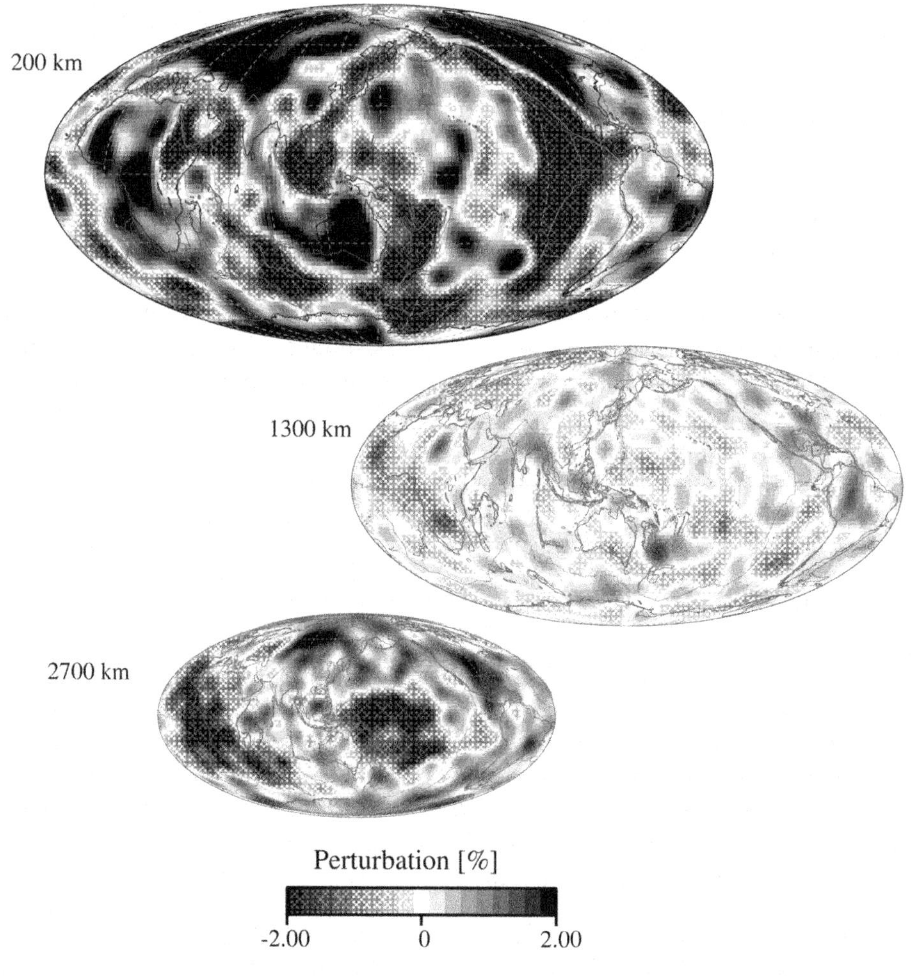

Figure 11.22. Maps showing the variation of shear wavespeed at depths of 200 km, 1300 km and 2700 km for the *SH* model SAW24B16 of Megnin & Romanowicz (2000) displayed as deviations from the PREM reference model (Dziewonski & Anderson, 1981). The ellipses are scaled in proportion to the radius of the section.

zone. The configuration of earthquake sources and, mostly, continental seismic stations restricts the sampling for body waves, unless multiply reflected body waves can be exploited. In those regions where different studies have achieved a comparable coverage, the major features of the tomographic images are in reasonable correspondence, even though the details vary. The highest levels of heterogeneity are found near the Earth's surface and near the core–mantle boundary (see Figure 11.22). More subtle features appear in the mid-mantle, including relatively narrow zones of elevated wavespeed that are likely to be produced by past subduction.

In the uppermost mantle, the ancient cores of continents stand out with fast wavespeeds, while the mid-ocean ridge system and orogenic belts show slow wavespeeds compared with the reference model. Below 400 km depth the high-wavespeed anomalies are mostly associated with subduction zones; in some regions they extend to around 1100 km depth, but in a few cases tabular fast wavespeed structures seem to extend to 2000 km or deeper. The base of the mantle shows long-wavelength regions of higher wavespeeds, most likely associated with past subduction, and two major regions of slow wavespeed beneath the central Pacific and southern Africa (see Figure 11.22) which may represent sites of upwelling of hotter material.

The early success of seismic tomography came from the striking images of large-scale three-dimensional structure and, later, of the details of subduction zones. The interpretation of such images is based firmly on the variations in seismic wavespeed. Thermal processes can be expected to play a major role, but chemical heterogeneity could also be important particularly in the regions with strong variability at the top and bottom of the mantle.

Results for a single wavespeed are not sufficient to indicate the nature of the observed anomalies. Recent developments in seismic imaging are therefore moving towards ways of extracting multiple images in which different aspects of the physical system are isolated. This may be from *P* and *S* images (preferably from common data sources) or via the use of the bulk modulus, shear modulus and density. Such multiple images of mantle structure encourage an interpretation in terms of processes and mineral physics parameters, since the relative variation of the different parameters adds additional information to the spatial patterns.

In principle, a significant increase in understanding of heterogeneity can be achieved if both *P* and *S* information are exploited. The *P* wavespeed α depends on both the bulk modulus κ and the shear modulus μ as

$$\alpha = [(\kappa + \tfrac{4}{3}\mu)/\rho]^{1/2}, \tag{11.4.1}$$

where ρ is the density. We can thus isolate the dependence on the shear modulus μ and the bulk modulus κ, by working with the *S* wavespeed

$$\beta = [\mu/\rho]^{1/2}, \tag{11.4.2}$$

and the bulk-sound speed ϕ derived from both the *P* wavespeed α and the *S* wave speed β

$$\phi = [\alpha^2 - \tfrac{4}{3}\beta^2]^{1/2} = [\kappa/\rho]^{1/2}, \tag{11.4.3}$$

which isolates the bulk modulus κ. This style of parameterisation has been employed in a number of studies (see Masters et al., 2000, for a comparison).

An unfortunate complication in the combined use of *P* and *S* information comes from the uneven geographic distribution of data. Whereas *S* wave data are available from globe circling paths; *P* wave information, dominantly from travel times, is dictated by the location of seismic sources and receivers. There is some component of *P* wave information in long period waveform data but this does not provide strong constraints on mantle structure. It therefore remains difficult to compare full global coverage from *S* with information derived from *P* waves with a much more limited geographic coverage. The inclusion of later phases and differential times helps but the *P* travel times still provide the dominant information.

An alternative approach is to restrict attention to paths for which both *P* and *S* information is available; the consequence is that sampling of the mantle is reduced but the reliability of the images is high and direct assessment can be made of the relative behaviour of either *P* and *S* wavespeed variations, or the variations in bulk-sound speed ϕ and shear wavespeed β (see, e.g., Kennett & Gorbatov, 2004).

In Figure 11.23 we show results from the model of Kennett & Gorbatov (2004) which was constructed with $2^\circ \times 2^\circ$ cells and 18 layers through the mantle, using a joint inversion of *P* and *S* arrival-time data with common source–receiver pairs, with light damping and a broad residual range designed to capture strong features in the uppermost and lowermost mantle. A linearised inversion is first performed for *P* and *S* separately, and then a joint inversion for ϕ and β is undertaken with three-dimensional ray tracing.

For the main shield regions, such as western Australia, there is a strong shear wavespeed anomaly (up to 6% or more) down to about 250 km accompanied by a somewhat weaker bulk-sound speed signature (Figure 11.23a). In contrast, the major orogenic belts from southern Europe to Iran (and also western N. America) show rather slow *S* wavespeeds, accompanied by fast bulk-sound speeds. A similar behaviour is evident for the Red Sea and E. African rifts. The anti-correlation of bulk-sound and shear wavespeed is pronounced, a large component can be thermal because of the very strong reduction of the shear modulus as the solidus is reached, but volatiles may also be significant. In eastern Asia we see some portions of subducted slab, e.g., the Ryukyu arc, standing out very clearly by their high shear wavespeeds from the lower background; these also have some expression in bulk-sound speed.

Only in a few places are there clear indications of slab-like behaviour extending below 1300 km depth, most notably in southern Asia and beneath the Americas. These two features have been linked to subduction in the past 80 Ma at the northern edge of the Tethys Ocean and of the now-extinct Farallon plate beneath

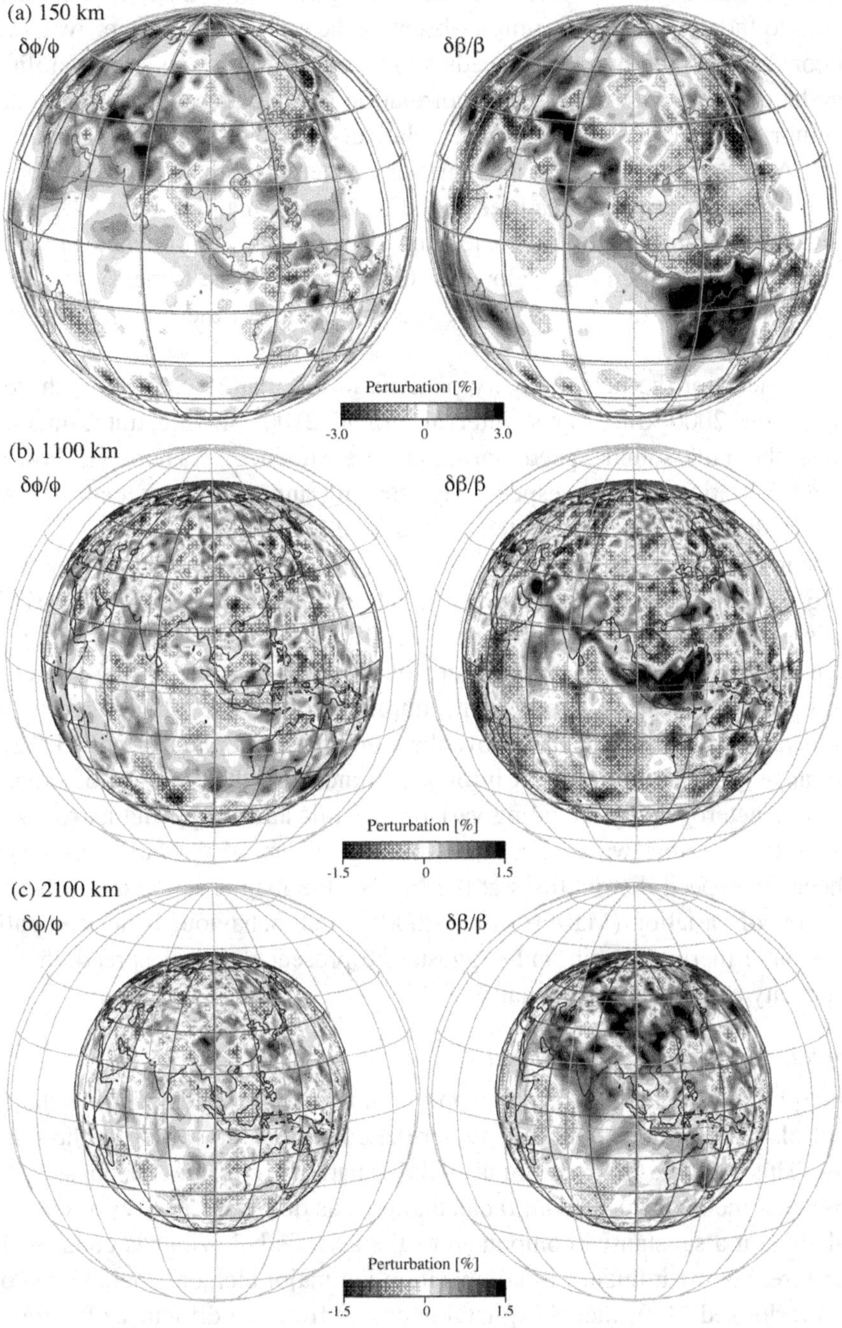

Figure 11.23. Variation of bulk-sound speed $\delta\phi/\phi$ and shear wavespeed $\delta\beta/\beta$ relative to the AK135 reference model. (a) layer from 100-200 km depth, (b) layer from 1000-1200 km depth, (c) layer from 2000-2200 km depth.

the Americas. In Figure 11.23(b) we see striking fast shear anomalies extending from Iran to Indonesia that are almost absent in the bulk-sound image. Away from the major features, the two wavespeeds show comparable levels of variability on intermediate scales, with a weak anti-correlation in the patterns that is compatible with minor thermal fluctuations. The relatively narrow, slab-like structures in southern Asia and the Americas become less coherent with depth and appear in places to link with drip-like features in the lowermost mantle. Cross-sections of these structures can tend to be misleading because of the influence of oblique cuts. Indeed, it is difficult to follow the behaviour in three dimensions because of the various factors that can influence the amplitudes of the imaged wavespeed variations.

The character of the heterogeneity patterns in the mantle changes with depth, notably below 2000 km. In the interval around 2100 km, illustrated in Figure 11.23(c), the bulk-sound speed variations are subdued. There is, however, significant variation in S wavespeed with striking anomalies, especially in Asia, with a very different pattern from that seen at 1100 km depth.

The character of the heterogeneity regime in the mantle undergoes further change as the core–mantle boundary is approached. The amplitude of bulk-sound variation, which is very low as we have seen near 2100 km, increases with depth towards the core–mantle boundary with a pattern of variation that increasingly becomes uncorrelated with shear wavespeed. The amplitude of the shear heterogeneity also increases rapidly with depth. Just above the core–mantle boundary, in the D'' zone, a wide range of different pieces of information indicates the presence of extensive but variable heterogeneity including variable seismic anisotropy and narrow zones with very low seismic wavespeeds. A striking feature of the bulk-sound speed and shear wavespeed distributions at the base of the mantle is the discordance in the patterns of variation (Masters et al., 2000). Such behaviour is not compatible with a simple thermal origin and suggests the presence of widespread chemical heterogeneity at the base of the mantle.

Interpretation of seismic heterogeneity

The interpretation of the results of seismic tomography depends on understanding the controls on seismic wavespeeds under the conditions prevailing in the Earth's mantle. This requires a strong input of information from mineral physics, both experimental methods and *ab initio* calculations, as discussed in Sections 9.4, 9.5. It is likely that a substantial component of the seismic behaviour is controlled by temperature, but the influence of composition via major element chemistry should not be overlooked. A further complication comes from the difficulty of extracting absolute velocity information from tomography, the influences of the damping and regularisation used in the inversion tend to lead to an underestimation of the amplitude of the anomalies, even though the spatial pattern may be appropriate.

In particular, we should note that the requirements for tomographic interpretation are somewhat different than for equations of state. In Chapter 6 we have discussed the development of material properties under hydrostatic pressure. The Eulerian (Birch–Murnaghan) formulation of finite strain is generally preferred because of the weak dependence on the pressure derivatives of the bulk modulus. This Eulerian approach gets us to a suitable reference state, but the information we now need on perturbations from that state is best described by an incremental Lagrangian treatment for non-hydrostatic strains, for which anelastic effects can be significant. We have already noted in Section 11.1.2 the rapid changes in shear modulus with elevated temperature as the solidus is approached. The consequence is a strongly non-linear dependence of the elastic moduli with temperature. To achieve the same size of change in seismic wavespeed will require a larger temperature contrast for fast anomalies than for slow anomalies where the temperature derivative is larger.

Because of the strong influence of shear wavespeed anomalies on P wavespeed images (e.g., Kennett & Gorbatov, 2004), the ratio of the P and S anomalies

$$R_{\alpha\beta} = \delta \ln \alpha / \delta \ln \beta \tag{11.4.4}$$

has little diagnostic value for the influence of composition, whereas the equivalent ratio for bulk-sound speed

$$R_{\phi\beta} = \delta \ln \phi / \delta \ln \beta \tag{11.4.5}$$

is more suitable for recognising the competing effects of temperature and composition.

11.5 Earthquakes and faulting

The description of an earthquake source as a propagating rupture across a fault plane captures much of the behaviour with respect to seismological observations. In reality there is a dynamic interaction between the propagating rupture front and the surrounding medium. Because the rheological properties vary with depth, the physical behaviour in a shallow event can differ significantly between the near surface and depth (Scholz, 1990).

For smaller earthquakes, the point moment tensor description introduced in Section 11.2 provides a good representation of the behaviour, and the radiation patterns for P and S waves can be used to infer the mechanism for the event. For shallow events the P waves overlap with the pP and SP arrivals reflected from above the source to give a complex interference packet, whose character depends on source depth. Waveform modelling procedures can then be used for the onset of the body waves to extract the mechanism, source depth and time history of the event (see, e.g., Stein & Wysession, 2003, Chapter 4).

For all events with magnitude greater than 5.5, a systematic procedure is applied to extract the centroid moment tensor by matching long-period seismograms from across the globe with modal synthetics for frequencies below 0.0125 Hz, with some allowance for major three-dimensional structure. The catalogue at http://www.globalcmt.org/CMTsearch.html covers the period from 1976 onwards.

Figure 11.24. ENVISAT interferogram of the Bam earthquake in Iran, 2003 December 26, Mw 6.6; the InSAR analysis was carried out using DORIS software. Each interferometric fringe represents 28.1 mm of line-of-sight deformation. The fault is predominantly strike-slip, with a peak slip of around 2 m. The elastic deformation cannot be explained by slip on a single planar fault (Funning et al., 2005). The areas of low coherence in the centre of the image are the heavily damaged Iranian cities of Bam and Bavarat. ENVISAT SAR data were made available through the European Space Agency (ESA). [Courtesy of J. Dawson.]

The number of high-quality broad-band stations across the globe has significantly increased in recent years, allowing more events to be characterised.

The seismological observations do not, by themselves, resolve the difference between the fault plane and the orthogonal auxiliary plane for a dislocation source. However, the patterns of ground deformation for shallow events provide additional information. As a consequence there has been a fusion of results from geodetic and seismic techniques to understand earthquake behaviour. Measurements of post-seismic displacements using space-geodetic methods such as GPS provide important constraints on possible fault behaviour. For very large earthquakes such as the M_w 9.3 Sumatra–Andaman event on 2004 December 26, perceptible deformation can be measured thousands of kilometres from the event itself.

In certain circumstances, a very direct view of ground deformation induced by an earthquake can be extracted from interferograms from satellite based synthetic-aperture radar (InSAR). The phase difference between images before and after the earthquake can be mapped into interference fringes that map the patterns of displacement, with particular sensitivity to vertical deformation. In particular for near-surface faulting in arid areas, the InSAR approach provides vivid images of ground deformation at fine resolution (~10 m) in areas 10–100 km wide, that can be matched to models of fault behaviour (Figure 11.24). For the 2003 earthquake that devastated the ancient town of Bam in Iran, InSAR images revealed the main fault trace that had been missed in the original post-event survey because of its limited surface expression (Funning et al., 2005).

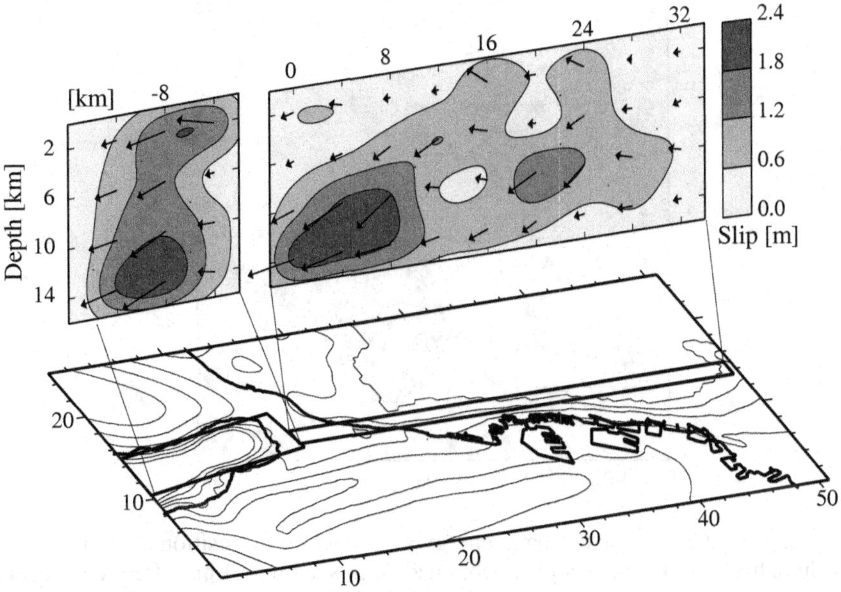

Figure 11.25. The slip distribution for the 1995 Kobe earthquake, Japan, proposed by Yoshida et al. (1996) related to a map view with contours for sediment thickness.

A major earthquake is not well described by a simple rupture, the displacement on the fault is spatially varying and episodic in time. The combination of near-field observations from strong ground motion instrumentation and teleseismic results from distant seismic stations can be used to constrain models of rupture behaviour, described as the superposition of the effects of many sub-faults. Characteristically, there are patches where relatively high moment release occurs, often referred to as *asperities*, and there may also be *barriers* where rupture is impeded or even stopped. Figure 11.25 illustrates the slip distribution inferred by Yoshida et al. (1996) for the 1995 Kobe earthquake from a wide range of information. There is bilateral slip on two segments with differing dip, and multiple concentrations of moment release along the extended fault trace.

Estimates of the slip distribution can now be made quite quickly following a major event. In such rapid analysis it is usual to adopt a single fault plane. More flexibly parameterised models, including multiple fault segments, tend to evolve as the full range of information with geodetic constraints becomes available.

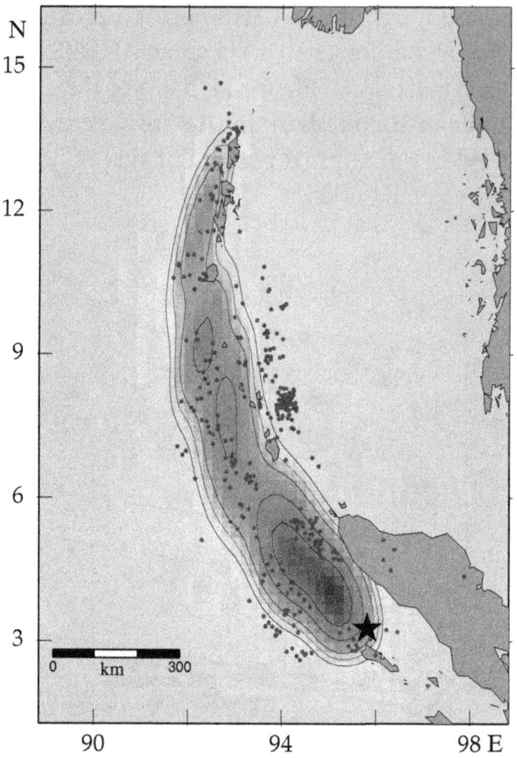

Figure 11.26. Image of radiated seismic energy as a function of position along the arc for the 2004 Sumatra–Andaman earthquake obtained from stacking of high frequency energy at the Japanese Hi-Net Array by Ishii et al. (2005). The changes in the intensity of the radiated energy have a close correspondence to structural features in the slab.

For a truly great earthquakes such as the M_W 2004 Sumatra-Andaman event, with a fault length exceeding 1300 km, the pattern of slip suggests a *domino effect* whereby the stress induced by rupture on one fault segment triggers slip on the next. The source time function includes at least two such stutters (Ammon et al., 2005), which correspond to distinct geographic points in the earthquake rupture. This great thrust event occurs where the Indo–Australian plate descends beneath the Asian plate and the Andaman micro-plate. The variable energy release along the Sumatra–Andaman arc is illustrated in Figure 11.26. It is likely that segmentation of the slip is associated with the prior structure of the subduction zone, through changes in the dip, strike and physical properties of the seismogenic part of the slab interface.

12

Lithospheric Deformation

12.1 Definitions of the lithosphere

The concept of a lithosphere arises from the presence of a zone of strength near the Earth's surface that is capable of transmitting and resisting stress. A consequence of the complex rheology of the Earth is that the apparent thickness of this zone becomes smaller as the time scale of the processes being considered becomes longer. This has lead to a variety of definitions of *lithosphere*, based on particular classes of observations, that are not necessarily mutually compatible.

We can think of the *mechanical lithosphere* as representing the outer part of the Earth where stress can be transmitted over geological time scales. This is linked to, but not identical with, the *thermal lithosphere*, which is that region where thermal energy is largely transferred by heat conduction (Figure 12.1).

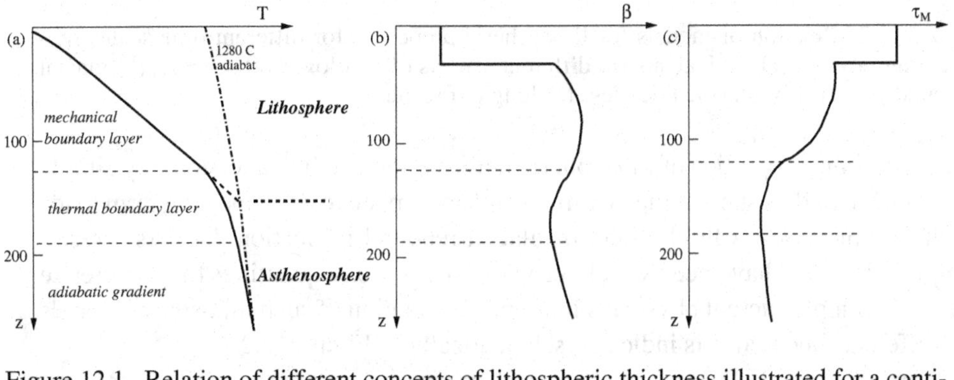

Figure 12.1. Relation of different concepts of lithospheric thickness illustrated for a continental situation: (a) thermal concept through the transition from a conductive geotherm to an adiabat, (b) seismic definitions based on the properties of the shear wavespeed distribution with depth, and (c) the viscous relaxation time as a function of depth.

In seismological studies the lithosphere is commonly taken to be associated with the region of elevated seismic wavespeeds (the 'lid'), with a base assigned either in the region of maximal vertical velocity gradient, or on entry into a zone of lowered shear wavespeed. Estimates of the temperature distribution associated with seismic wavespeed (e.g., McKenzie, Jackson & Priestley, 2005) suggest that this occurs in the same neighbourhood as the thermal transition (Figure 12.1). Although the seismic lid may have a relatively sharp base beneath the oceans, there are only a few observations suggesting very rapid velocity gradients in the continental environment.

A useful viewpoint for understanding the issue of the dependence of the apparent thickness of the lithosphere on the time scale of the phenomenon comes from the Burgers model for a viscoelastic medium introduced in Section 4.6. Such a system displays (i) instantaneous elasticity that can be associated with seismological observations, (ii) intermediate term viscoelastic response that can be used to describe the rebound after the removal of ice loading at the end of the ice age, and (iii) long term viscous flow. The time scale for the viscous process τ_M varies substantially with depth as indicated in Figure 12.1c.

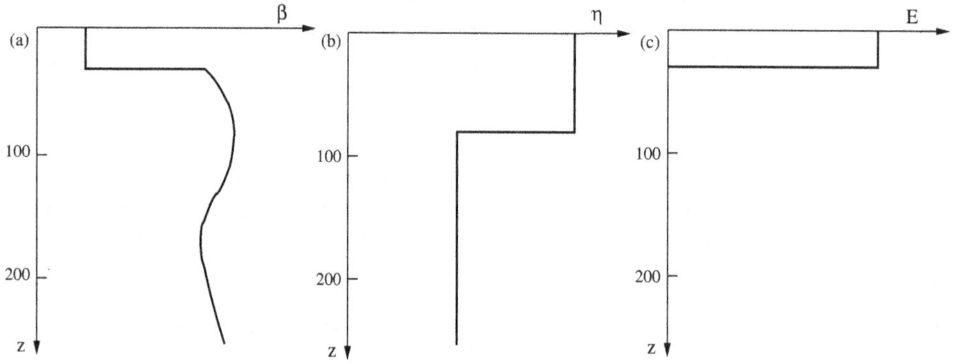

Figure 12.2. Relation of models for lithospheric properties for different time scales representing mechanical equivalents for different aspects of rheology: (a) seismic, (b) glacial rebound, (c) effective elastic thickness for long term loads.

Much of the properties of lithospheric deformation can be understood by simple mechanical models such as the flexure of an elastic plate to describe long term loads or the simple viscous fluid model of glacial rebound in Section 7.4. As a result simplified models have been widely used to extract proxy quantities to characterise the more complex actual physical situation. The relation of such lithosphere models for different time scales is indicated schematically in Figure 12.2.

12.2 Thermal and mechanical structure

12.2.1 Thermal conduction in the oceanic lithosphere

Oceanic lithosphere is formed at mid-ocean ridges, indicated by the double lines in Figure 12.7, and then is carried away from the ridges by the movement of the oceanic plates. Hot material from the mantle emerges at the axes of the mid-ocean ridges and forms juvenile oceanic crust, further basaltic material solidifies onto the base as the new oceanic lithosphere moves away from the axis. Hydrothermal circulation is important through the rapidly cooled materials created just at the ridge, and heat flux measurements are rather scattered. Heat flow stabilises for older lithosphere as hydrothermal effects become less important. As the lithosphere moves away from the mid-ocean ridge it cools and contracts with an increase of depth that is roughly proportional to the square root of age out to 65 Ma or so, but then deepens rather more slowly.

Much of this behaviour can be captured with relatively simple thermal models. Consider a two-dimensional situation, independent of the y-coordinate, with asthenospheric material ascending along the ridge axis at $x = 0$ with constant temperature T_M. With a plate transport speed v_x in the x-direction, the temperature is governed by (8.4.3)

$$\rho C_p \left[\frac{\partial T}{\partial t} + v_x \frac{\partial T}{\partial x}\right] = k\left[\frac{\partial^2 T}{\partial x^2} + \frac{\partial^2 T}{\partial z^2}\right]. \tag{12.2.1}$$

In a steady state $\partial T/\partial t = 0$, and we can make a further simplification if we assume that the conduction of heat in the horizontal direction is negligible compared with the advective contribution from $v_x \partial T/\partial x$ and vertical conduction. Thus, neglecting the $\partial^2 T/\partial x^2$ term in (12.2.1), we have

$$\frac{k}{\rho C_p}\frac{\partial^2 T}{\partial z^2} = \kappa_H \frac{\partial^2 T}{\partial z^2} = v_x \frac{\partial T}{\partial x}. \tag{12.2.2}$$

We now introduce the spreading time t_s so that $x = v_x t_s$, and then we recover the simple thermal conduction equation, in terms of t_s

$$\kappa_H \frac{\partial^2 T}{\partial z^2} = \frac{\partial T}{\partial t_s}. \tag{12.2.3}$$

The boundary conditions are $T = T_M$ at $x = 0$, and the top surface is maintained at zero temperature ($T = 0$ at $z = 0$); the requisite solution is provided by (8.4.28), so that

$$T(z, t_s) = T_M \operatorname{erf}\left(\frac{z}{\sqrt{4\kappa t_s}}\right), \tag{12.2.4}$$

with associated surface heat flow

$$Q_0(t_s) = -k\left.\frac{\partial T}{\partial z}\right|_{z=0} = -\frac{k T_M}{\sqrt{\pi \kappa t_s}}. \tag{12.2.5}$$

Thus for this half-space cooling model, the surface heat flux is proportional to $1/\sqrt{t_s}$. Isotherms, surfaces of constant temperature, will correspond to constant argument of the error function $z_c = 2c\sqrt{\kappa t_s}$, and hence the depth to a given temperature increases as the square root of the lithospheric age. It is conventional to take the base of the lithosphere as specified by such an isotherm, e.g., 1300°C.

The cooling lithosphere shrinks and the depth of the seabed can be estimated by assuming full isostatic compensation (Figure 12.3). We assume a compensation

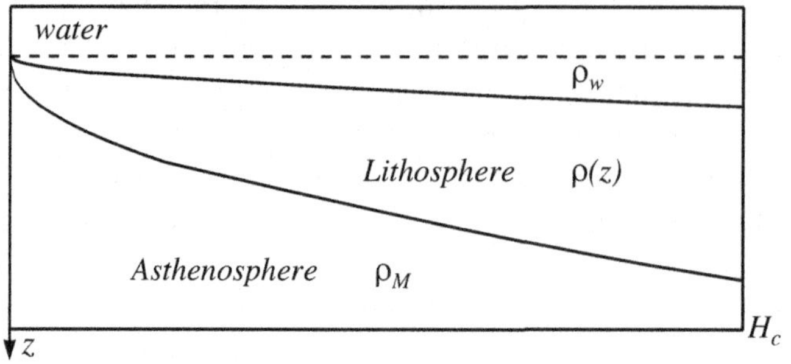

Figure 12.3. Oceanic depth profile associated with lithospheric cooling with full isostatic compensation

depth H_c in the mantle beneath the level of the deepest oceanic lithosphere and then require the mass in any vertical column extending to H_c to be the same.

We take the water depth to be w_0 at the ridge axis and the depth origin at the seafloor at the axis. The vertical traction at the level H_c at the axis

$$\sigma_{zz}(x=0, H_c) = g\rho_w w_0 + g\int_0^{H_c} dz\, \rho_M, \tag{12.2.6}$$

where ρ_w is the density of sea water and ρ_M is the density of the hot asthenosphere. Off the ridge axis, for lithosphere thickness L and additional water depth w,

$$\begin{aligned}\sigma_{zz}(x, H_c) &= g\rho_w(w_0 + w) + g\int_0^{H_c} dz\, \rho(z)\\ &= g\rho_w(w_0 + w) + g\int_w^{w+L} dz\, \rho(z) + g\int_{w+L}^{H_c} dz\, \rho_M.\end{aligned} \tag{12.2.7}$$

For the tractions to be equal at different horizontal positions we require

$$\rho_w w_0 + \rho_M H_c = \rho_w(w_0 + w) + \int_w^{w+L} dz\, \rho(z) + \rho_M(H_c - w - L), \tag{12.2.8}$$

and so the excess water depth is given by

$$w(\rho_M - \rho_w) = \int_0^L dz\, (\rho(z) - \rho_M). \tag{12.2.9}$$

The density profile $\rho(z)$ is controlled by the temperature variation through the effects of thermal contraction,

$$\rho(T) = \rho_M[1 - \alpha_{th}(T - T_M)] \tag{12.2.10}$$

in terms of the thermal expansion coefficient α_{th}, and so we can evaluate the integral in (12.2.9) to give

$$w(\rho_M - \rho_w) = \rho_M \int_0^L dz\, \alpha_{th}(T_M - T(z)). \tag{12.2.11}$$

With the temperature distribution (12.2.4), the excess water depth

$$w(t_s) = \frac{\rho_M}{\rho_M - \rho_w} \int_0^L dz\, \alpha_{th} T_M \left[1 - \operatorname{erf}\left(\frac{z}{\sqrt{4\kappa t_s}}\right)\right], \tag{12.2.12}$$

Once we are away from the axis the integral in (12.2.12) can be approximated by its asymptotic value as $L \to \infty$, $(4\kappa t_s/\pi)^{1/2}$, and so

$$w(t_s) \approx \frac{2\alpha_{th}\rho_M T_M}{\rho_M - \rho_w} \left(\frac{\kappa t_s}{\pi}\right)^{\frac{1}{2}}. \tag{12.2.13}$$

We can improve on the simple half-space model by taking a plate model in which the base of the rigid oceanic lithosphere derived from cooled asthenosphere is defined by an isotherm. In such a plate model the base of the lithosphere at $z = L$ is assumed to have the same temperature T_M as at the vertical ridge axis. We introduce the thermal Reynolds number $\mathfrak{R} = v_x L/2\kappa$ and can then find a series solution to (12.2.1). In the steady state when $\partial T/\partial t = 0$ we obtain a similar form to (8.4.29)

$$T(z, t_s) = T_M \left[\frac{z}{L} + \sum_{n=1}^{\infty} \frac{2}{n\pi} \sin\left(\frac{n\pi z}{L}\right) \exp\left(-[(\mathfrak{R}^2 + n^2\pi^2)^{1/2} - \mathfrak{R}]\frac{v_x t_s}{L}\right)\right]. \tag{12.2.14}$$

The isotherms initially deepen as the square root of age, but flatten out for older lithosphere as the steady-state thermal structure approaches a linear gradient in depth. The excess water depth can again be estimated from (12.2.11) using the isostatic assumption and has a similar age dependence to the isotherms.

12.2.2 Mechanical deformation

The lithosphere can be represented as a layered composite of differing rheological properties underlain by viscoelastic material with reduced viscosity associated with the asthenosphere. As noted in Section 12.1, the differing rheological attributes manifest themselves through a variation in resistance to load depending on the time scale of the imposed deformation (see Figure 12.1).

In the crust at shallow depth the dominant response is elastic, but the material is brittle, with finite strength, and failure occurs in localized zones. With increasing depth we pass into the ductile regime where flow is distributed with a temperature dependent relation between stress and strain, such as (9.3.3). The relative contributions of the different styles of rheological properties depend on a range of factors such as mineral properties and chemistry, pore geometries, stress conditions, and lithostatic and pore pressure. A useful summary of the nature of the transition from the brittle to ductile and plastic regimes and associated experimental evidence is provided by Evans & Kohlstedt (1995).

For deformation at seismic periods the dominant response comes from

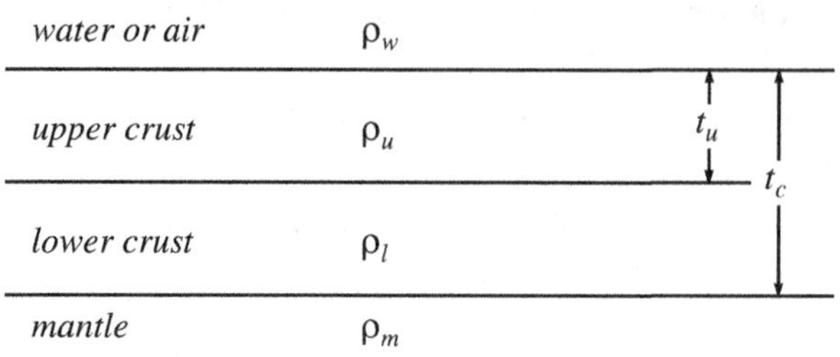

Figure 12.4. Model used to calculate crustal response to surface and internal loading. Loads are imposed: (1) at the surface by imposing a layer of thickness $s_1(x)$ and density ρ_u, (2) at the interface between upper and lower crust by moving the interface keeping the crustal thickness constant, and (3) at the Moho by adding a layer with a thickness $s_3(x)$ and density ρ_m.

instantaneous elasticity; whereas on the scale of millennia, associated with loading and unloading of ice in the glacial era, both the elastic and viscous components come into play. For long term loads, such as the support of major topography with time scales of many millions of years, the focus is again on elastic properties, but now for the upper part of the lithosphere in the crust and, perhaps, the uppermost mantle.

A relatively simple model that can capture many aspects of crustal deflection under long-term loads is provided by a thin sheet model with a laminated crust as in Figure 12.4. The upper crust is underlain by a higher density lower crust and loading is applied at the surface or Moho by the addition of layers with appropriate density, and in the interior by moving the interface between the upper and lower crust. The deflection w resulting from a load is assumed to be independent of depth in the thin sheet approximation.

When a layer of thickness $s_1(x)$ is added to the surface the resulting deflection w_1 is given by

$$\left[D\nabla^4 + g(\rho_m - \rho_w)\right] w_1 = -g(\rho_u - \rho_w)s_1, \qquad (12.2.15)$$

where D is the flexural rigidity of the plate, g the acceleration due to gravity, ρ_m the density of the mantle, ρ_u the density of the upper crust and ρ_w is the density of the fluid (water or air) overlying the crust. The hydrostatic restoring force $g(\rho_m - \rho_w)w_1$ arises because mantle with density ρ_m has been replaced through the deflection w_1 by the upper fluid of density ρ_w. The flexural rigidity of the plate D is related to the elastic thickness T_e by

$$D = \frac{ET_e^3}{12(1-\upsilon^2)}, \qquad (12.2.16)$$

where E is Young's modulus, and υ is Poisson's ratio. A simple derivation of (12.2.16) is provided by Watts (2001) §3.4.

If we take the Fourier transform of (12.2.15) with respect to the horizontal coordinate x we obtain

$$\left[Dk^4 + g(\rho_m - \rho_w)\right] \bar{w}_1 = -g(\rho_u - \rho_w)\bar{s}_1, \tag{12.2.17}$$

where the transformed variables are indicated by overbars and k is the horizontal wavenumber. The net surface topography $e_1(x)$ is the combination of the imposed load and the deflection so that

$$e_1(x) = s_1(x) + w_1(x). \tag{12.2.18}$$

The resulting gravity anomaly $\Delta\bar{g}_1$ for wavenumber k is given by (McKenzie, 2003)

$$\Delta\bar{g}_1 = 2\pi G\{(\rho_u - \rho_w)\bar{e}_1 + (\rho_l - \rho_u)e^{-kt_u}\bar{w}_1 + (\rho_m - \rho_l)e^{-kt_c}\bar{w}_1\}, \tag{12.2.19}$$

where G is the gravitational constant and, as shown in Figure 12.4, t_u is the thickness of the upper crust and t_c that for the whole crust. From (12.2.17) and (12.2.17) we have

$$\left[Dk^4 + g(\rho_m - \rho_u)\right] \bar{w}_1 = -g(\rho_u - \rho_w)\bar{e}_1, \tag{12.2.20}$$

and we can therefore express the gravity anomaly $\Delta\bar{g}_1$ in terms of the surface topography $\bar{e}_1$ through an admittance Z_1,

$$\Delta\bar{g}_1 = Z_1\bar{e}_1 = 2\pi G(\rho_u - \rho_w)(1 + Z_1')\bar{e}_1, \tag{12.2.21}$$

where Z_1' is given by

$$Z_1' = g\frac{(\rho_l - \rho_u)e^{-kt_u} + (\rho_m - \rho_l)e^{-kt_c}}{Dk^4 + g(\rho_m - \rho_u)}. \tag{12.2.22}$$

When the load is at the interface between the upper and lower crust the deflection $w_2(x)$ is determined by

$$\left[Dk^4 + g(\rho_m - \rho_w)\right] \bar{w}_2 = -g(\rho_l - \rho_u)\bar{s}_2, \tag{12.2.23}$$

with a corresponding gravity anomaly

$$\Delta\bar{g}_2 = 2\pi G\{(\rho_u - \rho_w)\bar{w}_2 + (\rho_l - \rho_u)e^{-kt_u}(\bar{s}_2 + \bar{w}_2) + (\rho_m - \rho_l)e^{-kt_c}\bar{w}_2\}. \tag{12.2.24}$$

The surface topography induced by this internal load $\bar{e}_2 = \bar{w}_2$ and the gravitational anomaly can again be expressed in terms of an admittance function Z_2,

$$\Delta\bar{g}_2 = Z_2\bar{e}_2 = 2\pi G(\rho_u - \rho_w)(1 + Z_2')\bar{e}_2, \tag{12.2.25}$$

with

$$Z_2' = \left[\frac{Dk^4 + g(\rho_m - \rho_w) - g(\rho_l - \rho_u)}{g(\rho_u - \rho_w)}\right] e^{-kt_u} + \left[\frac{\rho_m - \rho_l}{\rho_u - \rho_w}\right] e^{-kt_c}. \tag{12.2.26}$$

For an internal load placed at the Moho

$$\Delta\bar{g}_3 = Z_3\bar{e}_3 = 2\pi G(\rho_u - \rho_w)(1 + Z_3')\bar{e}_3, \tag{12.2.27}$$

where now

$$Z_3' = \left[\frac{\rho_l - \rho_u}{\rho_u - \rho_w}\right] e^{-kt_u} + \left[\frac{Dk^4 + g(\rho_l - \rho_w}{g(\rho_u - \rho_w)}\right] e^{-kt_c}. \tag{12.2.28}$$

The total surface topography will represent the sum of the effects of the three different types of loading and the total gravitational anomaly will again be the sum of the gravitational effects produced by the three classes of loads

$$\bar{e} = \sum_{i=1}^{3} \bar{e}_i, \qquad \Delta\bar{g} = \sum_{i=1}^{3} \Delta\bar{g}_i = \sum_{i=1}^{3} Z_i \bar{e}_i. \tag{12.2.29}$$

As in the work of Forsyth (1985), it is not unreasonable to assume that the surface and internal loads are uncorrelated with each other; in which case

$$\langle \bar{e}_1 \bar{e}_2^* \rangle = \langle \bar{e}_1 \bar{e}_3^* \rangle = \langle \bar{e}_3 \bar{e}_2^* \rangle = 0, \tag{12.2.30}$$

where an asterisk indicates a complex conjugate and the angle brackets indicate averages taken over a wavenumber band of width Δk centred on k.

The admittance Z relates the surface topography to the gravitational anomaly

$$\Delta\bar{g} = Z\bar{e} + \bar{n} \tag{12.2.31}$$

where $\bar{n}$ represents that part of the gravity anomaly $\Delta\bar{g}$ that is not coherent with $\bar{e}$ such as that due to loads that have no surface topographic expression. Thus

$$Z = \frac{\langle \Delta\bar{g}\, \bar{e}^* \rangle}{\langle \bar{e}\bar{e}^* \rangle} = \frac{\sum_{i=1}^{3} Z_i \langle \bar{e}_i \bar{e}_i^* \rangle}{\sum_{i=1}^{3} \langle \bar{e}_i \bar{e}_i^* \rangle}, \tag{12.2.32}$$

no cross terms appear because we have assumed that the loads are uncorrelated with each other (12.2.30).

The deflections due to a surface load can be determined directly from the solution of (12.2.15). The general solution takes the form,

$$w = e^{\zeta x}[A\cos(\zeta x) + B\sin(\zeta x)] + e^{-\zeta x}[C\cos(\zeta x) + D\sin(\zeta x)], \tag{12.2.33}$$

where A, B, C and D are integration constants. The flexural parameter ζ controls the amplitude and wavelength of the deformation

$$\zeta = \left(\frac{g(\rho_m - \rho_w)}{4D}\right)^{1/4}. \tag{12.2.34}$$

If we impose a line load P at $x = 0$, the solution in $x > 0$ cannot contain growing terms and, from symmetry $dw/dx = 0$ at $x = 0$, so

$$w = w_0 e^{-\zeta x}[\cos(\zeta x) + \sin(\zeta x)], \tag{12.2.35}$$

and from force balance $w_0 = P\zeta/2g(\rho_m - \rho_w)$. The deflection vanishes when $\cos(\zeta x) + \sin(\zeta x) = 0$ and thus when $\zeta x = 3\pi/4, 7\pi/4, \cdots$. In addition to the main downward deflection beneath the load, there is an upward peripheral bulge on either side of the central depression.

Watts (2001) §3.5.2 provides a detailed discussion of loading applied at or near a break using a formulation in terms of a semi-infinite beam for which

$$w = w_0 e^{-\zeta x} \cos(\zeta x), \tag{12.2.36}$$

with the maximum deflection $w_0 = 2P\zeta/[g(\rho_m - \rho_w)]$. There is again a bulge, but it is both narrower and of larger amplitude than for the continuous plate.

The thin-plate theory is based on the assumption that the thickness of the shell is thin compared with the radius of curvature of any deflection, and that the stress component σ_{zz} in the vertical direction can be neglected by comparison with the other stress components.

12.2.3 Estimates of the elastic thickness of the lithosphere

The thin plate formulation of the deflection of the lithosphere has been used by many authors to obtain estimates of the equivalent elastic thickness of the lithosphere by exploiting gravity anomalies, either by spectral methods or using forward modelling.

In the forward modelling approach the gravity anomaly due to a load, e.g. the surface load due to the topography of an oceanic island or a mountain chain, and the flexural compensation arising from the deflection of an elastic plate are calculated. The estimates of the gravity anomalies for different elastic plate thicknesses T_e are then compared with the observations and the value of T_e that secures the best fit between the observed and calculated gravity anomalies is selected.

The spectral methods are based on estimates of the admittance Z (12.2.32) between the gravity anomalies and surface topography. In principle it should be a relatively simple matter to determine the admittance through spectral division in the spatial Fourier domain, e.g.,

$$Z(k) = \frac{C(k)}{E_H(k)} = \frac{\Delta g(k) H^*(k)}{H(k) H^*(k)} \tag{12.2.37}$$

where $\Delta g(k)$ is the transform of the gravity anomaly and $H(k)$ the transform of the topography. However, in practice the situation is complicated by noise in the data and limitations in the geometry of data collection that are most pronounced in the distribution of ship tracks in the ocean. It is common practice to work with auxiliary information such as the *coherence*:

$$\gamma^2(k) = \frac{C(k) C^*(k)}{E_g(k) E_H(k)} \quad \text{with} \quad E_g(k) = \Delta g(k) \Delta g^*(k). \tag{12.2.38}$$

The coherence function derived from the observations is then compared directly with model predictions. At short wavelengths the coherence approaches 0 because the lithosphere appears strong irrespective of the value of T_e. Such small features would cause only minor flexure with little contribution to the gravity anomaly. At long wavelengths the coherence normally approaches 1, because the lithosphere has flexed under load. The presence of buried loads complicates the situation (Forsyth, 1985; McKenzie, 2003).

The two different approaches would be expected to give similar results for the effective elastic thickness of the lithosphere T_e and this is certainly true for the oceans. For example forward modelling for the gravity anomalies associated with the Hawaiian–Emperor seamount chain leads to an estimate of T_e for the Pacific Plate of 25 ± 9 km, whereas the spectral approach using free-air anomalies yields values of 20–30 km. In particular, the various T_e estimates provide the same dependence on age of the lithosphere at the time of loading: T_e increases from 2–6 km for young lithosphere to around 30 km for old lithosphere. Such values are somewhat thinner than the estimates of lithospheric thickness in the oceans derived from seismic observations, but reflect very different time scales for loading.

In the continents there are greater discrepancies between the two approaches to estimating T_e. The results of direct modelling are often smaller than those obtained by spectral transfer function methods. Improved techniques for spectral estimation (such as multi-taper techniques) have improved the spatial resolution of T_e. Forsyth (1985) introduced an allowance for the presence of internal loads using the coherence with the Bouguer anomalies, where corrections are made for the gravitational attraction of all material above sea level. However, McKenzie & Fairhead (1997) have argued that the Bouguer approach provides upper bounds on T_e rather than the true value. They proposed instead a free-air admittance method to argue that the effective elastic thickness of the continental lithosphere was less than 25 km. Subsequently McKenzie (2003) produced a refined coherence method to allow for the presence of internal loads, particularly when there is no associated surface topography. This approach is based on the model presented above (see Figure 12.4), but intriguingly the effective elastic thickness T_e derived by this method is smaller than the crustal thickness.

The forward modelling approach is commonly applied to foreland basins and in some cases such as, e.g., the Ganges basin has indicated T_e significantly in excess of the local crust thickness, 70 km compared with a crust of 40 km (Burov & Watts, 2006). Such a result suggests the presence of elastic strength in the mantle. The spectral methods have generally been derived for old cratons, but results have also been derived for orogenic belts and rifts. There is increasing evidence for relatively rapid horizontal variations of T_e on scales of a few hundred kilometres and also of anisotropy in the gravitational response (e.g., Simons et al., 2000). Yet, the theoretical development is based on a uniform isotropic plate, and so the values of T_e should be regarded as useful comparators between regions rather than having direct physical significance.

12.2.4 Strength envelopes and failure criteria

The clearest manifestation of the finite strength of the lithosphere comes from the presence of shallow earthquakes. It is rare for a major event to break fresh rock, so that the behaviour observed is dominantly that of a medium comprised of units linked together by the strong friction at the boundaries of the blocks. Experimental studies and computer simulations indicate that the friction laws are dependent on prior deformation history and slip-rate, so that strength can drop dramatically once relative motion starts. The friction is sufficient that for processes with large horizontal scale, e.g., earth tides and glacial rebound, the crust behaves as a single material. Yet the stress accumulating at places like convergent margins has eventually to be released, as in the great 2004 Sumatra-Andaman earthquake where a thrust front migrated over 1300 km along the plate boundary in a few minutes. The passage of a large earthquake on a fault redistributes local concentrations of stress that in turn are relieved by aftershocks, but can also change the stress distribution in the neighbourhood of another fault system, rendering that zone closer to the stress conditions required for failure.

The changing conditions from brittle failure at shallow depth to ductile flow at depth can be summarised though a *yield strength envelope* that attempts to provide a single strength profile through the lithosphere, based on the dominant constitutive law at each depth. As we have seen in Chapter 9, the details of the stress–strain (rate) relations depend strongly on material properties and temperatures, so that a single strength envelope for a particular regime is an oversimplification. Further, estimates of strength envelopes are based on data from experimental rock mechanics at much higher strain rates (10^{-4}–10^{-6} s^{-1}), than those in the Earth which rarely exceed 10^{-13} s^{-1} and where slower process such as dynamic recrystallisation might be important.

For the upper portion of the lithosphere, brittle deformation by frictional sliding on pre-existing fracture planes of suitable orientation can be described through the use of a Mohr–Coulomb fracture criterion (cf. Section 10.2.3), in these circumstances frequently known as Byerlee's law. The critical stress at which frictional sliding takes place σ_{sm} is then related to the normal stress σ_n and the ratio λ_f between the pore fluid pressure p_f and the lithospheric pressure p_{lith}, $\lambda_f = p_f/p_{lith}$, by

$$\sigma_{sm} = (M_1\sigma_n + M_2)(1 - \lambda_f), \tag{12.2.39}$$

where M_1 and M_2 are empirical constants, equivalent to μ and c in (10.2.6). The relation (12.2.39) incorporates the shift in effective stress state due to pore fluid pressure. For materials in compression, Brace & Kohlstedt (1980) suggest the use of

$$\begin{aligned} &M_1 = 0.85, \quad M_2 = 0\,\text{MPa}, \quad \text{when} \quad \sigma_n < 200\,\text{MPa},\\ &M_1 = 0.6, \quad M_2 = 60\,\text{MPa}, \quad \text{when} \quad \sigma_n > 200\,\text{MPa}. \end{aligned} \tag{12.2.40}$$

In tension, the differential stress needed to produce failure is significantly reduced (see Figure 10.10).

When material is fractured or porous, the controlling factor for brittle strength is pore fluid pressure p_f. In the upper crust a hydrostatic pore pressure gradient with a pore pressure coefficient $\lambda_f = 0.4$ works quite well. In the deep KTB borehole in Germany, hydrostatic pore pressure gradients persist to 9 km depth in metamorphic basement rock.

In the ductile regime, the creep can be described by the grain size dependent relation (9.3.3), that can be recast in a form where the stress σ is related to the strain-rate $\dot{\epsilon}$ as

$$\sigma = B \left[\frac{1}{\dot{\epsilon} G}\right]^n \left[\frac{d_0}{d}\right]^m \exp\left(-g\frac{T_M}{T}\right), \tag{12.2.41}$$

for grain size d and temperature T. Since the exponent n is generally greater than 1, the stress dependence on $(\dot{\epsilon})^{-n}$ is non-linear and highly temperature dependent.

Oceanic lithosphere

For the oceanic lithosphere the dominant mineralogy is olivine, and away from the ridge crest, the brittle regime extends beneath the oceanic crust. At greater depths the behaviour is ductile and the overall stress envelope is determined by the intersection of the linear yield stress relation from brittle failure and the nonlinear ductile law. At the *brittle–ductile transition* equivalent differential stress is predicted for the two different regimes. The resulting yield strength envelope, indicating the stress that can be supported by the lithosphere before it yields, has a configuration similar to a sail (Figure 12.5). The area inside the strength envelope can be regarded as a measure of the integrated strength of the lithosphere.

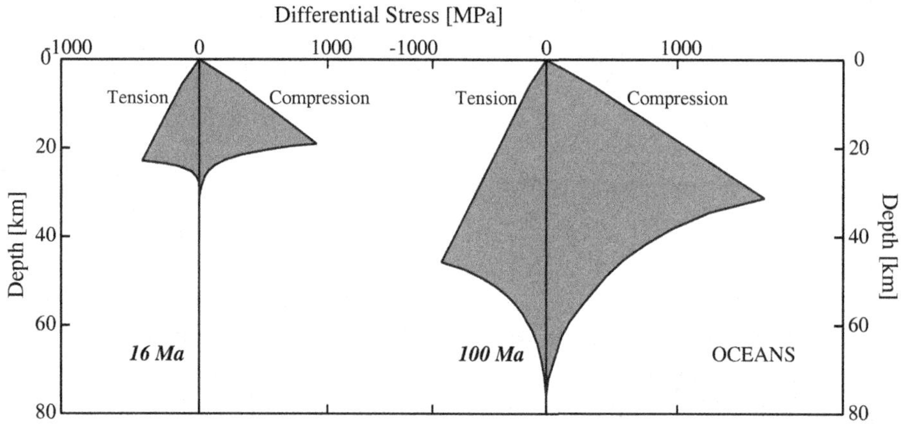

Figure 12.5. Yield strength envelopes for oceanic lithosphere, based on a strain rate of 10^{-14} s^{-1} and a half-space cooling model for the oceanic geotherm.

The typical strength envelope behaviour for young lithosphere (16 Ma) and older lithosphere (100 Ma) is shown in Figure 12.5 using a strain rate of 10^{-14} s^{-1} and a half-space model for the geotherm in the oceanic lithosphere (Section 12.2.1). For ages less than 100 Ma there would be little change using a cooling plate model.

The brittle–ductile transition corresponds approximately to the 450° isotherm in the oceanic lithosphere, and hence roughly to the estimates of T_e that show good correlation with the age of the oceanic lithosphere.

The majority of oceanic earthquakes are relatively shallow, but a few occur in the oceanic mantle, and are confined at depths above the 650°C isotherm. The maximum depth of earthquakes is thus somewhat deeper, but generally consistent with the apparent elastic thickness of the lithosphere T_e. Deeper events at the outer-rise of subduction zone trenches tend to indicate a compressional regime whilst the shallower events indicate tension. The separation in mechanism is not so distinct for the much rarer events within the oceanic plates.

Continental lithosphere

The details of the yield strength behaviour depend strongly on the particular mineralogy of the highly heterogeneous continental crust. A typical feature is a significant reduction in strength below 20 km depth, because the onset of ductility in quartz-rich rocks occurs at a shallower depth than for olivine under the same thermal gradients. There is the possibility of an increase in strength in the lower crust, and again beneath the Moho particularly when the uppermost mantle material is dry (Figure 12.6).

Evidence for a strong uppermost mantle has been adduced from occasional earthquakes below the Moho, but there could well be a residual influence of the assigned depth (33 km) used by international agencies for earthquakes for which a reliable depth cannot be found. Reinterpretation of possible mantle events at 70–85 km beneath Tibet, in light of improved constraints on crustal structure, suggests that they lie in the lower crust (Jackson, 2002), and may be associated with the transition from metastable granulite to eclogite in the presence of water.

Even very small amounts of water can have a dramatic effect on the strength distribution in a material, as can be seen from Figure 12.6. The conventional viewpoint has favoured a scenario with a wet lower crust and a dry uppermost mantle, in which case a considerable portion of the total strength of the lithosphere would reside in the mantle. This viewpoint has been challenged by Jackson (2002) building on the observations of deep lower crustal earthquakes and estimates of the effective elastic thickness T_e for the continents as less than that of the seismogenic layer. Jackson suggests that the normal continental scenario is both wet lower crust and uppermost mantle so that nearly all strength would reside in the seismogenic upper crust. In some regions, the lower crust could be dry and sustain earthquakes as in the Indian Shield material in the doubled crust under Tibet. Certainly most of the infrequent earthquakes in shield areas initiate above 15 km depth; though there is some evidence for lower crustal events in the Fennoscandian region associated with glacial unloading.

The concept that strength resides solely in the upper crust has been disputed by Burov & Watts (2006) who favour the retention of a strength contribution from the uppermost mantle. Simulations of lithospheric dynamics with weak or strong

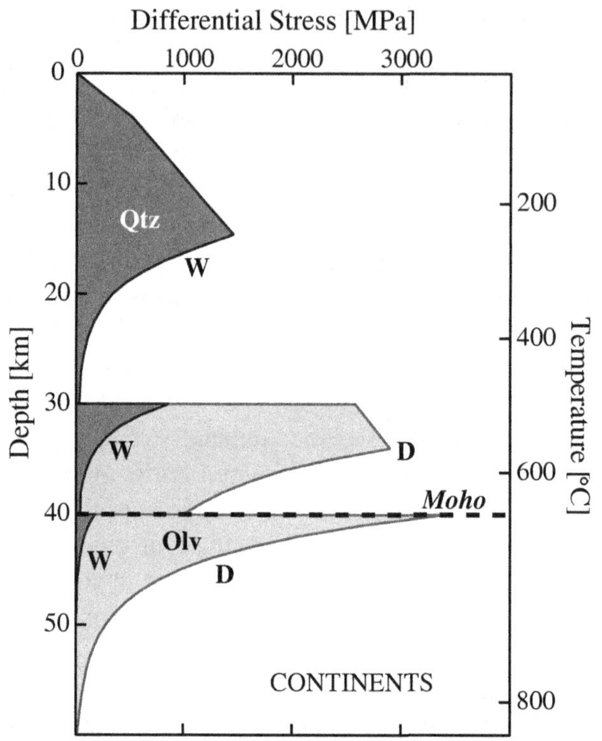

Figure 12.6. Yield strength envelope, in compression, for the continental lithosphere as a function of depth at a strain rate of 10^{-15} s^{-1} and a surface heat flow of 60 mW m^{-2}, based on experimental results for dry materials, and those containing small amounts of water. Dry conditions are indicated with grey lines and wet with black. The upper crust is taken as a wet quarzite, the lower crust as either a dry diabase (D) or undried granulite (W) and the upper mantle as wet or dry olivine. A wide range of strength conditions can be produced by different combinations of wet and dry elements. When both lower crust and mantle are wet, strength resides in the seismogenic upper crust. In contrast, with a dry lower crust, deep crustal earthquakes would be possible. If the whole crust is wet then most strength resides in the uppermost mantle. [Courtesy of J. Jackson.]

uppermost mantle (but similar effective T_e) show very significant differences in the stability of mantle structures. A weak uppermost mantle tends quite rapidly (<5 Myr) towards the development of drip structures sinking into the mantle, and the collapse of surface topography. Strength in the uppermost mantle allows topography induced by orogeny to be sustained for 100 Myr or more.

A major difficulty with resolving such fundamental issues is that most of the available information that bears on strength comes from regions that are clearly abnormal, because of the occurrence of sizeable earthquakes. The widespread development of seismic anisotropy in the upper part of the mantle lithosphere that shows preferred directions linked to past tectonic history provides evidence for preservation of structures over substantial time periods, which would be difficult to sustain without some strength in this zone.

12.3 Plate boundaries and force systems

12.3.1 Nature of plate boundaries

The lithosphere is sufficiently resilient to transmit stress over long distances and although there are some regions of internal deformation marked by sporadic earthquakes, the main features of plate tectonics are dominated by convergent and divergent plate boundaries (see Figure 12.7). The divergent boundaries mark the mid-ocean ridges where upwelling of hot material leads to the creation of new ocean crust and are shown by a double line in Figure 12.7. The rates of divergence vary from less than 10 mm/yr to more than 90 mm/yr, and in some cases there is some asymmetry in spreading rate on the two sides of the divergent boundary. The relative motion of the relatively rigid plates has to be accommodated by jogs in the divergent margins associated with transform faults. The traces of the transforms form small circles about the pole of rotation that describes the motion of the two separating plates. Most of the oceanic earthquakes occur on or near the transform zones.

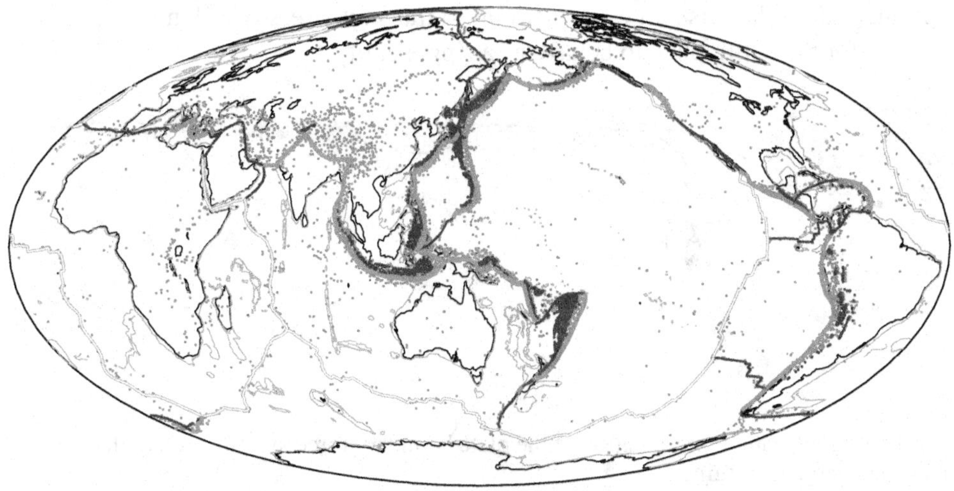

Figure 12.7. The earthquake distribution across the globe in relation to plate boundaries. Earthquakes below 150 km deep associated with subduction are shown in slightly darker tone. Divergent margins are indicated with a double line and convergent plate boundaries with a heavier line weight.

The addition of more oceanic material has to be accommodated by the re-assimilation of former oceanic lithosphere into the mantle at subduction zones, which are the major form of convergent margins. The convergent margin segments around the globe are indicated with heavier lines in Figure 12.7, and it is immediately apparent that they are accompanied by the majority of the world's earthquakes. The consumption of oceanic lithosphere in front of a continental component on the plate can lead to the scenario where, as under Tibet, an attempt is being made to force relatively buoyant continental lithosphere to depth.

The plate boundaries are not static but evolve over time, and only a limited number of possible triple junctions between plates are stable. When a plate overrides a speading centre, as appears to have happened for the extinct Farallon plate in North America, the resulting motions need still to be taken up and the modern San Andreas transform fault system links subduction in Mexico and Cascadia.

The modern-day configuration of plates does not have each plate with convergent and divergent boundaries. The African plate has no convergent margins, and the Philippine plate no current ridge system. The balance of forces associated with the motion of the plates will thus depend strongly on the geometry of the plate boundaries.

12.3.2 Plate boundary forces

Forsyth & Uyeda (1975) provided a useful description of the nature of the forces acting at the boundary of the lithospheric plates (Figure 12.8). As we have just seen the oceanic lithosphere is sufficiently strong to transmit stress over long distances and hence for the entire plate to behave as a single entity.

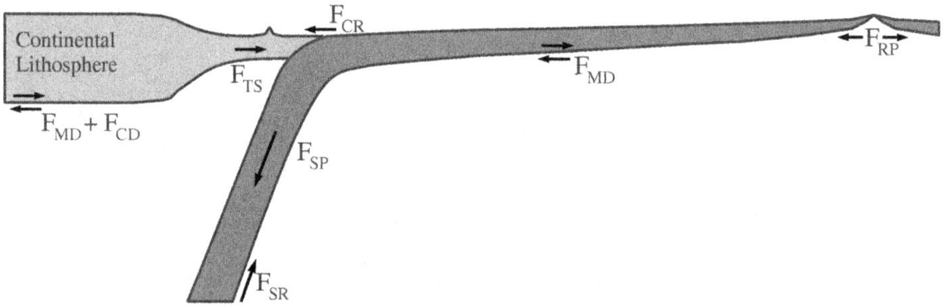

Figure 12.8. Schematic representation of the plate boundary force system, with both divergent and convergent margins

The various forces can be grouped as:

(i) Forces associated with divergent plate boundaries, associated with the gravitational effects of the relative elevation of the mid-oceanic ridge system (ridge push F_{RP}).

(ii) Forces arising from frictional resistance to motion at transform fault systems (transform fault resistance F_{TF} – not shown in Figure 12.8).

(iii) Forces linked to the complex processes at convergent plate boundaries: the negative buoyancy (*heaviness*) of the cooler subducting plate (slab pull F_{SP});

resistance of the mantle to slab penetration (slab resistance F_{SR}); forces associated with the impact of plate convergence (collisional resistance F_{CR}; and tensional forces in the over-riding plate resulting from the presence of the oceanic trench (trench suction F_{TS})

(iv) Forces related to the interaction of the plate and the underlying mantle (mantle drag F_{MD}). With the complexity and thickness variations of the continental lithosphere such drag effects are likely to be enhanced for portions of plates with continental material with an additional drag term F_{CD} compared with the oceanic portions of the plates.

In the original analysis Forsyth & Uyeda (1975) estimated the relative importance of the various forces (assumed to be the same for all plates) under the assumption that the sum of the torques acting on each plate must be zero. This leads to a set of linear equations in terms of the sizes of the forces, modulated by the configuration of the lithospheric boundaries. Some contributions are simply an integral along the length of the margin, but for the collisional resistance for convergent margins and the resistance at transform faults the direction of relative motion of the plates has to be taken into consideration. The inversion of the over-determined system of equations (36 simultaneous equations in 8 unknowns) can be accomplished using a singular value decomposition. The conclusion is that slab pull and slab resistance are an order of magnitude larger than the other forces, but that the net pull of downgoing slabs on the surface plates, $F_{SP} + F_{SR}$, is of the same order as the rest of the forces. This means that although slab pull is significant there can be a distinct contribution from ridge push.

A direct estimate of the ridge push per unit length of the ridge axis can be made by looking at the gravitational effect produced by the elevation of the ridge crest. Ritcher & McKenzie (1978) derive the simple form

$$F_{RP} = g w_e(\rho_m - \rho_w)\left(\tfrac{1}{3}L + \tfrac{1}{2}w_e\right), \tag{12.3.1}$$

in terms of the thickness L of the cooled plate and the elevation of the ridge axis above the equilibrium depth w_e. For a typical plate thickness of 85 km with a 3 km ridge elevation we obtain the estimate of 2×10^{12} N m^{-1}. The ridge push acts on the entire plate and not just as a line force at the ridge axis.

The slab pull force arises from the fact that the descending plate is cooler and denser than its surroundings. An estimate of the force F_{SP} at depth z can be obtained from models of the temperature distribution in the descending plate, so that, e.g.,

$$F_{SP}(z) \approx \frac{8g\alpha_{th}C_p T_M \rho_m^2 L^2 \mathfrak{R}}{\pi^4}\left[\exp\left(\frac{-\pi^2 z}{2c_p\rho_m\mathfrak{R}}\right) - \exp\left(\frac{-\pi^2(d_{UM}-L)}{2c_p\rho_m\mathfrak{R}}\right)\right], \tag{12.3.2}$$

for a plate of thickness L, with mantle temperature T_M, upper mantle thickness d_{UM}, α_{th} the coefficient of thermal expansion, C_p the specific heat and $\mathfrak{R}$ the thermal Reynolds number introduced in (12.2.14) expressed in terms of the rate of slab sinking. As discussed in Section 13.3 the elevation of the olivine to spinel phase transition within the slab, compared with the normal depth close to 410 km outside, provides an extra downward force because of the presence of denser material at shallower depth. The total contribution to F_{SP} is around 10^{13} N m^{-1}.

The descent of the plate is resisted by frictional forces acting on the sides of the plate and by the phase transition near 660 km. Seismic tomography gives evidence of impediments to penetration with stagnant slabs residing in the transition zone, particularly in regions where there has been considerable trench retreat. Most plates with a considerable length of subduction margin move with a similar speed which suggests that there is a terminal velocity for the plate achieved by the near balance of slab-pull and resistive forces.

The actual forces acting on any particular plate are likely to be more complex than indicated in Figure 12.8, because negative buoyancy (heaviness) of the subducting plate is age-dependent and the interaction of plates means that convergent margins may be in retreat or advance with a consequent change of the forces in the immediate neighbourhood of the descending slab. Modelling of the expected stress distribution in a plate based on the varying force contributions from its margins, taking into account age-dependence, can provide quite good agreement with the estimated stress patterns with the continent (see, e.g, the work of Reynolds, Coblenz & Hillis 2003, for the Indo-Australian Plate).

12.4 Measures of stress and strain

12.4.1 Stress measurements

A compilation of stress measurements from around the world is maintained by the World Stress Map project of the Heidelberg Academy of Science and Humanities. Nearly 16,000 reliable observations are included in their 2005 data release (http://www-wsm.physik.uni-karlsruhe.de). The data base is built up from

(i) earthquake focal mechanisms,
(ii) well-bore breakouts and drilling-induced features,
(iii) *in situ* stress measurements,
(iv) recent geological data such as fault slip analysis or alignments of volcanic vents

The primary indications of stress conditions in the lithosphere come from the source mechanisms of well-located earthquakes. For larger events the global centroid moment tensor catalogue (Section 11.5) provides an extensive suite of information for the period since 1976. The ambiguity between the faulting plane and the perpendicular auxiliary plane can only be resolved with additional information such as, e.g., geological constraints. The interpretation of the focal mechanism in terms of the principal stress directions requires the use of criteria for brittle failure, such as the Anderson criteria introduced in Section 10.2.3.

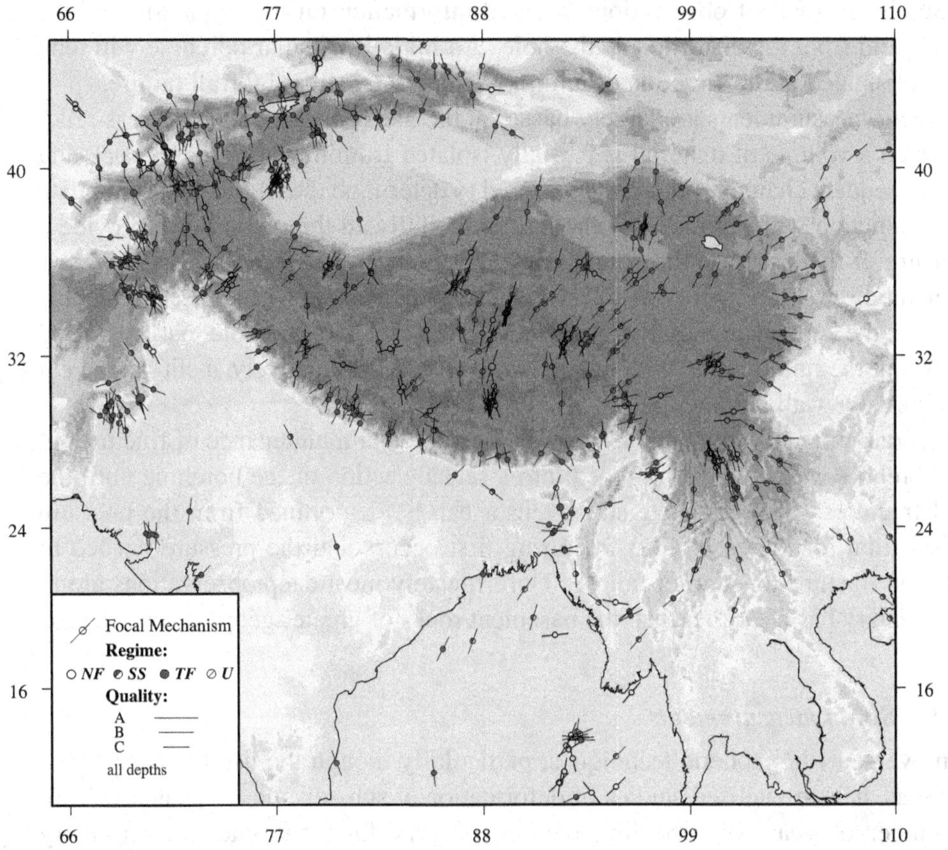

Figure 12.9. Stress conditions inferred from seismic focal mechanisms in central Asia. [Courtesy of O. Heibach.]

In areas of distributed deformation such as the region from the Himalayan front into Central Asia, the fault plane mechanisms for shallow events provide a remarkable view into the changes in stress regime. There is thrust faulting at the convergent margin that rapidly changes in the high Tibetan plateau to normal faulting associated with east–west rifting with also strike-slip faulting. The strike-slip events in particular show a dramatic rotation in direction around the syntaxis in northern Burma, indicating a flow of crustal material.

In regions with few seismic events, reliance has to be placed on information from smaller events, and then it is important to have as good an azimuthal distribution of stations as can be achieved to provide reliable results on source mechanism. As far as possible determinations should be based on the main event in a sequence since the mechanisms of aftershocks depend on the residual stress distribution rather than the ambient stress field, and there can be significant variations between nearby events.

Borehole breakout observations provide information on the orientation of the stress field from the distortion of the hole. An initially circular borehole will tend to be elongated in the direction of minimum horizontal stress by wall failure.

In-situ measurements are largely based on the use of the concept of stress relief in which a volume of material is partially isolated from the surrounding rock and the consequent change of shape is measured to determine the strain. The strain data in conjunction with estimates of the elastic moduli can then be used to provide a measure of the stress on the volume (see, e.g., Jaeger & Cook, 1979). A variety of methods are used to provide the stress relief. In over-coring the measurement is made of the deformation of a rock segment at the end of a borehole that is isolated by coring the final section with an annular drill bit. Alternatively, a slit can be cut in a borehole wall.

Hydraulic fracture methods rely on the creation and maintenance of fractures in the borehole wall. Fluid is pumped into a sealed section of the borehole until the wall fractures. The complete stress tensor can be determined from the pressure information for the case when fracturing first occurs and the pressure needed to hold the fractures open after failure. Unfortunately most deep observations are in sedimentary basins, rather than the basement rocks of greater geophysical interest.

12.4.2 Strain measurements

Improvements in geodetic techniques, particularly though the use of space-based methods, have revitalised studies of deformation at sub-seismic frequencies. With a number of years of recording the use of very long baseline interferometry (VLBI) has produced estimates of motion between the plates which are consistent with geological estimates, but continue to improve in accuracy as further data is accumulated.

On smaller spatial scales GNSS observations based on the analysis of signals from constellations of satellites designed to aid navigation (e.g., GPS, GLONASS, GALILEO) have become an increasingly important tool in the measurement of earth deformation. With a minimum of four satellites observed simultaneously the configuration of the station relative to the satellites can be determined with a knowledge of the satellite orbits. The orbit information transmitted by the satellites is sufficient for metre level accuracy, but precision observations at the centimetre level or better require the orbital parameters to be improved using observations from permanent ground stations and a set of corrections for other factors applied, such as the presence of water vapour that can change the apparent distance to a satellite. The GNSS observations can pick up coseismic signals, but are also well suited to capturing deformation occurring at time scales greater than a month. Such longer time scale phenomena are, e.g., the deformation associated with viscoelastic relaxation following an earthquake, and also decadal estimates of strain accumulation and plate motion, and their spatial variations.

GNSS observations indicate that the zone of deformation associated with the boundary between the North American and Pacific plates, with its main expression in the San Andreas fault system, is several hundred kilometres wide. This helps to explain the discrepancy between the deformation estimated from the cumulative slip of earthquakes and the predictions based on knowledge of the velocities between the plates. Increasing density of geodetic information means that seismological and geodetic information can be brought together to provide more detailed information on the deformation processes whose most obvious symptoms are earthquakes. There are indications that a number of different time scales of post-seismic deformation may be present that could, e.g., represent rapid afterslip on a time scale of a few weeks accompanied by slower viscoelastic relaxation of mantle and lower crust layers.

In areas of active deformation direct observations of strain accumulation can be obtained using appropriate instrumentation. The Plate Boundary Observatory component of the EarthScope project in the western U.S.A. is using borehole strainmeters that provide triaxial information and laser strainmeters for direct measurements of strain in particular directions. A long-baseline laser strainmeter measures the relative displacement between two end-piers separated by several hundred metres. Laser light is directed through a straight evacuated pipe to a reflector mounted on one pier that returns the light to a laser interferometer on the other pier. Changes in optical path length are detected by the interferometer in terms of the wavelength of light from a stabilized laser. The pipe is evacuated to suppress the variations in the index of refraction of the air that would otherwise affect the optical path-length measurement.

With high densities of permanent geodetic sites in many active source regions, direct constraints are available on the mechanisms of earthquakes from the deformation pattern. Models of the faulting process used in the analysis are tending to become more sophisticated with multiple fault segments and realistic Earth structure.

Multiple reoccupation of sites with short-term recording every few years provides a means of extending deformation information with a limited number of instruments. As an example, results from GPS studies across the complex micro-plates of the Papua New Guinea Region have shown that the present-day rates of change in relative position between stations are very consistent with the average rates over the past few million years (Tregoning et al., 1998). In such an active seismic region the patterns of observation can be disrupted on occasion by the occurrence of a major seismic event between the reoccupation of sites.

The interferometric radar technique (InSAR) introduced in Section 11.5 provides the opportunity for medium-term monitoring of deformation by the examination of differences between repeat satellite passes across the same region. Such information will be of particular value in detecting slow systematic vertical movements, which are less easily detected using GNSS methods.

Holt et al. (2000) present an approach to extracting horizontal strain and rotation components through the determination of a smooth vector rotation function $\mathbf{W}$ parameterising the horizontal velocity field

$$\mathbf{v}(\mathbf{x}) = r_e \mathbf{W}(\mathbf{x}) \times \mathbf{x}. \tag{12.4.1}$$

Both the rate of strain elements and the rotation rate can be represented in terms of the derivatives of $\mathbf{W}$. Attention can be focussed on the horizontal components, because even for rapid uplift (10 mm/yr), the contribution from vertical motion is around 1%. An inverse problem can be posed with an objective function composed of a sum of two parts, matching both direct observations of the rate of strain tensor from geological and seismological information and velocity vectors from geodetic measurements. The rate of strain field is required to satisfy St. Venant's compatibility relationships in the second derivatives of strain rate. Holt et al. (2000) provide a number of applications to, e.g., western North America, Central Asia and the Aegean.

An alternative approach for direct exploitation of geodetic information has been developed by Spakman & Nyst (2002) using the combination of many observations of differential velocities between pairs of stations. With observation of the horizontal velocity vector $\mathbf{v}_1$ at site 1 and $\mathbf{v}_2$ at site 2, using the same reference frame, the differential velocity

$$\mathbf{v}_1 - \mathbf{v}_2 = \int_{P_{12}} d\mathbf{x} \cdot \frac{\partial \mathbf{v}}{\partial \mathbf{x}} + \mathbf{f}_{12}, \tag{12.4.2}$$

where P_{12} is a suitable path between the sites and $\mathbf{f}_{12}$ is any contribution from faults between the sites. In many regions it is a plausible assumption that faults are locked at the surface, so that $\mathbf{f}_{12} = 0$. Then (12.4.2) forms the basis of a tomographic inversion for the horizontal components of the velocity gradient tensor, and hence the rate of strain and rate of rotation.

In Europe over 200 GPS observation sites provide more than 100,000 independent pairs of differential velocity, and the paths P_{12} can be chosen to optimise areal coverage. The resulting *deformation tomography* can be cast in terms of presentations of the principal components of rate of strain and rotation patterns as in Figure 12.10. The resolution is very good and the standard deviations in the model are generally much less than 5% of the imaged amplitudes. The results in Figure 12.10 are obtained by assuming that all faults in Europe are locked at the surface, and predominantly show the variation of the velocity gradient field across Europe. Notable features at this scale are the extension in Fennoscandia, the contraction in the Western Mediterranean and the distributed deformation in the Italian–Aegean–Anatolian–Zagros plate convergence zones. A rotation rate of 2.5×10^{-8}/yr corresponds to 1.4°/Myr; in the Aegean–Anatolian region the rotation rates are beyond the limits of the contouring scale and attain values of the order of 1.0×10^{-7}/yr (5.7°/Myr). Much more detail and systematic variations appear when attention is concentrated on specific regions.

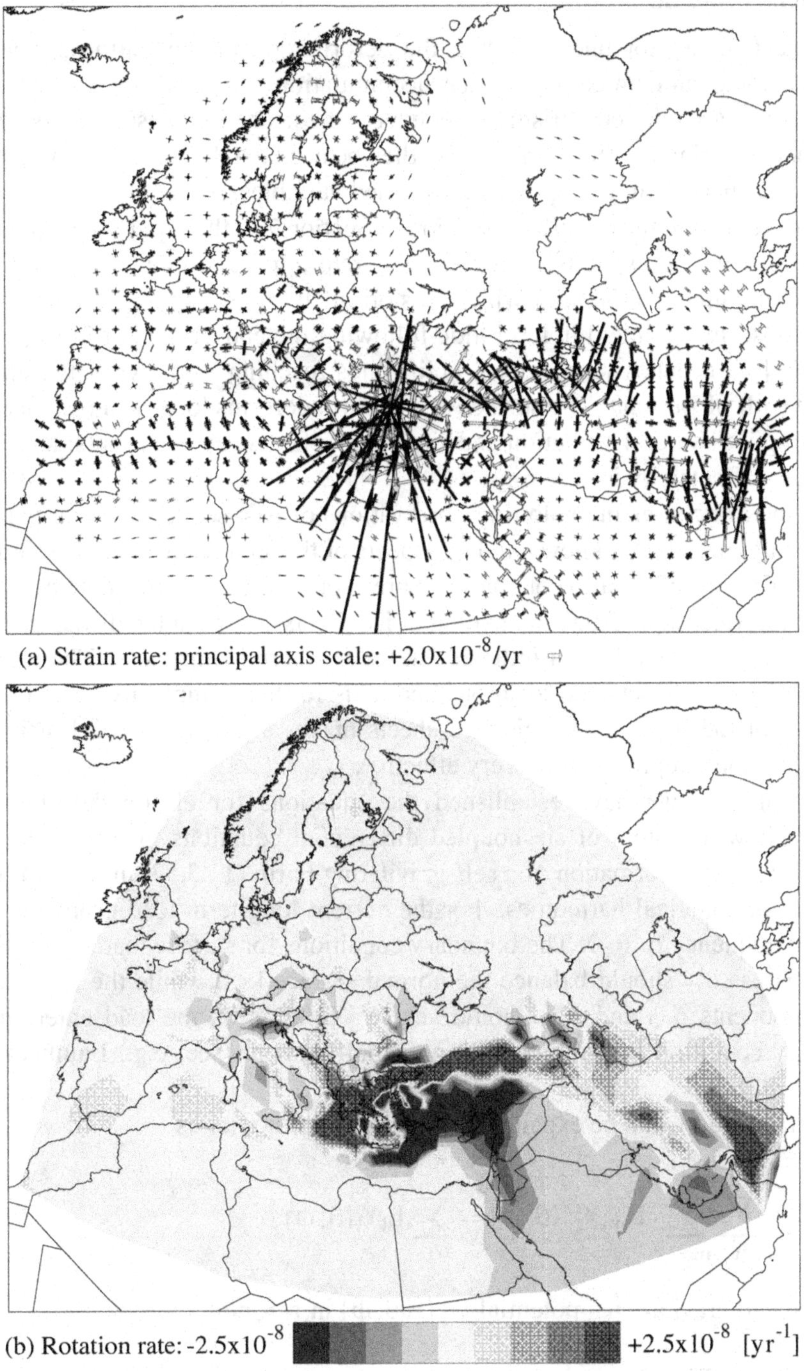

Figure 12.10. Velocity rates for Europe determined by deformation tomography: (a) Principal components of rate of strain, open vectors indicate extension, filled vectors compression; (b) rotation rates, dark shades represent counter-clockwise motion and patterning indicates clockwise motion. [Courtesy of G. Tanasescu, W. Spakman, B.A.C. Ambrosius.]

12.5 Glacial rebound

A convenient model for understanding the deformation of the Earth associated with the loading and unloading of ice accumulation during the recent glacial epoch is to use a model comprising a sequence of linear viscoelastic layers. The correspondence principle (Section 5.5) can then be used to take results from perfect elasticity and apply them to Laplace transformed variables.

As we have seen in the simple illustration in Section 7.4, the rapid removal of an ice load allows the lithosphere to progressively return to its original configuration, with maximum uplift in the zone which was originally most depressed. The uplift of the Fennoscandian shield in the Baltic following the end of the last ice age has produced raised beaches with current uplift rates reaching 9 mm/yr at the upper end of the Gulf of Bothnia. The surface of the Earth deforms locally in response to the changing ice–water distribution and significant effects can occur outside of the zone directly affected by ice-loading.

One of the difficulties in exploiting observations of past sea level is that the ice history is only imperfectly known and so aspects of the ice load have to be inferred at the same time as information on the structure supporting the load. There can also be significant water input from the melting of ice sheets at substantial distance, e.g., input from the Antarctic has an influence in the northern hemisphere. The timing of ice melting differs between locations and adds further complexity. The broad distribution of the loading of major ice sheets makes a global approach using a spherical harmonic representation very effective.

In Section 11.3 we have established the equations for elastic deformation and shown how a system of six coupled differential equations in radius can be established for the deformation of a self-gravitating Earth (11.3.57) in terms of the coefficients of spherical harmonics. For the current long-term load problems we can set the frequency $\omega = 0$. The boundary conditions for surface loading are that the radial stress σ_{rr} should balance the normal surface load, while the tangential stress components $\sigma_{r\theta}$ and $\sigma_{r\phi}$ should vanish. In addition the load enters into a continuity condition for gravitational potential gradient (see, e.g., Lambeck & Johnson, 1998).

The surface load L can be expanded in spherical harmonics as

$$L(\theta, \phi) = \sum_{l} \sum_{m=-l}^{l} L_{lm} Y_l^m(\theta, \phi) = \sum_{l} L_l Y_l(\theta, \phi) \tag{12.5.1}$$

with associated gravitational potential $\varphi_L(r, \theta, \phi)$ in $r > r_e$

$$\begin{aligned} \varphi^L(r, \theta, \phi) &= \sum_{l} \varphi_l^L(r) \left(\frac{r_e}{r}\right)^l Y_l(\theta, \phi), \\ &= 4\pi r_e G \sum_{l} \frac{1}{2l+1} \left(\frac{r_e}{r}\right)^l L_l Y_l(\theta, \phi). \end{aligned} \tag{12.5.2}$$

The new potential after deformation is

$$\varphi = \varphi^L + \varphi_d, \quad \text{where} \quad \varphi_d(r) = \sum_l k_l(r)\varphi_l^L(r), \tag{12.5.3}$$

and k_l is the Love number for gravitational potential.

The displacement vector $\mathbf{u}(r, \theta, \phi)$ is defined in terms of a spherical harmonic expansion by Love numbers h, ℓ as

$$\mathbf{u}(r, \theta, \phi) = \sum_l \frac{\varphi_l^L(r)}{g(r} \left(\frac{r_e}{r}\right)^l [h_l(r)Y_l(\theta, \phi)\hat{\mathbf{e}}_r + r\ell_l(r)\boldsymbol{\nabla}Y_l^m(\theta, \phi)] \tag{12.5.4}$$

with $\boldsymbol{\nabla}Y_l^m$ the horizontal gradient of the spherical harmonic function, and $g(r)$ the radial distribution of gravity. The displacement Love numbers h, ℓ define the radial and tangential deformation. The Love numbers are related to the variables U, V and Φ in (11.3.57) by

$$\begin{pmatrix} h_l(r) \\ \ell_l(r) \\ 1 + k_l(r) \end{pmatrix} = \frac{2l+1}{4\pi G L_l} \begin{pmatrix} g(r)U_l(r) \\ g(r)V_l(r) \\ \Phi_l \end{pmatrix}. \tag{12.5.5}$$

For linear viscoelastic stratification, a Laplace transform as in (5.5.14), (5.5.19) applied to the differential equations and boundary conditions reduces the problem to an equivalent elastic problem, where the equations depend on the Laplace transform variable p. In such a way a time-varying load can be introduced through its Laplace transform and the deformation history described through the inversion of the transformed results for displacement. Some care has to be taken in the numerical inversion of the Laplace transform.

Realistic solutions of the viscoelastic equation systems require a high-resolution of the coupled ocean–ice surface load. As ice melts load is transferred to the ocean through increasing sea level; the ocean surface must remain an equipotential and mass conservation is required between ice and ocean. The total change in the load from oceans and ice can be expressed as

$$L(\theta, \phi, t) = \rho_I \Delta I(\theta, \phi, t) + \rho_W \Delta W(\theta, \phi, t). \tag{12.5.6}$$

Here ΔI is the change in effective ice height and ρ_I its density, ΔW is the change in water depth of density ρ_W. For an iceshelf grounded below sea level a correction has to be made for the displacement of water by part of the ice load. The change in water level has to be confined to the area of the oceans and since coastline geometry is a function of water depth the spatial constraint varies in time. It is useful to introduce an ocean function $\mathcal{O}(\theta, \phi, t)$ equal to unity in the ocean and zero on land. Then the change in water depth at time t relative to the initial conditions at t_0 can be expressed as

$$\Delta W(\theta, \phi, t) = \int_{t_0}^{T} ds\, \mathcal{O}(\theta, \phi, s) \frac{\partial \zeta(\theta, \phi, s)}{\partial s}, \tag{12.5.7}$$

where $\zeta(\theta, \phi, t)$ is the relative sea level function.

To use the Love number formalism outlined above, the surface loads need to be expanded in terms of spherical harmonics as well as functions of time. Thus

$$L(\theta, \phi, t) = \sum_{l} \sum_{m=-l}^{l} [\rho_I \Delta I_{lm} + \rho_W \Delta W_{lm}], \tag{12.5.8}$$

where ΔI_{lm} and ΔW_{lm} are the spherical harmonic expansion coefficients. To ensure an accurate representation of the time varying coastlines the expansion needs to be carried to relatively high angular orders.

The primary information on rebound comes from estimates of past changes in sea level inferred from the heights or depths of shoreline features at a known time measured relative to present ocean surface (at time t_p). Although a wide range of observations can be exploited such as wave cut notches above current sea level and submerged wavecut platforms, such information is not instantaneous. Other proxies can only provide upper or lower bounds and good quality dating is essential for good constraints on behaviour.

Sea level moves up and down relative to the crust, which is itself deforming due to the changes in the loading from the combination of ice and water. The relative change in sea level ζ from the present time can be represented as

$$\Delta\zeta(\theta, \phi, t) = \zeta(\theta, \phi, t) - \zeta(\theta, \phi, t_p) \tag{12.5.9}$$

The sea level change in (12.5.8) includes *eustatic* change $\Delta\zeta^e$ and the contribution from the deformation of the crust and changes in the geoid as mass is displaced $\Delta\zeta^d$. The eustatic component is the mean change in relative sea level over the entirety of the oceans, and from mass conservation provides a measure of the volume of ice on the continents,

$$\Delta\zeta^e(\theta, \phi, t) = \int_{t_0}^{T} ds \, \frac{1}{\mathcal{O}_{00}(s)} \frac{dW_{00}}{ds} \approx -\frac{\rho_I}{\rho_W} \frac{\Delta I_{00}(t)}{\mathcal{O}_{00}(t)}, \tag{12.5.10}$$

since the zero-degree harmonic corresponds to an average value. The deformational components can be expressed in terms of the Love numbers as

$$\begin{aligned} \Delta\zeta^d(\theta, \phi, t) &= \frac{\varphi^d(r_e)}{g(r_e} - u_r(r_e) \qquad (12.5.11) \\ &= \frac{4\pi G r_e}{g(r_e)} \sum_{l=1}^{\infty} \frac{1}{2l+1} \left[1 + k_l(r_e, t) - h_l(r_e, t)\right] * L_l(t) Y_l(\theta, \phi), \end{aligned}$$

where $*$ denotes a convolution. To ensure mass conservation, a small contribution $d\zeta_0$ that is spatially uniform but time-dependent has to be added,

$$d\zeta_0 \approx -\frac{1}{\mathcal{O}_{00}} \sum_{l=1}^{\infty} \sum_{m=-l}^{l} \Delta\zeta^D_{lm} \mathcal{O}_{lm}. \tag{12.5.12}$$

The total relative sea-level change is then given by

$$\Delta\zeta(\theta, \phi, t) = \Delta\zeta^{e}(\theta, \phi, t) + \Delta\zeta^{d}(\theta, \phi, t) + d\zeta_{0}(\theta, \phi, t). \tag{12.5.13}$$

Lambeck & Johnson (1998) provide a number of examples of sea level behaviour and describe the way in which inferences can be made about mantle structure even though the ice history has major uncertainties. They demonstrate that there can be considerable trade-off between different aspects of the response of the Earth, e.g., between upper mantle viscosity and apparent lithospheric thickness. Nevertheless some very robust conclusions can be drawn, such as the requirement that the average lower mantle has a higher viscosity by at least a factor of 20 than that of the upper mantle ($\sim 5\times10^{20}$ Pa s). There is lateral variability in the viscosity of the upper mantle and the apparent thickness of the lithosphere on broad horizontal scales similar to those seen in seismic wavespeed and attenuation studies, but small wavelength features cannot be resolved.

In most studies of glacial rebound a simple Maxwell body (Figure 1.3) has been used for the viscoelastic constitutive equation. This is undoubtedly too simple a rheology to provide a full description of the deformation, but can provide a good fit to the limited available sea-level data. The estimates of viscosity are therefore best regarded as an "effective" viscosity subject to the assumption of linearity.

Recent studies in and around Antarctica comparing analysis of sea-level data and contemporary measures of uplift from direct space geodetic methods indicate differences in the viscosity estimates for the millennia and year time scales. Such results suggest that there may be a need to adopt a more complex rheology with a non-linear dependence of strain-rate on stress, that yields a stress-dependent effective viscosity. The flow in the mantle needed to sustain the rebound has to interact with the flow patterns associated with mantle convection. Since the rebound flow rates are small the two effects can reasonably be superimposed (even for a non-linear rheology).

12.6 Extension and convergence

The configuration of the lithosphere is modified through processes of extension or compression, with the most obvious manifestations in the crust through the influence on surface geology. Failure in tension is significantly easier than in compression (cf. Section 10.2.3) so that extension tends to impact on the whole lithosphere in a localised region. Convergent processes tend to drive mountain building with shallow zones of failure, but under certain circumstances gravitational instability can develop in the lithospheric root.

12.6.1 Extension

Much work on tensional processes has concentrated on the development of rifts as precursors to continental break-up. Many rifts appear to be linked to ancient zones

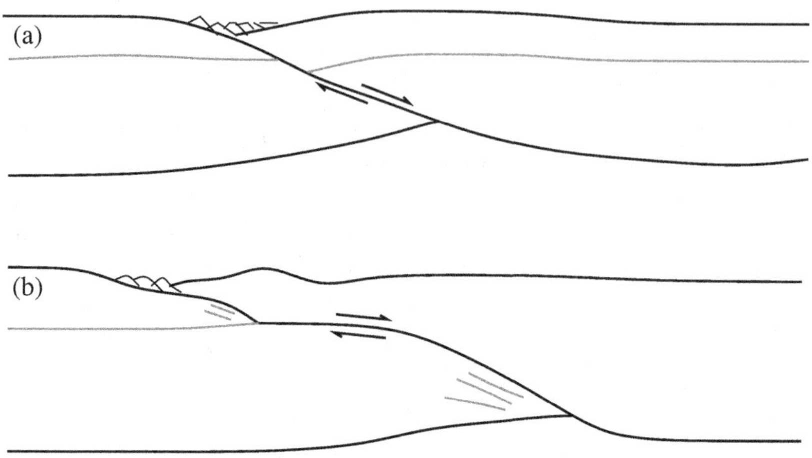

Figure 12.11. Lithospheric detachment models: (a) lithospheric wedge; (b) Delamination with ramps and flats

of contrasting properties, such as major lithospheric lineaments or the edge of the thick lithosphere beneath a craton as in the Tanzanian segments of the East African Rift.

Rifting can be accomplished by mechanical weakening of the lithosphere by stretching, the intrusion of heat and interaction with the dynamical asthenosphere beneath. Mechanical stretching accommodates the strain by the creation of faults with large offsets in brittle layers and localised ductile deformation in weaker regions. Depending on the specific rheology of the lithosphere, strain may be broadly distributed, or focussed on one large detachment fault. Rifting is frequently associated with volcanism, and the intrusion of magmatic material, either passively or actively, can be expected to enhance reduction of strength and localise strain. Extensional strain can be taken up by magma injection into the lower lithosphere, feeding dykes that penetrate the stronger lithospheric layers above. Small offset faults will be formed above such zones of magma injection. Buoyant melt can help to enhance stretching processes, whereas friction on fault surfaces will tend to impede the thinning of the lithosphere.

Lister, Etheridge & Symonds (1991) have pointed out major differences in continental margins that would have originally been contiguous before rifting and the formation of the intervening ocean. The asymmetry of the margins can be associated with their configuration relative to the detachment surface through the lithosphere. If strength is concentrated near the surface, the situation before break-up has a simple detachment surface with the development of a lithospheric wedge structure (Figure 12.11a). However, multiple zones of strength can lead to delamination with a set of ramps and flats (Figure 12.11b). The subsequent history will be significantly different for the upper and lower segments relative to the detachment.

The mathematical description of lithospheric deformation is challenging because of the strong temperature dependence of ductile rheologies, and the difficulty of providing an adequate representation of the behaviour of brittle material. To provide a continuum treatment, the brittle parts of the lithosphere can be assumed to have a dense distribution of fractures on length scales smaller than the size of lithospheric units so that a plastic rheology can be employed (see, e.g., Christensen, 1992). Such an approach cannot capture deformation on specific faults or in narrow shear zones, and so can only provide a generalised picture of structural evolution.

A simplified approach in which mechanical and thermal effects are decoupled was introduced by McKenzie (1978) in a study of the evolution of sedimentary basins. He considered the instantaneous stretching of the entire lithosphere followed by thermal recovery. Subsequent work used a similar assumption of pure shear with a uniformly strained lithosphere (with zero shear strain) represented through a thin viscous plate. The non-linear effective viscosity represents an average over the whole depth interval of the lithosphere. Such models provide valuable insights into coupled thermal and mechanical effects in deformation. For example, as the lithosphere thins in tension, conductive cooling will increase the strength and thereby inhibit further extension (Houseman & England, 1986). The changing configuration of the crust–mantle boundary and the lithosphere–asthenosphere boundary during deformation bring gravitational forces into play that themselves influence the extensional process.

As we have seen in Section 12.2.4, the strength profile of the continental lithosphere is strongly dependent on water content, and can, in certain circumstances, resemble a laminate of strong and weak components allowing differential deformation between the different zones. For such a configuration the use of a thin plate with an averaged viscosity will be somewhat misleading.

The treatment of the full complex rheology in deformation requires numerical methods that can accommodate substantial distortion of the original material configuration. This led Christensen (1992) to exploit an Eulerian formulation of the coupled thermo-mechanical deformation in terms of finite elements. In such a treatment the material flows through a fixed spatial grid and large deformation can be readily followed, which is very useful in following the later stages of rifting. Comparatively fine sampling may be needed to provide a good representation of material discontinuities in the Eulerian approach. If a Lagrangian approach is used the grid is fixed to the material and as deformation proceeds the mesh will be distorted, and remeshing is needed to keep the grid in a reasonable configuration. Braun & Sambridge (1994) have demonstrated how such remeshing can be accomplished dynamically to allow for the influence of faulting.

Whatever representation is used for the numerical framework, there is a need to suppress acoustic waves whose time scales are much shorter than the lithospheric deformation. This can be done with either the assumption of incompressibility, or the milder *anelastic approximation* $\partial\rho/\partial t = 0$. In a two-dimensional problem the representation of flow can be cast in terms of a stream function ψ (as in Chapter 7), but for three-dimensional simulations a full treatment in terms of velocities is normally used. Density changes induced by extension or thermal effects are unlikely to exceed 10%, and so the Boussinesq approximation (Section 7.5.2) in which variations of density are included only in the gravitational forces can generally be adequate. In the lithospheric case, because of the density contrasts between crust and mantle, density changes associated with, e.g., the deformation of the crust–mantle boundary must be included along with thermal effects.

A suitable rheological representation for ductile flow is

$$\dot{\epsilon}_{ij} = A \exp\left[-(E^* + p_L V^* + \mu^* C)/RT\right] \sigma^{n-1} \sigma_{ij}, \tag{12.6.1}$$

relating strain rate $\dot{\epsilon}$ to the stress-tensor $\boldsymbol{\sigma}$. In (12.6.1), E^* is the activation energy, V^* is the activation volume, μ^* is the chemical activation potential associated with changes in composition C, and p_L is the lithostatic pressure. The power-law dependence is introduced through the dependence on the second invariant of the stress tensor $\sigma = (\sigma_{ij}\sigma_{ij})^{1/2}$. The use of a chemical potential is a simplified description of the variation of creep strength with composition. The next level of complication would be to allow A and the index n to be functions of composition C and so allow different creep behaviour in crust and mantle. The constant A in the expression for the effective viscosity η_d can be eliminated by specifying a reference viscosity η_{ref} for a reference temperature, pressure, and strain rate $\dot{\epsilon}_{ref}$:

$$\eta_d = a \exp\left[-(E^* + p_L V^* + \mu^* C)/nRT\right] \dot{\epsilon}^{(n-1)/n}, \tag{12.6.2}$$

with

$$a = \eta_{ref} \exp\left[-(E^* + p_{ref} V^* + \mu^* C)/nRT\right] \dot{\epsilon}_{ref}^{(n-1)/n}. \tag{12.6.3}$$

In the cooler, shallow parts of the crust we have brittle deformation that can be approximated by a power law with a large stress exponent

$$\dot{\epsilon}_{ij} = B(p_L)\sigma^{q-1}\sigma_{ij}, \tag{12.6.4}$$

where, e.g., Christensen (1992) used $q = 30$. A brittle-plastic effective viscosity η_p can be extracted from (12.6.4), but again it is convenient to work in terms of reference values and so define the yield stress invariant σ_y at which the plastic strain equals $\dot{\epsilon}_{ref}$. The yield stress depends on lithostatic pressure as

$$\sigma_y = \sigma_y(0) + f p_L, \tag{12.6.5}$$

where the factor f represents the scaling between lithostatic pressure and pore-fluid pressure (cf. Section 10.2.3). The effective viscosity for the brittle (plastic) deformation is then

$$\eta_p = (\sigma_y \dot{\epsilon}_{ref}^{1/q}) \dot{\epsilon}^{-(q-1)/q}. \tag{12.6.6}$$

The mechanical deformation is coupled to temperature through the advection of

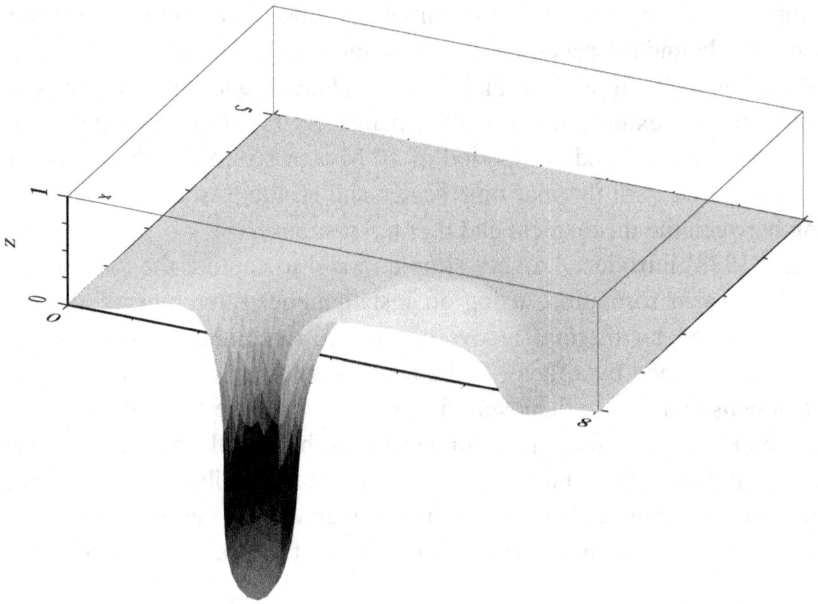

Figure 12.12. Illustration of numerical modelling of lithospheric deformation in three dimensions through a three-dimensional perspective view of a surface at the base of the mantle lithosphere, with extension in one region leading to gravitational instability of dense viscous mantle lithosphere just outside the extending zone. This model is used to explain the occurrence of deep continental seismicity beneath the SE Carpathian mountains. [Courtesy of P. Lorinczi and G. Houseman.]

heat and through heat generation processes. The temperature evolution equation can be cast in the form, cf. (7.1.15),

$$\rho C_p \left(\frac{\partial T}{\partial t} + \mathbf{v} \cdot \nabla T \right) = \nabla (k \nabla T) - h(C) - [\sigma \cdot \dot{\epsilon} - \rho_1 g \alpha_{th} v_z T] \qquad (12.6.7)$$

where c_p is the heat capacity, k the thermal conductivity, $h(C)$ represents the rate of volumetric heat production and ρ_1 is the original density profile. The two terms in square brackets represent the contributions from frictional heating and the effect of adiabatic compression. For consistency these two contributions must either be neglected or included together; commonly if the Boussinesq approximation is adopted they are neglected.

The difficulty with modelling extension processes is the choice of lithospheric parameters, since the coupled mechanical–thermal behaviour can be quite sensitive to the particular choice of parameters and approximations made (Christensen, 1992). The specification of boundary conditions plays an important role. There is a considerable difference between the situation where a side boundary moves with constant velocity or where it is fixed. Over a broad domain, extension in one region will have the effect of throwing the neighbouring zones into compression with consequent deformation (Figure 12.12).

In a simple stretching scenario, the initial response is a slight updoming of the crust-mantle boundary accompanied by a more pronounced upward shift in the boundary between lithosphere and asthenosphere. Once this step has been accomplished further extension can be comparatively rapid and extended crustal thinning can be accomplished in a period of 10 Myr or so. Such a time interval is quite short compared with thermal time scales and so there is a certain degree of decoupling between the mechanical and thermal response.

McKenzie (1978) introduced a very simple model to capture the basic features of sedimentary basin formation, using an instantaneous stretching of the entire lithosphere followed by thermal recovery. When a segment of lithosphere of thickness h_L is stretched by a factor β the new thickness is h_L/β, and to achieve isostatic compensation hot asthenospheric material will rise to take the place of the former thicker lithosphere. The thinned lithosphere will now have a steeper thermal gradient than before and will be out of thermal equilibrium. The stretched lithosphere will cool and thicken until it finally reaches its original thickness and temperature profile. The immediate effect is a drop of the surface elevation by an amount

$$S = \frac{h_L(\rho_M - \rho_L) + h_c(\rho_L - \rho_c)}{\rho_M - \rho_w}\left(1 - \frac{1}{\beta}\right), \tag{12.6.8}$$

where h_c is the initial thickness of the crust with density ρ_c, h_L is the initial thickness of the lithosphere with mantle density ρ_L, ρ_w is the density of water filling the subsidence, and the asthenosphere has density ρ_M. Further subsidence occurs as the lithosphere cools and an approximate solution of the equations of thermal evolution is

$$S(t) = \left[\frac{4h_L\rho_L\alpha_{th}T_M}{\pi^2(\rho_L - \rho_w)}\right]\frac{\beta}{\pi}\sin\left(\frac{\pi}{\beta}\right)\left(1 - \exp(-t/\tau_s)\right), \tag{12.6.9}$$

where the temperature of the asthenosphere is T_M and the relaxation time $\tau_s = h_L^2/\pi^2\kappa_H$. As the surface continues to subside there is room for sediment to be deposited. Since the sediment is denser than water the weight of sediment will also make a small extra contribution to the subsidence.

12.6.2 Convergence

As can be seen from Figure 12.13, the majority of great earthquakes (magnitude > 8) occur at convergent plate margins. Many of these earthquakes, like the 1960 Chilean event (M_w 9.6) and the 2004 Sumatra–Andaman event (M_w 9.3) have a large thrust component associated with slip on the interface between the subducting and over-riding plates. These thrust events are generally accompanied by tsunamis, since the sea-floor is raised locally along a long fault segment. A few of the events, e.g., the 1977 Sumbawa, Indonesia (M_w 8.1) and 2007 Kurile Islands (M_w 8.2) events represent normal faulting in the plate as it enters the trench.

Certain segments of the convergent margins have had many great events in the last century (e.g., Japan–Kuriles–Alaska, South America). The absence of such events elsewhere does not imply absence of risk, merely that the build up of stress has not yet reached critical levels. The 2004 Sumatra–Andaman segment had not had an earthquake of this magnitude for at least 500 years. The last major event on the Cascadia subduction zone (NW USA and Canada) occurred in January 1700, and the timing can be pinned down quite precisely from the arrival times of tsunami waves at sites in Japan.

The oceanic plate entering the subduction zone brings with it inherited structure, due to age variations or the presence of seamounts. The result is that a typical length of relatively homogeneous material entering the trench is of the order of 500 km and so can support a thrust earthquake of magnitude M_W up to around 8.7. Even larger events require multiple segments to fail through successive stress transfer. The presence of obstacles in the seismogenic zone such as a seamount being subducted beneath Shikoku in western Japan affects the slip pattern so that the Nankai earthquake of 1944 has most energy release to the east of the subducting seamount, but still a component to the west.

Large earthquakes along the continent–continent convergence zone in the Himalayas and in the Burma subduction segment (cf. Figure 12.9) are infrequent, but the population density in the region has grown rapidly over the last century. So now the potential for human disaster is much enhanced.

A further class of great earthquakes at plate boundaries is dominantly strike-slip events in transitional zones, e.g., Azores–Gibraltar or the Macquarie Ridge south

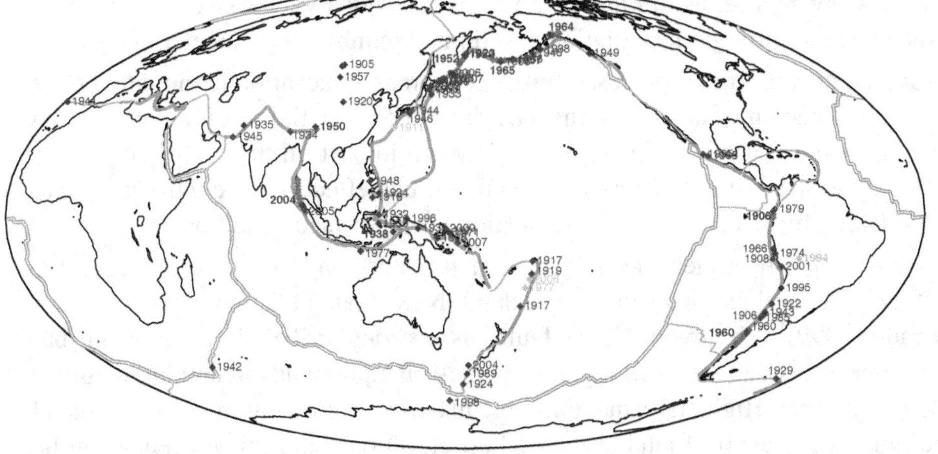

Figure 12.13. Great earthquakes since 1905 (magnitude > 8). Indicative rupture areas are shown for the largest events (dates in bold). The plate boundaries are represented with the same convention as in Figure 12.7. The majority of great earthquakes occur at convergent margins.

of New Zealand. Most events away from specific plate boundaries, e.g. in central Asia, can be linked to the regional compressional regime. Nevertheless, the tectonic settings of a few great earthquakes such as the 1998 event off Antarctica remain unclear. The 1994 Bolivian event occurred at a depth of 590 km and was felt in Toronto, Canada as a result of the radiation pattern from the source. This event appears to have broken through the whole thickness of the deep plate.

In the convergent zones, almost all the energy and most of the slip is taken up by the largest events and their major aftershocks. In between such events the faults can be treated as locked so that stress (and accompanying strain) accumulates in the vicinity of the plate boundaries. Because of the variations in the geometry of convergent plate margins and a significant oblique component in convergence for many subduction zones, stress concentrations can build up away from the main seismogenic zones. The resulting faulting can have very different character from the thrusts at the margin, e.g., the strike-slip faulting of the 1891 earthquake in the Neo valley, central Japan, illustrated in Figure 10.13.

Present-day faulting patterns can be strongly influenced by the prior configuration of deformation. An example occurs in the Zagros mountains of Iran where convergence is taken up by thrusting on listric faults, whose dip shallows with depth, that originated in an extensional regime. Listric fault systems are found in other thrust environments, as e.g., for the Chi-Chi earthquake in Taiwan whose structural setting is illustrated in Figure 10.19.

Modelling of convergent processes can be carried out with similar tools to those used for studying extension, subject to the appropriate boundary conditions. Rather than undertake a full dynamic model of subduction, a number of authors have used a kinematic prescription of the force distribution at the base of the lithosphere to drive the convergence, as in Figure 12.14. In this way a wide range of conditions can be explored through the variation of a limited number of parameters.

Not only do convergent processes produce increased elevation in mountain belts that can impinge on weather patterns, but also the differential erosion arising from prevailing wind and rainfall patterns can have an impact on the profile of surface relief and the stress/strain patterns in the crust. Such effects can be introduced into the modelling by an appropriate specification of the surface conditions.

Figure 12.15 illustrates the deformation produced in the convergence of the Pacific and Australian plates in South Island, New Zealand from the work of Batt & Braun (1999). The main Alpine fault has a strong strike-slip component, but the convergence leads also to significant vertical uplift and mountains of up to 4000 m elevation. High grade metamorphic material originating from lower crustal conditions is frequently found at the highest elevations, and this tendency can be explained by the flow pattern associated with the convergence.

Batt & Braun (1999) have used a rheological model in which the material behaves as a non-linear Maxwell viscoelastic body until the stress exceeds a critical value so that brittle failure occurs. The viscosity is stress-dependent and thermally

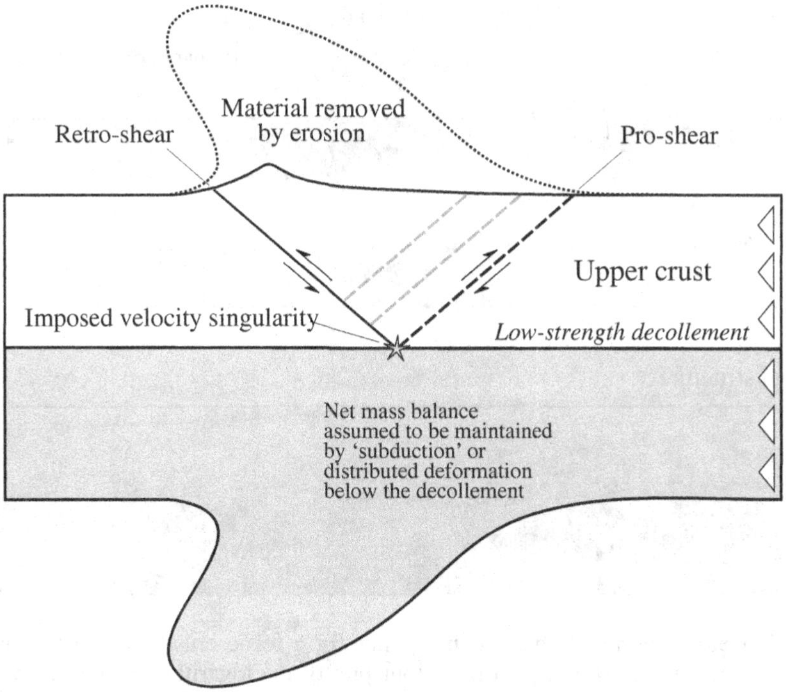

Figure 12.14. Kinematic model of convergence: the crust sits on a relatively strong mantle with finite flexural strength; deformation of the crust is driven by a velocity discontinuity at the base leading to an internal wedge structure whose geometry is determined by the internal strength of the crust and the strength of the decollement zone. [Courtesy of G. Batt]

activated with an Arrhenius form similar to (12.6.1) above, so that temperature can feed back to influence the mechanical properties of the model. The plastic failure criterion is of the form

$$J_{2D} + 12\, T_0 p = 0, \tag{12.6.10}$$

where J_{2D} is the second invariant of the deviatoric part of the stress tensor, T_0 the tensile strength and p the pressure, incorporating both lithostatic and dynamic effects due to deformation. A uniform rate of internal heating is employed, with a constant temperature condition at the base of the model. This approximation neglects the cooling effect of the subducting material and is appropriate for young orogens (<10 Myr duration). The calculations were carried out in a Lagrangian frame using dynamic remeshing (Braun & Sambridge, 1994).

During the initial stages of deformation, two conjugate shear zones form, linked at the velocity discontinuity at the base of the model, with dips close to 45°. The system evolves towards an assymmetric distribution of strain as the main deformation is accommodated on the retro-side where the subducted material descends. Material is rapidly advected through the other shear zone so that there is

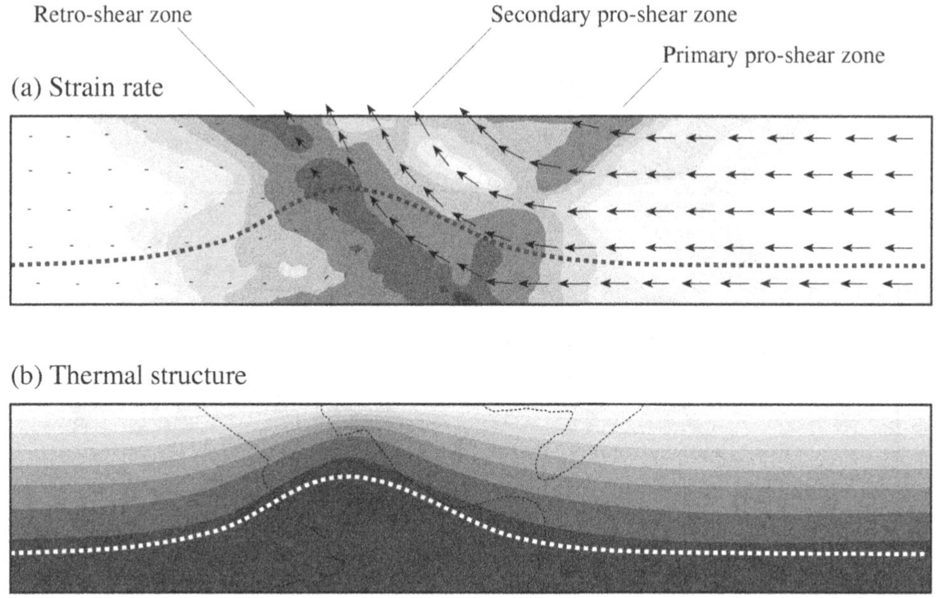

Figure 12.15. Deformation patterns in convergence for a felsic crust, with total erosion and a convergence rate of 10 mm/yr: (a) contour plot of the logarithm of the second invariant of the deviatoric part of the strain-rate tensor after 10 Myr of tectonic deformation (higher rates are darker), with superimposed particle velocities indicated by the arrows. The approximate location of the brittle–ductile transition is indicated by the dashed line; (b) corresponding thermal structure, the initial thermal profile is a gradient from 0°C to 600°C over the model depth of 30 km. [Courtesy of G. Batt.]

much less deformation on this side (Figure 12.15). The flow pattern brings material to the surface near the retro-shear zone. The rise of hotter material has the effect of decreasing viscosity and so enhancing the concentration in the neighbourhood of the retro-zone. The brittle-ductile transition is thereby brought closer to the surface. Strain softening in the natural environment is likely to be more localised than in the simulation.

In the circumstances where the mantle lithosphere is more dense than the underlying asthenosphere convergence will tend to thicken the lithosphere, with the possibility of gravitational instablity leading to drips into the lower viscosity zone beneath (cf. Figure 12.12). This Rayleigh–Taylor instability has been analysed by Houseman & Molnar (1997) to see how the stress dependence of viscosity might affect such convective instability for cold dense mountain roots. They employ both a perturbation analysis for the initial growth of disturbances and finite element analysis for the case of finite deformation. For a power-law rheology ($\dot{\epsilon} \propto \sigma^n$) the speed w at which the bottom of the dense layer descends grows with time as

$$w = \left[C \left(\frac{n-1}{n} \right) \frac{\Delta\rho\, g}{B} h^{1/n} (t_b - t) \right]^{n/(1-n)}, \qquad (12.6.11)$$

where $\Delta\rho$ is the density contrast between the dense layer and the underlying half space, h is the thickness of the layer and B is a measure of resistance to deformation. $C \sim 1$ is an empirical constant that depends on n, wavelength and the gradient of the density distribution in the dense layer. For $n > 1$ the speed w is small initially but increases significantly as t approaches t_b at which time the thickened portion of the dense layer detaches. The time t_b is related to the initial perturbation in the thickness of the layer z_o by

$$t_b = \left(\frac{B}{\Delta\rho g}\right)^n \left(\frac{n}{C}\right)^n \frac{(z_o/h)^{(1-n)}}{(n-1)}. \quad (12.6.12)$$

This model may help to explain why many mountain belts that are built by crustal thickening subsequently collapse with normal faulting and horizontal extension, since the removal of dense mantle lithosphere would provide potential energy to drive extension. The tongue of cool material projecting into the hotter asthenosphere will produce lateral temperature gradients that drive convection. To the sides of the descending material the lithosphere will thin. Once the dense material drops off, warmer material will replace the lithospheric root and uplift occurs. The process would be more effective for wet than dry materials and could remove between 50 and 75 per cent of the mantle lithospheric material over a period of about 10 Myr. The stress-dependent viscosity leads to a delay in the convective thinning that would not occur for a Newtonian fluid, and could explain the time gap between thrusting and normal faulting in the evolution of a number of current and former mountain belts.

13

The Influence of Rheology: Asthenosphere to the Deep Mantle

The contrast in behaviour between the lithosphere with stress transmission and the asthenosphere with flow linked to the mantle convection system represents one of the distinctions that can be made between different parts of the Earth through the nature of intrinsic rheology. However, we have to use indirect information to gain information on the change with depth.

A further major contrast in rheological properties in the near surface arises from the process of subduction where former oceanic lithosphere penetrates through the mantle transition zone to interact with the phase boundary at 660 km and the significant increase of mantle viscosity in the same neighbourhood.

A major control on rheology in Earth's mantle is likely to be the influence of temperature, but this will be accentuated by the effect of other variables such as grain size, which can at best be inferred indirectly.

Little is known of the rheology of materials under the pressure and temperature conditions prevailing at the base of the mantle, but the complexities of the D'' layer suggest that contrasts in physical properties comparable to those seen at the surface may well prevail. The D'' layer has distinct, but somewhat variable, seismic anisotropy that suggests a range of different deformation conditions.

13.1 Lithosphere and asthenosphere

As we have shown in Chapter 12, the definitions of lithosphere are time-dependent and also related to the phenomenon being investigated. The common feature is a transition at depth from a region in which stress is transmitted and heat transport is largely by conduction, to a new zone the *asthenosphere* where horizontal flow is expected in a mixed thermal regime.

This concept arose initially from interpretations of the dispersion of seismic surface waves in terms of a zone with reduced shear wavespeeds at depth. This *low-velocity zone* was associated with increased seismic attenuation that was ascribed to the presence of a small amount of partial melting of the mantle material. The low-velocity zone for S waves is well established in the oceanic environment (Figure 13.1), but is less clear beneath the older parts of continents.

As the concept of plate tectonics developed, the asthenosphere was invoked to accommodate the flows associated with the movements of the plates. In this sense, the lithosphere–asthenosphere boundary would represent the base of the zone that moves coherently in plate motion. This is close to the concept of the 'tectosphere' introduced by Jordan (1975), who suggested major differences between oceanic regions and the cratonic cores of continents based on the time differences in the passage of the seismic phase *ScS* reflected from the core–mantle boundary. Jordan (1978) added other evidence and geochemical arguments to suggest relatively deep roots to the continents that would have a significant influence on the way that plates containing continents might move.

The distinction that we wish to make between the lithosphere and the asthenosphere is dominantly rheological, based on a distinction between the time scales for deformation. Such information is not directly accessible, and so we must resort to a variety of proxy information to infer the presence of a change in rheology. Most information comes from seismology, since this provides a means of investigating present-day structure. The existence of a transition at depth is corroborated by electrical conductivity studies that indicate the presence of a more conducting layer. The logarithmic depth sensitivity of electrical sounding does not place close bounds on the location of this layer.

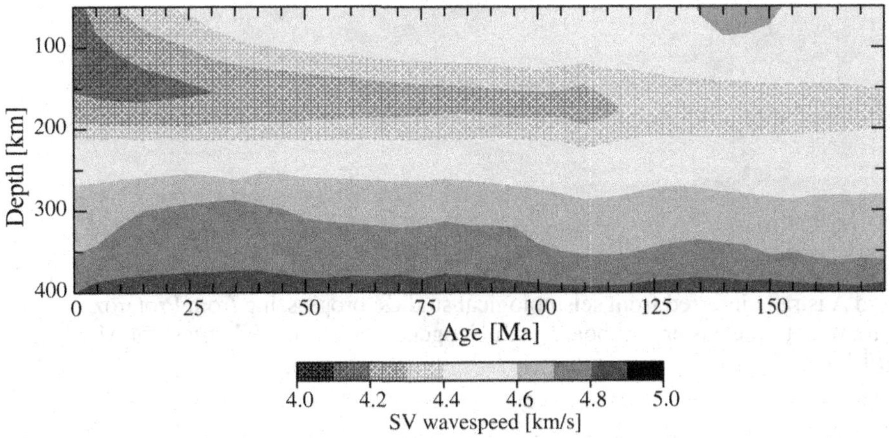

Figure 13.1. Averages of shear wavespeeds in the Pacific Ocean as a function of plate age, show a clear low velocity zone below a lithospheric lid. The averages are derived from a model obtained using surface wave tomography for the fundamental and first four higher modes. [Courtesy of K. Priestley.]

13.1.1 Seismic imaging

The zone of high seismic wavespeeds in the upper part of the mantle is often interpreted as an expression of the lithosphere (cf Figure 12.1). Much of this information comes from surface wave tomography exploiting frequencies in the range from 0.006 to 0.20 Hz, so that the attainable vertical resolution will be of the order of 25 km. On this scale, the base of the high-wavespeed zone (the seismic 'lid') frequently appears quite sharp in oceanic regions,

The situation is much less clear for the continents, where the base of the high-wavespeed region is not generally sharp. A variety of definitions for the 'seismic lithosphere' have been used by different authors, e.g., based on the point of strongest negative vertical gradient in velocity anomalies or where there is a drop in absolute wavespeeds. Studies with good path coverage suggest that high seismic wavespeed structures are mostly confined above 220 km for *SV* waves, with some faster zones at depth beneath Archaean regions. There are however indications of some level of polarisation anisotropy so that faster *SH* wavespeed structures may extend to greater depth than those for *SV* (Gung et al., 2004).

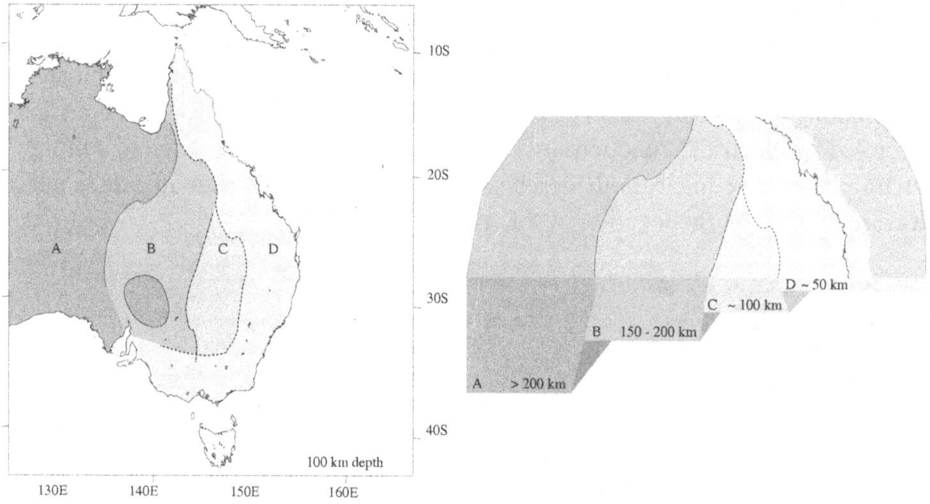

Figure 13.2. Schematic representation of the variation of lithospheric thickness in central to eastern Australia inferred from seismological studies, progressing from Proterozoic craton in the west to the eastern seaboard with Neogene volcanism. [Courtesy of M. Heintz and S. Fishwick.]

Although it may be difficult to get an absolute depth for the lithosphere–asthenosphere transition, relative depths may be extracted by using a consistent style of analysis for a tomographic model. Such results suggest that lithospheric thinning may be accomplished as a sequence of steps related to the history of continental assembly rather than a smooth gradient. Figure 13.2 shows the inferred progression of lithospheric thickness across the eastern half of the Australian continent derived from interpretation of surface wave tomography. The thickest lithosphere is associated with the oldest, cratonic material (early- to mid-Proterozoic 2.0-1.8 Ga). The sequence of steps to thinner lithosphere in the east appears to be related to the assembly of later belts of structure onto the cratonic core. The thinnest lithosphere in the east has experienced Neogene volcanism, after seafloor spreading that opened the Tasman Sea ceased around 80 Ma.

The distinct steps in lithospheric structure appear to have been resistant to thermal erosion over very long periods of time, given their clear association with different age bands at the surface. For resistance to thermal erosion to have persisted over such long intervals there needs to be some geochemical distinctiveness, most likely the presence of larger amounts of harzburgite in the older lithosphere. This mineral is both slightly less dense than other minerals and very refractory. The lower density allows the thicker lithosphere to float on the mantle beneath and still achieve isostasy.

The lithosphere–asthenosphere boundary between steps is likely to be somewhat undulating and local thinning of the lithosphere may provide points where kimberlitic volcanism can break through to the surface through a narrow pipe. Such specimens as we have of upper mantle material come from the *xenoliths* brought to the surface in the rapid ascent of the kimberlitic material, e.g., by removal of the wall rocks in the conduit. The dominant colour is undoubtedly pale green, but the extent to which the grain size and other characteristics of the xenolith material are representative of their original state is unclear.

13.1.2 Seismic attenuation

Some of the most direct seismic evidence for the existence of a significant change in properties at depth comes from seismic attenuation through the frequency content of seismograms. As shown by Gudmundsson et al. (1994) seismic wave propagation in ancient cratonic lithosphere shows very little loss with efficient transmission of high frequency energy for 2000 km or more. The P and S wave arrivals on seismograms show similar frequencies for epicentral distances to 16–18°, and then rather dramatically the high frequencies in S are eliminated for seismograms from sources at greater distances to the same stations. The loss of the high frequencies is a direct result of enhanced seismic attenuation, and the simple visual observations are reinforced by analysis of the spectral ratios of the P and S waves. A region of low Q and reduced shear wavespeed has to lie beneath the high wavespeeds of the Australian cratonic lithosphere.

The geometry of earthquakes is particularly convenient for such studies in Australia, but it is clear that significant shear attenuation has to be sited beneath the lithosphere for ancient continental regions. There are horizontal variations in attenuation as well, and the lithosphere beneath younger domains tends to be more attenuative than under cratons, but still has much less loss than for the asthenosphere below.

The low shear wavespeeds and increased attenuation in the asthenosphere have often been attributed to the presence of partial melt. However, geochemical evidence suggests that, even where melt is present, it is unlikely to be much more than 0.1%, which is insufficient to make a noticeable decrease in shear wavespeeds.

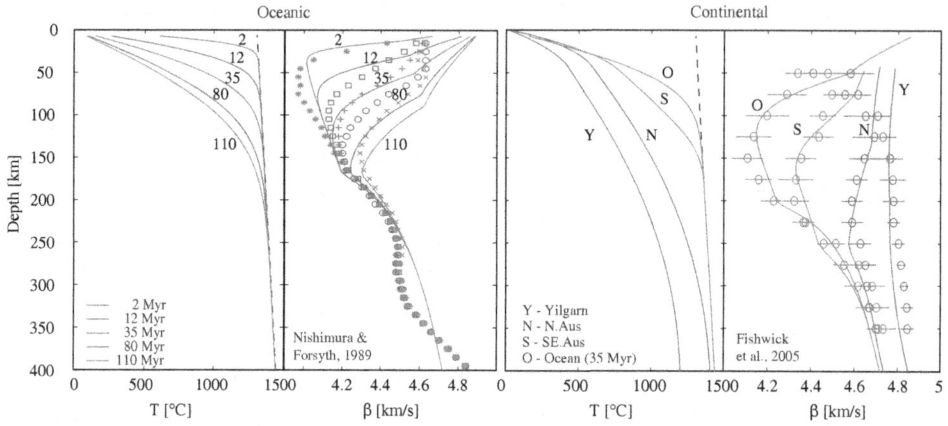

Figure 13.3. Fitting of the representation of seismic wavespeeds, including attenuation, from Section 11.1.2 to seismic velocity profiles from surface wave analysis. The ocean profiles are taken from Nishimura & Forsyth (1989) and the continental profiles from a study of the Australian region by Fishwick et al. (2005). The observational results are shown by symbols and the fitted velocities and associated temperature profiles are indicated by solid curves. A increase in grain size in the lower part of the asthenosphere is included in the fitting. [Courtesy of U. Faul & I. Jackson].

Laboratory measurements, such as those shown in Figure 11.2, indicate that there is a substantial reduction of shear wavespeed and dramatically increased shear attenuation for temperatures 100-200° below the solidus. The seismic properties of the asthenosphere can therefore be explained by thermal effects without needing to invoke significant melt. Figure 13.3 shows how the dependence of seismic shear wavespeeds on depth, derived from surface wave observations, can be well approximated using extrapolations of the results of laboratory experiments with suitable temperature profiles. The fits to the seismic velocities are achieved with the modified Burgers model discussed in Section 11.1.2, and include a systematic increase in grain size in the lower part of the asthenosphere.

The microstructural interpretation of the transition from lithosphere to asthenosphere is in terms of a change of deformation regime from dominant dislocation creep to a situation where diffusion creep, possibly enhanced by water, is the most important process. The dislocation creep in the lithosphere is associated with a power-law rheology, such as the $n = 3$ stress dependence in (9.3.9); whereas diffusion creep is close to linear in stress.

The seismological aspects of the transition from the lithospheric domain to the asthenosphere can certainly be accommodated within the modified Burgers model (11.1.8)–(11.1.11), by a shift in the importance of the different components induced by the increase in temperature with depth. Samples of upper mantle material from xenoliths suggest that there may be a reduction in grain size accompanying the change in thermal regime, enhanced by the presence of any traces of water; there might also be an accompanying change in the activation energy.

13.1.3 Seismic anisotropy

Further evidence for the influence of the asthenosphere comes from changes in the pattern of seismic anisotropy with depth. The orientation of minerals under strain fields (as in Section 10.1) leads to anisotropy in elastic moduli depending on the level of consistency in the crystallographic orientations. Major mantle minerals, such as olivine and clinopyroxene, show strong anisotropy in both P and S waves. Lattice preferred orientation (LPO) of crystals could arise from the active deformation of the asthenospheric mantle that accommodates the absolute plate motion, or be imposed during past deformation and subsequently 'frozen' in the lithosphere during post-tectonic thermal relaxation.

Since olivine is the most abundant and deformable mineral in the upper mantle, the direction of fastest S wavespeed is frequently taken as an indicator of the a-axis ([100]) of olivine, the dominant slip direction (Ismail & Mainprice, 1998).

There are two classes of rather different information that provide constraints on seismic anisotropy: (i) the azimuthal variations in the dispersion of surface waves can be interpreted in terms of the vertical distribution of angular variations in S wave speeds, and (ii) the patterns of splitting in the times of arrival of orthogonally polarised shear body waves as a function of the azimuth to the source can be interpreted in terms of models of seismic anisotropy.

Surface wave anisotropy

For surface waves, Rayleigh waves provides information on the angular variations in the wavespeeds of vertically polarised shear waves (SV) and Love waves provide information on the variations for horizontally polarised shear waves (SH). The local pattern of variation of the phase speed of the dispersed wavetrains as a function of azimuth ϑ depends on $\sin(2\vartheta), \cos(2\vartheta)$ and $\sin(4\vartheta), \cos(4\vartheta)$. For Rayleigh waves the dominant variation is with 2ϑ, whereas for Love waves the most important terms are those in 4ϑ that need much better path coverage to be resolved. As a result, most studies have concentrated on Rayleigh waves. Debayle, Kennett & Priestley (2005) have used Rayleigh wave dispersion on paths from 1000 km to 10 000 km length across the globe, measured from waveform matching techniques, to produce global maps of SV wave anisotropy as a function of depth (see Figure 13.4). Under most continents and the oceans the anisotropy decreases in amplitude between 100 km and 200 km depth, with little change in the directions of the fastest SV wave propagation. However, an exception is provided by the Australian continent which has the highest plate velocity of any continent of nearly 70 mm/yr. Under Australia the direction of fastest SV wave propagation shifts from roughly east–west at 100 km depth to more nearly north–south at 200 km depth. The shallow regime is consistent with anisotropy acquired in past deformation, whereas the orientation at 200 km is close to the direction of absolute plate motion. This suggests that shear on the fast-moving plate has influenced mineral orientation within a zone that

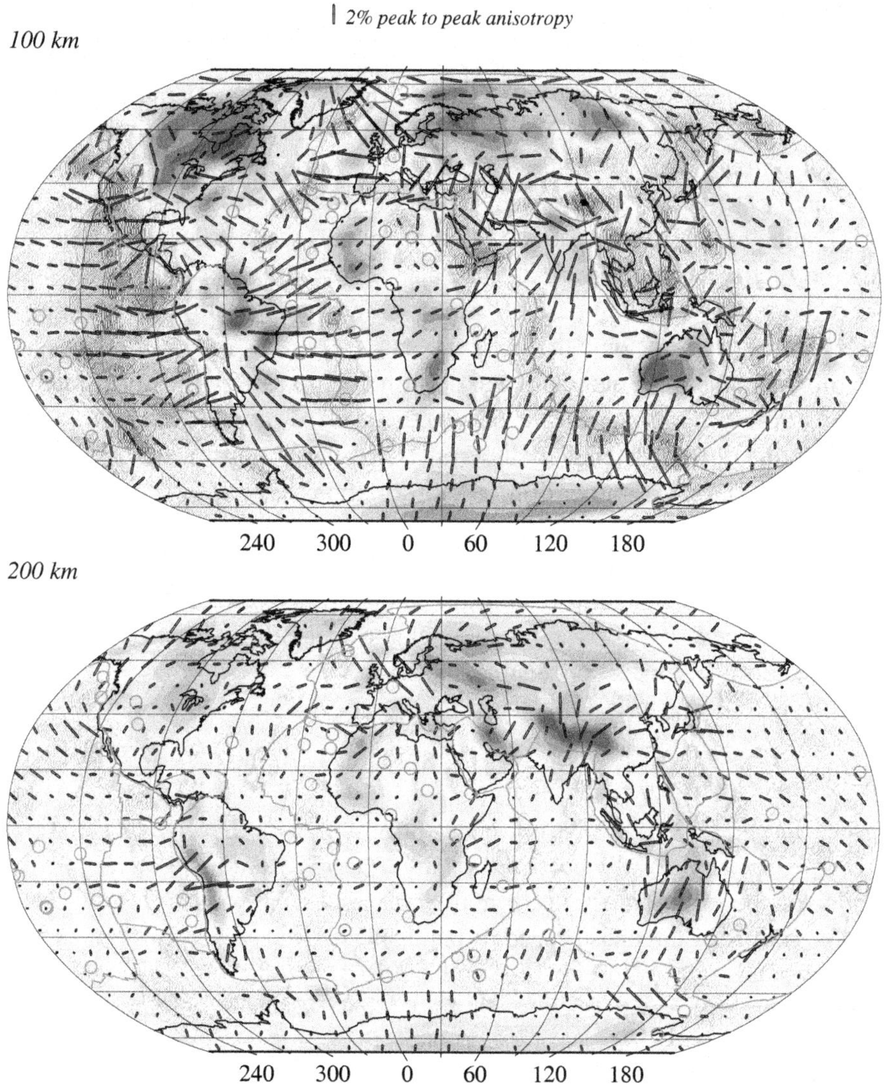

Figure 13.4. SV-wave heterogeneity and azimuthal anisotropy (black bars oriented along the axis of fast propagation) at 100 km and 200 km depth obtained from the inversion of 100779 Rayleigh waveforms (Debayle et al., 2005). Hot spot locations are indicated with circles. The positive S wavespeed anomalies are emphasised, and negative anomalies are shown with patterns. The length of the black bars is proportional to the maximum amplitude of azimuthal anisotropy. The reference velocity for SV wave variation is PREM (Dziewonski & Anderson, 1981).

still displays the fast shear wavespeeds that would normally be associated with the lithosphere.

The alignment of the fastest shear wave propagation with the direction of plate movement is normally regarded as a characteristic of the asthenosphere, but the example above indicates that the rate of strain may be important.

Body wave anisotropy

In the presence of anisotropy, time differentials arise between the arrivals of quasi-shear waves. The 'fast' and 'slow' polarisations are orthogonal to each other and the polarisation directions are characteristic of the anisotropy of the medium (Figure 13.5).

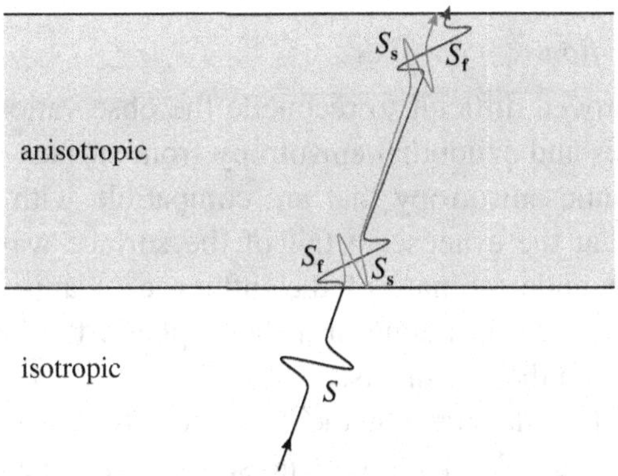

Figure 13.5. Passage of an S wave from an isotropic zone into an anisotropic region leads to the generation of two distinct quasi-S contributions (S_f, S_s) with orthogonal polarisation but different wavespeeds. As a result there is a separation in time, δt, between the arrivals of the two different polarisations.

For seismic body waves, measurements of the time separation between different S wave polarisations, *shear wave splitting measurements*, are commonly performed on shear-wave phases with a core refraction such as *SK(K)S* or *PK(K)S*. Such phases are generated with P to S conversion at the core–mantle boundary (CMB) and are thus polarized along the radial direction as they enter the mantle, i.e., along the great circle between source and receiver. These phases arrive at a station with a nearly vertical incidence, and energy appearing on the tangential component has to be acquired in passage through the mantle. Thus a difference in the shear wave times measured at the Earth's surface represents the vertically integrated effect of anisotropy from the CMB to the surface, with no indication of the depth location of the anisotropy source. Because conversion to *SV* waves occurs on the receiver side at the CMB, no contamination arises from structure near the source.

The measurement of shear wave splitting at a seismological station yields two parameters: φ, the orientation of the polarization plane of the faster S wave and δt, the delay between the arrival times of the fast and slow split waves. Petrophysical studies suggest that anisotropy will mostly be located in the upper mantle, between the 410 km olivine–spinel phase transition and the Moho. Some smaller contributions from the D$''$ layer, the lower mantle and the crust may also be present. The orientation of the polarization plane of the fast S-wave, φ is taken as a proxy for the orientation of the [100] axis of olivine in the upper mantle. The delay time δt will be a function of the intrinsic anisotropy, the thickness of the anisotropic layer, the orientation of the ray path with respect to the elastic matrix of the anisotropic medium, and the vertical coherence of the mantle fabric. Where measured orientations appear to be correlated with the superficial geological structures, the anisotropy should be shallow with a significant crustal component. Otherwise an origin for the shear wave splitting needs to be sought at greater depth.

13.1.4 Asthenospheric flow

In many areas it has proved difficult to reconcile the observations of shear wave splitting for body waves and azimuthal anisotropy from surface waves. Attempts to find models of seismic anisotropy that are compatible with both the sets of observations suggest that the evanescent tail of the surface wave displacements with depth leads to an underestimate of the influence of anisotropy within the asthenosphere. The presence of a zone of asthenospheric flow may therefore be a common feature between the continents.

The presence of such flow deflected by the structure at the base of the lithosphere may help to explain the complex pattern of hotspots tracks and volcanic edifices in southeastern Australia (W. Jason Morgan, personal communication). The asthenosphere flow is guided away from the deep rooted cratons in central Australia towards the east by the set of lithospheric steps (Figure 13.2). As the material rises there is consequent decompression and potential release of melt, which can reach the surface in zones of pre-existing weakness.

The patterns of shear wave splitting in eastern North America with a distinct change in the orientation of the fast direction of wave propagation in the New England region over a relatively short distance require a change in the controlling factor for the seismic anisotropy. Fouch et al. (2000) have suggested that this arises from the deflection of asthenospheric flow around the deep lithospheric keel beneath the cratonic region that appears to have a re-entrant feature near 200 km depth from surface wave tomography studies. Finite difference models of the flow pattern to be expected around a mantle keel indicate that a good fit to the observed patterns can be achieved with a segment ('divot') missing from the base of the keel. The mantle flows around the deep lithosphere but tends to enter into the gap, leading to rapid variations in the expected fast directions of shear waves over short distance scales. Beneath the craton root the fast directions are parallel to the plate motion, which is consistent with general patterns of observations.

13.1.5 The influence of a low-viscosity zone

The presence of a zone of low viscosity modifies the way in which fluid flow and convective behaviour is organised. Busse et al. (2006) have provided an analytical approach that allows the investigation of a region of lowered viscosity adjacent to material with an extended zone of uniform viscosity. The configuration of the system is sketched in Figure 13.6; it is convenient to take a symmetric scenario with both upper and lower viscosity layers. This provides a simple approximation to the presence of an asthenosphere and a lower mantle boundary layer.

We consider convection in a fluid layer of height h with constant kinematic viscosity ν_0 in most of the interior, but with a much lower value ν_b in thin layers of thickness δh ($\ll h$) next to the top and bottom boundaries (Figure 13.6). For an incompressible fluid ($\nabla \cdot \mathbf{v} = 0$), the Navier-Stokes equation (7.1.5) is

$$\rho\left(\frac{\partial}{\partial t} + v_k\frac{\partial}{\partial x_k}\right)v_j = \rho g_j - \frac{\partial}{\partial x_j}p + \frac{\partial}{\partial x_i}\left[\eta\left(\frac{\partial v_j}{\partial x_i} + \frac{\partial v_i}{\partial x_j}\right)\right], \tag{13.1.1}$$

with the linked thermal equation for material with constant thermal diffusivity κ_H (7.5.7),

$$\left(\frac{\partial}{\partial t} + v_k\frac{\partial}{\partial x_k}\right)T = \kappa_H\frac{\partial}{\partial x_k}\frac{\partial}{\partial x_k}T, \tag{13.1.2}$$

where we have neglected the work done by the weak flow and assumed no internal heat sources.

In the Boussinesq approximation, as in Section 7.5.2, we include thermally induced density variations in the buoyancy term (ρg_j) but neglect such variations elsewhere. We can absorb the hydrostatic pressure in the reference state as in (7.5.6) by introducing

$$P = p - \rho_0 g z. \tag{13.1.3}$$

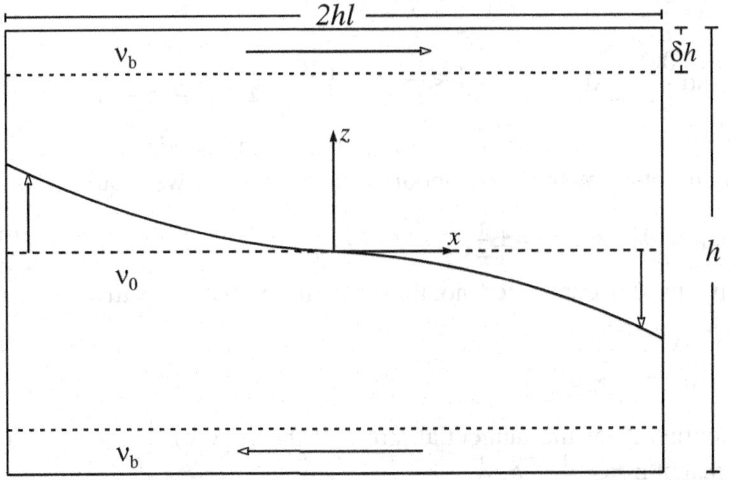

Figure 13.6. Configuration of a convection system with zones of lowered viscosity, flow directions are indicated with open arrows.

The buoyancy contribution is then $-\rho_0\alpha_{th}[T - T_0]g(\hat{\mathbf{e}}_z)$.

It is convenient to work with non-dimensional quantities and we use the height of the layer h as the length scale, the thermal timescale h^2/κ_H, and the difference between the top and bottom temperatures $(T_2 - T_1)$ as the temperature scale. In terms of non-dimensional quantities the Navier-Stokes equation takes the form

$$\frac{1}{Pr}\left(\frac{\partial}{\partial t} + \nu_k\frac{\partial}{\partial x_k}\right)\nu_j = -\frac{\partial P}{\partial x_j} + Ra\,\Theta(\hat{\mathbf{e}}_z)_j + \frac{\partial}{\partial x_i}\left[\frac{\nu}{\nu_0}\left(\frac{\partial \nu_j}{\partial x_i} + \frac{\partial \nu_i}{\partial x_j}\right)\right], \tag{13.1.4}$$

where we have introduced the Prandtl number Pr (7.1.18) and the Rayleigh number Ra (7.5.1)

$$Pr = \frac{\nu_0}{\kappa_H}, \qquad Ra = \frac{\alpha_{th} g(T_2 - T_1)h^3}{\nu_0\kappa_H}. \tag{13.1.5}$$

The equation for the non-dimensional temperature Θ has the simple form

$$\left(\frac{\partial}{\partial t} + \nu_k\frac{\partial}{\partial x_k}\right)\Theta = \frac{\partial}{\partial x_k}\frac{\partial}{\partial x_k}\Theta. \tag{13.1.6}$$

We examine two-dimensional solutions of (13.1.5), (13.1.6) in the circumstances when the Prandtl number is very large. As in Sections 7.4, 7.5.2 we introduce a stream function ψ such that

$$\nu_x = -\frac{\partial \psi}{\partial z}, \quad \nu_z = \frac{\partial \psi}{\partial x}. \tag{13.1.7}$$

We can eliminate the term from (13.1.5) by taking the curl of the equation and then from the x-component we find the following equation system for the stream functions ψ_0 in the interior and ψ_b in the boundary layers

$$\nabla^4\psi_0 - Ra\frac{\partial\Theta}{\partial x} = 0 \quad \text{in} \quad -\tfrac{1}{2} + \delta < z < \tfrac{1}{2} - \delta, \tag{13.1.8}$$

$$\nabla^4\psi_b - \beta Ra\frac{\partial\Theta}{\partial x} = 0 \quad \text{in} \quad -\tfrac{1}{2} \le z \le -\tfrac{1}{2} + \delta, \tfrac{1}{2} - \delta \le z \le \tfrac{1}{2}. \tag{13.1.9}$$

Here β is the ratio of the interior viscosity to that in the boundary layers ν_0/ν_b.

With the assumption of stress-free boundaries at $z = \pm\frac{1}{2}$ we require

$$\psi_b = \partial^2_{zz}\psi_b = 0 \quad \text{at} \quad z = \pm\tfrac{1}{2}, \tag{13.1.10}$$

where we have used a contracted notation for the partial derivative. We restrict attention to one convection cell so

$$\psi_b = \psi_0 = 0 \quad \text{at} \quad x = \pm l. \tag{13.1.11}$$

We require continuity of the tangential stress at the edge of the zone of lowered viscosity, so that at $z = -\frac{1}{2} + \delta, \frac{1}{2} - \delta$

$$\psi_b = \psi_0, \quad \partial_z\psi_b = \partial_z\psi_0, \quad [\partial^2_{zz} - \partial^2_{xx}]\psi_b = \beta[\partial^2_{zz} - \partial^2_{xx}]\psi_0. \tag{13.1.12}$$

In the limit of very large Rayleigh numbers we will get a periodic array of alternate rising hot plumes and descending cold plumes with a very narrow width. We restrict attention to the interval $-l \le x \le l$ with a hot plume at $x = -l$ and a cold plume at $x = l$. Away from these regions the horizontal derivative $\partial_x\Theta$ will be very small and can be neglected in (13.1.8), (13.1.9). Without loss of generality we can set the mean temperature to zero and so the temperature boundary condition is

$$\Theta = \mp\tfrac{1}{2} \quad \text{at} \quad z = \pm\tfrac{1}{2}. \tag{13.1.13}$$

Stream function

We assume symmetry about the mid-plane of the layer $z = 0$ and concentrate attention on the upper zone. We need to include the condition of continuity of normal stress at the viscosity interface $z = \frac{1}{2} - \delta$. To a good approximation the effect of the normal viscous stress in the interior layer acts on the less viscous layer as a compensating pressure term at $z = \frac{1}{2} - \delta$,

$$\partial_x p_b(x) = \beta\left[3\partial^3_{xxz} + \partial^3_{zzz}\right]\psi_0 \tag{13.1.14}$$

In the thin low-viscosity layer this pressure drives a shear flow specified by

$$\nabla^2\partial_z\psi_b = \partial_x p_b(x) = \beta\left[3\partial^3_{xxz} + \partial^3_{zzz}\right]\psi_0\Big|_{z=\frac{1}{2}-\delta} = \beta\Psi_b. \tag{13.1.15}$$

Since the thickness δ of the layer is small we can replace the Laplacian by ∂_{zz} and integrate with respect to z to obtain

$$\psi_b = \tfrac{1}{6}\beta\Psi_b\bar{z}^3 + f(x)\bar{z}, \quad \text{with} \quad \bar{z} = z - \tfrac{1}{2}, \tag{13.1.16}$$

where $f(x)$ is a function to be determined; this form for ψ_b satisfies the continuity conditions (13.1.10). The continuity conditions for tangential stress, (13.1.12), require that at $z = \frac{1}{2} - \delta$ the interior stream function ψ_0 satisfies

$$\psi_0 = -\delta^3\tfrac{1}{6}\beta\Psi_b - \delta c(x), \tag{13.1.17}$$

$$\partial_z\psi_0 = \delta^2\tfrac{1}{2}\beta\Psi_b + c(x), \tag{13.1.18}$$

$$\left[\partial^2_{zz} - \partial^2_{xx}\right]\psi_0 = \delta\beta\left[3\partial^3_{xxz} + \partial^3_{zzz}\right]\psi_0. \tag{13.1.19}$$

We can extract $f(x)$ from (13.1.17), (13.1.18) as

$$f(x) = -\tfrac{1}{2}\left[\frac{3}{\delta}\psi_0 + \partial_z\psi_0\right]_{z=d}, \quad \text{with} \quad d = \tfrac{1}{2} - \delta. \tag{13.1.20}$$

The solution of the biharmonic equation $\nabla^4\psi_0 = 0$ that meets the horizontal boundary conditions (13.1.11) is

$$\psi_0 = \tfrac{1}{2}C(x^2 - l^2) + \sum_{m=1}^{\infty} \cos(a_m x)\{A_m\cosh(a_m z) + B_m\sinh(a_m z)\}, \tag{13.1.21}$$

where

$$a_m = (2m-1)\frac{\pi}{2l}. \tag{13.1.22}$$

The solution for A_m, B_m exploiting the continuity conditions (13.1.17) – (13.1.19) is straightforward but algebraically complex (for details see Busse et al., 2006). To a good approximation the solution for ψ_b can be written as

$$\psi_b = \sum_{m=1}^{\infty} \tfrac{1}{2}\Gamma_m \cos a_m x \left[\left(\frac{\bar{z}^3}{\delta^3} - 3\frac{\bar{z}}{\delta} \right) (2 + G_m(\delta)) + ... \right] \tag{13.1.23}$$

where $\Gamma_m = (-1)^m C/(a_m^3 l)$ and

$$G_m(\delta) = \frac{\delta^3 \beta a_m^4 d - 3\cosh^2(a_m d)}{\delta^3 \beta a_m^3 [\cosh(a_m d)\sinh(a_m d) - a_m d] - \frac{3}{2}\cosh^2(a_m d)} \tag{13.1.24}$$

The expression (13.1.23) can be well approximated by

$$\psi_b = \tfrac{1}{4} C(x^2 - l^2) \left[\frac{\bar{z}^3}{\delta^3} - 3\frac{\bar{z}}{\delta} \right] (2 + G_1(\delta)), \tag{13.1.25}$$

since terms with $m \geq 2$ make only a minor contribution.

Thermal boundary layer

The stream function (13.1.25) with a parabolic profile describes the dominantly horizontal flow in the upper layer with low viscosity, and can now be used to consider the thin thermal boundary layer close to $z = \frac{1}{2}$. The temperature equation (13.1.6) can be written in terms of the stream function as

$$\nabla^2 \Theta = \partial_z \psi_b \partial_x \Theta - \partial_x \psi_b \partial_z \Theta. \tag{13.1.26}$$

With the introduction of the coordinates

$$\xi = \int_{-l}^{x} d\tilde{x}\, \partial_z \psi_b(\tilde{x}), \qquad \zeta = \tfrac{1}{2} \partial_z \psi_b\, \bar{z}, \tag{13.1.27}$$

the temperature equation can be transformed into a diffusion equation

$$\frac{\partial \Theta}{\partial \xi} = \frac{1}{4} \frac{\partial^2 \Theta}{\partial \zeta^2}. \tag{13.1.28}$$

A solution which satisfies the transformed boundary conditions

$$\Theta = -\tfrac{1}{2} \quad \text{at} \quad \zeta = 0, \qquad \Theta = 0 \quad \text{for} \quad \zeta \to \infty, \tag{13.1.29}$$

can be found from the results in Section 8.4.2 as

$$\Theta = -\tfrac{1}{2} - \frac{1}{\sqrt{\pi}} \int_0^{\zeta/\sqrt{\xi}} dy\, e^{-y^2}. \tag{13.1.30}$$

The temperature distribution (13.1.30) in the boundary layer does not link directly to the rising plume at $x = -l$, but numerical solutions indicate that this does not represent a problem for positions close to the upper boundary.

The local Nusselt number $\mathrm{Nu}_{\mathrm{loc}}$ can be found from the normal derivative of the temperature at $z = \frac{1}{2}$ since the heat transport in the absence of convection is unity

$$\mathrm{Nu}_{\mathrm{loc}} = -\frac{\partial}{\partial z}\Theta(\xi(x)) = \frac{3}{4}\left(\frac{C(2+G(a_1)}{\pi\delta}\right)^{\frac{1}{2}} \frac{l-x}{(2l-x)^{1/2}}. \qquad (13.1.31)$$

The regular Nusselt number is the average of (13.1.31) over the interval $-l \leq x \leq l$

$$\mathrm{Nu} = \frac{1}{2}\left(\frac{C(2+G(a_1)}{\pi\delta}\right)^{\frac{1}{2}}. \qquad (13.1.32)$$

The heat transport arriving at the upper boundary has to equal that carried by the rising plume

$$2l\mathrm{Nu} = \int_{-l-\epsilon}^{-l+\epsilon} dx\,\Theta\langle[-\partial_z\psi]_{x=-l}\rangle = lC[1+H(a_1)]\int_{-l-\epsilon}^{-l+\epsilon} dx\,\Theta, \qquad (13.1.33)$$

where the angle brackets denote the average over the z-dependence. The function

$$H(a_1) = \frac{4}{\pi^2}\left[\frac{\delta^3 2\beta a_1^2(\sinh^2(a_1 d) - 2(6\sinh(a_1 d)\cosh(a_1 d) - a_1)}{\delta^3 2\beta a_1^3(\sinh(a_1 d)\cosh(a_1 d) - a_1 d) + 3\cosh^2(a_1 d)}\right]; \qquad (13.1.34)$$

terms with $m \leq 2$ make a negligible contribution. The constant C can be evaluated by balancing the tangential stress on the plane $x = -l$ with the buoyancy of the rising plume,

$$C = \frac{1}{2}\mathrm{Ra}\int_{-l-\epsilon}^{-l+\epsilon} dx\,\Theta = \mathrm{Ra}^{2/3}\left(\frac{[2+G(a_1)]l}{4\pi[1+H(a_1)]^2\delta}\right)^{1/3}. \qquad (13.1.35)$$

Hence the Nusselt number can be expressed as

$$\mathrm{Nu} = \frac{1}{2}\mathrm{Ra}^{1/3}\left(\frac{[2+G(a_1)]^2 l^2}{2\pi[1+H(a_1)]\delta^2}\right)^{1/3}. \qquad (13.1.36)$$

For a constant viscosity layer the maximum heat transport occurs for an aspect ratio $2l$ of order unity, whereas with the presence of the low viscosity layer the maximum of the Nusselt number occurs for somewhat higher aspect ratios. For a viscosity ratio of 1000 the horizontal cell size doubles. The approximate treatment gives the correct dependence of Nu on l but somewhat underestimates the actual values.

This simple model indicates that a region of reduced viscosity in the upper mantle as indicated from studies of glacial rebound (Section 12.5) has important consequences, particularly with respect to the aspect ratio of possible convection cells. The aspect ratios needed for a convective origin for plate motion are far more consistent with the presence of a zone of lowered viscosity than the constant-viscosity case.

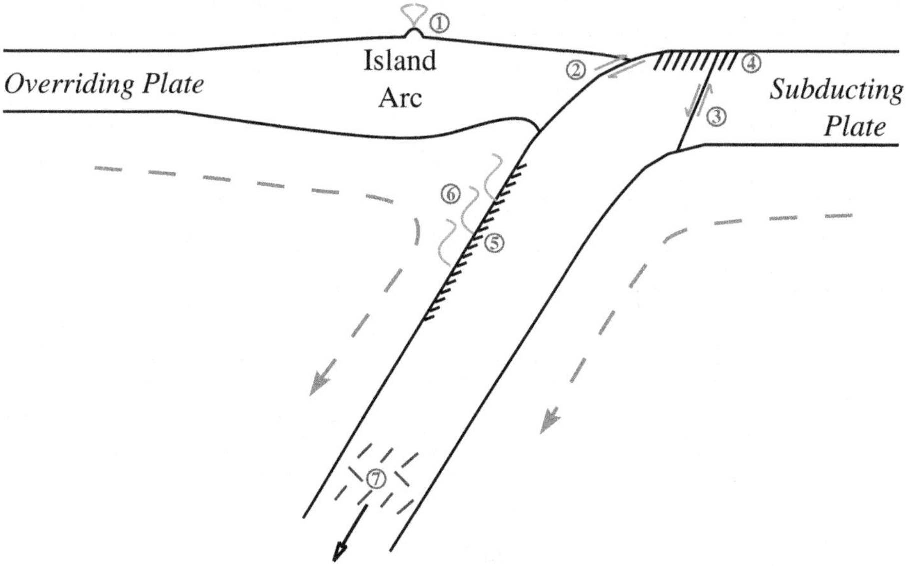

Figure 13.7. Cross-section through a subduction zone system indicating the nature of flow, the stress state of the slab through observed earthquake mechanisms and the typical location of the volcanic arc associated with subduction. (1) volcanic arc, small earthquakes occur in the over-riding plate in the forearc; (2) seismogenic zone associated with thrust earthquakes, e.g., the great 2004 Sumatra–Andaman event (Mw 9.3); (3) normal faulting through the ocean lithosphere as in the 1997 Sumbawa event (Mw 7.7); (4) fracturing of plate as it approaches trench and starts to bend; (5) dehydration reactions in former oceanic crust, intermediate earthquakes near the top of the slab with largely down-dip tension; (6) volatiles released from the downgoing slab; (7) deep earthquakes in slab under compression.

13.2 Subduction zones and their surroundings

The mechanical strength of lithospheric material is of particular importance in subduction zones. We have seen in the analysis of forces acting on the lithosphere in Section 12.3.2 that *slab-pull* is a major component of the system. The mass circulation achieved by the recirculation of cold former lithosphere into the deeper mantle forms a very important part of the mantle convection process, which we address in Chapter 14. Here we concentrate on the way in which subducted material enters the mantle and the consequent flow fields in their vicinity.

Figure 13.7 displays the general features associated with subduction in an oceanic environment, with a deep ocean trench where the subducted lithosphere enters the mantle and mantle flow above and below the plate. Earthquake activity generally occurs within the subducted plate except for great thrust events such as the 1960 Chile (Mw 9.6) and 2004 Sumatra–Indonesia (Mw 9.3) earthquakes which propagate along the interface between the subducting and overriding plate. Most subduction is accompanied by volcanism in an island arc lying back somewhat from the trench on the overriding plate, linked to volatiles released from the former oceanic lithosphere.

13.2.1 Configuration of subduction zones

The dominant behaviour in subduction is of a slab of oceanic lithosphere bending at the entry into the mantle, and then descending to depth with a dip typically in the range 35–60° (Figure 13.7). The particular circumstances depend on the relative motion of the subducting and over-riding plates. Although the dominant behaviour can be extracted from two-dimensional models in which subduction occurs perpendicular to the surface expression of the plate boundary, many systems have a significant oblique component. For example, at the northern end of the Sumatra–Andaman arc, along which the great 2004 earthquake (Mw 9.3) occurred, the plate motion is only at an angle of 10° to the boundary. The earthquake initiated thrusting that propagated northward along the arc for some 1300 km (see Figure 11.26), with variations in character associated with changes in the strike of the plate boundary and consequent modifications of the physical state of the upper plate surface.

Considerable variability can occur in the geometry of subduction, even for a single plate boundary. Figure 13.8 illustrates the morphology of the subduction zone in the Izu–Bonin–Mariana arc to the south of Japan, where the old Pacific plate is subducting beneath the younger Philippine plate in a purely oceanic environment. The shape of the subducted slab is inferred from seismic tomography studies. A very substantial change in the geometry of the deeper part of the slab is inferred to occur between the Izu–Bonin and Mariana segments. The near-horizontal segment in the mantle transitions zone for the Izu-Bonin arc may well be a consequence

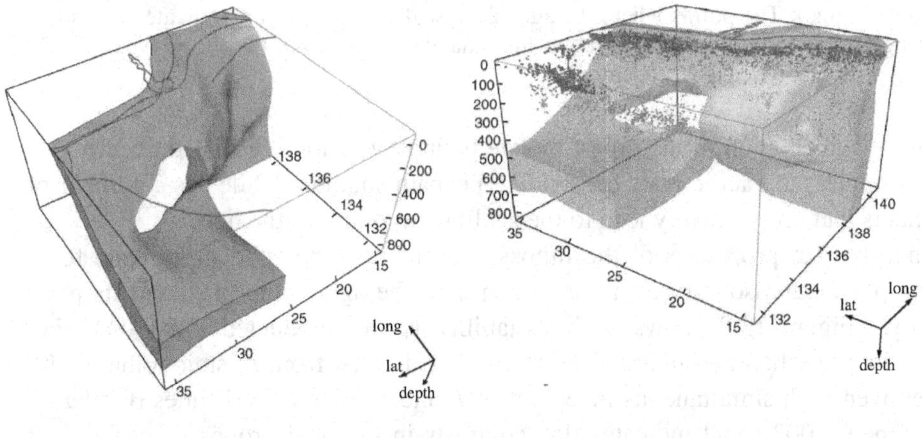

Figure 13.8. Interpreted morphology and geometry of the subducting Pacific slab beneath the Izu–Bonin–Mariana arc from P wave tomography images (Miller et al., 2005). The hole corresponds to a region where there is change in seismic properties. Earthquakes from the NEIC catalogue for events from 1967–1995 are superimposed on the slab configuration.

of rapid slab *roll-back* in the evolution of the plate boundary. The change to the much steeper dips in the Mariana arc occur where the Marcus–Nesker ridge enters the subduction system, with at present the Ogasawara plateau at the trench. There is a zone of much reduced contrast in *P* wavespeed at depth, that gives rise to the apparent 'hole' in Figure 13.8. This zone may indicate the development of a tear in the slab linked to the oblique component of subduction since it is marked by pronounced earthquake activity.

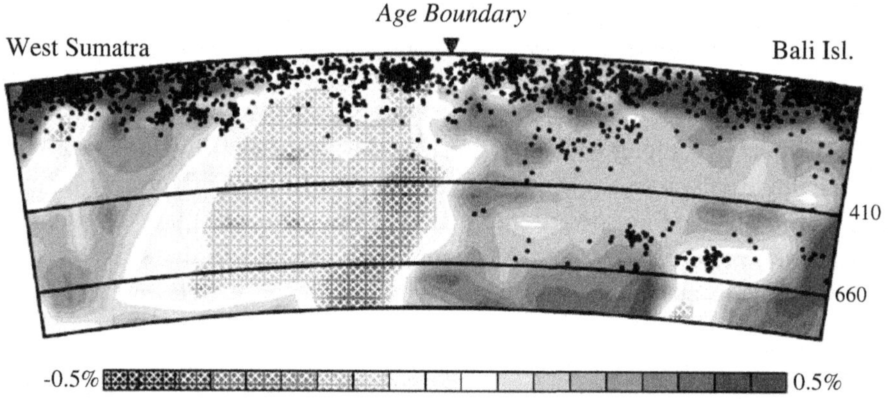

Figure 13.9. Variations in the relative amplitude of bulk-sound speed and shear wavespeed anomalies along the subducted plate in the Java–Sumatra subduction zone (Gorbatov & Kennett, 2003). The cells with high-velocity anomalies associated with the subducted plate were isolated and averages of the difference between the wave-speed variations from the AK135 reference model were made through the thickness of the plate. The bulk-sound speed values are subtracted from the shear wave-speed values, and the results are then projected on the vertical section. Thus large bulk-sound velocity anomalies appear as negative values (patterned grey tone), while high shear wave speed is presented as positive anomaly (solid grey tone). The points where the age of the seafloor changes are indicated and show a strong correlation with the changes in the balance between the two wave types. Solid black dots are earthquake hypocentres.

Such segmentation of the dip of a plate boundary is not an isolated occurrence. For example, the Pacific plate descending beneath southern Chile has a number of segments that progressively lead to the shallowing of dip to the north.

The physical properties of the lithosphere entering the subduction zone for a single plate can also vary as a consequence of the age of the plate and its prior history. Figure 13.9 shows such variability in the subducted Indo-Australian plate along the Indonesian arc. The image is extracted from seismic tomography undertaken with simultaneous inversion of *P* and *S* wave arrival times (Gorbatov & Kennett, 2003), and indicates that contrasts in physical properties can persist to significant depth. The dominance of shear wavespeed anomalies occurs for lithosphere older than 85 Ma as it enters the trench, such oceanic lithosphere may well have reached thermal maturity.

These examples illustrate that the subduction process carries its own complexity linked to the history of plate motions, so that there can be variations in the physical properties of the subducted slab with segmentation on scales of 500 km or so. Nevertheless, we can capture the dominant behaviour associated with the descent of cool, dense material into the mantle with simple two-dimensional conceptual models.

13.2.2 Flow around the slab

Many of the characteristics of the flow around a subduction zone can be represented via a simple mechanical model (Figure 13.10) in which the stationary overriding plate and the subducting plate both have thickness a, with convergence at speed V_S. The subducting plate descends with dip δ_S, and the relative motion of the two plates drives flow in the mantle wedge above the subducting plate. The motion of the subducting plate also leads to an underside flow.

The flow of mantle material outside the cold slabs is represented through the behaviour of an incompressible viscous fluid, in a steady state, so the velocity of the fluid must satisfy

$$\eta\nabla^2\mathbf{v} + \rho\nabla\varphi = \nabla p, \quad \text{with} \quad \nabla\cdot\mathbf{v} = 0. \tag{13.2.1}$$

Here we have neglected spatial variations in viscosity associated with variation in temperature, φ is the gravitational potential and p is the pressure. For this incompressible viscous fluid the vorticity $\boldsymbol{\varpi} = \nabla\times\mathbf{v}$ satisfies $\nabla^2\boldsymbol{\varpi} = 0$.

The convergence of many subduction zones includes an oblique component, but we can capture the main behaviour with the two-dimensional approximation that

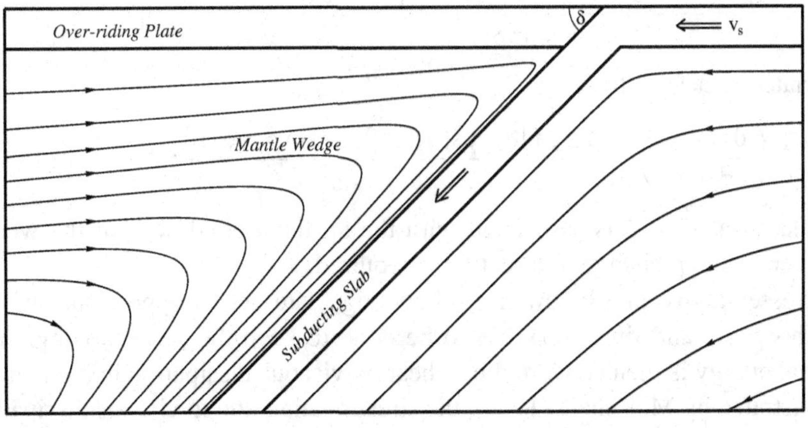

Figure 13.10. Streamlines for viscous flow around a subducting slab, driven by the motion of the sinking slab and its attached plate, using a corner flow representation neglecting any thermal convection effects. The lithosphere is taken as 50 km thick and the dip angle δ_S is 45°.

convergence is normal to the overriding plate. As in Section 7.4 we introduce a stream function ψ such that the velocity field $\mathbf{v}$ in the fluid is expressed as

$$\mathbf{v} = \nabla \times \psi \, \hat{\mathbf{e}}_\perp \quad \text{with} \quad \nabla^4 \psi = 0, \tag{13.2.2}$$

where the unit vector $\hat{\mathbf{e}}_\perp$ lies perpendicular to the plane of convergence. It is convenient to adopt cylindrical polar coordinates in each of the wedges so that

$$\mathbf{v} = \left[v_r, v_\theta, 0\right] = \left[-\frac{1}{r}\frac{\partial \psi}{\partial \theta}, \frac{\partial \psi}{\partial r}, 0\right]. \tag{13.2.3}$$

We want to find velocity fields that have fixed values on radial vectors from the wedge corners and seek solutions in the form (Batchelor, 1967)

$$\psi(r, \theta) = r f(\theta), \tag{13.2.4}$$

so that from (13.2.2) we require

$$\frac{1}{r^3}\left(\frac{\partial^4 f}{\partial \theta^4} + 2\frac{\partial^2 f}{\partial \theta^2} + f\right) = 0, \tag{13.2.5}$$

with general solutions of the form

$$f(\theta) = A \sin\theta + B \cos\theta + C\theta \sin\theta + D\theta \cos\theta; \tag{13.2.6}$$

the constants A, B, C, D have to be found from the specific boundary conditions.

In the mantle wedge zone we require

$$\mathbf{v} = \mathbf{0} \quad \text{on} \quad \theta = 0; \qquad v_r = V_s, \; v_\theta = 0 \quad \text{on} \quad \theta = \delta_s, \tag{13.2.7}$$

where we measure θ anticlockwise from the base of the overriding plate. The conditions are satisfied when (McKenzie, 1969)

$$\psi^w = \frac{r V_s[(\delta_s - \theta)\sin\delta_s \sin\theta - \theta\delta_s \sin(\delta_s - \theta)]}{\delta_s^2 - \sin^2\delta_s}, \tag{13.2.8}$$

the associated shearing stress

$$\sigma^w_{r\theta} = \frac{\eta}{r}\left(\frac{\partial^2 f}{\partial \theta^2} + f\right) = \frac{2V_s\eta}{r}\frac{[\delta_s \cos(\delta_s - \theta) - \sin\delta_s \cos\theta]}{\delta_s^2 - \sin^2\delta_s}; \tag{13.2.9}$$

the singularity at $r = 0$ is associated with the artificial conditions at the wedge corner where a sharp change in direction is required.

The stresses above and below the subducting plate act to oppose the motion of the lithosphere and thus work has to be done to keep the plate moving. The mechanical energy is then converted into heat by viscous dissipation in the mantle. As demonstrated by McKenzie (1969), the stress heating due to viscous dissipation in the flow $H^w = \sigma^2_{r\theta}/\eta$, and so the contours of the stress heating function are the same as those for the shearing stress. The high stresses near the wedge corner would lead to a modification of the viscous constitutive relation, but the solutions (13.2.8) and (13.2.9) provide a reasonable representation for $r > 50$ km.

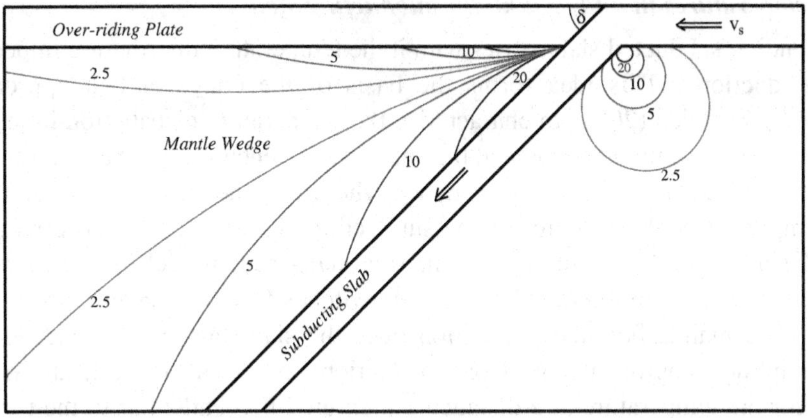

Figure 13.11. Shear stresses in MPa associated with the flow in Figure 13.10, for a convergence rate of 100 mm/yr and mantle viscosity of 3×10^{20} Pa s. The forces act counter to the descent of the slab material and are larger beneath the subducting plate than in the mantle wedge.

For the corner flow beneath the subducting plate, the boundary conditions are now

$$v_r = -V_s,\ v_\theta = 0 \quad \text{on} \quad \theta = 0; \qquad v_r = V_s,\ v_\theta = 0 \quad \text{on} \quad \theta = \delta_s, \quad (13.2.10)$$

with θ measured clockwise from the horizontal base of the subducting plate. The corresponding stream function is

$$\psi^b = -\frac{rV_s[(\pi - \delta_s - \theta)\sin\theta + \theta\sin(\delta_s + \theta)]}{\pi - \delta_s + \sin\delta_s}, \quad (13.2.11)$$

with associated shear stress

$$\sigma^b_{r\theta} = -\frac{2V_s\eta}{r}\frac{[\cos\delta_s + \cos(\delta_s + \theta)]}{\pi - \delta_s + \sin\delta_s}. \quad (13.2.12)$$

The streamlines of constant ψ and the contours of shear stress can be mapped out by sweeping across the relevant span of angles. The resulting patterns are illustrated in Figures 13.10, 13.11 for a dip angle of 45°, a convergence velocity of 100 mm/yr and a lithosphere thickness of 50 km. The mantle viscosity in Figure 13.11 is taken as 3×10^{20} Pa s both in the mantle wedge and below the subducting plate.

The streamline pattern in Figure 13.10 shows that the sinking slab drags fluid down as it descends, as might be expected. Yet, in the mantle wedge, the path of much of the fluid heads upwards towards the corner before turning to descend with the slab. The stress field is contoured in Figure 13.11 and we see that the stresses beneath the subducting slab are much more concentrated than in the mantle wedge.

13.2.3 Temperatures in and around the subducting slab

The descending subducted slab carries with it the temperature distribution imposed before subduction. This idea forms the basis of the conveyor belt approach employed by Frohlich (2006) to characterise the temperature distribution in terms of the age of the oceanic lithosphere at the entry to the trench t_{age} with an additional time component in subduction $t_{sub} = h/V_z$, where h is the current depth and V_z is the component of the velocity of the subducting slab in the vertical direction ($V_z = V_s \sin \delta_s$). Frohlich employs the thermal half-space model for the oceanic lithosphere discussed in Section 12.2.1, and switches from an oceanic boundary condition to a mantle boundary condition once the slab starts to descend. This highly simplified approach, based on conduction processes alone, leads to an estimate for the temperature at a distance $\bar{z}$ measured perpendicular to the top of the slab of

$$T(\bar{z}) = T_M + (T_O - T_M)\left(\mathrm{erf}\left[\frac{\bar{z}}{2\sqrt{\kappa t_{tot}}}\right] - \mathrm{erf}\left[\frac{\bar{z}}{2\sqrt{\kappa t_{sub}}}\right]\right), \qquad (13.2.13)$$

where T_M is the mantle temperature, T_O is the oceanic temperature, and the total time

$$t_{tot} = t_{age} + t_{sub} = t_{age} + h/V_z. \qquad (13.2.14)$$

The coolest temperature in the slab occurs at

$$\bar{z}_c = 2\kappa t_{age}\frac{h}{\Phi}\left(1 + \frac{h}{\Phi}\right)\ln\left(1 + \frac{\Phi}{h}\right), \qquad (13.2.15)$$

where the thermal parameter for the slab $\Phi = t_{age}V_z$. The temperature contrast between this coldest point (T_c) and the mantle temperature (T_M) is controlled by Φ/h, and thus the residence times in contact with ocean and mantle since

$$\frac{\Phi}{h} = \frac{t_{age}}{t_{sub}}. \qquad (13.2.16)$$

The temperature contrast is approximately linear in $\log(\Phi/h)$ (Frohlich, 2006)

$$\frac{T_c - T_M}{T_O - T_M} \approx 0.18 + 0.31\log\left(\frac{\Phi}{h}\right), \qquad (13.2.17)$$

a result which can be applied even if the slab does not have a simple geometry. Compared with numerical estimates of the temperature distribution (Kirby et al., 1996) this simple model including just conduction effects tends to overestimate temperature in the slab at shallower depths and to underestimate below 300 km.

A comparable approach can also be built from the plate thermal model (12.2.14) as used by McKenzie (1969), but the series representation is much less amenable to analytic simplification.

The previous results ignore the details of the behaviour of the fluid in the mantle wedge with imposition of a constant temperature condition on the top of the subducting slab. The corner flow will in fact maintain an advective boundary layer on the top of the slab that will control the temperature at the interface between the slab and the wedge (England & Wilkins, 2004); this temperature T_s is depth-dependent.

Although the dominant direction of flow in the mantle wedge above the slab is down the subducting plate, there is a component $v_\perp$ perpendicular to the slab that can advect heat. In terms of the coordinate $\bar{z}$ perpendicular to the top of the slab, the temperature equation is

$$\kappa_H \frac{d^2T}{d\bar{z}^2} + v_\perp(\bar{z}) \frac{dT}{d\bar{z}} = 0, \tag{13.2.18}$$

in $\bar{z} < 0$ above the top of the slab. The thickness of the advective boundary layer L_a is determined by the requirement that the time scale for the diffusion of heat through this layer, $L_a^2/(\pi^2 \kappa_H)$, is comparable to the timescale for heat to be carried the same distance by the flow, L_a/U, where U is a characteristic velocity. Thus we require

$$\frac{L_a^2}{\pi^2 \kappa_H} \frac{U}{L_a} \sim 1, \quad \text{so} \quad L_a \sim \frac{\pi^2 \kappa_H}{U}, \tag{13.2.19}$$

and so the Péclet number introduced in Section 7.2.4 is given by

$$Pe = \frac{UL}{\kappa_H} \sim \pi^2 \tag{13.2.20}$$

England & Wilkins (2004) examine the case of corner flow in the mantle wedge in some detail and derive approximate expressions for the thickness of the boundary layer in terms of the dip angle δ_s and convergence speed V_s. For the corner flow in the mantle wedge

$$L_a \approx \left(\frac{5 \kappa r^2 \delta_s}{2\sqrt{\pi} V_s \xi} \right)^{1/3}, \tag{13.2.21}$$

where r is the radial distance from the wedge corner and ξ is a slowly varying function of δ_s with a value around 0.55. Similar power-law dependences could be expected in the more complex real situation. For typical slab parameters, the advective boundary layer is 10–20 km thick, with a temperature contrast of more than 600°C between the wedge and the slab, corresponding to a heat flux of around 0.1–0.2 W m^{-2}.

The temperature in the advective layer is approximately

$$T(\bar{z}) = T_s + (T_w - T_s) \operatorname{erf}\left(\frac{-\bar{z}}{L_a} \right), \tag{13.2.22}$$

where T_s is the temperature at the top of the slab and T_w the maximum temperature in the mantle wedge on a perpendicular profile to the slab. At the interface between the mantle wedge and the subducting plate, conservation of heat flux requires the temperature gradients in the slab and in the advective layer to be equal (under the assumption of common thermal conductivity), so that

$$T_s \approx \left[T_w + \frac{\sqrt{\pi} L_a T_M}{2a} \operatorname{erf}\left(\frac{a}{2\sqrt{\kappa T / V_s}} \right) \right] \left[1 + \frac{\sqrt{\pi} L_a}{2\sqrt{\kappa t_{sub}}} \right]^{-1}, \qquad (13.2.23)$$

where a is the plate thickness.

These simple models are in good general agreement with the results of full numerical modelling (England & Wilkins, 2004). Dissipation can also make a significant contribution to heating near the wedge corner. Once the slab has penetrated 100–200 km into the mantle, the dominant influence on temperature structure comes from advection by flow in the mantle wedge.

The corner flow model describes the dominant mantle motions but does not represent the undoubtedly important movement of volatiles from the slab to the surface where they help to create the volcanic arc. This secondary process has much smaller mass-transport. The temperature at the top of the subducting plate does not vary strongly with subduction parameters and so the negative correlation of the distance of the volcanic front from the trench with convergence rate and slab dip suggests that the positions of the volcanoes in the island arc are controlled by a strongly temperature-dependent process taking place in the wedge (England & Wilkins, 2004).

The situation on the underside of the subduction zones may well be somewhat more complex than in the simple corner flow models. For a number of subduction zones, seismic tomography results reveal a concentrated zone of lowered velocities beneath the subducted slab at depths around 400 km that is not easily explained, but might involve some counter flow.

Although two-dimensional numerical simulations of the evolution of a subduction zone can provide a good indication of behaviour, the details have a strong dependence on the rheology assigned to the different materials. Frequently a visco-plastic formulation is employed for the plate material, but this does not provide an adequate description of fault failure and the initiation of subduction. Most current subduction zones have some component of oblique convergence, and in some cases, such as the Andaman Arc, the oblique component can be dominant. This means that three-dimensional effects cannot be ignored in the understanding of subduction systems.

13.2.4 Subduction and orogeny

Since the recognition of current-day subduction zones and their related features, plate tectonic concepts have been extensively employed in the explanation of past patterns of orogeny. The gravitational drag of the slab has a tendency to induce a retreat of the hinge where the slab bends. Such slab retreat appears to have played a significant role in some regions where there has been major realignment of plate

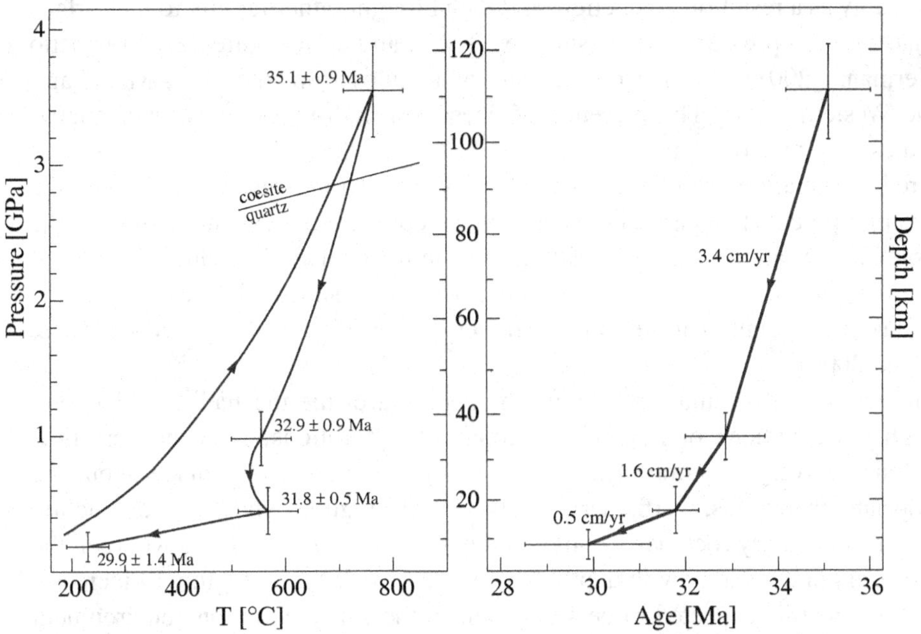

Figure 13.12. Pressure–temperature and depth–age paths estimated for ultra-high-pressure rocks in the Dora–Maira unit in the Western European Alps. The age estimates were obtained using in situ U-Th-Pb dating on titanite samples. The pressure and temperatures were estimated using a variety of different methods. [Courtesy of D. Rubatto & J. Hermann.]

boundaries, such as the Izu–Bonin region near Japan and the Tonga–Kermadec arc. For the trench line now near Tonga, the line of subduction has pivoted about the North Island of New Zealand over the last 40 Ma. The hinge retreat has led to a rotation of the Fiji Island block by 90°, determined from palaeomagnetic measurements. The rapid slab retreat has enabled asthenospheric upwelling to produce back-arc spreading centres in the Lau basin behind the Tonga volcanic arc. Images from seismic tomography indicate that in regions where rapid slab roll-back has occurred, 'stagnant' slabs lie horizontally on top of the 660 km discontinuity with a connection to the present subduction. Such a situation can be seen for the Izu–Bonin arc in Figure 13.8.

Within the generally convergent regimes associated with subduction and the consumption of oceanic lithosphere by passage into the mantle, offers the possibility of an extensional regime. Some time can be expected to elapse after the initiation of subduction before roll-back becomes significant. The insertion of extensional episodes in a convergent environment helps to reconcile some of the puzzling geological structures in former orogens.

Within many mountain belts there are isolated occurrences of ultra-high-pressure minerals whose characteristics require that material has been carried to great depth, presumably as a result of subduction, and then brought rather rapidly to the surface so that the high-pressure phases survive. The example in Figure 13.12 (Rubatto & Hermann, 2001) is based on analysis of nodules from the Dora–Maira unit in the Western Alps. The presence of coesite, a high-pressure form of quartz, indicates that material has reached depths of more than 100 km. The rates of inferred exhumation are of the order of cm/yr and so not much slower than the subduction process that carried the material to depth. Numerical simulations of the complex deformation of the over-riding plate in the course of subduction indicate that rather complex material trajectories can arise, with substantial variations in the paths taken by points that are initially rather close (see, e.g., Gerya, Stockhert & Perchuk, 2002).

Such numerical simulations require the coupling of the mechanical behaviour, through the equations of motion and continuity conditions, with the heat flow equations. A very important role is played by the assumptions made about the appropriate rheologies, particularly in the brittle regime where the distinction between wet and dry rock has a marked influence on behaviour. For a wet rock the brittle strength decreases with depth, but for a dry rock the strength will increase. Diffusion creep is expected to be significant in the subduction zone environment and the interplay between temperature and grain size in (9.3.4) leads to only modest variations in viscosity. An effective rheology can be found by combining creep and quasi-brittle behaviour controlled by a yield stress σ_y. A 'brittle' viscosity η_b can be defined as

$$\eta_b = \tfrac{1}{2}\sigma_y/(\dot{e}_{II})^{1/2} \quad \text{with} \quad \sigma_y = (M_1 p_{\text{lith}} + M_2)(1-\lambda_f), \tag{13.2.24}$$

where $\dot{e}_{II}$ is the second invariant of the strain rate tensor $\frac{1}{2}\dot{e}_{ij}\dot{e}_{ij}$. The total effective viscosity η_e is then determined by the relative size of the yield stress and the stress associated with creep (Gerya et al., 2002)

$$\eta_e = \begin{cases} \eta_{\text{creep}} & \text{when} \quad 2(\dot{e}_{II})^{1/2} < \sigma_y, \\ \eta_b & \text{when} \quad 2(\dot{e}_{II})^{1/2} > \sigma_y. \end{cases} \tag{13.2.25}$$

A further complexity comes from the potential release of fluid from the subducted slab with migration of a hydration front into the mantle wedge introducing weak serpentenized mantle. Many of the necessary physical properties are poorly constrained by conventional laboratory experiments, and in consequence there is considerable latitude available in the choice of material parameters that can lead to significant differences in the overall mechanical behaviour.

Continent–continent collisions can produce major mountain chains that are susceptible to the influence of erosion. The elevation of mountains can have the effect of changing weather patterns, e.g., the rise of the Himalayas appears to have caused a major modification of the monsoon system around 8 Ma ago. A number of authors have demonstrated that erosion has the potential to modify the deformation pattern and material trajectories in a collisional environment. For example, Pysklywec (2006) provides a number of examples of the way in which surface erosion patterns can influence the configuration of the lithosphere, and indeed the nature of an orogeny, when buoyant continental material is brought into the subduction environment.

13.3 The influence of phase transitions

Evidence for the composition of the mantle comes from xenoliths, from slices of mantle rocks emplaced at the surface in a few locations, from the composition of mantle-derived melts and from cosmochemical arguments, supplemented by geophysical results. The general consensus is that the mantle resembles a form of peridotite with around 60 per cent olivine by volume. Most observations are consistent with a single chemical composition through the bulk of the mantle, for which the physical properties are modified through the progressive effect of high pressure phase transitions over the depth range from 300–800 km (Figure 13.13).

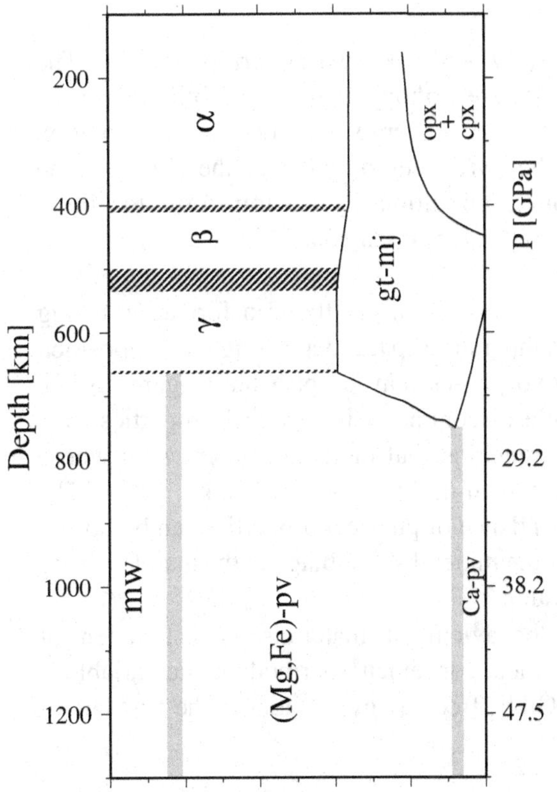

Figure 13.13. Depth-varying proportions of different mineral phases for a 'pyrolite' model of the mantle. The phases represented are: (α) olivine, (β) wadsleyite, (γ) ringwoodite, (opx) orthopyroxene, (cpx) clinopyroxene, (gt-mj) garnet-majorite, (mw) magnesiowüstite, (Me,Fe-pv) ferromagnesian silicate perovskite, and (Ca-pv) calcium silicate perovskite. [Courtesy of C. Bina].

The olivine component (α) transforms to wadsleyite (β) and then to ringwoodite (γ-spinel) with increasing pressure, finally ending up as a mixture of silicate

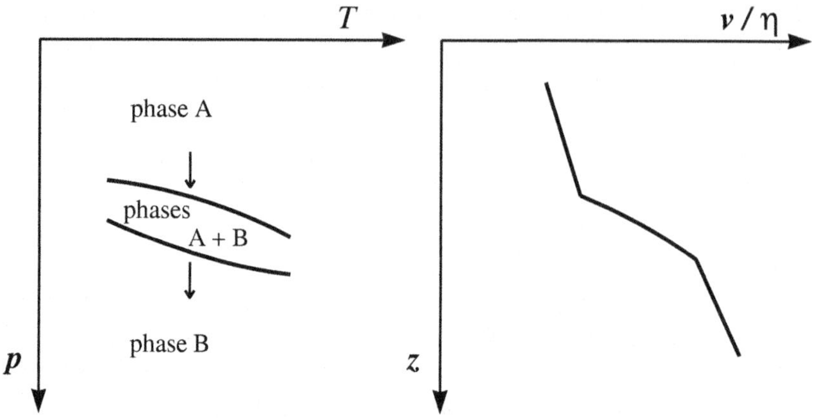

Figure 13.14. Where the transition from phase A to phase B does not take place immediately there will be a gradient in the physical parameters linking the behaviour associated with the two individual phases.

perovskite (pv) and magnesiowüstite (mw - also known as ferropericlase). The remaining components of the peridotite are orthopyroxene (opx), clinopyroxene (cpx) and garnet (gt). These components undergo a more gradual evolution under the influence of pressure as the pyroxenes dissolve in the garnet. The resulting garnet-majorite solid solution (gt-mj) ultimately also transforms to silicate perovskite, but this transition extends to greater pressure, and hence depth, than for the olivine component.

Although a simple picture of a phase transition has the transformation taking place at a line boundary in pressure–temperature space, depending on composition some degree of coexistence of the two phases may be possible (Figure 13.14). In this case there will not be a simple discontinuity in physical properties such as seismic velocity across the phase transition, but rather a gradient zone whose properties will depend on the way in which the transformation takes place. The situation with regard to the depth dependence of physical properties can be further complicated by several constituents of the mineral assemblage in the mantle having transitions in a similar pressure interval.

When we wish to examine the behaviour of materials as a function of temperature (T) and pressure (p), the most convenient thermodynamic variable is the Gibbs free-energy. The specific Gibbs free-energy G satisfies the first law of thermodynamics in the form

$$dG(T, p) = -S\,dT + \nu\,dp. \tag{13.3.1}$$

where ν is the specific volume.

We will consider a simple univariant phase transformation for which there is no mixed domain. Then, across the transition between the two phases A,B there can be

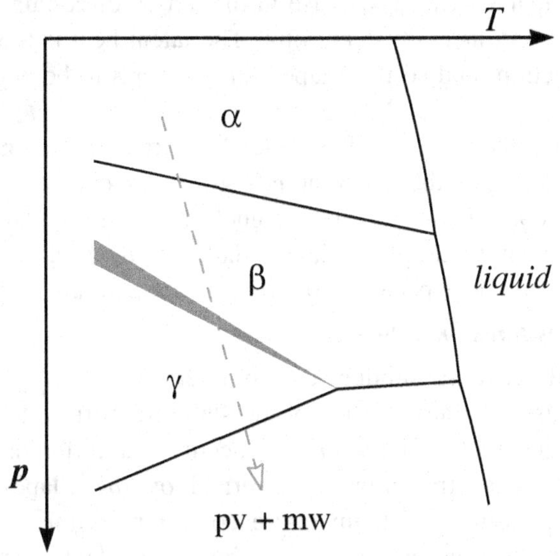

Figure 13.15. Schematic representation of the pressure–temperature relations for olivine in the upper mantle, with ultimate transformation to perovskite and magnesiowustite (pv + mw). The Clapeyron slope of the $\alpha \rightarrow \beta$ transition associated with the 410 km seismic discontinuity is positive; whereas for the γ-spinel to perovskite transition the Clapeyron slope is negative. The $\beta \rightarrow \gamma$ transition is not simple and as indicated takes place over a range of pressure at a particular temperature. The dashed grey line indicates typical mantle conditions; hotter or cooler trajectories will encounter the phase transitions at different pressures.

no net change in G. Thus, considering the situation at a particular p, T combination on the boundary, we require

$$\Delta G_{B,A} = (\nu_B - \nu_A)dp - (S_B - S_A)dT = 0. \tag{13.3.2}$$

The slope of the phase boundary, the Clapeyron slope γ_c, in pressure-temperature space is given by

$$\gamma_c = \left.\frac{dp}{dT}\right|_{bdy} = \frac{\Delta S}{\Delta \nu} = \frac{L}{T\Delta\nu} = \frac{\rho^2 L}{T\Delta\rho}, \tag{13.3.3}$$

where L is the latent heat associated with the transition, $L = T\Delta S$. For the upper mantle minerals the higher pressure phase is always denser $\Delta\rho > 0$, and so $\Delta\nu$ is always negative. Latent heat may be absorbed or evolved depending on the particular transition.

For most phase transitions, the transformation pressure increases with increasing temperature, this means that the Clapeyron slope γ_c is positive and the reaction is exothermic $L < 0$. The $\alpha \rightarrow \beta$ transition near 410 km depth has this character (Figure 13.15). However, in the γ-spinel to perovskite transition associated with the 660 km seismic discontinuity, the change in coordination of the silicon atoms from 4 to 6 is accompanied by an increase in the length of certain atomic bonds with a consequence of an increase in entropy. The latent heat is then positive for this endothermic reaction, and so the Clapeyron slope has to be negative (Figure 13.15).

A further complication around 660 km depth is transformation of garnet to perovskite that, depending on the temperature regime, can occur over an extended depth interval. The combination of the γ-spinel and garnet transitions leads to complex changes in the physical properties, so that, e.g., the seismic reflectivity of the 660 km transition can be expected to be strongly temperature-dependent.

Impact of phase transitions on subduction

The material in a descending subducted slab interacts with the mineralogical transitions, but because the thermal state of the slab differs from the surrounding mantle, phase changes within the slab will occur at a different depth. The displacement of the phase transition is governed by the Clapeyron equation (13.3.3), under the pressure and temperature conditions within the slab. The Clapeyron relation can be recast as a representation for the depth change

$$\frac{dz}{dT} = \frac{\gamma_c}{\rho g}, \tag{13.3.4}$$

with the sign of the deflection controlled by the Clapeyron slope γ_c. The exothermic $\alpha \rightarrow \beta$ transition, with positive Clapeyron slope, will be shallower in the cool core of the slab compared with external conditions. In contrast, the endothermic γ-spinel to perovskite transition would be expected to produce an enhanced depth for the phase boundary in the colder slab. The 410 km discontinuity tends to enhance the subduction potential since the presence of denser material at shallower depth increases the downward forces, whilst the 660 km transition tends to impede subduction. Careful seismological studies of slabs are consistent with the predictions from the thermodynamic arguments, but the detailed shapes of the deflected boundaries are difficult to estimate in this heterogeneous environment.

The Clapeyron theory describes an equilibrium phase transition, but the conditions within the slab are conducive to the preservation of metastable states. Metastable material can be expected if the high-pressure phase is produced by nucleation on grain boundaries with ultimate growth to consume the crystals of the lower-pressure phase. Within the coldest slabs it is likely that a wedge of metastable olivine can persist to depths approaching 660 km in a zone concentrated towards the upper slab interface (Stein & Rubie, 1999). It has been suggested that phase changes within this metastable zone may act as the trigger for deep earthquakes via transformational faulting in which slip occurs along thin shear zones where metastable olivine transforms to denser spinel. An alternative model is that decomposition of hydrated minerals leads to the presence of water that can cause reduction of effective stresses on faults in the slab. Deep earthquakes have the characteristics of slip on faults that in some cases, such as the 1994 Mw 8.4 deep Bolivian earthquake, appear to cut through the whole slab.

Mineral physics studies in recent years have demonstrated that many nominally anhydrous silicate phases can carry significant amounts of water to depth. As a result the hydration potential adds another variable in the description of behaviour, since even small amounts of free water lead to a major reduction of the effective viscosity of materials.

13.4 The deeper mantle

Inferences on the long-term mantle viscosity have been made from two major classes of observations: (a) patterns of sea-level change related to the adjustment of the Earth to the effects of the ice loading and unloading during the glacial cycles of the late Pleistocene glacial cycles, dominated by isostatic adjustment; (b) a suite of surface geophysical observables linked to convective processes, such as long wavelength free-air gravity anomalies, the horizontal divergence of tectonic plate motions, and dynamic surface topography.

Models that satisfy both classes of data are characterised by a significant increase in viscosity within the lower mantle relative to the upper mantle by a factor of 30–100. The position of the increase has been a matter of some debate, and although it is often modelled as a sharp change the data would be entirely compatible with a transition spread over some hundreds of kilometres. For example, there could be a ramp in viscosity from the mean upper mantle of around 3×10^{20} Pa s to 10^{22} Pa s below 1000 km depth. Some models such as that of Mitrovica & Forte (2004) invoke a thin layer of significantly reduced viscosity just above the 660 km discontinuity and thereby reduce the absolute value of viscosity required in the lower mantle. Such a zone would have the effect of partially decoupling flow patterns between the upper and lower mantle.

What might be happening to give such a significant change in viscosity within the upper part of the lower mantle? Calculations of the properties of the mineral assemblage for the mantle suggest that the proportion of garnet should diminish with depth and drop to zero by 900–1000 km leaving just perovskites such as $(Mg,Fe)SiO_3$ and ferropericlase $(Mg,Fe)O$. A strong gradient in viscosity could therefore arise from the diminishing influence of the more deformable garnets.

13.4.1 Viscosity variations in the mantle and the geoid

A fluid body under rotation assumes the shape of an oblate spheroid as a result of the balance between self-gravitation and the centrifugal effects of rotation. The internal variations in density mean that the figure of the Earth departs from a solid of revolution. Density anomalies in the mantle contribute to the gravity field of the Earth in two ways: a local mass anomaly (excess or deficiency) has a direct effect on the gravitational field, but also causes flow in the mantle that leads to distortion of the surface and the core–mantle boundary as well as any internal

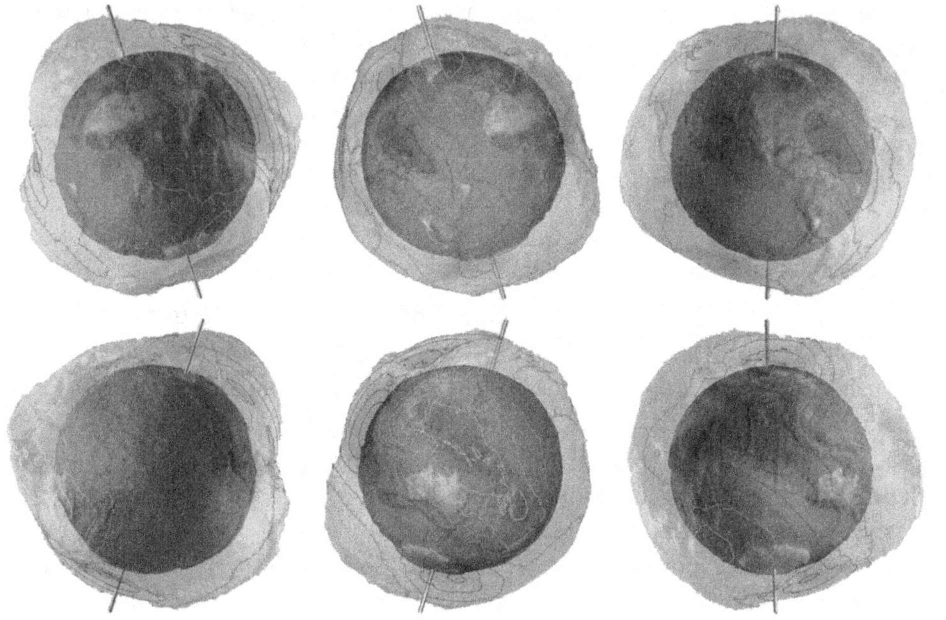

Figure 13.16. Configuration of the long-wavelength components of the geoid, displayed as an outer surface with an exaggerated scale of deformation from a spherical shell. The inner spherical surface is present to provide a geographical reference. [Courtesy of M. Sandiford].

density interfaces. The flow-induced distortions lead to additional mass anomalies that can contribute to the gravitational field and its potential.

The surface of constant gravitational potential closest to mean sea level is termed the *geoid* (Figure 13.16). Across the oceans the geoid lies close to the actual surface. In the continents the geoid has no direct relation to topography, but would be close to the water level cut across a continent to connect two oceans directly. Because gravitational potential decays as R^{-1} with distance R from a mass anomaly, the geoid is much more sensitive to the mass configuration at depth than gravity with an R^{-2} dependence.

The topographic variation of the geoid from the rotational ellipsoid is of the order of about 100 metres with a profound depression to the south of India (Figure 13.16). The large-scale variations in gravitational potential φ can be well represented through a spherical harmonic expansion with the origin placed at the centre of mass

$$\varphi = \frac{GM_e}{r}\left\{1+\sum_{l=2}^{\infty}\left(\frac{r_e}{r}\right)^{l} P_l^m(\cos\theta)\sum_{m=0}^{l}[C_l^m\cos(m\phi)+S_l^m\sin(m\phi)]\right\}, \quad (13.4.1)$$

where G is the gravitational constant and M_e is the mass of the Earth. The $l = 0$ term is extracted from (13.4.1) as the central mass contribution, and the $l = 1$ term vanishes with the choice of origin. The number of contributions to the expansion increases rapidly with increasing degree l.

The oblate ellipsoid associated with rotation, with mean radius r_e and ellipticity e, takes the form

$$\frac{r(\theta)}{r_e} = \left(\frac{(1-e)^2}{(1-e)^2\sin^2\theta + \cos^2\theta}\right)^{1/2} \approx (1+\tfrac{1}{3}e) + \tfrac{2}{3}e\,P_2(\cos\theta), \qquad (13.4.2)$$

the higher-order spherical harmonic contributions decrease steadily but the P_2 term alone does not represent an ellipsoidal surface.

The shape of the geoid depends on the mass distribution within the Earth and the influence of self-gravitation and flow that tend to deform the Earth's surface and the core–mantle boundary as well as any internal boundary surface. The deviations of the geoid from the reference ellipsoid depends on instantaneous force balance rather than flow rates and hence are sensitive to the variations of viscosity rather than the absolute values.

Dynamic compensation

Following Hager (1984) we consider a two-dimensional flow system as illustrated in Figure 13.17(a) consisting of a plane layer of unit depth with uniform viscosity and density overlying an inviscid fluid half-space of greater density. A surface density contrast $\Delta\rho_m$ is placed at the midpoint of the layer with an amplitude

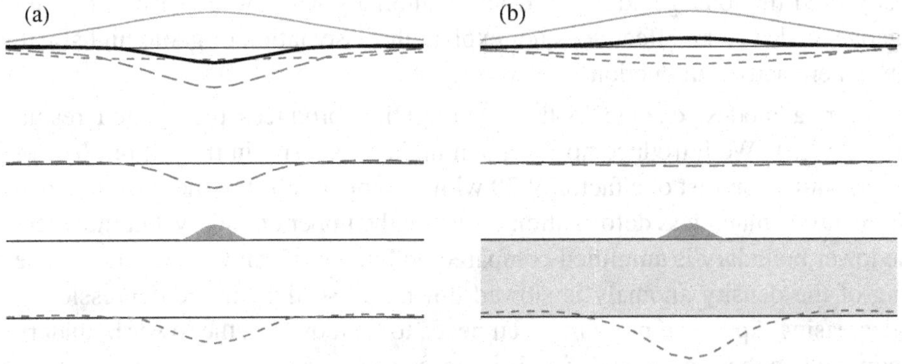

Figure 13.17. Illustration of the components of the geoid anomaly due to a density contrast in the form of cosine bell at the centre of a layer: (a) a layer with uniform viscosity η, (b) with the bottom half of the layer having a viscosity 30 times larger than that of the upper half. In each case the total geoid anomaly (heavy solid line) is the sum of contributions from the density contrast (light solid line), the dynamic deformation of the upper boundary (long dashes) and the lower boundary (short dashes). In case (a) the total geoid anomaly is negative for a positive density contrast, but in case (b) the corresponding geoid anomaly is negative.

specified by a cosine bell, the effect of this anomaly is to lead to an equal mass displacement at both upper and lower boundaries.

At the top and bottom of the viscous layer the normal and shear tractions will be continuous. The introduction of a sudden density contrast will lead to the boundaries deforming until a steady state is reached. A sinking positive density contrast will push material downwards and pull material from behind to create a depression of the upper surface. The surface density contrast due to the depression $\Delta\rho_t$ then acts to set up a secondary flow tending to fill in the topography in a way that is precisely analogous to post-glacial rebound, with the same time constant. In the steady state the rate at which the sinking heavy blob pulls material away from the surface is balanced precisely by the rate at which flow tends to fill in the depression. A similar behaviour occurs for the bulge at the base of the layer with associated equivalent surface density contrast $\Delta\rho_b$. $\Delta\rho_t$ and $\Delta\rho_b$ will have the opposite sign to $\Delta\rho_m$.

The amplitude of the surface depression depends on the size and depth of the density contrast and the thickness of the layer, and is unaffected by the value of the viscosity in the layer. However, the time needed to reach steady state is proportional to the viscosity. The displacement of the density contrast during this interval is similar to the final surface deformation.

The total geoid anomaly is the sum of contributions from $\Delta\rho_m$, $\Delta\rho_t$, and $\Delta\rho_b$ as indicated at the top of Figure 13.17(a). The surface density anomaly $\Delta\rho_t$ can contribute at short spatial wavelengths because of its proximity to the measuring point, but the influences of $\Delta\rho_m$ and $\Delta\rho_b$ will shift to longer wavelengths due to the geometric decay of gravitational potential from a mass anomaly. In a uniform viscosity fluid the total geoid effect from an anomaly with an increase in density is negative, and such an effect does not explain the association of geoid highs with regions where active subduction is occurring.

However, a modest change in the configuration produces the desired result, Figure 13.17(b). We introduce now a much higher viscosity in the bottom half of the layer, with a contrast of a factor of 30 with the upper half, For the same positive density contrast much less deformation occurs at the upper boundary, but the effect on the lower boundary is amplified compared with the uniform viscosity case. The sinking of the density anomaly is slowed down and so the surface depression is less, decreasing $\Delta\rho_t$; now more material needs to be moved at the lower boundary to accommodate the motion and $\Delta\rho_b$ is increased.

The total geoid anomaly is now positive, but only a fraction of that due to the mass contrast $\Delta\rho_m$ alone which is unchanged from the previous case. The increase in the effect of $\Delta\rho_t$ overcomes that from $\Delta\rho_t$, which is smaller than in the uniformly viscous case.

These two models illustrate that the sign of the geoid anomaly for a given density contrast is not immediately obvious, but depends on the viscosity variation with depth. Both distance and viscous flow act as filters to determine the magnitude of the total anomaly. The behaviour can be expected to be quite complex for features such as the density contrasts associated with subduction zones.

Spectral representation

For spherical harmonic degree l and order m, the contribution to the geoid anomaly $\delta\varphi_l^m$ can be found as an integral over the corresponding spherical harmonic load contribution from density $\delta\rho_l^m$ in the form (Richards & Hager, 1984)

$$\delta\varphi_l^m = \frac{4\pi G r_e}{2l+1}\int_{r_c}^{r_e} dr\, K(r,\eta)\delta\rho_l^m, \tag{13.4.3}$$

where $K(r,\eta)$ is the geoid response kernel calculated from the deformation produced by viscous flow due to a surface density contrast at radius r, including the boundary deformation contributions.

The issue then is how to find the requisite density contributions. A common assumption is to use some class of scaling relation between seismic wavespeed heterogeneity determined from seismic tomography and density variations. The usual choice is based on a simple scaling from the shear wave heterogeneity

$$\frac{\delta\rho}{\rho} = C(r)\frac{\delta\beta}{\beta}, \tag{13.4.4}$$

where $C(r)$ is a scaling factor that is allowed to vary with radius. Such a relation would be appropriate when the only effective influence is temperature; however, we can expect somewhat different behaviour when compositional variations come into play. Consider the relative variation in shear wavespeed as a function of shear modulus G, density ρ, temperature T and some compositional component X,

$$\frac{\delta\beta}{\beta} = \tfrac{1}{2}\frac{1}{G}\frac{\partial G}{\partial T}\delta T - \tfrac{1}{2}\frac{1}{\rho}\frac{\partial\rho}{\partial T}\delta T + \tfrac{1}{2}\frac{1}{G}\frac{\partial G}{\partial X}\delta X - \tfrac{1}{2}\frac{1}{\rho}\frac{\partial\rho}{\partial X}\delta X. \tag{13.4.5}$$

We cannot expect a similar scaling of the influences of temperature and composition and so a simple relation as in (13.4.4) is likely to fail. There are indications that such scaling works reasonably well for the longest-wavelength features of heterogeneity where temperature may well dominate. However, compositional effects can be expected to be localised with a strong effect for shorter wavelength and also to be more prominent near the boundaries of the mantle.

A significant component of the geoid signal comes from subducted slabs and once this has been removed using plausible slab models, based on tomographic imaging and past plate configurations, the remaining component is well described using a two-layer mantle with a factor of 30 contrast in viscosity in the neighbourhood of 1000 km deep. Such a change at depth is compatible with the observations of glacial rebound, even though such data can be adequately represented with hardly any change. The available data provide little support for the concept of any internal density boundaries apart from those associated with the 410 km and 660 km seismic discontinuities.

13.4.2 Material properties

In the deep Earth the effect of pressure can be very significant and change perceptions of material properties based on attainable laboratory conditions. Many experiments and *ab initio* calculations refer to isothermal conditions rather than the adiabatic state appropriate to seismic waves. The relation between the adiabatic bulk modulus K_S and the isothermal bulk modulus K_T is (6.1.18)

$$K_S = K_T(1 + \alpha_{th}\gamma_{th}T), \tag{13.4.6}$$

in terms of the thermal expansion coefficient α_{th}, Grüneisen constant γ_{th} and temperature T. The derivatives with respect to temperature are related by

$$\frac{\partial K_S}{\partial T} = \frac{\partial K_T}{\partial T}(1 + \alpha_{th}\gamma_{th}T) + K_T\left(\alpha_{th}\gamma_{th} + T\frac{\partial \alpha_{th}\gamma_{th}}{\partial T}\right). \tag{13.4.7}$$

The thermal expansivity α_{th} decreases with increasing pressure but increases with temperature. The Grüneisen parameter γ_{th} will vary only slightly under lower mantle conditions. The isothermal bulk modulus varies monotonically with pressure and temperature with $\partial K_T/\partial p > 0$ and $\partial K_T/\partial T < 0$. At lower mantle conditions $(\partial K_T/\partial T)_V \sim$ -0.002 GPa K^{-1}, $\alpha_{th} \sim 2\times10^{-5}$ K^{-1}, $\gamma_{th} \sim 1.5$ with $K_T \sim$ 500 GPa, and so the second and third terms in (13.4.7) are significant. Indeed for the derivatives at constant volume $(\partial K/\partial T)_V$ these extra terms can outweigh the first at higher temperatures, and so reverse the sign of the adiabatic derivative compared with the isothermal. However, the influence of pressure on volume is such that this will not occur for the derivatives at constant pressure $(\partial K/\partial T)_P$, even though the adiabatic temperature derivative will be somewhat smaller than the isothermal.

13.4.3 The lower boundary layer

Below a depth of about 2100 km the amplitude of seismic wavespeed heterogeneity inferred from different styles of seismic tomography shows a steady increase and near the core–mantle boundary the size of the variations is comparable to those close to the surface (see, e.g., Figure 11.22). The increase in level of heterogeneity is common to both the bulk modulus and shear modulus, even though the major anomalies in the Pacific and beneath southern Africa near the core–mantle boundary have opposite sign for the anomalies in bulk-sound and shear wavespeeds. This anti-correlation of the different material properties has frequently been used as an argument for the presence of chemical heterogeneity in the D$''$ layer at the base of the mantle (e.g., Masters et al., 2000), since it cannot be achieved by purely thermal means.

The discovery of the perovskite to post-perovskite phase transition (Murakami et al., 2004), with an increase in density, has raised a range of new questions about the nature of structures at the base of the mantle. The phase transition occurs for pressures comparable to those at the core–mantle boundary, and it seems likely that the presence of minor components brings the transition into the mantle regime.

The estimates of the Clapeyron slope for the pure Mg end member are of the order of 8–10 MPa K^{-1}, more than twice the magnitude, and of opposite sign, to the γ-spinel to perovskite transition near 660 km depth ($\gamma_c \sim -3\ K^{-1}$). As a result the influence of temperature on the topography of this exothermic transition will be magnified compared with the deflections of phase boundaries within subducted slabs, discussed above in Section 13.3. A further complication comes from the influence of the iron (Fe) content of the perovskite and the admixture of oxide components. The topographic undulations of the phase boundary are controlled by the change in the phase transition pressure

$$\delta p = \left(\frac{\partial p}{\partial T}\right)_X \delta T + \left(\frac{\partial p}{\partial X}\right)_T \delta X = \gamma_c \delta T + \left(\frac{\partial p}{\partial X}\right)_T \delta X, \tag{13.4.8}$$

where, as in (13.4.5), X represents some additional chemical component. The actual configuration will depend on the balance between thermal buoyancy and compositional buoyancy. When the effect of including an additional component is to increase the transition pressure ($\partial p/\partial X > 0$), then temperature and composition work together and the phase transition moves to greater depth. However, if the presence of a component is such that $\partial p/\partial X < 0$ the depression of the phase boundary will be reduced and could even lead to an elevated boundary in a region of upwelling.

The presence of undulations in the phase boundary associated with mixed materials leads to localized forces in the momentum equation (8.1.4), due to the density differences arising from the phase change. An additional constraint has also to be placed on the continuum equations to ensure conservation of the different chemical species. The phase boundary can also be expected to involve some change in viscosity that will influence flow patterns.

The complex mixture of components at the base of the mantle means that the simple analysis for a single phase change has to be supplemented by a consideration of mixture properties. Spera, Yuen & Giles (2006) have looked at the influence of Fe on the perovskite to post-perovskite transition and conclude that there is likely to be a mixed phase region perched above a post-perovskite layer. For reasonable molar proportions Fe/(Fe+Mg) ~ 0.05–0.20, and temperatures appropriate to the region around the core-mantle boundary (3800–4400 K), a perched layer 200–400 km thick would start about 200–400 km above the core-mantle boundary. The mixed layer would be displaced upwards for regions with lower temperatures, such as where former slab material reaches the D″ zone. The postulated mixed phase zone occupies the region where seismic heterogeneity increase rapidly with depth and the base might well be identified as a seismic discontinuity.

14
Mantle Convection

The Navier–Stokes equation introduced in Chapter 7 provides the basis for understanding the behaviour of mantle convection. Because the mantle is highly viscous, the dominant contributions to the momentum balance come from viscous and buoyancy forces. This means that a scaling analysis from boundary layer theory can be used to compute mantle convection velocities from estimates of mantle viscosity and buoyancy forces. The peculiar geometry, or planform, of mantle convection cells, with subduction zones separated by distances several times the mantle depth, results from a strong increase in mantle viscosity with depth. The internal mantle temperature distribution departs considerably from the adiabat, because the mantle overturns slowly in a time comparable to the internal heat production time scale and the rate of secular cooling. The topography associated with cooling of the oceanic lithosphere and with hot spot swells gives a useful reference to constrain the ratio of internal mantle heating relative to the heat flux coming from the core, although the topography associated with plumes must be corrected for non-adiabatic effects. The Mesozoic and Cenozoic circulation of the mantle can be inferred from plate motion histories, although a limitation is the lack of information on initial conditions.

14.1 Convective forces

We start by providing a simple treatment of convection based on boundary layer theory, and we will see that this provides velocities of the size associated with plate motion (Figure 14.1).

14.1.1 Boundary layer theory

The Earth has much weaker heat diffusion than momentum transport, so that the Prandtl number is large ($\sim 10^{23}$ as discussed in Section 7.1.2). If we take the limit of infinite Prandtl number the momentum equation represents a balance between viscous forces and the forces associated with thermal and other buoyancies. This insight puts us into a position to derive a a simple analytic model of convection

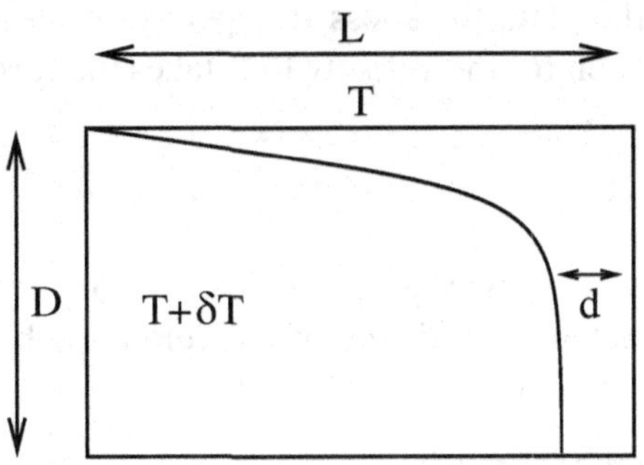

Figure 14.1. Simple convection model with force balance associated with a downgoing slab of lower temperature than its surroundings.

where the force balance is computed from estimates of thermal buoyancies and the mantle viscosity. The approach was first used by Turcotte & Oxburgh (1967) to show that it is possible to obtain a reasonable approximation to the velocity of tectonic plates. Consider the weight of a subducting slab of length D:

$$F_W = g\, D d\, \rho \alpha_{th}\, \Delta T, \tag{14.1.1}$$

where α_{th} is the thermal expansion coefficient (6.1.11), d is the thickness of the slab and D the depth of penetration. ΔT represents the average temperature difference between the mantle and the slab, which is approximately $\Delta T = -T/2$, where T is the interior temperature. Hence

$$F_W = -g\, D d\, \rho \alpha_{th}\, T/2. \tag{14.1.2}$$

The thickness of the slab is

$$d = (k t_s)^{1/2}, \tag{14.1.3}$$

where k is the thermal conductivity and t_s is the age of the slab material (see Section 12.2). In terms of the distance from the ridge axis to subduction L and the plate velocity v, the slab thickness

$$d = (kL/v)^{1/2}. \tag{14.1.4}$$

The viscous resistance to the penetration of the slab into the mantle is given by the force per unit area,

$$\sigma = \eta 2v/D. \tag{14.1.5}$$

We get the force per unit length by multiplying σ by the vertical length D:

$$F_R = D\sigma = 2\eta v. \tag{14.1.6}$$

The force balance between the weight of the slab F_W and the resistive force F_R yields the downgoing velocity as

$$v = -g D d \rho \alpha_{th} T/4\eta, \tag{14.1.7}$$

but we must recall that the plate thickness d depends on the velocity v through (14.1.4). The final expression for the velocity thus takes the form

$$v = \left(\frac{g\rho\alpha_{th} T k^{1/2} L D^2}{4\eta}\right)^{\frac{2}{3}}. \tag{14.1.8}$$

In simple convection systems the aspect ratio is close to unity so it is reasonable to take $D \approx L$. We can obtain an estimate of the velocity v by using appropriate parameters for the Earth's mantle:

$$\begin{aligned} &D = 3000 \text{ km}, && \rho = 4000 \text{ kg m}^{-3}, \\ &\alpha_{th} = 2\times10^{-5} \text{ K}^{-1}, && T = 1400°\text{C}, \\ &k = 10^{-6} \text{ m}^2\text{s}^{-1}, && \eta = 10^{22} \text{ Pa s}, \end{aligned}$$

we find $v = 2.8\times10^{-9}$ m s^{-1} = 90 mm yr^{-1}. Even if we assume that slabs do not penetrate further than half the mantle so that $D \approx L/2$, we obtain $v \approx 30$ mm yr^{-1}. These two values bracket most of the inferred plate speeds.

14.1.2 Basic equations

We wish to focus attention on longer time scale features in fluid motion and so need to suppress acoustic waves. We can do this by making the *anelastic approximation* $\partial\rho/\partial t = 0$, which yields the equation of mass conservation in the form

$$\nabla \cdot (\rho \mathbf{v}) = 0. \tag{14.1.9}$$

A further restriction to an incompressible medium leads to the vanishing of the divergence of the velocity field ($\nabla \cdot \mathbf{v} = 0$).

The equation of motion for fluid flow in the absence of bulk viscosity takes the form (7.1.4):

$$\rho\frac{\partial}{\partial t}v_j + \rho\left(v_k\frac{\partial}{\partial x_k}\right)v_j = \rho g_j - \frac{\partial}{\partial x_j}p + \frac{\partial}{\partial x_i}\left[\eta\left(\frac{\partial v_j}{\partial x_i} + \frac{\partial v_i}{\partial x_j}\right) - \tfrac{2}{3}\eta\delta_{ij}\frac{\partial v_k}{\partial x_k}\right]. \tag{14.1.10}$$

If we allow a temperature dependent rheology the viscosity η will vary with position through the variations in temperature and so we need to retain the spatial derivatives of η in (14.1.10).

The fluid flow equations are linked to the behaviour of temperature through the transport of heat in convection. The temperature equation in the presence of flow is given by (7.1.15):

$$\rho\left[\frac{\partial}{\partial t}(C_p T)\right] = \rho v_k \frac{\partial}{\partial x_k}(C_p T) + \frac{\partial}{\partial x_k}\left[k\frac{\partial}{\partial x_k}T\right] + 2\eta\hat{D}_{ij}\hat{D}_{ij} + h, \tag{14.1.11}$$

where k is the thermal conductivity, C_p is the specific heat at constant pressure and h is the heat generation per unit mass due to radioactivity; $\hat{D}_{ij}$ is the deviatoric strain-rate tensor. The dominant thermal behaviour is described by the *thermal diffusivity* (7.1.17), $\kappa_H = k/\rho C_p$. For most of the mantle the conductive term will dominate the viscous dissipation by a factor of 50 or more; however, significant dissipation may arise in regions of strong shear such as the edges of thermal boundary layers or rising plumes. In (14.1.11) we have retained the possibility of spatial variations in specific heat C_p and thermal conductivity k. We can expect slow variations of these quantities with depth, but lateral variations induced by temperature would normally be small. The additional effects come from the spatial gradients of C_p and k and so will be concentrated in regions of sharp temperature changes.

Reference state

To complete the picture we need to understand the relation of the density to temperature and composition. We envisage a reference state in which the density has an adiabatic radial profile in hydrostatic equilibrium, so that

$$-\boldsymbol{\nabla} p_0(r) + \rho_0(r)\mathbf{g}(r) = 0. \tag{14.1.12}$$

The variations in density associated with deviations from the equilibrium temperature $T_0(r)$ profiles due to convective flow can then be approximated by

$$\rho = \rho_0(r)\{1 - \alpha_{th}[T - T_0(r)]\}, \tag{14.1.13}$$

where the thermal expansivity α_{th} (6.1.11) can be expressed as

$$\alpha_{th} = \frac{1}{\rho}\left(\frac{\partial \rho}{\partial T}\right)_P. \tag{14.1.14}$$

14.1.3 Boundary conditions

The viscosity of the mantle is much larger than that of the core or the air and water above. Thus the appropriate boundary conditions for the velocity are *free-slip*, i.e., zero normal velocity and zero shear stress as in (8.2.4), at the core–mantle boundary (r_c), and the free surface boundary (r_e) :

$$v_r = 0, \quad \sigma_{r\theta} = \sigma_{r\phi} = 0, \qquad \text{at} \quad r = r_c, r_e. \tag{14.1.15}$$

The corrections needed to allow for dynamic distortions of the boundaries are small and generally ignored.

The temperature at the outer boundary will be dictated by the temperature at the ocean bottom or the continental surface, whereas, at the core–mantle boundary the temperature is practically isothermal and equilibrated to that of the core. If core heat flux has strong geographic variation, the isothermal surface and the core–mantle boundary will not coincide.

14.1.4 Non-dimensional treatment

In our discussion of fluid flows in Chapter 7, we have seen how considerable insight into the physical character of the system can be obtained by working with non-dimensional quantities, so that laboratory experiments and computer simulations can be compared with conditions in the mantle through a set of dimensionless numbers. The choice of non-dimensionalisation should be designed to bring the main terms to comparable size.

A sensible choice for the length scale for mantle convection is the depth of the mantle. $L_M = r_e - r_c$, i.e., 2890 km. There are two choices for the time scale based on thermal diffusion or the advection time. We will here use the advection time, $T_a = L_M/u$, where u is the velocity of plates of the order 5 cm per year. The corresponding flow velocity $U_M = L_M/T_a$ is around 10^{-6} m/s, and the effective flow and diffusion time scales are then comparable. The set of main scaling factors are thus:

$$L_M = r_e - r_c, \qquad T_a = \frac{L_M}{u}, \qquad U_M = \frac{L_M}{T_a}. \tag{14.1.16}$$

Following scaling by the non-dimensional quantities we can write the equation governing the behaviour of the convecting fluid in the form

$$\begin{aligned} Re_0 \left[\frac{\partial}{\partial t} v_j + \left(v_k \frac{\partial}{\partial x_k} \right) v_j \right] = \\ \rho_0 g_j - \frac{\rho_0}{\rho} \frac{\partial}{\partial x_j} p + \frac{\rho_0}{\rho} \frac{\partial}{\partial x_i} \left[\frac{\eta}{\eta_0} \left(\frac{\partial v_j}{\partial x_i} + \frac{\partial v_i}{\partial x_j} \right) - \frac{2}{3} \frac{\eta}{\eta_0} \delta_{ij} \frac{\partial v_k}{\partial x_k} \right]. \end{aligned} \tag{14.1.17}$$

where ρ_0 and η_0 are the values for the reference distribution. The Reynolds number Re_0 is also constructed using these reference values

$$Re_0 = \frac{\rho_0 U_M L_M}{\eta_0} = \frac{U_M L_M}{\nu_0}. \tag{14.1.18}$$

As we have seen in Section 7.2.1 the value of Re_0 for the mantle of the Earth is very small, around 10^{-20} and so the inertial forces on the left-hand side of (14.1.17) pale into insignificance relative to the viscous forces on the right-hand side. Viscosity variations of even 10^{10}, equivalent to temperature changes of 1000 K, will not affect this argument.

In consequence we have a Stokes flow condition in the presence of gravitation

$$0 \approx \rho_0 g_j - \frac{\rho_0}{\rho} \frac{\partial}{\partial x_j} p + \frac{\rho_0}{\rho} \frac{\partial}{\partial x_i} \left[\frac{\eta}{\eta_0} \left(\frac{\partial v_j}{\partial x_i} + \frac{\partial v_i}{\partial x_j} \right) - \frac{2}{3} \frac{\eta}{\eta_0} \delta_{ij} \frac{\partial v_k}{\partial x_k} \right] \tag{14.1.19}$$

in terms of the relative variations of density and viscosity.

The elliptic nature of the equation (14.1.19) has the consequence that the balance of momentum is global and instantaneous. That is deformations in some place (i.e., plate motion) have the consequence of instantaneous stress variations throughout the volume. This character of creeping flow is important in a number of ways:

(i) for attempts to infer the flow throughout the mantle from a knowledge of plate motion (i.e. the insight of Hager & O'Connell, 1979),
(ii) for attempts to tackle the inverse problem of mantle convection where flow is inferred back in time (through the adjoint problem) knowing the history of plate motion, and
(iii) for discussions of a decoupling layer in the asthenosphere; for reasonable values of the low-viscosity channel, a factor of 100 or so relative to the deeper mantle, the nature of the equation (14.1.19) prevents such decoupling.

An alternative development for the non-dimensionalisation using the thermal diffusion time, rather than the advective time, leads to a representation in terms of the Rayleigh number Ra rather than the Reynolds number Re (Landau & Lifshitz, 1987, §57).

14.1.5 Computational convection

Having derived a mathematical representation of mantle flow in the form of the Navier–Stokes equations, we require numerical techniques for their solution. We wish to focus on three-dimensional spherical convection and seek a discretisation of the sphere. An obvious, but ultimately unsuitable, choice is a grid based directly on longitude and latitude. Such grids are at a disadvantage, because there are many more grid points near the poles than at the equator so that the polar regions are oversampled, and because the convergent latitude and longitude lines result in two singularities at the poles so that exceedingly small time steps are required to ensure numerical stability. Collectively these difficulties are known as the *pole problem*.

We can avoid the pole problem if we discretise the sphere with a triangular mesh derived from the icosahedron. This so called *geodesic* grid starts from projecting the regular icosahedron, a platonic body with twenty equilateral triangles as its faces, upon the sphere. The mesh is then constructed recursively by splitting nodal distances in half and inserting new nodes at the midpoints. Each time the process is repeated the lateral resolution in the mesh is doubled, and so a mesh of arbitrary refinement on the sphere can be found. Recursive geodesic mesh refinements taken from the three-dimensional spherical TERRA mantle dynamics code of Bunge & Baumgardner (1995) are shown in Figure 14.2. The volume of the mantle can be discretised by replicating the icosahedral mesh along the radial direction. The mesh then subdivides a thick spherical shell into elements of almost uniform size.

Once the mantle volume is discretised, we need to find a local representation of the variables and their derivatives. Many mantle convection studies adopt a finite element or finite volume approach, because the local character of these techniques is well suited for parallel computers and accommodates large local variations in

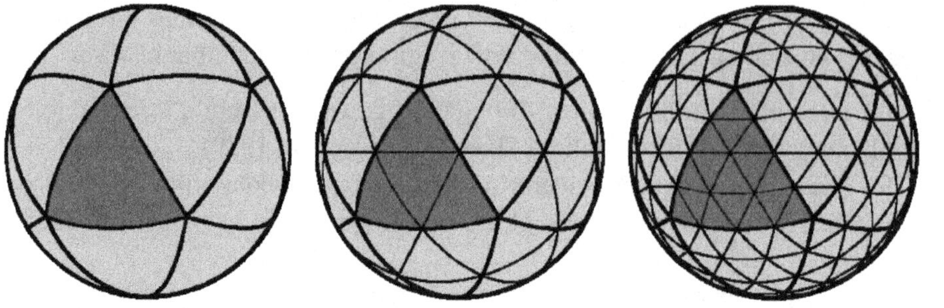

Figure 14.2. Successive refinements of icosahedral grid used to represent spherical behaviour

viscosity. In finite elements one adopts a variational, or weak formulation of the problem and represents the solution locally through low-order shape, or basis functions. These functions are chosen in such a way that they vary between zero and one within an elementary element of integration but vanish elsewhere, that is they have a local support. A comprehensive survey of the finite element method in many circumstances in fluid mechanics is provided by Zienkiewicz et al. (2005).

We illustrate the basic finite element procedure through the equation of motion, and start by representing velocity and pressure locally through

$$\mathbf{v} = \sum_{l=1}^{n} \Psi_l v'_{lj}, \quad j = 1,2,3\,, \tag{14.1.20}$$

where the Ψ_l are the finite element basis functions. The weak form of the equation of motion is derived by multiplying the momentum equation by appropriate trial functions (ϕ) and then integrating the resulting system over the mantle domain Ω,

$$\int_V \phi \left(\rho g_j - \frac{\partial}{\partial x_j} p + \frac{\partial}{\partial x_i} \left[\eta \left(\frac{\partial v_j}{\partial x_i} + \frac{\partial v_i}{\partial x_j} \right) - \tfrac{2}{3}\eta\delta_{ij} \frac{\partial v_k}{\partial x_k} \right] \right) d\Omega = 0, \tag{14.1.21}$$

so that we enforce the balance of momentum in an integral (weak) sense. A further step is taken by using integration by parts on the viscous term in (14.1.21)

$$\begin{aligned} \int_V \phi \left(\frac{\partial}{\partial x_i} \left[\eta \left(\frac{\partial v_j}{\partial x_i} + \frac{\partial v_i}{\partial x_j} \right) - \tfrac{2}{3}\eta\delta_{ij} \frac{\partial v_k}{\partial x_k} \right] \right) d\Omega &= \\ - \int_V \eta \frac{\partial \phi}{\partial x_i} \left[\left(\frac{\partial v_j}{\partial x_i} + \frac{\partial v_i}{\partial x_j} \right) - \tfrac{2}{3}\delta_{ij} \frac{\partial v_k}{\partial x_k} \right] d\Omega & \\ + \int_S \eta\, \phi \left[\left(\frac{\partial v_j}{\partial x_i} + \frac{\partial v_i}{\partial x_j} \right) - \tfrac{2}{3}\delta_{ij} \frac{\partial v_k}{\partial x_k} \right] dS, & \end{aligned} \tag{14.1.22}$$

which reduces the order of differentiation for this term. It is common to take trial functions (ϕ) and finite element basis functions (Ψ) as identical, in the *Galerkin* approach to the finite element technique. Upon inserting the local representation of trial functions and velocity (14.1.20), the volume integral on the right-hand side of

(14.1.22) reads

$$-\int_V \eta \frac{\partial \Psi_m}{\partial x_i} \left[\left(\frac{\partial \Psi_l}{\partial x_i} v'_{lj} + \frac{\partial \Psi_l}{\partial x_j} v'_{li} \right) - \tfrac{2}{3} \delta_{ij} \frac{\partial \Psi_l}{\partial x_k} v'_{lk} \right] d\Omega = A'_{limj} v'_{lj} \quad (14.1.23)$$

where A' is the volume part of the viscous finite element tensor operator for a compressible fluid with variable viscosity

$$A'_{limj} = -\int_V \eta \left(\frac{\partial \Psi_l}{\partial x_j} \frac{\partial \Psi_m}{\partial x_i} - \frac{2}{3} \frac{\partial \Psi_l}{\partial x_i} \frac{\partial \Psi_m}{\partial x_j} + \delta_{ij} \frac{\partial \Psi_l}{\partial x_k} \frac{\partial \Psi_m}{\partial x_k} \right) d\Omega,$$
$$\text{for} \quad l, m = 1, \ldots, n, \quad n, i, j = 1, 2, 3. \quad (14.1.24)$$

The expression (14.1.24) shows us that viscosity variations enter the local finite element operator through the nodal values of the viscosity field η. In the special case of an incompressible fluid with constant viscosity (14.1.24), reduces to the scalar Laplacian operator

$$S_{lm} = -\int_V \eta \frac{\partial \Psi_l}{\partial x_k} \frac{\partial \Psi_m}{\partial x_k} d\Omega. \quad (14.1.25)$$

We may make a number of choices for the finite element basis functions (Ψ); these include linear, quadratic or cubic functions. However, the integration by parts in the weak form of the momentum balance has the effect of reducing the order of differentiation for the viscous term from second to first order, so that it is sufficient to consider piecewise linear basis functions in the elemental operator. The TERRA code adopts piecewise linear basis functions for velocities and piecewise constant basis functions for pressure. Its elementary element of integration $d\Omega$ on the sphere is that of a spherical triangle.

We have seen that the momentum balance in the mantle is instantaneous and global due to the Stokes flow condition for flow at extremely low Reynolds number. We express this condition in the elliptic form of the momentum balance. The discrete representation of the momentum equation is an algebraic *sparse* matrix system, with entries clustered around the diagonal, where the sparse character results from the local support of finite elements. Numerical schemes exist to solve such sparse matrix systems directly, for example through *Gaussian elimination*.

However, when fine resolution is required the elliptic equation systems become rather large and are best tackled through iterative schemes. The key to solve such systems efficiently lies in the *multigrid* procedure (see, e.g., Briggs, 1987). This approach is optimal in the sense that the computational cost scales linearly in the number of unknowns, that is it costs the same per grid point to solve small or large matrices. This makes multigrid competitive with spectral methods based on the Fast Fourier Transform (FFT), without incurring the penalties associated with such global methods when handling large viscosity variations or in implementation on parallel computers. Multigrid methods excel because they rely on a hierarchy of nested computational grids, so that near- and far-field components of the global momentum balance are effectively solved at once. The nested structure of the icosahedral grid is particularly well suited for multigrid. A number of mantle dynamics models, such as the TERRA code, rely on multigrid methods for their performance.

We also need to include the condition for mass conservation, which enters the momentum balance as a constraint on the velocity field $\mathbf{v}$. We take the algebraic weak form of the momentum and the mass conservation equation in the anelastic (14.1.9) limit:

$$\mathbf{G}^{\mathsf{T}}\rho\mathbf{v} = 0, \qquad \mathsf{A}\mathbf{v} - \mathbf{G}\mathsf{p} = \mathsf{f}, \tag{14.1.26}$$

where $\mathbf{G}$ is a gradient matrix and A is the full tensor operator built from the operator elements A' (14.1.22) that revert to the scalar Laplacian (14.1.25) for incompressible flow without viscosity variations. The coupled system (14.1.26) is a saddle-point problem, and we seek to solve it for a velocity field $\mathbf{v}$ that satisfies the divergence-free constraint for incompressible, anelastic flow. Pressure is the Lagrange multiplier associated with this constraint. A standard approach to solve (14.1.26) is through a conjugate gradient algorithm, often referred to as the *Uzawa* method for incompressible flow, where an inner loop solves the elliptic problem by means of multigrid for a velocity field $\mathbf{v}$, while an outer loop enforces the divergence constraint on the velocity through a conjugate gradient search. This method has been adopted in the TERRA code.

The parabolic nature of the energy equation (14.1.11) differs from the elliptic form of the momentum balance, and finite volume methods provide a straightforward way for its solution. Remember that temperature variations ($\partial \mathsf{T}/\partial \mathsf{t}$) arise from advection ($-\mathbf{v} \cdot \boldsymbol{\nabla}\mathsf{T}$) through a divergence term, and from diffusion ($\nabla^2\mathsf{T}$) through a scalar Laplacian (assuming isotropic thermal conductivity) in addition to the heat source term. The *finite volume* method, like the finite element method, is a technique for solving the differential equation as a system of algebraic equations. The name finite volume refers to small volumes surrounding each node point on a mesh. Volume integrals that contain a divergence term are converted to surface integrals, using the divergence theorem. The requisite terms are then evaluated as fluxes at the surfaces of each finite volume. The Péclet number of the mantle is large, so mantle heat transport is controlled primarily by advection. Thus, outside of thermal boundary layers, the divergence term dominates, and the finite volume formulation is very effective. The finite volume method can also easily be adapted to unstructured meshes, such as the icosahedral grid. Further, because the flux entering a given volume is identical to that leaving the adjacent volume, the finite volume method is conservative. This approach has been adopted in the TERRA code, with a simple finite difference scheme used to handle diffusion. The energy balance, and thus the temporal evolution of the temperature field ($\partial \mathsf{T}/\partial \mathsf{t}$), is computed from an explicit second-order Runge–Kutta marching method in time.

It is common to use parallel computers for mantle convection simulations. The hardware in use includes traditional super computers, groups of individual machines connected into so-called *clusters* and new *multi-core* machines. *Explicit message-passing* in the MPI implementation is the standard programming approach to accommodate these diverse platforms. Finite element, volume, and difference techniques are all well suited for parallel computers because their local support offers inherent parallelism. Typically one subdivides the computational domain into smaller regions through an approach known as *domain decomposition*. We illustrate this for the icosahedral grid, and show a subdivision of the grid for 4 and 16 processors in Figure 14.3. Most of the work within each sub-domain is performed independently of the others, and boundary data are exchanged as required among them.

Figure 14.3. Domain decomposition for the icosahedral grid. The spherical shell is shown with one portion moved into the foreground. (a) The grid is partitioned into four sub-domains and a greytone scheme shows mapping to individual processors, with different tones used for processors 1, 2, 3, 4. (b) A recursion strategy allows for flexibility in the number of processors employed. In a second decomposition step, each sub-domain itself is split into a further set of four, to yield a total of sixteen subsidiary domains. In this way one can accommodate parallelism with many thousands of processors.

14.2 Convective planform

We describe the geometry of convection cells through their *planform*. The term originates from fluid dynamic tank experiments, where flow patterns are imaged by projecting collimated light (i.e., with parallel rays) through a tank filled with a transparent fluid such as corn syrup. Because the refractive index of light depends on the local fluid temperature, light rays are bent away from hot regions and converge toward colder regions. In the projected view upwellings show up as dark and downwellings as bright areas.

Much of the original studies of planform for mantle convection were done in the laboratory, and showed that the horizontal spacing of upwellings and downwellings is comparable to the depth of the convecting fluid. This agrees well with the theoretical predictions from linear stability derived in Chapter 7; see Whitehead (1988) for a comprehensive review.

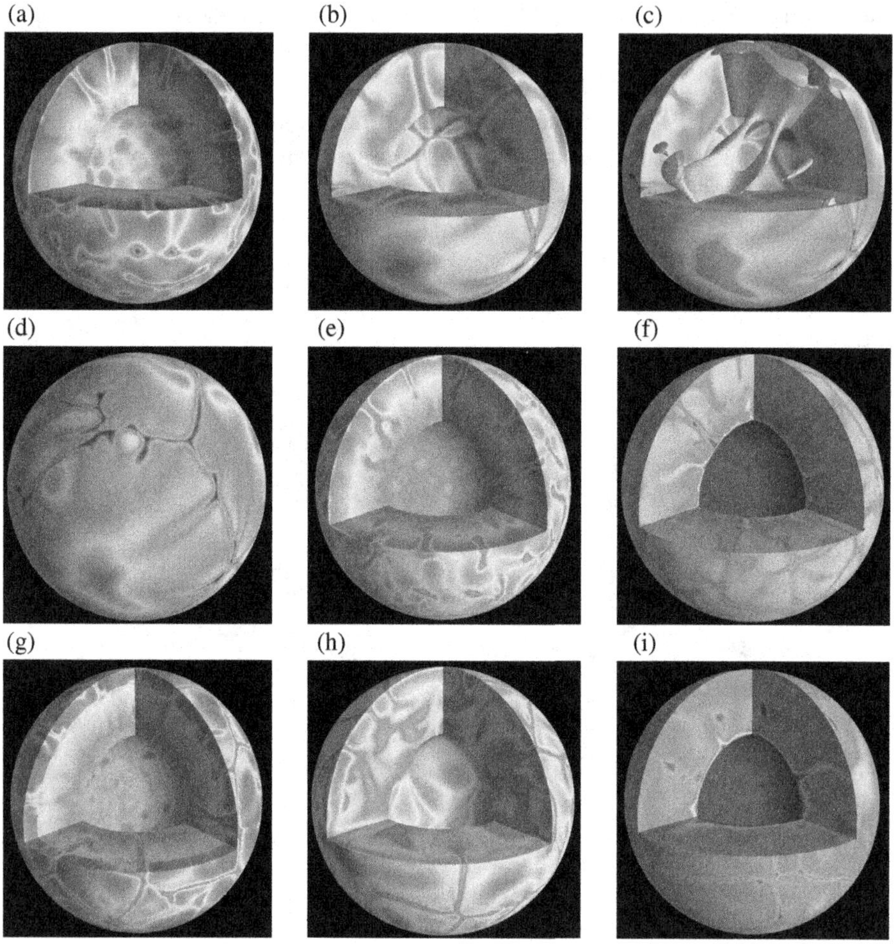

Figure 14.4. (a) A snapshot in time of the temperature field for a simulation of an incompressible (Boussinesq) and only internally heated mantle flow (zero Reynolds number). The Rayleigh number Ra_H based on internal heating is 4×10^7. The upper 200 km (roughly the depth of the upper thermal boundary layer) is removed, to permit a view on the convection planform below the boundary layer. The isoviscous reference model shows point-like downwellings from the upper boundary layer and relatively short wavelength convection cells, quite unlike the Earth. (b,c,d) same as (a) but with the viscosity of the lower mantle set thirty times larger than that for the upper mantle. (e) The superadiabatic temperatures for the compressible, purely internally heated, isoviscous reference mantle convection calculation with $Ra_H = 1.1\times10^8$. Isolated downwelling plumes from the upper boundary layer dominate the planform. (f) Same as (e) except for the addition of 38% bottom heating from an isothermal core. Convection is influenced by largely axisymmetric upwellings (plumes) from the lower thermal boundary layer. (g) Same as (e) except for the addition of an endothermic phase change at 670 km depth with $\gamma_c = -4$ MPa K^{-1}. Downwellings pause in the transition zone, before entering the lower mantle. (h) Stratified viscosity case: same as (e) except that the viscosity of the lower mantle has been increased by a factor of 30. The planform is dominated by long, linear downwelling sheets. (i) combination of depth-dependent viscosity and bottom heating.
The dominant features in (a),(e),(f),(g),(h),(i) are cold downwellings. Hot upwellings are prominent in (b),(c). Case (g) has a very warm upper mantle.

However, in looking at the Earth it is clear that the characteristic horizontal scale of the planform is much larger than the mantle depth, as evidenced by the spacing between subduction zones. The difference is particularly striking for the Pacific plate, where the plate width (12,000 km) exceeds the mantle depth (3000 km) by a factor of four. In Section 13.1.5, we examined the linear stability of a fluid with a low viscosity asthenosphere, and found that the dominant unstable convective mode is shifted to longer wavelengths. This shift favours convection cells that are much wider than they are deep.

Growing computer power means that convective planforms can now be studied through numerical simulations. Such computer simulations are particularly useful to understand convective systems where physical properties vary with depth, which are difficult to examine in the laboratory. The most pronounced variation in mantle properties with depth is arguably the large viscosity jump between the upper and lower mantle. But other parameters, such as thermal conductivity and thermal expansivity are known to vary with depth. A wide range of geodynamic studies on the influence of these parameters has yielded the following conclusions: an elongated mantle convection planform is associated with a reduction of the Rayleigh number with depth. In other words, a low thermal expansivity or a high thermal conductivity reduces the convective vigour of the deep mantle much like an increase in lower mantle viscosity. We present a group of examples of convective planforms for three-dimensional simulations of mantle convection in Figure 14.4 to illustrate the differences between different modes of heating and viscosity variations.

We take isoviscous, incompressible convection with pure internal heating as the most simple representation of flow in the mantle. This reference model, Figure 14.4(a), produces isolated, point like downwellings from the cold, upper thermal boundary layer. The convective planform for this simple case is dominated by short length scales, which agrees with laboratory results as we have noted above.

However, a dramatic change in the planform is produced by a single parameter variation from the reference model, with an increase of the lower mantle viscosity relative to that for the upper mantle. Sinking plumes in Figure 14.4(a) give way to sheetlike downwellings much like the subduction dominated planform of the Earth, Figure 14.4(b)–(d): the three panels show different aspects of the temperature distribution for the same convection scenario.

Compressibility effects do not significantly alter the character of the flow. Compressible, isoviscous, purely internally heated convection, Figure 14.4(e), produces point-like downwellings from the upper thermal boundary layer much like in the incompressible reference model. The parameters that do affect the planform include bottom heating from the core, and the buoyancy-associated mineral phase transitions in the transition zone. Core heating adds a hot thermal boundary layer at the bottom of the mantle. Because the surface area of the core–mantle boundary is only a quarter of that at the outer surface of the Earth, the geometrical spacing of the upwelling plumes increases the horizontal distance between upwellings and downwellings, Figure 14.4(f). Otherwise the planform with additional bottom heating remains similar to the case of convection with pure internal heating.

The introduction of phase transitions into the mantle produces minor planform effects. In the model shown in Figure 14.4(g) we use a Clapeyron slope of $\gamma_c = -4$ MPa K^{-1} for a phase transition at 670 km depth. This value is already probably too large to be representative of the γ-spinel to perovskite phase change and results in closely spaced, sheetlike downwellings that pause in the transition zone.

In Figures 14.4(h),(i) we show compressible convection with an increase in lower mantle viscosity, with either pure internal heating or a mix of internal and bottom heating. These cases are dominated by the increase in viscosity in the lower mantle.

14.3 Thermal structure and heat budget

The thermal structure of the mantle is characterised by steep thermal gradients concentrated in narrow thermal boundary layers and a nearly adiabatic temperature profile in the intervening regions. The presence of internal heat sources combined with the very long time it takes to cycle the volume of the mantle through the oceanic plates (of the order of 1–2 billion years) results in departures from adiabaticity, because the assumption of constant entropy employed to calculate the adiabat no longer holds for internally heated convection. The amount of heat entering the mantle from the core can be inferred from buoyancy studies over mantle hot spots.

14.3.1 Thermal boundary layers and the geotherm

The thermal diffusivity of the mantle is small, and so heat transport by conduction is restricted to thermal boundary layers where temperatures increase rapidly with depth. In the intervening well mixed regions the temperature distribution is nearly adiabatic. The overall temperature profile through the Earth with adiabatic and boundary layer regions is called the *geotherm*; a comprehensive discussion has been provided by Jeanloz & Morris (1986). Figure 14.5 shows the basic convective geotherm for a whole-mantle convection model of Earth-like convective vigour.

Because of their large vertical temperature gradients, the thermal boundary layers control, via thermal conduction, the influx (through the bottom) and outflux (through the top) of heat. Such boundary layers also dominate the dynamics of the mantle through their large temperature and buoyancy contrasts, as we saw from the boundary layer analysis in Section 14.1.1. Gravitationally unstable material from the boundary layers must eventually either sink (if in the cold top boundary layer) or rise (if in the hot bottom boundary layer), thus feeding upwelling or downwelling convective currents.

In the mantle we identify the lithosphere and tectonic plates as the cold upper thermal boundary layer. A hot thermal boundary layer also exists at the bottom of the mantle, where it separates the mantle from the core.

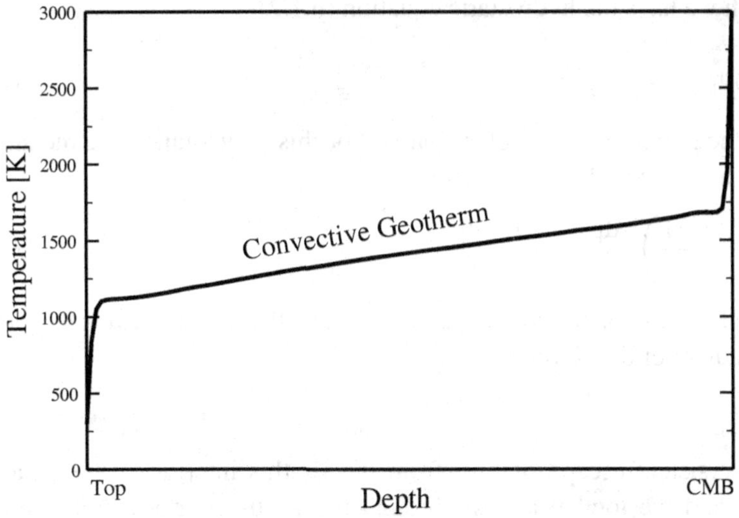

Figure 14.5. Basic convective geotherm for a whole mantle convection model.

Adiabatic gradient

For a mantle that is well mixed in an isentropic state, the geotherm away from thermal boundary layers is that of an adiabatic temperature gradient, which is given by:

$$\left(\frac{\partial T}{\partial p}\right)_S = \left(\frac{\partial T}{\partial S}\right)_p \left(\frac{\partial S}{\partial p}\right)_T, \tag{14.3.1}$$

from Maxwell's relation

$$\left(\frac{\partial S}{\partial p}\right)_T = \left(\frac{\partial V}{\partial T}\right)_p, \tag{14.3.2}$$

and so the isentropic temperature gradient with respect to pressure is

$$\left(\frac{\partial T}{\partial p}\right)_S = \left(\frac{\partial T}{\partial S}\right)_p \left(\frac{\partial V}{\partial T}\right)_p. \tag{14.3.3}$$

We recall the definition of the thermal expansion coefficient α_{th} (6.1.11), and the specific heat at constant pressure C_p (6.1.14):

$$\alpha_{th} = \frac{1}{V}\left(\frac{\partial V}{\partial T}\right)_p, \quad C_p = T\left(\frac{\partial S}{\partial T}\right)_p. \tag{14.3.4}$$

The adiabatic derivative of temperature T with respect to pressure p from (14.3.3) can then be recast as

$$\left(\frac{\partial T}{\partial p}\right)_S = \frac{T\,\alpha_{th}}{\rho\, C_p}. \tag{14.3.5}$$

For the Earth we have the hydrostatic equation (6.1.2)

$$\frac{dp}{dr} = -g\rho, \tag{14.3.6}$$

in terms of the gravitational acceleration g. For this hydrostatic case the adiabatic temperature increase with depth is given by

$$\left(\frac{\partial T}{\partial r}\right)_S = \left(\frac{\partial T}{\partial p}\right)_S \frac{dp}{dr} = \frac{T\alpha_{th}}{\rho C_p}\frac{dp}{dr}. \tag{14.3.7}$$

Using typical values for thermal expansivity, density, and heat capacity yields an adiabatic gradient of 0.5 K km^{-1}.

Heat sources

There is a persistent escape of heat from the Earth's interior to the oceans and atmosphere, and the total heat loss through the Earth's surface is 44 TeraWatts (TW). About 31 TW of this heat is lost through the oceanic plates; another 13 TW comes from the continents. Roughly half the continental heat flux (about 7 TW) has its source in heat production from within the continental crust. The rest is supplied via conduction from the mantle beneath. Thus the total mantle heat loss is about 37 TW.

Heat is produced in the mantle by the decay of radioactive isotopes. The most important contributors are uranium (^{238}U, ^{235}U)), thorium (^{232}Th) and potassium (^{40}K). Such isotopes are believed to be disseminated throughout the mantle, but are expected to be concentrated in the thermal boundary layers at the top and bottom of the mantle.

There is a tendency for U, Th, K to be segregated via melting processes and to stay with one of the fractions. As a result the continental lithosphere that represents a residue from multiple phases of melting has high concentrations of heat producing isotopes. Crustal concentrations reach parts per million (ppm) for U and Th. Potassium is much more abundant (of the order of per cent), but the radioactive isotope (^{40}K) represents only 10^{-4} of total potassium and so the contribution to total heat production is less than for uranium and thorium. Inside the mantle the radioactive elements are thought to be about two orders of magnitude less abundant than in the crust. Typically U, Th and K occur in similar proportions relative to one other in the crust and mantle, even though their absolute concentrations can vary greatly. The average abundance ratio Th/U is about 3.5–4, while the ratio K/U is usually $1\text{–}2\times10^4$.

A rough upper bound on the concentration of heat-producing elements in the mantle can be made, if we assume the mantle is in a thermal steady state. In this case the heat loss from the mantle must be balanced entirely by heat production from radioactive decay. With a mantle mass of 4×10^{24} kg, we arrive at an upper bound for heat production of 11×10^{-12} W kg^{-1}. Correcting for the heat produced in the continental crust, the upper bound on mantle heat production is 9×10^{-12}

$W\,kg^{-1}$. This upper limit is substantially larger than the heat production rate inferred from geochemical studies of upper mantle rocks. For example, Jochum et al. (1983) estimate that the heat production rate of upper mantle rocks is only about $0.6\text{–}1.0\times10^{-12}\ W\,kg^{-1}$. If such heat productivity applied throughout the mantle it would at most result in a surface heat flux of about 4 TW, that is only one tenth of the total surface heat loss.

We can account for some of the deficit by assuming that the mantle is not in a thermal steady state, so that we assume that the mantle undergoes secular cooling, so that the interior temperature profile is progressively lowered by 50 to 100 K per billion years. Mantle secular cooling of $70\ K\,Gyr^{-1}$ would result in a surface heat loss of 10 TW. Still it is not clear whether there is enough radioactivity inside the mantle to account for the observed heat loss.

Subadiabaticity

Jeanloz & Morris (1987) appear to have been the first to notice that secular cooling and the presence of internal heat sources must act to move the mantle geotherm away from adiabaticity. An easy way to see this is to focus on the effect of internal heating, and to follow a mantle volume element on its way from the core–mantle boundary to the surface. As the volume element rises there is cooling due to adiabatic decompression, but at the same time the temperature increases in response to the internal heat released from radioactive decay. The net result is to make the radial temperature change smaller than the adiabatic gradient, so that the geotherm is subadiabatic. A subadiabatic model geotherm derived from three-dimensional spherical convection modelling is shown in Figure 14.6.

Analytic and computational studies suggest departures from the adiabat by as much as 300–500 K in the mantle. The heat equation (14.1.11) allows us to derive a simple scaling argument. Outside of thermal boundary layers, where thermal gradients are necessarily large, we ignore the diffusive term $k\nabla^2 T$. This simplification means nothing else than that heat is transported primarily by advection, or alternatively that the Péclet number in the mantle is large, as we have seen before. We also ignore the effects of shear heating. With these assumptions (14.1.11) reduces to static heating by internal heat generation:

$$\frac{DT}{Dt} = \frac{h}{C_p}, \qquad (14.3.8)$$

where DT/Dt is the total time derivative of temperature T moving with the volume element. For internally heated convection all material must eventually cycle through the upper thermal boundary layer (the lithosphere in mantle convection) to lose its heat. Note that this behaviour differs from purely bottom-heated fluids, where it is sufficient for material from the bottom boundary layer to cycle through the upper thermal boundary layer. We can derive an estimate of the circulation time scale (Δt) from the geometry of plate tectonics. Taking the total length of the oceanic spreading system as 65 000 km and the average plate velocity as $5\ mm\,yr^{-1}$,

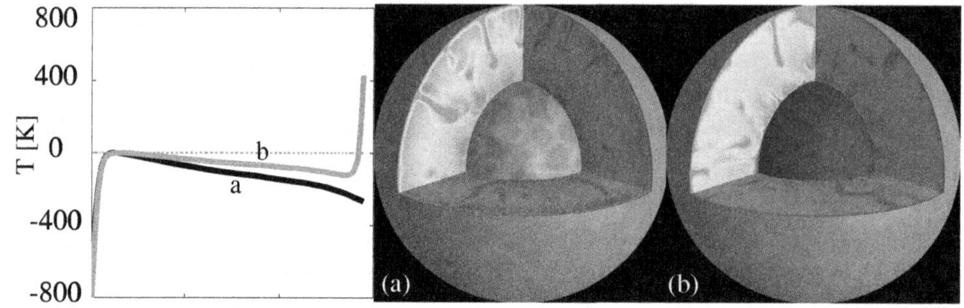

Figure 14.6. Convective geotherms with the reference adiabat removed, for (a) pure internally heated convection, and (b) convection with 50% core heating. The effect of core heating is to make the geotherm closer to adiabatic.

some 3 km^2 of ocean floor is created each year. The same amount must enter the mantle through subduction. With the assumption that slabs and the material entrained with them are 200 km thick we infer that some 600 km^3 material with mass 2×10^{15} kg enters the mantle (with total mass 4×10^{24} kg) each year. At this rate the entire mantle circulates through the upper thermal boundary layer on a time scale Δt_{mantle} of order 2 billion years (Gyr), which together with an assumed internal heating rate h of 5–10×10^{-12} W kg^{-1} and a mantle heat capacity C_p of 1000 J kg^{-1} K^{-1} implies a non-adiabatic component of order 250–500 K in the mantle geotherm.

14.3.2 Plates

Tectonic plates are the cold upper thermal boundary layer of the mantle. These plates are the surface expression of mantle convection and accomplish the primary mantle heat loss, that is they cool the mantle. In Section 12.2.1 we have derived the thermal structure for such plates from a half-space cooling model. The plates differ from the rest of the mantle fluid, because they are cooler and thus stronger than the mantle, although, as we have seen in Section 12.2.4, their strength is not unlimited and with sufficient stress they yield in the form of brittle fracture. Thus the defining feature of plates is their combination of strength and mobility.

Heat transport

Plates carry the majority of the mantle heat loss. The heat transport can be estimated from equation (12.2.5), which relates the heat flux to the plate age

$$Q(t) = -\frac{kT_M}{\sqrt{\pi\kappa}}\frac{1}{\sqrt{t_s}} = -\frac{a}{\sqrt{t_s}}, \tag{14.3.9}$$

where we have set $a = kT_M/\sqrt{\pi\kappa}$. The total seafloor area A is 3.1×10^8 km^2 and new seafloor is formed at a rate of 3 km^2 per year. At this rate the entire ocean floor would be replaced in about 100 Myr and plates at the time of subduction (t_s) would be on average 100 million years old. Taking the heat flux (q_s) for 100 Myr old seafloor as 50 mW m^{-2} allows us to evaluate the integral heat flux over the whole seafloor from

$$Q = \int_0^{t_s} dt\, Sq = 2Sa\sqrt{t_s} = 2Aq. \tag{14.3.10}$$

We find that the average heat flux through the oceans is about 100 mW m^{-2} and that the total heat flux is 31 TW in agreement with observations. Thus plates account for 75% of the total heat loss of the Earth and nearly 90% of the heat lost by the mantle.

Influence on mantle flow

Plates affect the mantle through their mechanical strength, since they remain nearly rigid over geological time and thereby stabilise the upper thermal boundary layer. The plate effect is particularly evident in the subduction process. Whereas a fluid boundary layer becomes gravitationally unstable through local Rayleigh–Taylor instabilities, the high viscosity of plates ensures that downwellings enter the mantle only at a few major subduction zones. The stabilising effect of plates is taken to the extreme in the case of Venus and Mars, where there is only one single plate with little apparent movement and where the planet convects in a so-called *rigid-lid regime*.

The Earth is in the *mobile-lid regime*, that is it has a strong yet mobile lithosphere (plates). Thus the long-wavelength planform of the mantle reflects both the influence of depth-dependent viscosity and the great mechanical strength of plates. The essence of this effect has been captured by saying that plates *organise* the flow.

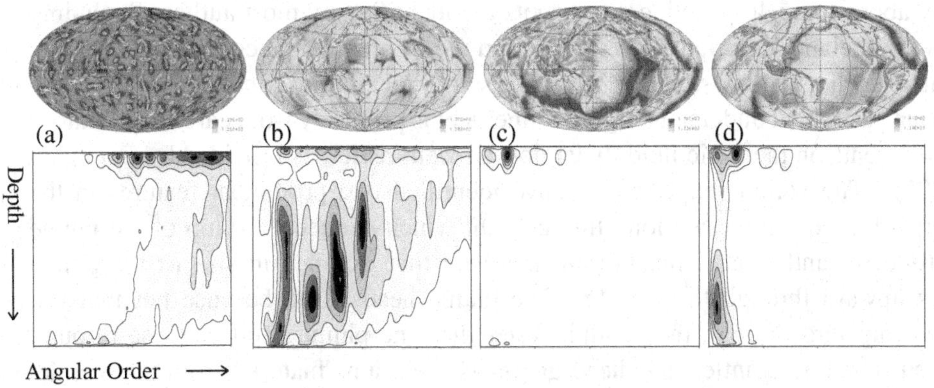

Figure 14.7. Near surface temperature (100 km depth) and the spectrum of thermal heterogeneity as a function of depth and angular order for three-dimensional spherical computer models: (a) isoviscous, (b) stratified viscosity convection models without the stabilising effects of plate motion and a stiff lithosphere; (c) isoviscous, (d) stratified viscosity with a relatively stiff lithosphere

In Figure 14.7 we show results from the models of Bunge & Richards (1996) to illustrate the plate effect for global incompressible, internally heated mantle flow. The planforms and the thermal heterogeneity spectra are shown for isoviscous convection and convection with a high viscosity lower mantle, with and without the stabilising effect of the plates. Both isoviscous convection with a plate effect and stratified viscosity convection without the plate effect are unable by themselves to reproduce the observed planform of the mantle. The dominance of long-wavelength heterogeneity in mantle convection results from a combination of the two features: a significant increase in mantle viscosity with depth, and the existence of strong surface plates. The large-scale character of mantle structure thus arises from independently established physical mechanisms in otherwise simple mantle convection models.

14.3.3 Hot spots and plumes

Up to this point we have encountered the cold, upper thermal boundary layer of mantle convection in the form of plates. However, there must also be a hot, lower thermal boundary layer at the bottom of the mantle associated with the conductive heat loss from the core. Thermal instabilities arising from this hot lower boundary layer are known as plumes, and constitute the other important mode of mantle convection.

Tuzo Wilson introduced the concept of plumes in order to explain the anomalous volcanic activity of hot spot regions, which are areas of extensive intra plate volcanism not associated with plate margin volcanism. Later Jason Morgan envisioned vertical conduits of hot thermal plumes rising through the mantle and accomplishing a major component of heat transport through the mantle. Although between 20 and 100 different hot spots have been proposed at different times, only about some 40 prominent hot spots are identified by most authors, including Hawaii, Iceland, Réunion, Cape Verde, and the Azores. Large continental volcanic centres such as Yellowstone or Afar are also attributed to hot spots.

Our theoretical understanding of plumes stems primarily from careful laboratory experiments on the basic fluid dynamics; a recent review is provided by Campbell (2007). The various experiments have identified two prominent features in the morphology of plumes: a long thin conduit which connects the top of the plume to its base, and a large mushroom like head that expands in size as the plume rises upward through the mantle. The plume heads form because hot material moves upward through the conduit faster than the plume itself can rise through the surrounding mantle. The head grows as it entrains material from the mantle. Plume heads in the Earth can reach substantial size, exceeding $500 \times 500 \times 500$ km to produce a total volume of melt in excess of 10^7 km^3.

When plume heads encounter the base of the lithosphere they spread sideways, and undergo decompression melting to form enormous volumes of basaltic magma, called flood basalts. Radiometric dating puts the durationof eruptions from the plume heads at less than a few hundred thousand years. At this rate an area the size of California or Germany could be covered by 4 km of basalt in less than 1 million years to form a continental flood basalt (if it erupts through continental crust) or an oceanic plateau (if the eruption takes place in the oceans).

The location of prominent flood basalts is shown in Figure 14.8. Examples of continental flood basalts are the Deccan traps and the Rajmahal traps in India, as well as the Siberian traps of Asia. Other examples include the Karoo basalts in South Africa, and the Parana basalts in South America which were linked with the Etendeka basalts in Africa, prior to the opening of the South Atlantic. Oceanic flood basalts include the Ontong Java plateau of the southwest Pacific and the Maniheken plateau of the Indian ocean. A comprehensive review of flood basalt eruptions and their possible relation to mass extinctions is provided by Courtillot & Renne (2003).

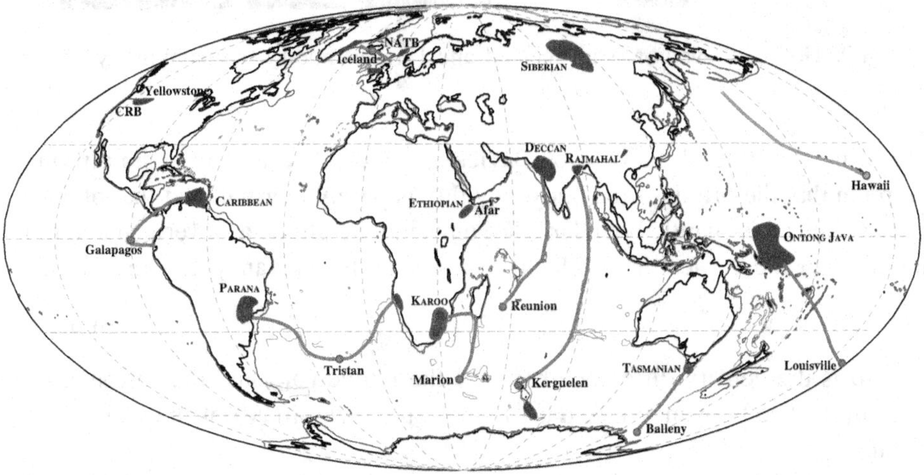

Figure 14.8. Map of hot spots with well defined tracks and flood basalts at their origins (indicated by small capitals).

Hot spot swells

Most hot spots are associated with distinctive bathymetric swells, where the ocean floor is anomalously shallow for its age. Figure 14.9 shows the swells associated with Hawaii and Iceland, through perspective views of ocean-floor topography. An example of topographic effects on the continents is the elevated region around Afar, in eastern Africa. Hot spot swells must be maintained by buoyancy forces associated with the plume. If a strong plume is located beneath a fast-moving plate, such as the Pacific plate, large amounts of anomalous topography can be created each year. In the case of Hawaii, an area 1000 km wide is elevated by 1 km at a velocity of 10 cm per year (the speed with which the Pacific plate travels across the hot spot). Rigorous studies of hot spot bathymetry have been presented by Davies (1988) and Sleep (1990) to place constraints on the amount of core heat entering the mantle.

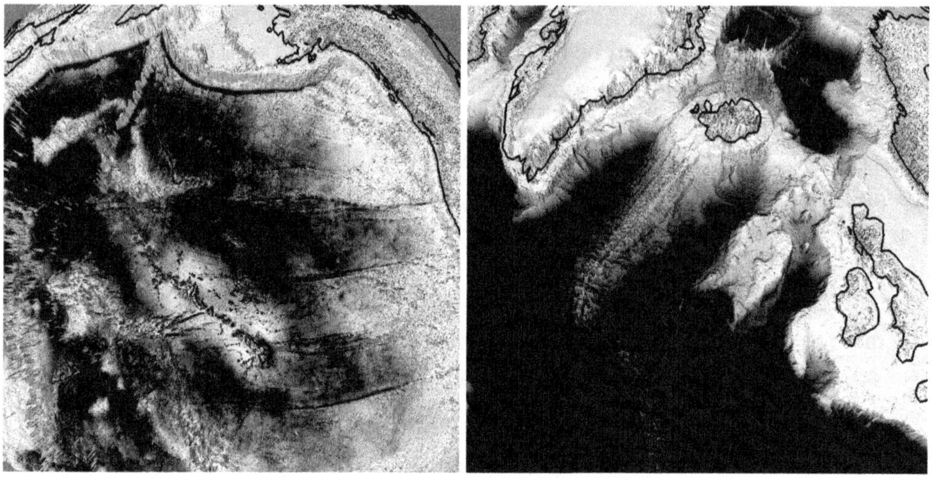

Figure 14.9. Hot spot swells around Iceland and Hawaii modify the bathymetry of the oceans.

Following the style of analysis of Davies and Sleep, we make the simplifying assumption that the origin of the buoyancy flux is purely thermal, and that plumes come from the core–mantle boundary. We envision the plume as a vertical conduit with radius (r) and vertical flow velocity (v), so that the buoyancy flow rate (b) is

$$b = g\,\Delta\rho\,\pi r^2 v, \tag{14.3.11}$$

where $\Delta\rho$ is the density difference between the plume (ρ_p) and the surrounding mantle (ρ_m). The buoyancy flux must balance the weight (W) of the anomalous topography

$$b = W = g\,(\rho_m - \rho_w) w_s e_s u_{plate}. \tag{14.3.12}$$

where $\rho_m - \rho_w$ is the density difference between mantle and overlying sea water, w_s and e_s are the width and elevation of the swell, and u_{plate} is the plate velocity. For Hawaii this yields a flux of 7×10^4 N s^{-1}.

For a purely thermal origin we equate the heat flux with the plume flux, since both depend on the excess temperature $\Delta T = T_p - T_M$ of the plume relative to surrounding mantle, the resulting density difference

$$\Delta\rho = (\rho_p - \rho_m) = \rho_m \alpha_{th}\,\Delta T, \tag{14.3.13}$$

The plume heat flux is also controlled by the excess temperature ΔT,

$$Q = \pi r^2 v\,\rho_m\,C_p\,\Delta T, \tag{14.3.14}$$

and is related to the buoyancy flux b by

$$Q = \left(\frac{C_p}{g\alpha_{th}}\right) b. \tag{14.3.15}$$

The heat flux is proportional to the buoyancy flux weighted by heat capacity, gravity and the thermal expansivity; yet the expression (14.3.15) does not involve the excess temperature ΔT. Thus, the same buoyancy flux b could be maintained by a small material flux with large excess temperature or a large material flux with a small excess temperature.

For Hawaii, using typical values for the physical parameters ($C_p = 1000$ J kg^{-1} K^{-1}, $\alpha_{th} = 3\times10^{-5}$ K^{-1}) we obtain a heat flow $Q = 2\times10^{11}$ W. This flow represents about 0.5% of the global heat flux.

Similar studies of plume flux have been presented by Davies and by Sleep for all major hot spots. Although there are about 40 hot spots, all of them are weaker than Hawaii and many are very much weaker. The best estimate of the total plume heat flux derived this way is about 2.3×10^{12} W (2.3 TW), which is about 6% of the total global heat flow. In addition to the heat carried by plume tails, we must account for the heat transported by plume heads. From the frequency of flood basalt eruptions this has been estimated at about 50% of the heat carried by plume tails. Thus the total heat flux transported by plumes would be around 3.5 TW, less than 10% of the total global heat flow rate, but still an important component of the Earth's heat budget.

Hot Spot reference frame

A common feature of many hot spots is a well defined track made up of chains of volcanoes and underwater sea mounts aligned in the direction of motion of the overriding plate. The best example is the Hawaiian–Emperor chain, which extends nearly 4000 km from the big island of Hawaii to the Aleutians near Alaska and shows a clear progression with the age of submerged volcanoes increasing with distance from Hawaii. The rate of development of the hot spot track is about 90 mm yr^{-1} for the past 40 Ma, which indicates the average velocity of the Pacific plate relative to Hawaii during this period.

When we correct for the motion of the overriding plates, hot spots appear to be relatively stationary with respect to each other and the mantle. Of course, there must be some relative motion, due to mantle convection. For example, hot spots in the Pacific seem to move relative to those in the African hemisphere. Detailed studies for Hawaii have detected relative motion with respect to the Indian and Atlantic hot spots of the order of a few mm per year. The clearest evidence for hot spot motion has emerged for Hawaii, which seems to have originated significantly further north from its present location, and subsequently drifted south. Tarduno et al. (2003) provide a summary of the relevant palaeomagnetic data.

Nevertheless, the overall motion among hot spots appears to be smaller by about an order of magnitude than the average motion associated with plate tectonics, making hot spot fixity a useful first order approximation for time periods of 50-100 million years. This observation has been used to define a global reference frame tied to the deep mantle through hot spots, the *hot spot reference frame*.

Plume excess temperature and global mantle heat budget

A longstanding observation about plumes is that they have low excess temperature relative to normal mantle (e.g., Schilling, 1991). Most volcanic rocks associated with hot spots have basaltic composition with major element chemistry similar to that of mid-ocean ridge basalt. The petrology of hot spot lavas can be explained with an excess temperature of the order of 200–300 K. This low value for the excess temperature would appear at odds with independent considerations that suggest a much larger temperature change across the core–mantle boundary of the order of 1000 K. If plumes originate from a thermal boundary layer at the core–mantle boundary the difference in temperature contrasts is difficult to understand.

One way to reduce the excess temperature is through a chemically dense layer at the core–mantle boundary. Such a layer could buffer the temperature contrast between the core and the mantle. Yet, two other considerations are probably more important. Firstly, one expects a lowering of the excess temperature associated with plumes because the mantle has internal heat sources. As we have seen before,

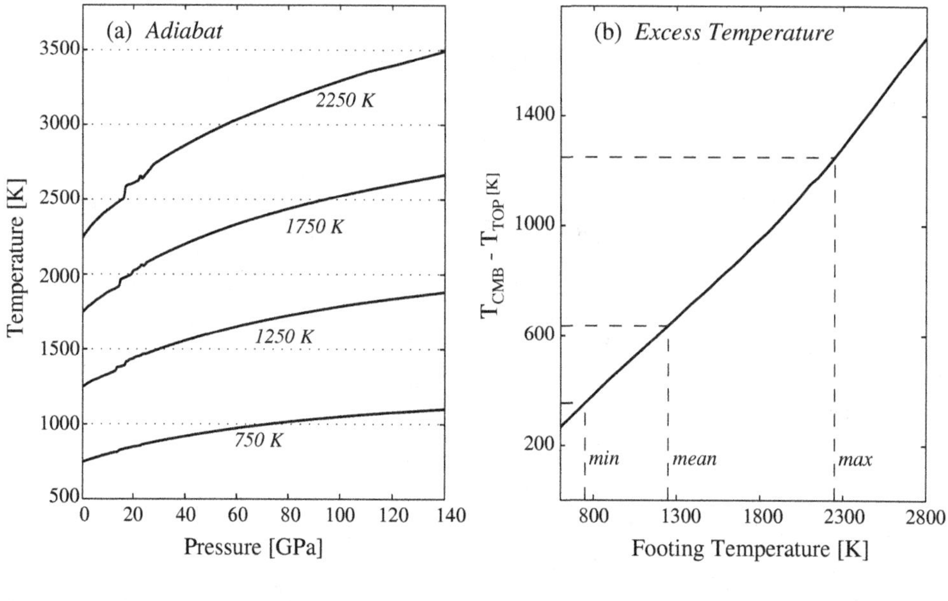

Figure 14.10. (a) Curves of constant entropy adiabats for footing temperatures of 750, 1250, 1750, 2250 K (the adiabatic projection of the mantle temperature up to the Earth's surface). The slope of the adiabat increases with footing temperature, and generally decreases with depth. Temperature jumps corresponding to the discontinuities in entropy of the phase transitions are less evident in the colder profiles. (b) Total adiabatic temperature variation from surface to core–mantle boundary plotted against footing temperature. There is an increase in adiabatic temperature variation with footing temperature. Also shown are the footing temperatures corresponding to mean, minimum and maximum radial temperatures in a spherical convection model. Hot thermal upwellings (plumes) undergo a stronger adiabatic cooling relative to surrounding mantle, so that their excess temperature decreases systematically in the mantle from the bottom to the top.

internal heating in combination with the slow overturn of the mantle results in the mantle geotherm departing from the adiabat by as much as 250–500 K. Plumes, however, rise through the mantle relatively quickly, on a time scale of the order of 100 million years. The plume temperature distribution is therefore expected to be nearly adiabatic. We can express this result in a different way: the normal geotherm decreases less rapidly than the adiabatic gradient in moving from the deeper to the shallower mantle, whilst the temperature inside the plumes follows the adiabat closely. The net result is a systematic loss of excess temperature in the plume as it moves away from the core–mantle boundary.

A second consideration is evident from equation (14.3.7), which states that the adiabatic temperature change depends on temperature. For example, from Figure 14.10 we see that an isentrope tied to a footing temperature of 2000 K, at zero pressure, undergoes a temperature increase with depth nearly twice that of an adiabat footed at 1000 K. For plumes the adiabatic temperature drop therefore is about 200 K larger than that of normal mantle, due entirely to their higher temperature.

Our observations bear on the heat loss of the core, which as we have seen can be constrained from buoyancy studies over hot spots. If we correct for the steeper adiabatic gradient in plumes (about 250 K) and the effects of subadiabaticity in the mantle geotherm (about 250 K) the excess temperature in a plume and the associated buoyancy flux at the core–mantle boundary could be larger by a factor of three than the valued inferred from surface observations. The total heat flux transported by plumes could thus be in the range of 10–12 TW, which represents about 30% of the total mantle heat budget. Together with a secular cooling contribution of 70 $\mathrm{K\,Gyr^{-1}}$ (10 TW) and a radioactive heating rate of $4\text{–}6\times10^{-12}$ $\mathrm{W\,kg^{-1}}$ amounting to 16 TW, the mantle heat budget would be balanced.

14.4 Circulation of the mantle

It is common to use the term circulation to describe the motion of the mantle, in analogy to the general circulation in the oceans and the atmosphere. The direct evidence we possess for mantle circulation comes from plate tectonics, because the viscosity of the mantle is sufficiently large to ensure coupling between plate motion and mantle flow. Geodynamic calculations suggest upper mantle viscosities as low as 10^{17} Pa s are required before plate motion and deep mantle convection are effectively decoupled. Thus the kinematics of plate motion models provides first-order constraints on the dynamics of the mantle. A close linkage between plate motion and mantle circulation must also be inferred from the requirement to conserve mass, because mass transfer associated with plate motions at the surface must be balanced by internal mass displacements in the mantle. Plate motions change over time, and these variations reveal important temporal variations in the pattern of the internal circulation.

The most important data used to construct plate motion models come from the magnetic isochron record of oceanic plates. Thus reliable reconstructions of plate motion are restricted to the past 100–150 Myr, the age of the oldest ocean floor. Early in the chapter we saw that the primary mantle heat loss occurs through the oceanic lithosphere. This means that a close linkage between plate motion and mantle circulation must also be inferred from thermal considerations, and that the large scale thermal structure of the mantle must be dominated by past subduction.

The combination of convection models with reconstructions of past plate motion allows us to explore the evolution of the Mesozoic and Cenozoic mantle in *mantle general circulation models.* Such models solve the mantle convection equations subject to boundary conditions on surface velocities taken from the history of plate motion. Mantle circulation models explain some deep mantle heterogeneity structures imaged by seismic tomography, especially those related to past subduction. However, their lack of initial condition information is a significant limitation. The problem of unknown initial conditions can be addressed using an inverse approach to convection, based on the exploitation of the adjoint equations for mantle convection.

14.4.1 Present-day and past plate motion models

Models of present-day plate motion are given in terms of rotation poles and angular velocities. We start from Euler's displacement theorem which states that rigid body motion at the surface of the Earth can be expressed as a rotation about an axis passing through the Earth's centre. This was evident to Alfred Wegener when he described the drift of North America relative to Eurasia as the rotation around an axis passing through Alaska. Euler's theorem gives us a compact and quantitative way to describe the Earth's surface velocity field. In particular, plate velocities can be described by an angular velocity vector and then standard vector algebra can be applied to combine motions. Thus if both the angular velocity of plate A relative to B, ${}_{\mathrm{A}}\omega_{\mathrm{B}}$, and that of B relative to C, ${}_{\mathrm{C}}\omega_{\mathrm{B}}$, are known, we obtain the motion of C relative to A from vector addition:

$$ {}_{\mathrm{C}}\omega_{\mathrm{A}} = {}_{\mathrm{C}}\omega_{\mathrm{B}} + {}_{\mathrm{B}}\omega_{\mathrm{A}}. \tag{14.4.1} $$

Angular velocities can be deduced from a variety of observations that include the orientation of active transform faults between two plates. Since relative motion between two plates sharing a mid-ocean ridge is parallel to transform faults, the fault arcs must lie on small circles. The rotation pole then lies somewhere on the great circle perpendicular to the transform faults. With information from two or more transform faults the position of the rotation pole can be determined (see, e.g., Morgan, 1968). Another way to examine the relative motion is to map spreading rates along a mid-ocean ridge, because the magnetic anomaly pattern varies as the sine of the angular distance from the rotation pole. A third method to compute the relative motion between two plates is the use of fault plane solutions (focal mechanisms) from earthquakes along plate boundaries, although this method provides just the location of the pole, and not the spreading rate. A global model for current plate motions based on these types of data has been constructed by De Mets et al. (1990, 1994).

Today plate motions are measured with great accuracy using space geodetic methods (cf. Section 12.4.2). Very long baseline interferometry (VLBI) exploits very distant quasars as source signals and terrestrial radio telescopes as receivers. The difference in distance between two telescopes is then tracked over a period of years. The use of Global Navigation Satellite Systems (GNSS) such as the Global Positioning System (GPS) is another common method to measure plate motions. A good review on the application of such space geodetic methods in a geological context is provided by Dixon (1991). GNSS measurements allow real-time monitoring of plate motions; a survey time of about 1–2 years is normally sufficient to obtain a meaningful estimate of the rate and direction of motion at a site. A plate motion model based only on GPS data was constructed by Sella et al. (2002).

Models of past plate motion are constructed primarily by matching magnetic anomalies and fracture zones of the same age, corresponding to patterns of palaeo-ridge and palaeo-transform segments at a given reconstruction time. The plates then need to be restored to their palaeo-positions through a series of finite rotations, often referred to as stage poles. Bullard, Everett & Smith (1965) used finite rotations based on similarities in the Atlantic coastlines to publish the first set of computer-generated plate reconstructions describing the opening of the Atlantic. Figure 14.11 presents a modern summary of reconstructions of plate motion over the past 130 Ma, a period that has seen significant shifts in the relative positions of the continents.

Palaeomagnetic plate reconstructions constrain the palaeo-latitude and angular orientation of a plate, but leave its palaeo-longitude unknown. To minimise the uncertainty in longitude one is required to specify a reference frame. One choice is the selection of a slow-moving plate, such as Africa, as the common reference against which the motion of all other plates is restored. Africa's lack of elongated hot spot tracks is commonly taken as evidence that the plate has moved by less than 15 degrees over the past 100 Myr. An alternative choice is the use of the *no-net-rotation reference frame*, where one assumes that the mean lithosphere velocity integrated over all plates must vanish. The inherent difficulty with this reference frame rests in the fact that it is not obvious whether the mean velocity of plates should vanish. Thus a third and physically more plausible reference frame is often used by relating plate reconstructions directly to the deep mantle through hot spots in the *hot spot reference frame*, as we have seen before. Steinberger & O'Connell (1998) have tried to improve on the accuracy of the hot spot reference frame by accounting explicitly for errors associated with relative hot spot motion using global mantle flow models. In general, plate motion models based on a hot spot reference frame result in minimal longitudinal motion of Africa (compared

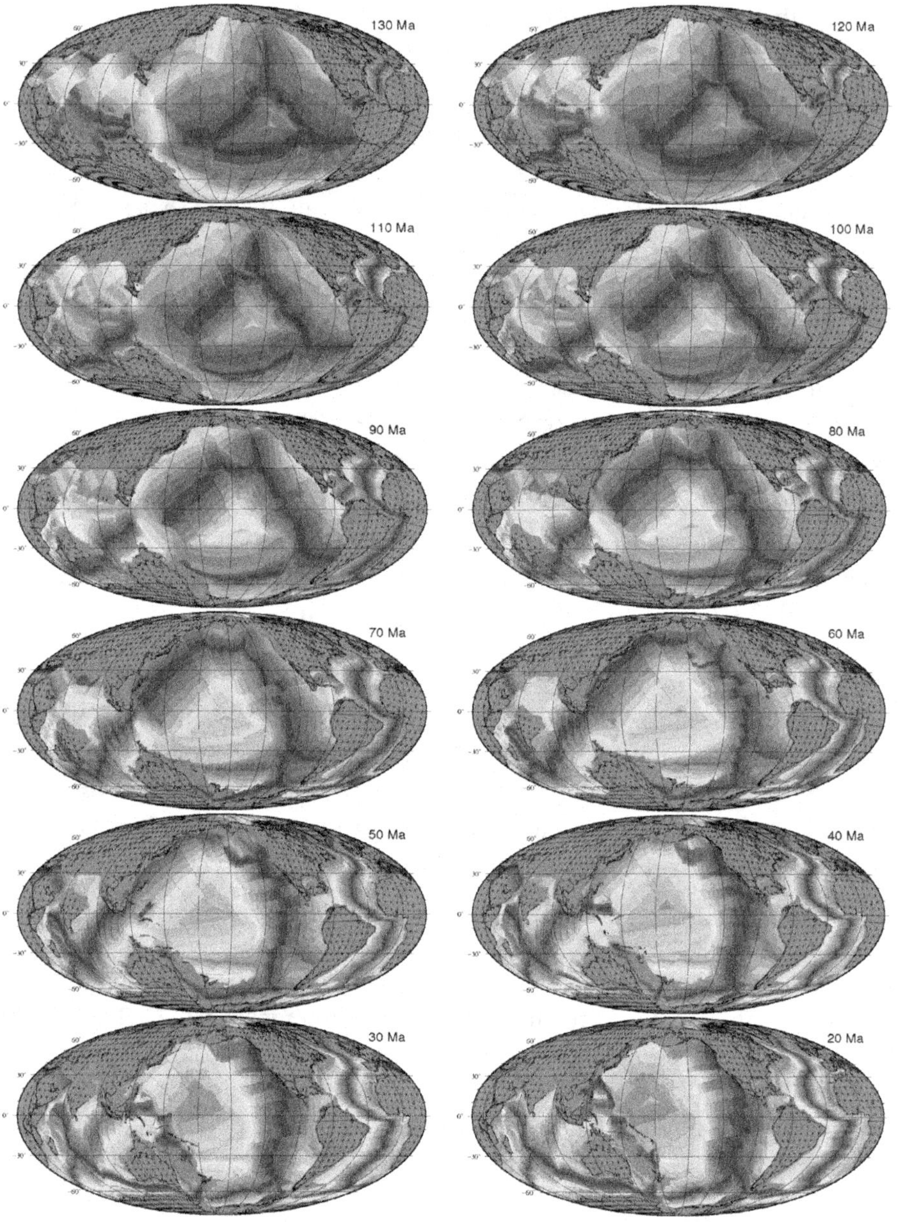

Figure 14.11. Reconstruction of plate motions for the past 130 Ma. [Courtesy of D. Mueller.]

with most other plates), and thus confirm the relative fixity of the African plate. Finite rotation poles have been published for most plates describing their late Cretaceous/Tertiary history of motion. A widely used plate motion model in the hot spot reference frame for the Cenozoic is that of Gordon and Jurdy (1986).

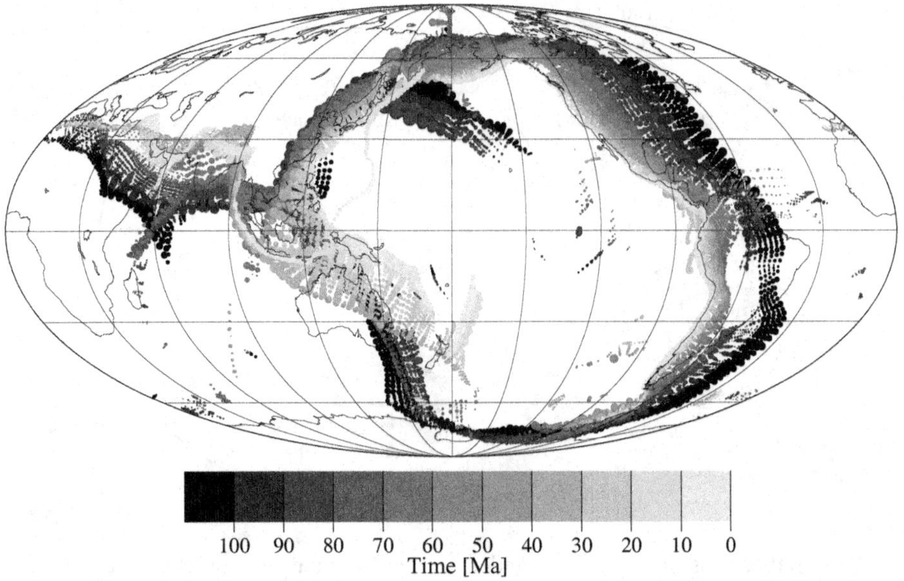

Figure 14.12. Subduction history for the last 110 Ma, where lighter tones are used for progressively younger subduction. Note the significant differences in current and past subduction in the northwest and southwest Pacific. [Courtesy of B. Steinberger.]

A consequence of the changes in the motion of the plates is that the configuration of subduction zones has changed, as can be seen from Figure 14.12 where the position of subduction is coded by age, with lightest tones used for the most recent period. For the last 100 Ma, the perimeter of the Pacific has been the location of most of the subduction zones. The current sites of subduction show up clearly in seismic tomography (Section 11.4.4) for the upper mantle. Clear indications of the former Farallon plate beneath the Americas and the subduction beneath South Asia at the northern edge of the Tethys ocean are found in the upper part of the lower mantle (see, e.g., Figure 11.23 at 1100 km).

14.4.2 Implications of plate motion models for mantle circulation

We can use plate motion models to make a number of direct predictions about structure and dynamics of the mantle. For example, the temperature variations in the mantle should be dominated by heterogeneities associated with past subduction reflecting the fact that vast amounts of old oceanic lithosphere descended into the mantle over the past 100 Myr (see Figure 14.12). In a classic paper, Chase & Sprowl (1983) point to a rough correspondence between major geoid lows and Cretaceous subduction zones, while Anderson (1982) noted the complementary correspondence of the African geoid high with the ancient supercontinent Pangea. Section 13.4.1 showed us that the geoid is dominated by these broad highs and lows, and as a result the geoid spectrum peaks at spherical harmonic degrees l of 2 or 3. The time lag between Mesozoic plate tectonics and modern geoid

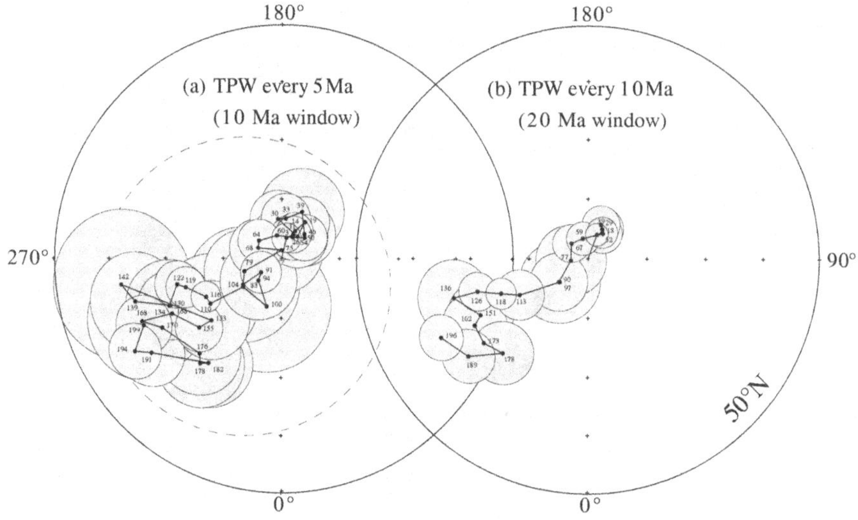

Figure 14.13. Indications of true polar wander derived from palaeomagnetic observations. [Courtesy of J. Besse, V. Courtillot.]

anomalies, effectively the age of the geoid, is due to mantle convection and reflects the time it takes for slabs to sink through the mantle, as explained by Davies (1984). A formal correspondence between large-scale mantle heterogeneity and the time-integrated history of subduction was made by Richards & Engebretson (1992), and a quantitative model of mantle heterogeneity derived from subduction history was later developed by Ricard et al. (1993).

We can infer also that the mantle's circulation pattern must vary in time. In other words, as can be seen from the boundary layer analysis in Section 14.1.1, long-term changes in plate motion on time scales of the order of 10–100 Myr must be matched by long term changes in the convective forces of the mantle. Independent observational evidence for time variations in mantle circulation comes in the form of palaeomagnetic studies of the secular motion of the Earth's rotation axis relative to the surface. This *true polar wander* accommodates changes in the Earth's moment of inertia due to internal mass redistributions as the mantle convects, and is required to conserve angular momentum. The well known true polar wander model of Besse & Courtillot (2002), shown in Figure 14.13, suggests secular motion of the Earth's rotation axis of the order of 10 degrees over the past 100 Myr.

The fact that mantle circulation varies in time means that convectively generated *dynamic topography* is a transient feature. While the existence of dynamic topography as a result of vertical stresses associated with mantle convection is commonly accepted, its magnitude is poorly understood, being overshadowed entirely by the first-order topographic signal of the continent–ocean contrast of approximately 4 km. Buoyant mantle upwellings, such as hot spots, produce

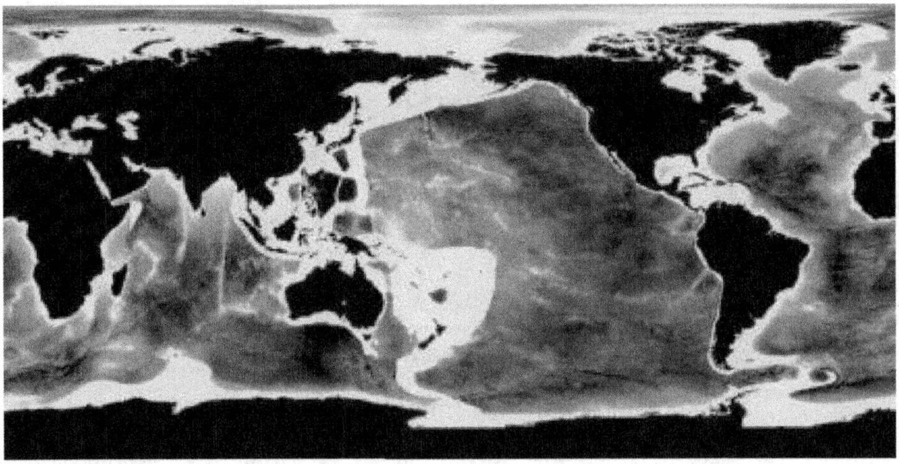

Figure 14.14. Residual topography beneath the Earth's oceans with the removal of the effect of features due to the lithosphere; the variations in topography are of the order of 500 m. [Courtesy of M. Sandiford.]

topographic highs, while downwellings go along with depressions of the surface. We can estimate the range of dynamic topography once the dominant shallow, isostatic mass anomalies in the crust and lithosphere (i.e., the thickness variations in the crust and lithosphere, sediment loading and age related thermal subsidence) are removed from the observed topography. A map obtained this way for the oceans is shown in Figure 14.14, revealing that large areas in the western Pacific are anomalously shallow for their age. Much of this area lies above mantle regions with rather slow seismic wavespeeds (see Figure 11.23) that are presumably buoyant. Furthermore the western Pacific has been strongly affected by hot spot activity. Shallow ocean floor in this region is thus not entirely surprising, and should not be taken as evidence against the half-space cooling model presented in Section 12.2.1.

The dynamic topography of the sea floor along the global mid-ocean ridge system is particularly informative. Since ridges by definition represent ocean floor of the same (zero) age, their depth below sea level should be the same. Yet prominent variations in the height of ridge crests occur, for example, at Iceland (above sea level) and south of Australia. The pronounced Australia–Antarctic discordance coincides with mantle downflow over an old sinking slab, and suggests a signal from dynamic topography of the order of several hundred meters. On a regional scale Lithgow-Bertelloni & Silver (1998) confirm this inference for the African plate, showing that the anomalously high stand of southern Africa (by a kilometre or so) correlates with lower mantle heterogeneity and the geoid. Husson (2006) provides a discussion of the timing and magnitude of dynamic topography over retreating subduction zones. The presence of dynamic topography means that the various continents should differ in their relative sea level curves as recorded in sediments on continental platforms, a view that is still somewhat alien to the

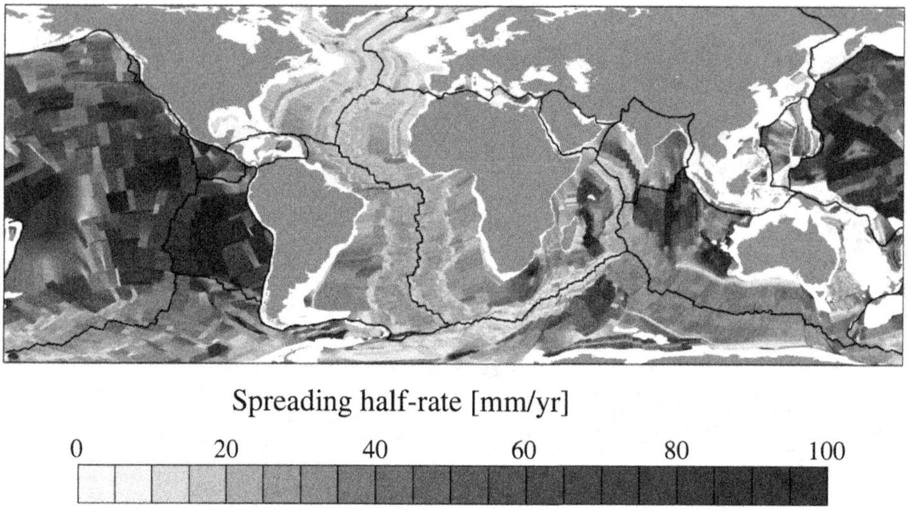

Figure 14.15. Ocean half spreading rates for the last 100 Myr.

conventional sedimentological perspective that such changes reflect only the global eustatic signal of the oceans.

Plate motions change on a variety of time scales. While long-term variations on the order of 10–100 Myr are clearly driven by mantle-related processes, there are quite rapid plate movement changes superimposed on this which occur over periods of 10 Myr or less. A well known feature is the Pacific plate motion change recorded in the Hawaiian–Emperor bend, which is commonly believed to reflect the reorientation of the Pacific plate to more westward motion around 40 Myr ago.

A recent compilation of oceanic spreading half rates over the past 100 Myr is shown in Figure 14.15, which documents frequent short-term velocity variations for all major plates during the Mesozoic and Cenozoic. An important new source of information on plate motion has come from geodetic plate motion models, and a detailed comparison of such models with those derived from palaeomagnetic reconstructions shows that while the rates and direction of plate movement averaged over several years agree well with those deduced from averages over millions of years, there are also small but important differences. For example, present-day Nazca–South America motion from the REVEL model (Sella et al., 2002) derived from geodetic measurements is slower than predicted by the 3 million year average in NUVEL-1A (De Mets et al., 1994), as shown in Figure 14.16; the light arrows from REVEL are somewhat shorter than the dark arrows representing NUVEL-1A. Such rapid variations cannot arise from changes in convective processes in the mantle, which operate on a longer time scale of the order of 100–150 Myr. One must thus appeal to shallow forces in the lithosphere operating along plate boundaries. Examples include the initiation or termination of subduction zones, the loss of a ridge due to subduction, variations in plate coupling along subduction zones due to variations in sediment thickness, or the rise of high topography associated with mountain building along convergent margins. For example, Iaffaldano et al. (2006) explain the rapid reduction of plate convergence between the Nazca and South American plates as associated with the recent uplift of the Andes. Short-term variations in plate boundary forces especially along convergent margins may therefore play an important role in modulating the short-term evolution of plate motion.

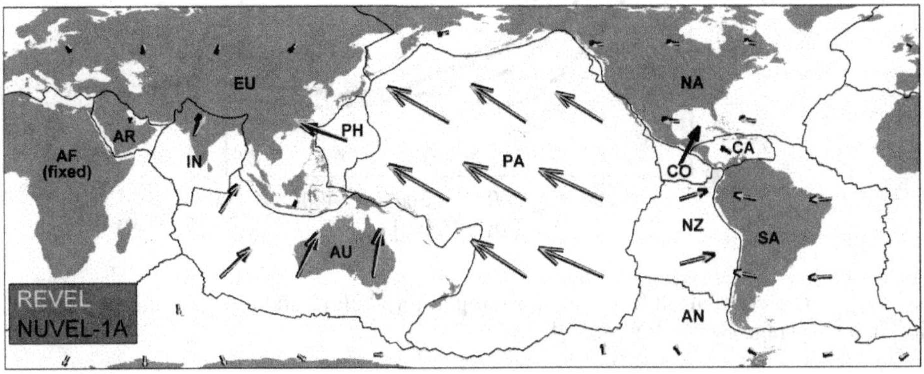

Figure 14.16. Comparison of the plate motion patterns for the NUVEL-1A model of DeMets et al. (1994) shown by dark arrows, with the REVEL model of Sella et al. (2002) derived just from recent space geodetic methods shown with light arrows.

14.4.3 Mantle circulation models

We have seen that a careful analysis of past plate motion models provides important information about structure and dynamics of the mantle. One can exploit this insight in a formal way through mantle circulation modelling. The circulation models solve the Navier–Stokes equations in the Stokes limit (14.1.19), that is they tie past plate movement directly to the history of deep mantle flow through the elliptic nature of the momentum balance. The circulation models are also required to satisfy the energy equation (14.1.11). This combination of flow and temperature requirements gives rise to the development of thermal boundary layers at the surface and the core–mantle boundary. Thus the temperature heterogeneity of circulation models is also tied to past plate motion, primarily through the location of cold downwellings along current and past subduction zones.

In mantle circulation modelling one faces a fundamental hurdle: namely it is impossible to know the initial state from which to start the circulation sometime in the past. Missing initial condition information precludes a deterministic approach to modelling the history of mantle flow. The *initial condition problem* is shared with circulation models of the ocean and the atmosphere. The difficulty reflects the fact that typically there are far more degrees of freedom in the model than available

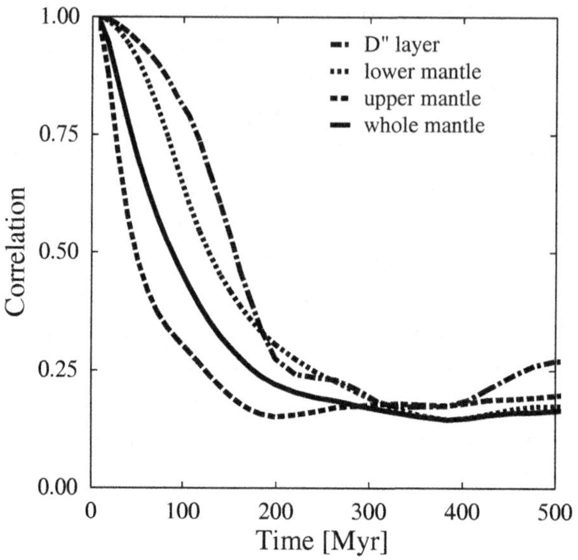

Figure 14.17. The correlation functions for temperature before and after a change of plate motion in a two-stage convection model.

observations. Ocean and atmospheric modellers overcome this shortcoming by a method known as *data assimilation*; through a variety of techniques computed values from the model are replaced by observations whenever possible. In this way the flow and dynamics in a model can be adapted to the observations at hand provided the assimilation period is long enough.

Plate motion histories take the form of data in a mantle circulation model. The plate motions enter the calculation as a set of time-dependent boundary conditions on the surface velocity field, which is required to match plate geometries and velocities at any given time. In other words, when the models are integrated forward in time from some assumed initial state, the computed surface velocities are replaced by velocities corresponding to the plate motion at a particular time in the past. As a result mantle circulation models assimilate past plate motions, and we need to ensure that plate motion histories are imposed for a sufficiently long time that the *memory* of initial conditions is lost.

Figure 14.17, based on Bunge et al. (1998), illustrates how initial condition information is lost in a mantle circulation model. The results are taken from a whole mantle convection model similar to the one shown in Figure 14.4. The scenario includes compressible flow, with a constant-temperature core supplying about 20% of the mantle heat flux (the remainder coming from internal heating), and a factor of 40 increase in mantle viscosity through the transition zone. A regime of mid-Cretaceous plate motions is imposed at first, and when quasi-steady-state is reached a sudden switch is made to present-day plate motion. The adjustment time between the two plate motion regimes can be monitored through the global correlation between model temperatures before and after the plate motion change. The four curves in Figure 14.17 show the correlations of the temperature patterns for the entire mantle, the upper mantle, the lower mantle, and the core–mantle boundary (D″) region; all correlations begin at unity. The whole mantle curve levels off at a correlation of about 0.2 after an elapsed time of about 175 Myr. The other curves behave similarly, with the most rapid adjustment occurring, not surprisingly, in the upper mantle, where plate-related flow is most directly felt, and the slowest in the D″ layer. The results of Figure 14.17 suggest the following conclusions:

(a) the adjustment time is comparable to the vertical transit time of these models, which is controlled by the radial viscosity structure and is of the order of 150–200 Myr.
(b) the adjustment time is also comparable to the time over which global plate reconstructions are feasible (about 150 Myr), which is no coincidence: old seafloor represents, roughly, the vertical transit time in the mantle, and a faster mantle overturn would merely serve to lower the mean age of the lithosphere.
(c) the imposition of plate motion models for the past 150 Myr is not sufficient to overcome initial condition information in the deeper mantle and near the core–mantle boundary, where the memory to earlier flow regimes is retained for closer to 200 Myr. For the mantle this observation is supported by seismic models showing slabs of early Jurassic or late Permian age, but not older, buried beneath Siberia (e.g., van der Voo et al., 1999).

Heterogeneity pattern

Seismic tomography provides us with increasingly clear images of mantle heterogeneity, which can, in principle, be compared with heterogeneity structure predicted from circulation modelling. Figure 14.18 shows thermal heterogeneity in two standard circulation models of whole mantle flow from Bunge et al. (2002). The only difference between the models lies in their amount of bottom heating, which is kept at less than 1% of the global surface heat flux in model one (at the top), while model two (below) has 35% core heating reflecting the uncertainty in the range of core heat flux that we have discussed before.

We represent compressible flow through the anelastic approximation, with a thermodynamic background configuration based on a Murnaghan equation of state. The reference density increases radially from 3500 $\mathrm{kg\,m^{-3}}$ at the surface to 5568 $\mathrm{kg\,m^{-3}}$ at the core–mantle boundary, in compliance with the Preliminary Reference Earth Model (PREM). The thermal expansivity decreases from a surface value of $4.0\times10^{-5}\ \mathrm{K^{-1}}$ to $1.2\times10^{-5}\ \mathrm{K^{-1}}$ at the core–mantle boundary. The heat capacity is kept constant at $1.1\times10^{3}\ \mathrm{J\,kg^{-1}K^{-1}}$, and thermal conductivity at 6.0 $\mathrm{W\,m^{-1}K^{-1}}$. Internal heating at a rate of $6.0\times10^{-12}\ \mathrm{W\,kg^{-1}}$ is applied, comparable to a value suggested by chondrite meteorites. The upper mantle viscosity of 8.0×10^{21} Pa s exceeds the value inferred from post-glacial rebound by about an order of magnitude. We must make this choice in order to reduce the thermal Rayleigh number and thereby lower the computational burden. The viscosity increases from

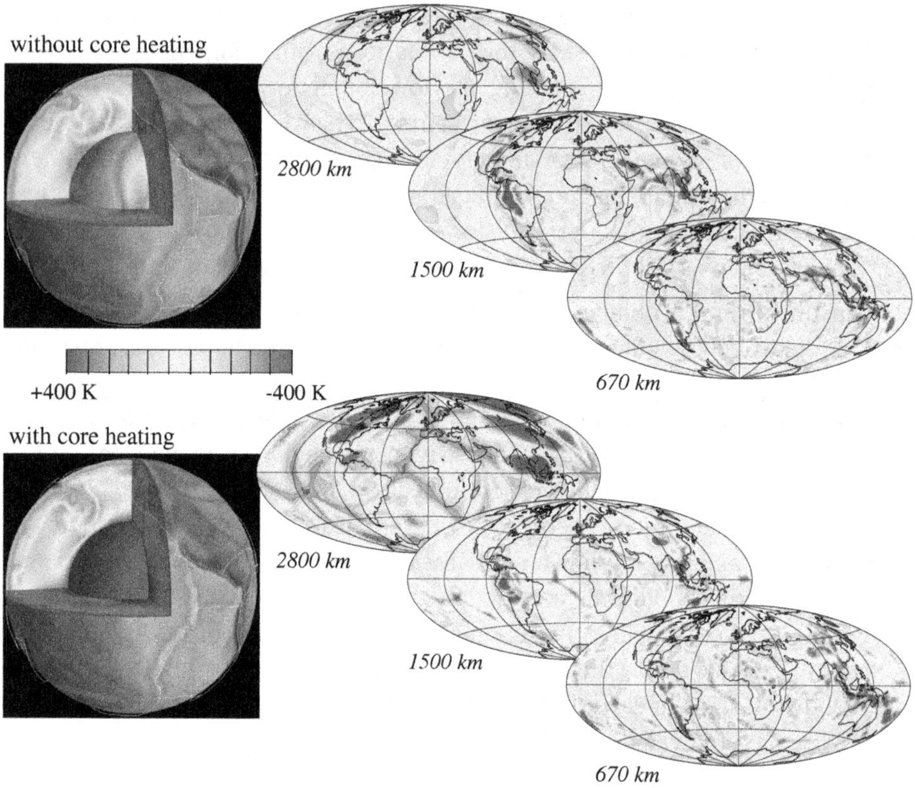

Figure 14.18. Mantle circulation models calculated with and without core heating shown as cross-sections at 2800 km, 1500 km and 670 km depth, as well as in a spherical view.

the upper to the lower mantle by a factor of 40. A mechanically strong lithosphere results from raising the viscosity in the upper 100 km of the model by a factor of 100 relative to the upper mantle. The convective vigour measured by the Rayleigh number for internal heating $\mathrm{Ra_H}$ is 10^8 based on the upper mantle viscosity.

We have seen before that one of the most uncertain assumptions in these models is the initial state. We adopt the following choice: we approximate the unknown mid-Cretaceous mantle heterogeneity as a quasi-steady-state of mantle flow with plate motions corresponding to the mid-Cretaceous (119 Myr ago), that is we apply mid-Cretaceous plate motion for all earth history (4.5 billion years) prior to that time. We account for the reduced convective vigour in these models by scaling the root mean square (RMS) plate velocities such that convective motion is neither increased nor reduced by the assimilated velocities, i.e., we scale the RMS plate velocities to match the RMS surface velocity of convection with a free slip surface. In this way the model Péclet number remains unchanged.

There is a dominant heterogeneity in the circulation models associated with past subduction (Figure 14.18). In the transition zone (670 km depth slice) cold downwellings beneath North and South America correspond to subduction of the

Farallon and Nazca plates. Likewise the cold anomalies under India and the Western Pacific arise from Cenozoic subduction of the Indian and Pacific plates beneath Eurasia. The model without strong core heating lacks hot active mantle upwellings, and the temperatures are nearly uniform away from the downwellings, whereas the core heated model (35%) shows additional hot upwellings, or plumes. These are located under the Pacific and the Atlantic. The heterogeneity otherwise is similar to the internally heated case.

The picture of slab-dominated mantle heterogeneity does not differ for the cross-section much deeper in the mantle (1500 km depth slice). Cold anomalies persist under eastern North America, the Caribbean, the northernmost part of South America, and the Alpine–Himalayan mountain belt due to past subduction. There are also two strong hot thermal anomalies in the core heated model in the mid-Atlantic and the Eastern Pacific region.

A general observation we can draw from Figure 14.18 is that the location of cold downwellings agrees reasonably well with fast seismic velocity anomalies imaged by tomography, while agreement in the location of hot upwellings is poor. The difference is particularly striking for the strong low seismic velocity anomaly imaged by many tomographic studies under Africa. Its counterpart in the core heated circulation model is located some 5000 km further west in the Mid and South Atlantic.

Heterogeneity near the core–mantle boundary in the two circulation models differs through the existence of a hot thermal boundary layer. The strength of thermal heterogeneity is low in the internally heated model, whilst there are prominent hot upwellings in the model with core heating. The heterogeneity pattern of the models at the core–mantle boundary does not resemble seismic images from the region just above the core–mantle boundary. Rather it is an artefact of the model initialisation, made evident upon comparison with the 120 Ma mid-Cretaceous plate reconstruction (Figure 14.19b), which we used to initialise the model. We note that heterogeneity at the core–mantle boundary corresponds closely to the location of spreading and subduction margins. In other words, the deep mantle and core-mantle boundary heterogeneity in a circulation model is entirely controlled by the assumed initial condition. The artificial nature of heterogeneity near the core–mantle boundary is also evident from the recent circulation model by McNamara & Zhong (2005) which resembles the heterogeneity structure in Figure 14.18 closely.

Adjoint mantle circulation models

We have seen that it is impossible to overcome the initial condition problem with circulation models running forward in time from some unknown state in the past. However, the general character of the mantle convection equations suggests a different approach to get around the problem. There is an absence of inertial effects in the momentum balance of the mantle, which means that the equation of motion is instantaneous and does not involve time. Time enters mantle convection only through the energy equation (14.1.11), where irreversible processes are associated with heat diffusion. We recall that heat diffusion in the mantle is small, and that it can be neglected outside of thermal boundary layers. It is then feasible to reverse time in the energy equation provided we assume zero heat diffusion. In that case one may start from a model of present mantle heterogeneity and step convection backward in time.

There is an error associated with ignoring thermal diffusion effects, and we may quantify this error by estimating the length scales of diffusive and advective mantle heat transport. Thermal disturbances diffuse in the mantle by about 100 km - the thickness of the thermal boundary layer - over 100 Myr. Over the same time the disturbances travel by advection over distances exceeding 1000 - 5000 km. The error we make then in ignoring thermal diffusion is reasonably small. In practice one will not want to set the thermal diffusion term to exactly zero in the time-reversed energy equation. Rather, for reasons of numerical stability one will choose a very small negative diffusion coefficient, so that heat diffuses still from hot to cold regions when we step back in time.

The calculations of Steinberger & O'Connell (1997) illustrate the capabilities of this approach. They consider the advection of a mantle density field back in time; the density variations are inferred from seismic tomography as discussed in Section 11.4. The changes of large-scale mantle density obtained in this way predict temporal variations in Earth's rotation axis which are in reasonable agreement with paleomagnetically derived inferences of true polar wander for the past 60 Myrs. There are other geologic observables, such as dynamic topography and relative hot spot motion that one can infer as well, and Steinberger and colleagues have done so in subsequent work.

There is a formal way to infer mantle flow back in time that requires the formulation of an inverse problem. We seek optimal initial conditions that minimise, in a weighted least squares sense, the difference between what a mantle convection model predicts as heterogeneity structure in the mantle and the heterogeneity that is actually inferred from, say tomography. In general, this class of problems is known as *history matching*, for example in hydrology. It is also often referred to as *variational data-assimilation*, for example in meteorology, meaning that model parameters are inferred from a variational principle.

In mantle circulation modelling the images of seismic tomography take on the form of data and the assimilation is posed as an inverse problem with the initial conditions as the variables to be determined. The misfit between current day structure and the predictions for specific initial conditions is specified through a cost function F. The necessary condition for a minimum of F, that the variation $\boldsymbol{\nabla}\mathsf{F} = 0$, produces a set of equations which involve the usual mantle convection equations coupled to the corresponding *adjoint equations*. The adjoint equations are nearly identical to the forward model except for forcing terms and a change of sign from positive to negative in the diffusion operator. Reversing the sign of the

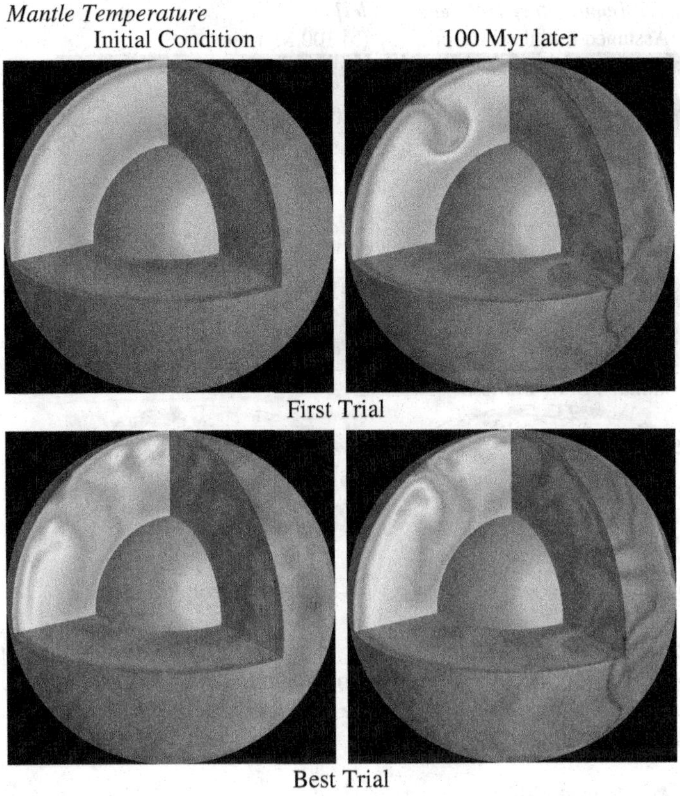

Figure 14.19. The refinement of the estimate of the initial conditions in the mantle, 100 Myr ago using the adjoint method. The first trial is a simple stratified model, but after 100 iterations a complex initial state is achieved.

diffusion operator makes the adjoint equations unconditionally stable to integration backwards in time.

There are a number of ways that can be used to derive adjoint equations for the initial condition problem. Here we illustrate the basic approach to which further complexities can be added (Bunge et al., 2003). We start with the definition of the cost function F to measure the misfit between predicted and actually observed heterogeneity for a specific initial condition. We take the general form

$$\mathsf{F} = \mathsf{F}_{\mathrm{I}} + \mathsf{F}_{eq}, \tag{14.4.2}$$

where F_{I} depends on the initial temperature distribution $\mathrm{T}_{\mathrm{I}}(\mathbf{x})$, and F_{eq} allows for contributions from the model equations used to map the initial conditions to a final state that can be compared with tomography. The component F_{I} is given by

$$\mathsf{F}_{\mathrm{I}} = \int_V d^3\mathbf{x} \int_V d^3\mathbf{x}'\, j(\mathbf{x}) W_j(\mathbf{x}, \mathbf{x}') j(\mathbf{x}'), \tag{14.4.3}$$

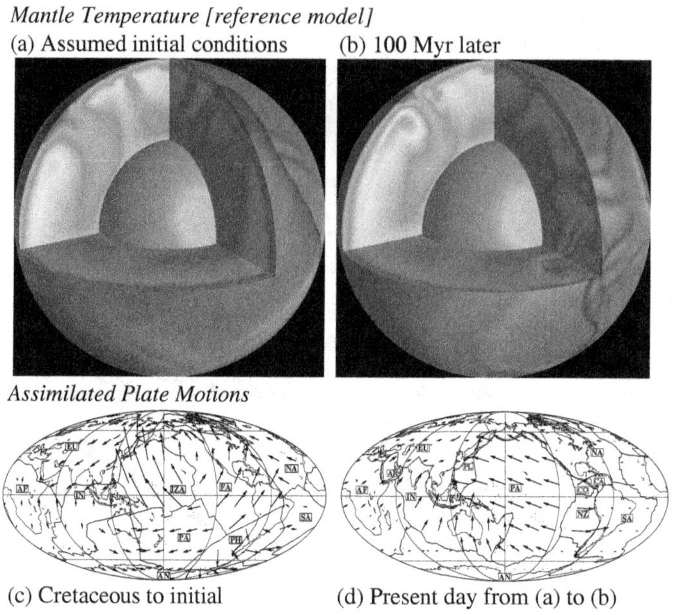

Figure 14.20. Reference model calculated using data assimilation, the initial state (a) was created by imposing the Cretaceous plate motion shown in (c) up to 100 Myr ago and then switching to the modern plate motion (d) to give a final convective pattern (b) that is much more like the modern Earth.

with a dependence on the initial temperature distribution through the residual $j(\mathbf{x})$,

$$j(\mathbf{x}) = T(\mathbf{x}, t_0) - T_I(\mathbf{x}), \tag{14.4.4}$$

i.e., the difference between the *trial* initial condition for temperature $T_I(\mathbf{x})$ and the optimal condition $T(\mathbf{x}, t_0)$, integrated over the model space V with appropriate weighting functions W_j.

There are two components to the model error F_{eq}. Such errors can arise from the numerical inaccuracies of a convection code, i.e., the computational approximations imposed in solutions of the Navier–Stokes equations, or can reflect errors in the model equations themselves. For instance, the heat source term h in the energy equation is often taken as constant in mantle circulation modelling, although a full treatment requires time-dependence $h(t)$. Model errors can be included collectively as non-zero terms in the convection equations, and we define residuals for divergence (D), momentum ($\mathbf{M}$) and temperature (Θ). For an incompressible flow, the convection equations take the form

$$\boldsymbol{\nabla} \cdot \mathbf{v} = D(\mathbf{x}, t), \tag{14.4.5}$$

$$\boldsymbol{\nabla}(\eta \boldsymbol{\nabla} \mathbf{v}) + Ra\, \Delta\theta\, \hat{\mathbf{e}}_r - \boldsymbol{\nabla} p = \mathbf{M}(\mathbf{x}, t), \tag{14.4.6}$$

$$\frac{\partial \theta}{\partial t} + \mathbf{v} \cdot \boldsymbol{\nabla} \theta - \nabla^2 \theta - h = \Theta(\mathbf{x}, t). \tag{14.4.7}$$

We have here adopted a non-dimensionalisation based on the thermal diffusion time, so that the momentum equation involves the Rayleigh number Ra and the contrasts in non-dimensional temperature $\Delta\theta$.

The model error functional F_{eq} is constructed as a weighted integral of the squared model residuals over the volume V and the entire time interval that separates the initial condition and final state, the assimilation period I:

$$\begin{aligned} F_{eq} = \int_I dt \int_V d^3\mathbf{x} \int_I dt' \int_V d^3\mathbf{x}' \Big\{ & D(\mathbf{x},t) W_D(\mathbf{x},t,\mathbf{x}',t') D(\mathbf{x}',t') \\ & + \mathbf{M}^T(\mathbf{x},t) W_{\mathbf{M}}(\mathbf{x},t,\mathbf{x}',t') \mathbf{M}(\mathbf{x}',t') \\ & + \Theta(\mathbf{x},t) W_{\Theta}(\mathbf{x},t,\mathbf{x}',t') \Theta(\mathbf{x}',t') \Big\}. \end{aligned} \qquad (14.4.8)$$

Considerable simplification of the cost function F_{eq} can be achieved by introduction of *adjoint variables*, $\chi(\mathbf{x},t)$, $\boldsymbol{\phi}(\mathbf{x},t)$ and $\tau(\mathbf{x},t)$ as weighted integrals over the model errors:

$$\chi(\mathbf{x},t) = \int_I dt' \int_V d^3\mathbf{x}' \, W_D(\mathbf{x},t,\mathbf{x}',t') D(\mathbf{x}',t'), \qquad (14.4.9)$$

$$\boldsymbol{\phi}(\mathbf{x},t) = \int_I dt' \int_V d^3\mathbf{x}' \, W_{\mathbf{M}}(\mathbf{x},t,\mathbf{x}',t') \mathbf{M}(\mathbf{x}',t'), \qquad (14.4.10)$$

$$\tau(\mathbf{x},t) = \int_I dt' \int_V d^3\mathbf{x}' \, W_{\Theta}(\mathbf{x},t,\mathbf{x}',t') \Theta(\mathbf{x}',t'). \qquad (14.4.11)$$

The adjoint variables have no particular physical meaning, although one sometimes refers to them loosely as adjoint pressure $\chi(\mathbf{x},t)$, velocity $\boldsymbol{\phi}(\mathbf{x},t)$ and temperature $\tau(\mathbf{x},t)$ by analogy with the familar forward variables.

In terms of the adjoint variables the full cost function F takes the form

$$\begin{aligned} F = \int_I dt \int_V d^3\mathbf{x} \Big\{ & [\nabla \cdot \mathbf{v}] \chi(\mathbf{x},t) + [\nabla(\eta \nabla \mathbf{v}) + \mathrm{Ra}\,(\bar{\theta} - \theta)\,\hat{\mathbf{e}}_r - \nabla p]^T \boldsymbol{\phi}(\mathbf{x},t) \\ & + [\partial_t \theta + \mathbf{v} \cdot \nabla\theta - \nabla^2\theta - h]\, \tau(\mathbf{x},t) \Big\} \\ + \int_V d^3\mathbf{x}\, [\theta(\mathbf{x},t_0) - \theta_I(\mathbf{x})] & \int_V d^3\mathbf{x}' \, W_j(\mathbf{x},\mathbf{x}') [\theta(\mathbf{x},t_0) - \theta_I(\mathbf{x})], \end{aligned} \qquad (14.4.12)$$

where $\bar{\theta}$ is the mean non-dimensional temperature.

A sure approach to finding the first variation of F is through a classical perturbation procedure. We introduce a small scalar quantity ϵ and consider the functions $\theta(\mathbf{x},t) + \epsilon\theta$ and $\mathbf{v}(\mathbf{x},t) + \epsilon\mathbf{v}$ as arguments for F, where $\epsilon\theta$ and $\epsilon\mathbf{v}$ satisfy all required initial and boundary conditions. Taking the derivative of F with respect to ϵ and then looking at the special case $\epsilon = 0$ we arrive at the first variation of F:

$$\begin{aligned} \delta F = \int_I dt \int_V d^3\mathbf{x} \Big\{ & [\nabla \cdot \delta\mathbf{v}] \chi(\mathbf{x},t) + [\nabla(\eta \nabla \mathbf{v}) + \mathrm{Ra}\,\delta\theta\,\hat{\mathbf{e}}_r - \nabla \delta p]^T \boldsymbol{\phi}(\mathbf{x},t) \\ & + [\partial_t(\delta\theta) + \delta\mathbf{v} \cdot \nabla\theta + \mathbf{v} \cdot \nabla\delta\theta - \nabla^2\delta\theta]\, \tau(\mathbf{x},t) \Big\} \\ + \int_V d^3\mathbf{x}\, \delta\theta(\mathbf{x},t_0) & \int_V d^3\mathbf{x}' \, W_j(\mathbf{x},\mathbf{x}') [\theta(\mathbf{x},t_0) - \theta_I(\mathbf{x})]. \end{aligned} \qquad (14.4.13)$$

We take an important step in the derivation of the adjoint equations when we carry out integration by parts in both space and time on the elements of δF in (14.4.12). Our aim is to isolate the terms involving $\delta \mathbf{v}$, $\delta\theta$ and then rearrange the differential operators to act upon the adjoint variables.

A necessary condition of the misfit function F is that the first variation δF be zero at the extremum. This criterion requires that the forward and adjoint variables satisfy a set of adjoint equations and boundary conditions coupled to the forward model through the model error terms (Bunge et al., 2003):

$$\nabla \cdot \boldsymbol{\phi} = 0, \tag{14.4.14}$$

$$\nabla \cdot (\eta \nabla \boldsymbol{\phi}) + \tau \nabla \theta = \nabla \chi, \tag{14.4.15}$$

$$-\frac{\partial \tau}{\partial t} - \nabla \cdot (\tau \mathbf{v}) + Ra\, \hat{\mathbf{e}}_r \cdot \boldsymbol{\phi} = \nabla^2 \tau + \delta(\mathbf{x}, t - t_1)[\theta(\mathbf{x}, t_0) - \theta_I(\mathbf{x})], \tag{14.4.16}$$

where $\delta(\mathbf{x})$ is the Dirac delta function, and $\mathbf{v}$, θ are the velocity and temperature fields that couple the adjoint equations to the forward model.

The diffusion term and the time derivative in the adjoint energy equation (14.4.16) have changed sign from positive to negative so the equation can be stepped backwards in time. Both of these changes arise from the action of the integration by parts on the elements of the misfit function F. The adjoint variables have to satisfy the following set of boundary conditions, for the full assimilation period I. At the Earth's surface S and the inner boundary ∂C,

$$\boldsymbol{\phi} = 0, \quad \mathbf{x} \in S, \qquad \frac{\partial \boldsymbol{\phi}}{\partial n} = 0, \quad \mathbf{x} \in \partial C, \tag{14.4.17}$$

since the plate velocities are specified at the surface. Over both boundaries (∂V) we require

$$\boldsymbol{\phi} \cdot \mathbf{n} = 0, \qquad \chi = 0, \qquad \tau = 0. \tag{14.4.18}$$

There is also a final time condition for the adjoint system,

$$\tau(\mathbf{x}, t_1) = 0. \tag{14.4.19}$$

The adjoint equations are integrated backwards in time from t_1 to t_0 from the final time condition (14.4.19) on the adjoint variable τ, which has the role of coupling the adjoint and forward sets of equations.

Coupling between the forward equations and the adjoint equations can be ignored if we can assume that our mantle convection model captures all relevant physical processes, and the numerical error is insignificant. This simplification, which makes the computation far more tractable, amounts to taking the limit as the relevant model weights in the cost function become infinite. It is equivalent to taking the model as a constraint in the minimisation of F.

The physical significance of the adjoint equation is readily apparent. Instead of advecting mantle density heterogeneity directly back in time, the adjoint equations advect heterogeneity differences, that is the misfit between predicted and current heterogeneity, into the past. Advection follows the flowlines of the forward calculation, so that adjoint modelling is not unlike one style of tomographic inverse problem in seismology, where travel time residuals are projected back along the ray path. Numerically, adjoint modelling takes the form of an iterative procedure. One starts by solving the forward equations from a first trial for the initial condition. In the next step the misfit between predicted and observed heterogeneity is carried back in time through the adjoint equations along the lines of flow computed in the forward run. The trial initial condition is then updated by adding the heterogeneity residual, and the procedure repeats itself. As is customary in inverse problems, one must consider issues of suitable damping and regularisation. The close correspondence of forward and adjoint equations makes it possible for essentially the same computer code to be used to solve the forward and the adjoint system.

Figure 14.19 represents a modelling experiment to test the adjoint procedure. The figure shows the progress of the iterative optimisation scheme through illustrations of the *first-trial* initial condition state of the temperature field and that achieved after 100 iterations. The initial condition assumption is that of a simple stratified mantle which nevertheless evolves over 100 Myr to produce a more complex structure. Through repeated refinements of the initial condition temperature distribution by the adjoint method we arrive at a *best-trial* estimate of the initial condition which may be compared with the correct initial condition for this case in Figure 14.20.

14.5 Mantle rheology

In Chapter 9 we encountered the wide range of mechanisms that control the mechanical properties of silicate minerals through, e.g., diffusion and dislocation creep. Each of these mechanisms has an exponential dependence on temperature through the Arrhenius factor (9.3.3)–(9.3.5), so that apparent viscosity can alter by many orders of magnitude for a few hundred degrees temperature change. The inverse dependence on temperature as T^{-1} in the Arrhenius exponent for all the different mechanisms renders viscosity most sensitive to temperature fluctuations at lower temperatures, so that the cold, upper thermal boundary layer of convection can acquire very high viscosity, or strength, akin to plates.

The viscosity of the lithosphere is around 10^{25} Pa s and may drop to as low as 10^{17} Pa s in the asthenosphere before recovering to 10^{21} Pa s in the lower part of the upper mantle (see Chapter 13). The viscosity may change by as much as 7 orders of magnitude in the top 200 km of the mantle. The low viscosity of the asthenosphere has a significant effect on the patterns of mantle flow (as discussed in Section 13.1.5), and encourages the formation of convection cells with large lateral extent compared with their depth.

Plate tectonics is associated with strong lateral reductions in effective viscosity that concentrate deformation into narrow regions, referred to as zones of shear localisation. These regions cannot arise within linear rheologies such as elasticity, viscoelasticity or Newtonian viscous flow, where an increase in deformation (or rate of deformation) goes along with greater resistance to deformation rather than a loss of strength. Plate-like behaviour in convection models therefore requires nonlinear rheologies and feedback mechanisms, so that loss of strength is controlled by processes that are themselves a function of deformation.

14.5.1 Temperature dependence

We start our consideration of the influence of rheology with the effects of temperature on viscosity. The main role of temperature dependent viscosity when put into mantle convection models is to stiffen the upper thermal boundary layer and to make it stronger than the mantle beneath. A straightforward application of laboratory values for the activation energy (Karato & Wu, 1993) yields viscosities sufficiently large to inhibit motion of the underlying fluid, which convects almost as if it were in a state of isoviscous convection with an imposed no-slip top boundary condition.

At the same time there is a strong asymmetry in the horizontally averaged temperature profile between the top and bottom of the fluid, in response to the inherent strength and associated large temperature drop across the cold upper boundary layer. The interior temperature of the fluid layer is strongly hotter than the average temperature of the top surface. The large temperature contrast across this surface facilitates small-scale convection, so that the flow removes heat from beneath the boundary layer with no need for the entire boundary layer to return into the mantle. The resultant stagnant lid regime provides a good description for the tectonic style of Mars, Mercury, the Moon and perhaps Venus, where the entire planetary surface is made up from a single plate. However, this convective style is a poor description of plate tectonics on Earth.

Tozer (1972) identified an additional role of temperature by considering the thermal evolution of terrestrial planets. If viscosity were too high for convection to occur, then a planet would heat up until the viscosity reduced sufficiently for convection to begin. All larger terrestrial bodies in the solar system msut therefore be in a state of convection due to the strong temperature dependence of the silicates making up their mantles, and their internal temperatures must be in a state of *self-regulation.* The long-term evolution of the plate-mantle system must therefore take account of the sensitivity of heat flow to internal temperature via viscosity variability. Thus in the early Earth, when temperatures were higher, the vigour of convection can be expected to have been higher, and the circulation time for a particle through the mantle somewhat shorter than at present.

Temperature-dependent viscosity provides a feedback mechanism capable of producing shear localisation, from the coupling of viscous heating and viscosity. The feedback works in such a way that shear zones undergo dissipative heating and associated weakening of the material. The weak zones then focus deformation, leading to a further rise in temperature and weakening of the material. Although simple, this mechanism is unlikely to account for the generation of plate tectonics, partly because thermal diffusion is too efficient to maintain narrow hot zones over extended times, and also because observations suggest that the majority of lithospheric faults are characterised by rather low heat flow.

14.5.2 Strain dependence

A more effective feedback between rheology and flow than that of temperature-dependent viscosity involves the non-linear dependence of stress upon strain rate. In Chapter 9 we encountered the general relation between stress and strain rate for silicate minerals in (9.3.3), and this power-law rheology is one of the simplest that produces shear localisation. In this rheology the strain rate is proportional to stress raised to some power $n > 1$. The resultant non-Newtonian flow behaviour is typical for silicate polycrystals at high temperature and low stress (with T approaching the melting temperature and σ in the range of 10–100 MPa), and it is likely that much of the upper mantle deforms by dislocation creep with experimentally determined creep laws having power-law exponents in the range 2–5.

The feedback mechanism of power-law rheology works such that rapidly deforming regions soften, which facilitates further deformation in such zones. From (9.3.5) we recall the apparent viscosity at constant strain rate, which we write here as:

$$\eta_{\dot{\epsilon}} = \frac{\sigma(\dot{\epsilon})}{\dot{\epsilon}} \propto \dot{\epsilon}^{\frac{1}{n}-1}. \qquad (14.5.1)$$

Thus in the limit of large n the effective viscosity scales inversely with strain rate. This class of non-Newtonian rheology has been implemented into numerous two-dimensional convection models (commonly with $n = 3$), but it produces a poor plate-like behaviour even when one takes large values ($n > 7$) in the exponent of the power law.

A considerable improvement in shear localisation is obtained if we invoke an exotic rheology by setting the power law exponent to a value of $n = -1$. The effective viscosity then scales inversely with the square of the strain rate. This rheology, which is not observed in the laboratory and lacks a clear physical interpretation, is known variably as *pseudo-stick-slip*, *self-lubrication* or *self-weakening*. Models with this rheology produce narrow regions of deformation and behaviour with a strong plate-like component, as one would expect. This type of rheology is difficult to implement numerically due to the strong reduction in local viscosity. Bercovici & Karato (2002) provide a good overview of this class of models.

We have seen that the coldest parts of the lithosphere are in the brittle regime in which deformation occurs through slip on faults. This behaviour can be represented in the continuum limit by modelling temperature-dependent viscosity combined with a plastic *yield stress*. The temperature dependence causes the cold upper boundary layer (lithosphere) to be strong, whilst the yielding allows the boundary

Figure 14.21. Three-dimensional spherical convection model with temperature dependent viscosity and yield stress; grid resolution is about 50 km. The contours are for the logarithm of viscosity, with the darkest hues corresponding to 10^{21} Pa s and mid-grey to 3×10^{23} Pa s. High-strain rate zones (lighter tones) separating plates show reduced viscosity due to plastic yielding. A zone of reduced viscosity by a factor of 1000 lies beneath the lithosphere from 140 to 400 km depth, and is visible as the nearly black zone. The yield stress is set at 140 MPa.

to fail locally in regions of high stress. With the use of this style of rheology Moresi & Solomatov (1998) identified three styles of convection:

(1) the *stagnant lid* regime, which we encountered before, in which the cold upper boundary later is immobile,

(2) the *mobile lid* regime at low yield stress in which ubiquitous plastic failure of the lithosphere yield a highly mobile upper boundary layer, and

(3) a transitional *sluggish lid* regime that switches episodically between the frozen and mobile regimes.

Regime 1 applies to Mars and the Moon, while regime 3 may describe the distributed deformation and episodic overturn of the lithosphere on Venus.

An example of this coupled temperature and yield-stress rheology is shown in Figure 14.21 from Richards et al. (2001). The rheology produces a reasonably good plate-like behaviour when used in conjunction with an imposed low-viscosity asthenosphere.

There is abundant evidence that the lithosphere is pervasively fractured and that faults play a fundamental role in controlling the strength of the lithosphere. This result lies at the heart of mantle convection models that incorporate mobile faults,

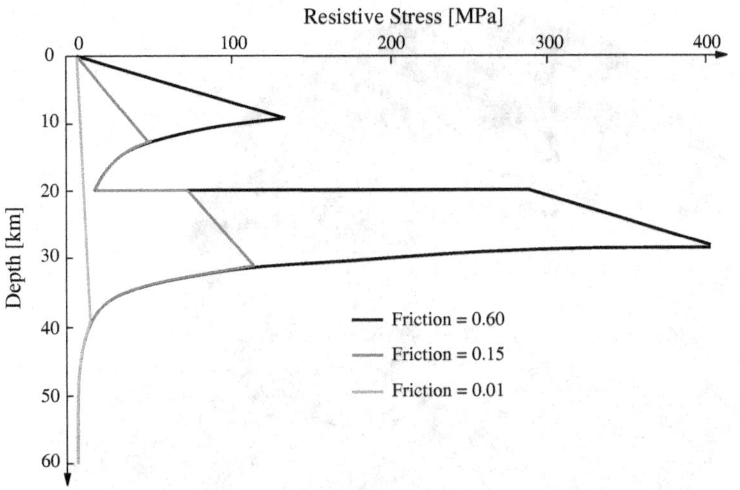

Figure 14.22. Yield strength envelopes for varying coefficients of friction corresponding to faulted and non-faulted material.

e.g., Zhong & Gurnis (1996). The strength envelope of the lithosphere is controlled by brittle failure down to considerable depth, especially if one assumes that faults are weak (Figure 14.22). This means that the coefficient of friction for fault failure is much lower than the value ($M_1 = 0.8$) that would be taken for Byerlee's law for unfaulted material (see Section 12.2.4). The reduced coefficient of friction agrees with observations of stress indicators and low heat flow that are characteristics of many major fault systems.

It is only with the use of such exotic, self-lubricating rheologies that a single continuum treatment can be used to generate realistic plate-like behaviour. However, we already know that there is a rheologically distinct lithosphere, and that plate boundaries represent features that must cut through the entire lithosphere zone with different plates on either side. Lithospheric rheology is sufficiently different from the rest of the mantle that, at our present state of understanding, it is appropriate to treat it separately.

14.6 Coupled lithosphere–mantle convection models

The distinction between the lithosphere and the rest of the mantle is largely rheological, as we have seen in Chapter 13. Although we can treat mantle circulation through a viscous flow, we need to take account of more complex properties and shorter length scales in the lithosphere. Some aspects of lithosphere behaviour, particularly in subduction, can be captured by a visco-plastic formulation, but a full description requires a combination of rheological processes whose computational requirements are rather different from those for the fluid mantle.

However, there is an effective way to link lithospheric modelling with mantle

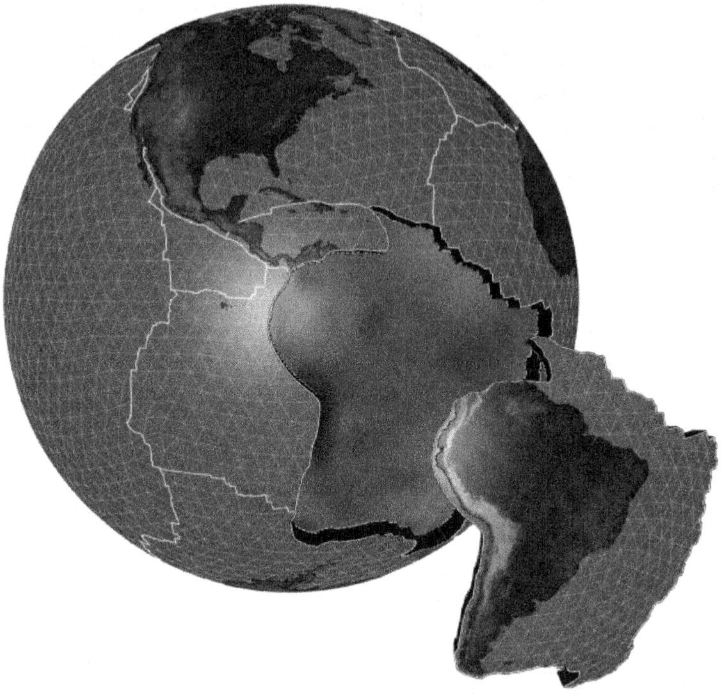

Figure 14.23. Lithospheric models can be linked to mantle convection through the imposition of suitable driving conditions at plate boundaries and the base of the lithosphere.

circulation through the driving stresses at the base of the lithosphere (Figure 14.23). Iaffaldano et al. (2006) have shown how a global thin-plate representation of the lithosphere (SHELLS, Kong & Bird, 1995) can be coupled to a full mantle circulation model. The thin-plate approach reduces the lithospheric problem from three to two dimensions by employing the concepts of isostasy and vertical integration of lithospheric strength (cf. Section 12.6). In the SHELLS model the lithosphere is treated by a finite element method with a representation of rock strength employing frictional sliding on faults in the shallow brittle zone and dislocation creep in the ductile zone at depth. The major plate boundaries are introduced by faults, represented through finite element interfaces, with dips controlled by seismological results.

Iaffaldano et al. (2006) have considered the impact of raising the topography of the Andes on the global mantle flow. The extra load results in an increase in frictional forces on the plate boundary by about 1×10^{13} N m^{-1}, which is comparable to the shear stresses in the mantle. The increased elevation of the mountains leads to a slowing of the rate of convergence of the Nazca plate and South America. This result is consistent with the decline from a convergence rate around 100 yr^{-1} at 10 Ma, estimated from plate reconstructions (Gordon & Jurdy, 1986), to about 70 mm yr^{-1} in the geodetic model of Sella et al. (2002) as illustrated in Figure 14.16. The influence on mantle flow is significant, but largely confined to the zones beneath the Nazca and South America plates. These results indicate that the raising of topography has the potential of consuming a significant amount of the driving forces available from the mantle.

The work of Iaffaldano et al. (2006) combines two different styles of well-established modelling, and demonstrates the potential of a coupled lithosphere–mantle approach. The restriction to a uniform-thickness thin plate means that a common treatment is applied to both oceanic and continental environments and that subduction is taken up by flow in the mantle rather than direct feed from the lithosphere above. We can envisage more complex three-dimensional lithosphere models coming into play as computational resources improve. Such coupled calculations allow the separation of the main flow and represent an excellent test-bed for concept testing, without the need for the incorporation of fine-scale complex rheology in three-dimensional global mantle circulation models.

14.7 Thermochemical convection

In addition to the dynamical state of the mantle, we would like to know how chemical components are distributed in the mantle and the way in which this distribution has evolved.

A number of classes of questions can be posed that require somewhat different treatment. The simplest approach is the use of passive tracers in a computational convection scheme, so that the evolution of the mantle from some initial configuration can be tracked and mixing patterns determined. Such studies are consistent with geochemical estimates of a mantle residence time of 1–2 Gyr. The fate of a particular component, e.g., the former oceanic crust, can be followed with the use of tracers whose properties, such as density, are modified by the current thermal state and passage through phase transitions.

A similar concept can be used to associate tracers with the concentration of chemical species through local segregation and assimilation processes governed by appropriate partition coefficients. Most studies, as in the work of Tackley & Xie (2002), have used laboratory estimates of volumetric partition coefficients. The nature of mantle mixing means that surface melt extraction is likely to be more important than volumetric effects, and these would be described by somewhat different partition coefficients. In addition the partition coefficients can be influenced by conditions such as the oxidation state that are difficult to capture in a numerical code.

Over time most of the mantle is well stirred and initially contiguous volumes tend to be spun into thin ribbons. Although some pockets of material with high viscosity may be less well mixed, it is difficult to find conditions suitable for the preservation of large volumes of relatively undisturbed mantle as envisaged in the geochemical concept of a *primitive mantle reservoir* that has preserved the geochemical abundances for the different elements imposed at the time of the formation of the Earth.

A full treatment of thermochemical issues requires keeping track of both major element chemistry and the minor elements that are important for their geochemical signatures. Tackley & Xie (2002) employed an initial state with two components 'basalt' and 'harzburgite' with their own distinct densities. The 'basalt/eclogite' tracers, initially set at 25% of the total, are allowed to melt with a change in composition as the mantle convection system evolves. Once differentiation takes the composition of a tracer to the 'harzburgite' end-member no further melting is allowed.

The melting process needs to be assessed in each grid cell for each time-step by comparing the calculated temperature with a solidus temperature that depends on pressure and composition derived from experimental results. Sufficient melt is extracted from the tracer to return the temperature to the solidus. The tracer is then assigned the composition of the residuum. The melt component is immediately transferred to the surface, under the assumption that melt percolation through the mantle and eruption occur on time scales short compared with the time-step for the mantle convection run. The melt forms 'crust' with compaction of the material beneath. In the process of melting trace elements (e.g., Pb, U, He, Ar, K, Th) are partitioned between the melt and residuum, ensuring mass conservation. The noble gases are assumed to largely escape to the atmosphere from the melt crust.

In addition to assumptions about the initial state of the major elements, a starting distribution of trace elements has to be imposed (e.g., allowing for continental crust differentiation). The complex tracer system provides a useful way of investigating the dependence of a simplified geochemical system on different assumptions about the initial state. Many aspects of the results resemble the geochemical signatures determined from the limited available samples of melt products at the Earth's surface.

Because of the level of computational resources needed for such a thermochemical convection scheme, two-dimensional simulations have often been used, such as a cylindrical geometry approximating a cross-section through the Earth (e.g., Tackley & Xie, 2002).

A major complication in the treatment of thermochemical issues is the rather different scales associated with the main flow patterns and the filamentary structures that may carry specific thermochemical signatures. A similar scenario is encountered in oceanographic models where fine-scale structures, such as eddies and jets, are very important to detailed weather patterns, whilst the climate is determined by the main flow. New styles of oceanographic calculations are being developed where the representation of the model is adaptive to the flow, and so computational effort is concentrated in regions of rapid spatial variation. Similar concepts would be attractive for the thermochemical scenario, since they would allow a more direct representation of melt processes and melt transport without the need for parameterisation of sub-grid scale features.

15

The Core and the Earth's Dynamo

Passage from the Earth's mantle to the core marks a transition from a silicate mineralogy to a liquid metallic region dominated by iron. The principal evidence for composition comes from shock-wave equation of state measurements (Section 9.4), reinforced by studies of meteorites. Some lighter component than iron is required in the outer core to match seismological estimates of density. The density deficit is of the order of 10% compared with a liquid iron–nickel alloy. O'Neill & Palme (1998) provide a careful discussion of the merits of the main candidates for a light element component: oxygen, silicon and sulphur. No single element is able to produce the desired density variation and meet geochemical constraints, but a mixture of several elements (Si, O, S, C) might be appropriate. The seismological results are compatible with an adiabatic state through most of the outer core with a change of density of about 10% from the top of the core to the inner core boundary.

The outer core contains 30% of the mass of the Earth and behaves as a fluid on seismic time scales, preventing the passage of seismic shear waves. Slower motions in the liquid core are inferred to be the origin of the Earth's magnetic field through some class of dynamo action. The effective mixing through the convective motion in the core is responsible for the adiabatic density profile with radius. In contrast, the inner core appears to be solid, with a large density jump from the outer core to the inner core associated with a nearly pure iron composition. As mentioned in Chapter 1, the properties of the inner core are rather complex, with evidence for seismic anisotropy that is somewhat weaker near the surface. A significant difficulty is that uncertainty as to the behaviour of other parts of the Earth tends to be transferred via different observations to the properties of the inner core, which is only 5% of the total mass.

The arguments for a solid inner core come most forcibly from the frequencies of free oscillation of the Earth (Masters & Shearer, 1990). A number of efforts have been made to identify seismic phases with a shear leg (J) in the inner core. The most promising result from Duess et al. (2000) uses stacking methods and synthetic seismogram analysis to identify a likely SKJKP phase consistent with a mean shear wavespeed of 3.6 km/s.

The boundary between the inner and outer cores is sufficiently sharp to reflect high-frequency seismic energy (PKiKP). Gradients in seismic properties below the boundary have suggested the possibility of some fluid percolation in the outer region as progressive freezing takes place.

The complex interaction of fluid flow and electromagnetic effects in the core sustains a geodynamo that generates the internal magnetic field of the Earth. This magnetic field has to penetrate the much more poorly conducting silicate mantle to reach the surface, and in consequence short-wavelength components are suppressed.

Sources of energy to power the geodynamo can come from gravitational energy released by compositional convection due to expulsion of the lighter component as the inner core solidifies and from the cooling and contraction of the core as heat is extracted, latent heat release will accompany the solidification and there may also be internal radioactive heat sources (such as ^{40}K). The idea that the precessional motion of the Earth powers the dynamo has strong advocates, but is not generally accepted as a likely mechanism.

Computations of a magnetohydrodynamic dynamo for the core have yet to reach a fully realistic Earth-like state. Nevertheless some simulations display many features of the Earth's magnetic field, including polarity reversals in the field; a comprehensive review is provided by Kono & Roberts (2002).

15.1 The magnetic field at the surface and at the top of the core

Most of the magnetic field measured at the Earth's surface comes from sources within the Earth, but there are also external influences from the ionosphere. We can build a model of the contribution to the magnetic field from internal sources in terms of a magnetic potential φ_m with a spherical harmonic expansion as

$$\varphi_m(r) = r_e \sum_l \sum_{m=-l}^{l} \left(\frac{r_e}{r}\right)^{l+1} Y_l^m(\theta, \phi) G_l^m, \tag{15.1.1}$$

where, as in (11.3.38), Y_l^m is a normalised spherical harmonic. The magnetic field $\mathbf{B}_m$ above the core is derived from the gradient of φ_m under the assumption that the conductivity of the mantle is negligible,

$$\mathbf{B}_m = -\nabla\varphi_m, \quad \text{with} \quad \nabla^2\varphi_m \approx 0. \tag{15.1.2}$$

At the core–mantle boundary all the components of the magnetic field must be continuous, so that the field on the underside of the boundary will satisfy

$$\mathbf{B} = \mathbf{B}_m \quad \text{at} \quad r = r_c. \tag{15.1.3}$$

Inside the core the requirement that $\nabla \cdot \mathbf{B} = 0$, from Maxwell's equations, can be met by representing the solenoidal field $\mathbf{B}$ in terms of toroidal and poloidal contributions,

$$\mathbf{B}(\mathbf{x}, t) = \nabla \times [T(r, t)\mathbf{x}] + \nabla \times \nabla \times [S(r, t)\mathbf{x}], \tag{15.1.4}$$

where T is the defining scalar for the toroidal contribution and S that for the poloidal part. In spherical polar coordinates, the components of the magnetic field are then

$$\mathbf{B} = \left[B_r,\ B_\theta,\ B_\phi\right] = \left[\frac{1}{r}\mathfrak{L}^2 S,\ \frac{1}{r}\frac{\partial}{\partial\theta}\frac{\partial}{\partial r}(rS) + \frac{1}{\sin\theta}\frac{\partial}{\partial\phi}T,\ \frac{1}{r\sin\theta}\frac{\partial}{\partial\phi}\frac{\partial}{\partial r}(rS) - \frac{1}{r}\frac{\partial}{\partial\theta}T\right]. \quad (15.1.5)$$

where $\mathfrak{L}^2$ is the angular momentum operator,

$$\mathfrak{L}^2 = \frac{\partial}{\partial r}\left(r^2\frac{\partial}{\partial r}\right) - r^2\nabla^2 = -\frac{1}{\sin\theta}\frac{\partial}{\partial\theta}\left(\sin\theta\frac{\partial}{\partial\theta}\right) - \frac{1}{\sin^2\theta}\frac{\partial^2}{\partial\phi^2}, \quad (15.1.6)$$

so that $\mathfrak{L}^2 = -r^2\nabla_H^2$, where ∇_H is the horizontal derivative operator.

We can convert the poloidal and toroidal representation for the magnetic field into an expansion in terms of spherical harmonics, introduced in (11.3.38). Thus, for the poloidal contribution we write

$$S(r,t) = \sum_l \sum_{m=-l}^{l} S_l^m(r,t) Y_l^m(\theta,\phi). \quad (15.1.7)$$

The spherical harmonics Y_l^m are eigenfunctions of the angular momentum operator

$$\mathfrak{L}^2 Y_l^m(\theta,\phi) = l(l+1)Y_l^m(\theta,\phi) = \mathcal{L}^2 Y_l^m(\theta,\phi). \quad (15.1.8)$$

Further, from (15.1.5) we can cast the magnetic field in terms of the vector spherical harmonics introduced in (11.3.41) as

$$\mathbf{B}(\mathbf{x},t) = \sum_l \sum_{m=-l}^{l} \left\{ \frac{\mathcal{L}^2}{r} S_l^m(r,t)\, \mathbf{P}_l^m(\theta,\phi) + \frac{\mathcal{L}}{r}\frac{\partial}{\partial r}\left(r S_l^m(r,t)\right) \mathbf{B}_l^m(\theta,\phi) + \mathcal{L} T_l^m(r,t)\, \mathbf{C}_l^m(\theta,\phi) \right\}, \quad (15.1.9)$$

with the poloidal contribution in the core specified by S_l^m and the toroidal by T_l^m.

The magnetic field in the mantle is purely poloidal and so at the core–mantle boundary $r = r_c$ we require the toroidal component to vanish

$$T(r,t) = 0, \quad T_l^m = 0 \quad \text{at} \quad r = r_c. \quad (15.1.10)$$

The continuity condition (15.1.3) for the radial component B_r leads to

$$\frac{l}{r_c} S_l^m(r_c) = G_l^m \left(\frac{r_e}{r_c}\right)^{l+2}, \quad (15.1.11)$$

and, by eliminating G_l^m between the radial and θ components, we require

$$\frac{\partial}{\partial r}(r S_l^m) + l S_l^m = 0, \quad \text{at} \quad r = r_c. \quad (15.1.12)$$

We have therefore a way of transferring information on the magnetic field at the surface into the poloidal part of the magnetic field in the core.

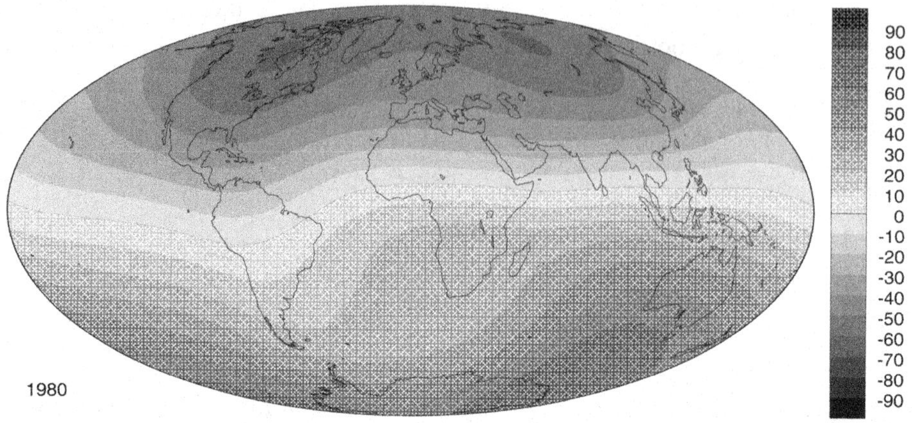

Figure 15.1. Radial magnetic field at the Earth's surface in 1980. [Courtesy of A. Jackson.]

Satellite observations such as those from the CHAMP and Oersted satellites, specifically designed to make magnetic field measurements have allowed the construction of maps of the large scale magnetic field at the Earth's surface (see Figure 10.1). The dominant component of the internal field is dipolar and it is this property that is used in the standard compass. However, there are significant non-dipole components.

The spectrum of the surface magnetic field shows a sharp decline with the angular order of the spherical harmonics l out to order 14, corresponding to a wavelength of the surface of around 3000 km. For short wavelengths, and thus larger angular orders, the spectrum shows a slight increase. This is interpreted as being the result of contributions from the lithosphere, both remanent magnetism from rocks that have cooled in the magnetic field prevailing at their time of formation and fields induced by the main core field.

The contribution to the surface magnetic field displayed in Figure 15.1 for angular orders l up to 12 is consistent with an origin at the core–mantle boundary. We can therefore extrapolate the magnetic field back to this boundary using the potential function φ_m from (15.1.1). From (15.1.11) the coefficients of the expansion of the magnetic potential at the core–mantle boundary are amplified by a factor $(r_e/r_c)^{l+2}$ compared with G_l^m. Since r_e is 6371 km and r_c is about 2890 km, $r_e/r_c \approx 2.2$ and hence there is a substantial enhancement of higher angular orders at the top of the core relative to the surface. The result is that the field at the core–mantle boundary appears to have much more power at shorter wavelengths (Figure 15.2) than at the surface. In fact what is happening is that the higher angular orders generated in the core are severely damped in passage through the nearly insulating mantle and so make little contribution at the surface.

Direct use of (15.1.12) to create field maps has a tendency to produce unstable results. The problem of the creation of the magnetic map at the core surface is

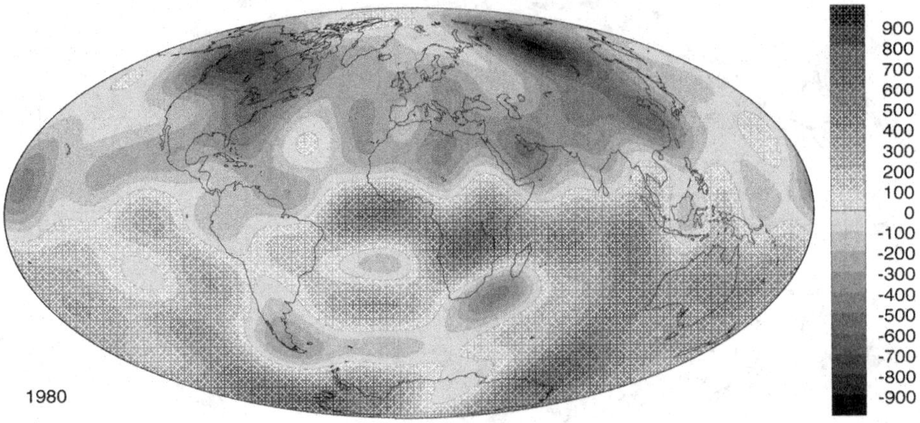

Figure 15.2. Magnetic field at the core–mantle boundary reconstructed from the data in figure 15.1, the projections of the continents are indicated for reference. [Courtesy of A. Jackson.]

frequently formulated as constraints on the coefficients S_l^m, with a solution through the minimisation of a mean-square error.

Although the general pattern of the magnetic field at the core–mantle boundary remains similar over time there is evidence for secular variation in time. For example, the patches of magnetic field in the Atlantic Ocean near the equator have been moving steadily westwards over the period for which measurements are available - producing an apparent westward drift of the north magnetic pole for stations in the northern hemisphere. Other features of the field are more static, such as the field concentrations at higher latitudes (around $\pm 70°$).

The toroidal component of the core magnetic field does not escape from the core, and as a result auxiliary assumptions have to be invoked to deduce the flow field in the core from the time variation of the magnetic field at the surface, based on the very high conductivity of the core fluid. It is necessary to assume that the flow is large-scale and invoke some approximations as to the relation of the magnetic field and the flow.

The frozen flux approximation discussed in Section 8.3.6 reduces to a relation for the time derivative of the radial component of the magnetic field B_r in terms of the horizontal derivative of the product of the velocity field $\mathbf{v}$ and B_r,

$$\frac{\partial}{\partial t} B_r = -\boldsymbol{\nabla}_H \cdot (\mathbf{v} B_r). \tag{15.1.13}$$

This assumption is not sufficient to determine the flow field uniquely. The ambiguity can be reduced by assuming that the horizontal Lorentz forces from the magnetic field play a minor role in the dynamics of the flow at the core surface. This assumption of 'tangential geostrophy' leads to a further equation for the horizontal derivative of $\mathbf{v}$,

$$\boldsymbol{\nabla}_H \cdot (\mathbf{v} \cos\vartheta) = 0, \tag{15.1.14}$$

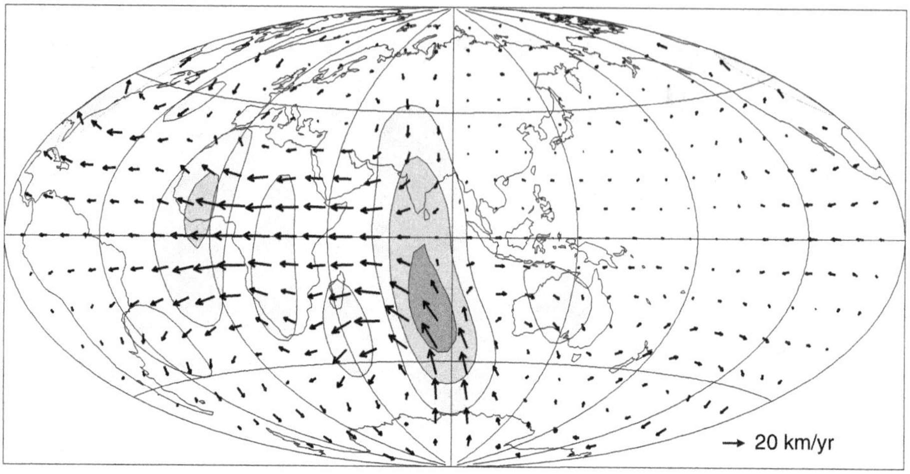

Figure 15.3. Flow at the top of the core deduced from the radial component magnetic field at the core–mantle boundary and its rate of change. [Courtesy of A. Jackson.]

where ϑ is the colatitude. With observations of B_r and its time derivative from secular variation the flow field $\mathbf{v}$ can be estimated from (15.1.13), (15.1.14) with suitable smoothness assumptions (see, e.g., Jackson, 1997).

The model displayed in Figure 15.3 represents the flow field just below the core–mantle boundary deduced from the magnetic field and secular variation information. The scale of the flow is much faster than those of plate tectonics, with typical velocities of 20 km/yr; this is five orders of magnitude faster than plate motions of cm/yr.

15.2 Convection and dynamo action

The time scales of the variations in the Earth's internal magnetic field are of the order of decades, although the core is subjected to the daily rotations of the Earth. To provide a description of the coupled magnetic and fluid behaviour that can give rise to dynamo action, we need therefore to combine the treatment of a rapidly rotating fluid from Section 7.6 and the description of magnetic fluid dynamics from Section 8.3.6.

15.2.1 Basic equations

The equation of mass conservation, (8.1.1), reduces in the case of an incompressible medium ($D\rho/Dt \equiv 0$) to the vanishing of the divergence of the velocity field

$$\nabla \cdot \mathbf{v} = 0. \tag{15.2.1}$$

The slightly less restrictive requirement that $\partial\rho/\partial t = 0$ provides the anelastic approximation (cf. Section 14.1.2):

$$\nabla \cdot (\rho \mathbf{v}) = 0. \tag{15.2.2}$$

In each case the removal of the time derivative of density suppresses the possibility of acoustic disturbances with much faster time scales than the rest of the motion.

The equation of motion for a rotating fluid takes the form (7.6.3)

$$\frac{\partial \mathbf{v}}{\partial t} + (\mathbf{v} \cdot \nabla)\mathbf{v} = -\frac{\nabla p}{\rho} - \boldsymbol{\Omega} \times (\boldsymbol{\Omega} \times \mathbf{x}) - 2\boldsymbol{\Omega} \times \mathbf{v} + \nu \nabla^2 \mathbf{v} + \frac{1}{\rho}\mathbf{F}, \tag{15.2.3}$$

where $\nu = \eta/\rho$ is the kinematic viscosity.

The force term $\mathbf{F}$ in the core arises from gravitational effects and the magnetic force

$$\mathbf{F} = \rho \mathbf{g} + \mu_0 \mathbf{j} \times \mathbf{B}; \tag{15.2.4}$$

following the treatment in Section 8.3.6 we neglect contributions from the displacement current $\partial \mathbf{D}/\partial t$ and thereby suppress electromagnetic wave phenomena as well as acoustic waves. In this *magnetohydrodynamic* approximation, the current $\mathbf{j}$ is given by

$$\mathbf{j} = \frac{1}{\mu_0} \nabla \times \mathbf{B}. \tag{15.2.5}$$

Thus from (8.3.49) the core force term becomes

$$\mathbf{F} = \rho \mathbf{g} + (\nabla \times \mathbf{B}) \times \mathbf{B} = -\rho \nabla \varphi + (\mathbf{B} \cdot \nabla)\mathbf{B} - \tfrac{1}{2}\nabla B^2, \tag{15.2.6}$$

where we have introduced the gravitational potential φ.

As in Section 7.6 we can represent the centrifugal force as a gradient

$$\boldsymbol{\Omega} \times (\boldsymbol{\Omega} \times \mathbf{x}) = -\nabla(\tfrac{1}{2}\rho|\boldsymbol{\Omega} \times \mathbf{x}|^2), \tag{15.2.7}$$

to absorb the influence of the centrifugal force. Thus we can write the equation of motion as

$$\begin{aligned}\frac{\partial \mathbf{v}}{\partial t} + (\mathbf{v} \cdot \nabla)\mathbf{v} = &-\frac{\nabla p}{\rho} - \nabla(\tfrac{1}{2}\rho|\boldsymbol{\Omega} \times \mathbf{x}|^2) - 2\boldsymbol{\Omega} \times \mathbf{v} + \nu\nabla^2 \mathbf{v} \\ &-\rho \nabla \varphi + (\mathbf{B} \cdot \nabla)\mathbf{B} - \tfrac{1}{2}\nabla B^2.\end{aligned} \tag{15.2.8}$$

When the density is constant, or very slowly varying in space, the various gradient terms can be combined in terms of an augmented pressure

$$P = p - \tfrac{1}{2}\rho|\boldsymbol{\Omega} \times \mathbf{x}|^2 - \rho\varphi, \tag{15.2.9}$$

with a consequent form of the Navier–Stokes equation for the rapidly rotating magnetic fluid

$$\frac{\partial \mathbf{v}}{\partial t} + (\mathbf{v} \cdot \nabla)\mathbf{v} = -\frac{\nabla P}{\rho} - 2\boldsymbol{\Omega} \times \mathbf{v} + \nu\nabla^2 \mathbf{v} + \frac{1}{\rho}(\nabla \times \mathbf{B}) \times \mathbf{B}. \tag{15.2.10}$$

From Maxwell's equations (8.3.1), the magnetic field has to satisfy the requirement of vanishing divergence

$$\nabla \cdot \mathbf{B} = 0, \tag{15.2.11}$$

and the induction equation (8.3.48)

$$\frac{\partial \mathbf{B}}{\partial t} = \nabla \times (\mathbf{v} \times \mathbf{B}) + \frac{1}{\breve{\sigma}\mu_0}\nabla^2 \mathbf{B}, \tag{15.2.12}$$

where we have assumed that spatial variation in the electrical conductivity $\breve{\sigma}$ can be neglected. The magnetic diffusivity $\lambda = (\mu_0 \breve{\sigma})^{-1}$.

For the toroidal and poloidal representation for the core magnetic field introduced in (15.1.4), the current is to be found from the curl of the magnetic field, $\nabla \times \mathbf{B}$, which has the representation

$$\nabla \times \mathbf{B} = \nabla \times [-\nabla^2 S\, \mathbf{x}] + \nabla \times \nabla \times [T\, \mathbf{x}]. \tag{15.2.13}$$

Thus the toroidal current system determines the poloidal magnetic field and vice versa.

In the highly conducting metallic fluid of the outer core we will have a small amount of Joule heating ($\mathbf{j}^2/\breve{\sigma}$) and also some viscous heating effects. We can extract an equation for the evolution of temperature from (8.3.53) in a similar form to (7.1.15):

$$\rho \left[\frac{\partial}{\partial t}(C_p T) + v_k \frac{\partial}{\partial x_k}(C_p T) \right] = h + \frac{\partial}{\partial x_k}\left[k \frac{\partial}{\partial x_k} T \right] + 2\eta \hat{D}_{ij}\hat{D}_{ij} + \frac{1}{\breve{\sigma}\mu_0^2}(\nabla \times \mathbf{B})^2, \tag{15.2.14}$$

Here k is the thermal conductivity, C_p is the specific heat at constant pressure and h is the heat generation per unit mass due to radioactivity; η is the viscosity, $\hat{D}_{ij}$ is the deviatoric strain-rate tensor and the last term on the right hand side is the Joule heating contribution. The dominant thermal behaviour is described by the *thermal diffusivity* (7.1.17), $\kappa_H = k/\rho C_p$.

The effects of compositional buoyancy in the core can be captured with a simple model of a binary alloy of iron and a lighter constituent with mass fraction ζ. The evolution of the mass fraction is described by a diffusion equation similar to that for temperature

$$\frac{\partial}{\partial t}\zeta + \mathbf{v} \cdot \nabla \zeta = h_\zeta + \kappa_\zeta \nabla^2 \zeta, \tag{15.2.15}$$

where we have made the simplifying assumption of uniform diffusivity κ_ζ. h_ζ is a source term that represents the influx of the lighter component into the fluid outer core as it is frozen out from the growth of the inner core. The molecular diffusivity κ_ζ is very small and so direct transport of the light component will be rather slow. However, we can regard (15.2.14), (15.2.15) as summarising the influence of small-scale motions for which the diffusivities will be enhanced. Such small-scale motions are not required to be isotropic in character, and hence the diffusivities should strictly be tensor quantities (Braginsky & Roberts, 1995).

Reference state

To complete the picture we need to understand the relation of the density to temperature and composition. We envisage a reference state in which the density has an adiabatic radial profile in hydrostatic equilibrium with no magnetic field, so that

$$-\nabla P_0(r) + \rho_0(r)\mathbf{g}(r) = 0. \tag{15.2.16}$$

The variations in density associated with deviations from the equilibrium temperature $T_0(r)$ and compositional $\zeta_0(r)$ profiles due to convective flow can then be approximated by

$$\rho = \rho_0(r)\{1 - \alpha_{th}[T - T_0(r)] + \alpha_\zeta[\zeta - \zeta_0(r)]\}, \tag{15.2.17}$$

where the thermal expansivity α_{th} and the influence of composition α_ζ are determined by compositional effects

$$\alpha_{th} = \frac{1}{\rho}\left(\frac{\partial\rho}{\partial T}\right)_{p,\zeta}; \qquad \alpha_\zeta = \frac{1}{\rho}\left(\frac{\partial\rho}{\partial\zeta}\right)_{p,T}. \tag{15.2.18}$$

15.2.2 Boundary conditions

The fluid outer core lies in the region between the solid inner core at radius r_i and the core–mantle boundary at r_c. From the discussion in Section 8.2 the interface condition between the viscous fluid and the solids requires *no slip* as in (8.2.4). If the radial boundaries are stationary in the rotating reference frame, the fluid velocity must vanish at r_i, r_c:

$$\mathbf{v} = 0, \qquad r = r_i, r_c. \tag{15.2.19}$$

However, if there is differential rotation between the inner and outer core then the condition at the inner core boundary needs to be modified to

$$\mathbf{v} = \boldsymbol{\omega}_i \times \mathbf{x}, \qquad r = r_i, \tag{15.2.20}$$

where ω_i is the angular velocity of the inner core relative to the outer core (and the mantle).

The temperature at the inner core boundary will be dictated by the freezing of the iron alloy, whereas, at the core–mantle boundary we need continuity of radial heat flux, which is expected to translate into an approximately constant temperature. If the heat flux at the core–mantle boundary is not uniform, the equi-temperature surfaces no longer coincide with radii. The state of zero motion, $\mathbf{v} = 0$, does not then satisfy the equation of motion and so the fluid must move. The resultant flow is termed *free convection*.

The constant temperature condition at the inner core boundary is accompanied by a release of latent heat in freezing and a flux of the light constituent. There will be no mass fraction flux into the mantle, so

$$(\nabla\zeta)_r = 0, \qquad r = r_c. \tag{15.2.21}$$

All components of the magnetic field are continuous at both the inner core boundary and the core–mantle boundary. The mantle has a much lower electrical conductivity than the metallic core, and to a first approximation can be treated as an insulator with no sources of magnetic field. The magnetic field in the mantle $\mathbf{B}_m$ can then be determined from

$$\mathbf{B}_m = -\nabla\varphi_m, \qquad \nabla^2\varphi_m = 0, \tag{15.2.22}$$

with a harmonic solution that vanishes as $r \to \infty$, as in (15.1.1). All the components of the magnetic field should match at the boundary:

$$\mathbf{B} = \mathbf{B}_m, \qquad r = r_c. \tag{15.2.23}$$

There is likely to be limited electromagnetic coupling between the mantle and the outer core, concentrated in the D″ region, which will slightly modify the form of solution.

The inner core is expected to have a similar conductivity to that of the outer core and so we need to impose the condition that the horizontal components of the electric field should be continuous across $r = r_i$ (8.3.14).

15.2.3 Interaction of the flow with the magnetic field

In order for a dynamo to function there must be sufficient energy generated to overcome dissipation. We need therefore to examine the way in which energy is distributed between the flow and the magnetic field. From (15.2.10) we can extract the rate of change of the kinetic energy, by taking the scalar product with $\mathbf{v}$, as

$$\frac{D}{Dt}\left(\tfrac{1}{2}\rho v^2\right) = \mathbf{v}\cdot(\mathbf{j}\times\mathbf{B}) - \mathbf{v}\cdot(\rho\nabla\varphi + \nabla P), \tag{15.2.24}$$

The term $\mathbf{v}\cdot(\mathbf{j}\times\mathbf{B})$ represents the rate at which the magnetic energy is converted into kinetic energy through the agency of the Lorentz force. The corresponding rate of change of magnetic energy, comes from (15.2.12)

$$\frac{D}{Dt}\left(\frac{1}{2}\frac{B^2}{\mu_0}\right) = -\mathbf{v}\cdot(\mathbf{j}\times\mathbf{B}) - \frac{J^2}{\breve{\sigma}}. \tag{15.2.25}$$

In a dynamo the rate of growth of magnetic field energy from $-\mathbf{v}\cdot(\mathbf{j}\times\mathbf{B})$ over the whole core needs to exceed the integrated Joule heating $J^2/\breve{\sigma}$ from the degradation of magnetic energy by ohmic resistance. Local balance is not essential, but sufficient of the core has to be building magnetic energy to overcome the total ohmic dissipation.

Lines of magnetic force will tend to be carried with the flow field in the electrically conducting fluid in the core. An azimuthal shear with rate ω_s will tend to shear the lines of forces in the radial and latitudinal directions ($\mathbf{B}_M$) to create a component in the azimuthal direction (B_ϕ). This ω-effect creates a zonal field B_ϕ from a meridional field $\mathbf{B}_M$. For a successful dynamo there has to be a converse loop that creates $\mathbf{B}_M$ from B_ϕ, and as demonstrated by Cowling in 1934 no such process exists if the system is totally axisymmetric.

The eddies in a rotating fluid undergoing turbulent convection are expected to resemble the patterns of atmospheric circulation. These cyclonic (or anticyclonic) disturbances will deform, and may amplify the magnetic field in a conducting fluid. Consider then a decomposition of the magnetic field into a large-scale component $\langle \mathbf{B} \rangle$ and a much more rapidly varying small-scale field $\mathbf{b}'$, with a similar form for the flow field $\mathbf{v}$. There will be an electromotive force (emf) $\boldsymbol{\epsilon} = \mathbf{v}' \times \mathbf{b}'$ that will have a large-scale component

$$\langle \boldsymbol{\epsilon} \rangle = \langle \mathbf{v}' \times \mathbf{b}' \rangle. \qquad (15.2.26)$$

When the turbulence is sufficiently complex the average over the small scales in (15.2.26) does not vanish and the net emf $\langle \boldsymbol{\epsilon} \rangle$ is approximately proportional to the large-scale component of the magnetic field,

$$\langle \boldsymbol{\epsilon} \rangle = \alpha \langle \mathbf{B} \rangle. \qquad (15.2.27)$$

In general, α would be a tensor. If the small-scale motions are statistically independent of direction, but do not have mirror symmetry, then α is approximately a scalar. The azimuthal component $\langle \boldsymbol{\epsilon} \rangle_\phi = \alpha \langle \mathbf{B} \rangle_\phi$ is able to generate a meridional field $\langle \mathbf{B}_M \rangle$; this is known as the α-effect.

Dominantly axisymmetric large-scale fields can therefore be sustained by a simple feedback loops. In the $\alpha\omega$ dynamo, the α-effect creates $\langle \mathbf{B}_M \rangle$ from $\langle B_\phi \rangle$ and in turn the ω-effect produces $\langle B_\phi \rangle$ from $\langle \mathbf{B}_M \rangle$ to give a self-sustaining magnetic field. The size of the zonal flows deduced from secular variation suggests that the $\alpha\omega$ dynamo is a plausible candidate for the Earth's core. An alternative, the α^2 dynamo, relies on the α-effect to generate both $\langle \mathbf{B}_M \rangle$ from $\langle B_\phi \rangle$ and $\langle B_\phi \rangle$ from $\langle \mathbf{B}_M \rangle$. Such a dynamo model is more likely to lead to steady solutions. Depending on the choice of α and the differential rotation in the flow the dominant solutions can have either dipolar or quadrupolar symmetry. Reversal in the magnetic field would occur when the form of internal motion changes, leading to a modification of the preferred symmetry of the field configuration. Both α- and ω-effects are likely to be active, and the classification as $\alpha\omega$ or α^2 dynamos will depend on which effect is more significant in generating $\langle \mathbf{B}_M \rangle$.

15.2.4 Deviations from the reference state

The reference state introduced in (15.2.16) has purely radial variations of pressure and temperature, whereas in the flow field there will be additional components with spatial and temporal variations; we therefore write

$$P = P_0(r) + P'(r, \theta, \varphi, t), \qquad T = T_0(r) + \Theta(r, \theta, \varphi, t), \qquad (15.2.28)$$

To first order in the pressure and temperature deviations (P', Θ) the equation of motion takes the form

$$\frac{\partial \mathbf{v}}{\partial t} + (\mathbf{v} \cdot \nabla)\mathbf{v} = -\frac{\nabla P'}{\rho_0} - 2\boldsymbol{\Omega} \times \mathbf{v} + \nu\nabla^2\mathbf{v} + \frac{1}{\mu_0\rho_0}(\nabla \times \mathbf{B}) \times \mathbf{B}; \qquad (15.2.29)$$

and the equation for the temperature evolution is

$$\frac{\partial \Theta}{\partial t} + \mathbf{v} \cdot \nabla\Theta = h + \kappa_H\nabla^2\Theta - \mathbf{v} \cdot \nabla T_0 + d_q, \qquad (15.2.30)$$

where d_q represents the combined effect of Joule and viscous heating.

In many numerical simulations a somewhat simplified scenario is investigated. The variation in density across the outer core is ignored and a Boussinesq approximation is adopted in which just the density effects due to temperature (and perhaps composition) are included (cf. the treatment of convection in Section 7.5.2). The resulting equations are similar to (15.2.29), (15.2.30), but the energy dissipation d_q cannot be included in a fully consistent fashion.

15.2.5 Non-dimensional treatment

In our discussion of fluid flows in Chapter 7, we saw how considerable insight into the physical character of the system can be obtained by working with non-dimensional quantities, so that different configurations can be compared through a set of non-dimensional numbers. A similar approach can be made for the magnetic fluid dynamics of the core. A wide variety of different choices have been made in studies of the core dynamo and a convenient summary and inter-comparison is provided by Kono & Roberts (2002).

The choice of non-dimensionalisation is designed to bring the main terms to comparable size. There are two plausible choices for the length scale, either the radius of the outer core, r_c, or the span of the outer core from the mantle to the core-mantle boundary $L_s = r_c - r_i$, i.e., 2260 km. There are more choices for the time scale based on viscous, thermal or magnetic diffusion times. We will use the magnetic diffusion time $T_s = L_s^2/\lambda$, where $\lambda = 1/(\mu_0\breve{\sigma})$ is the magnetic diffusivity; T_s is of the order of 65,000 years. The corresponding flow velocity $U_s = L_s/T_s$ is around 10^{-6} ms^{-1}, and the effective flow and diffusion time scales are then comparable. A scaling for the magnetic field is provided by $B_s = (\Omega\rho\mu_0\lambda)^{1/2}$, which ensures that the Lorentz and Coriolis force terms have a similar size.

The main scaling factors are thus:

$$L_s = r_c - r_i, \qquad T_s = \frac{L_s^2}{\lambda} = L_s^2\mu_0\breve{\sigma}, \qquad U_s = \frac{L_s}{T_s},$$

$$B_s = (\Omega\rho\mu_0\lambda)^{1/2} = \left(\frac{\Omega\rho}{\breve{\sigma}}\right)^{1/2}. \qquad (15.2.31)$$

The Ekman number (7.6.9), the ratio of the viscous and Coriolis forces, is given by

$$\mathrm{Ek} = \frac{\nu}{\Omega L^2} \approx 10^{-12}, \tag{15.2.32}$$

based on molecular diffusivity in the core. If an eddy diffusivity is employed, representing the effects of small scale flow within the core, then the Ekman number is increased by a factor of around 10^3. The viscosity of the fluid outer core is not well constrained, but is assumed to be comparable to that of water.

The magnetic Rossby number, representing the balance between magnetic and Coriolis forces, is defined as

$$\mathrm{Ro_M} = \frac{\lambda}{\Omega L^2} = \frac{1}{\mu_0 \breve{\sigma} \Omega L^2} \approx 10^{-8}. \tag{15.2.33}$$

The Coriolis forces from the rapid rotation overwhelm both the viscous and magnetic forces.

The ratio of the thermal to magnetic diffusion times defines the Roberts number

$$q = \frac{\kappa_H}{\lambda}, \tag{15.2.34}$$

which plays an analogous role to the Prandtl number $\mathrm{Pr} = \nu/\kappa_H$ that describes the relative significance of momentum and heat transport.

For the rapidly rotating fluid it is convenient to use a modified Rayleigh number to include the angular velocity Ω

$$\widetilde{\mathrm{Ra}} = \frac{g \alpha_{th} \beta L_s^2}{\Omega \kappa_H}, \tag{15.2.35}$$

where g is the acceleration due to gravity, α_{th} the thermal expansion coefficient and β is the radial temperature gradient.

Following scaling by the non-dimensional quantities we can write the equations governing the behaviour of the magnetic fluid, in the absence of internal heating, in the form

$$\frac{\partial}{\partial t}\mathbf{B} = \nabla^2 \mathbf{B} + \nabla \times (\mathbf{v} \times \mathbf{B}), \tag{15.2.36}$$

$$\mathrm{Ro_M}\left(\frac{\partial}{\partial t} + \mathbf{v}\cdot\nabla\right)\mathbf{v} + 2\hat{\mathbf{e}}_z \times \mathbf{v} = -\nabla P + \mathrm{Ek}\,\nabla^2\mathbf{v} + (\nabla \times \mathbf{B}) \times \mathbf{B} + \widetilde{\mathrm{Ra}}\, q\Theta\mathbf{x}, \tag{15.2.37}$$

$$\left(\frac{\partial}{\partial t} + \mathbf{v}\cdot\nabla\right)\Theta = q\nabla^2\Theta, \tag{15.2.38}$$

$$\nabla \cdot \mathbf{B} = 0, \tag{15.2.39}$$

$$\nabla \cdot \mathbf{v} = 0, \tag{15.2.40}$$

where Θ is the non-dimensional temperature perturbation.

Induction equation

Consider the equation for the magnetic field (15.2.36), but neglect initially the diffusion term in $\nabla^2\mathbf{B}$. Then

$$\frac{\partial}{\partial t}\mathbf{B} = \nabla \times (\mathbf{v} \times \mathbf{B}). \tag{15.2.41}$$

This equation is analogous to the vorticity equation of fluid dynamics

$$\frac{\partial}{\partial t}\varpi = \nabla \times (\mathbf{v} \times \varpi), \tag{15.2.42}$$

which states that vortex lines evolve as material lines. Thus, just as vortex lines can be stretched and thereby be amplified, so may the magnetic field lines. The presence of the neglected term from (15.2.36) will ultimately lead to diffusion of the field behaviour.

Geostrophy and Taylor Columns

The inertial term in (15.2.37) appears with the very small magnetic Rossby number Ro_M and the viscous term with the not much larger Ekman number Ek so we can anticipate that as in the treatment of a rotating fluid in Section 7.6 we could make a geostrophic approximation to yield a reduced equation

$$2\hat{\mathbf{x}} \times \mathbf{v} = -\nabla P + (\nabla \times \mathbf{B}) \times \mathbf{B} + \widetilde{\mathsf{Ra}}\, q\theta\mathbf{x}. \tag{15.2.43}$$

The appropriate boundary condition on the flow at the core–mantle boundary is then the inviscid condition $\mathbf{v} \cdot \mathbf{n} = 0$. In the absence of the magnetic field there would be no variation in the velocity field in the direction parallel to the rotation axis. However, such a free geostrophic solution is not possible for (15.2.43) unless the Lorentz force obeys the very restrictive Taylor condition

$$\int_{C(s)} (\nabla \times \mathbf{B} \times \mathbf{B})_\phi dz d\phi \equiv T(s) = 0, \tag{15.2.44}$$

where (s, ϕ, z) are cylindrical polar coordinates and $C(s)$ is the cylinder of radius s aligned with the rotation axis within the core.

The problem can be overcome by working with a nearly geostrophic state including viscous terms, either through the reinstatement of $\mathsf{Ek}\nabla^2\mathbf{v}$ in (15.2.43) or by inclusion of viscous effects in thin Ekman boundary layers at the top and bottom of the fluid outer core. Nevertheless, the forms of flows displayed by many numerical dynamos with significant magnetic fields have a strong columnar component (*Taylor columns*) as suggested by the simple geostrophic approximation.

15.3 Numerical dynamos

The advent of fast computers has meant that it is feasible to attempt full three-dimensiona; calculations of the dynamo with coupled velocity and magnetic fields. Viscosity is retained in the calculations and the Ekman number Ek made as small as possible. However, the parameter values are still a long way from being geophysically reasonable even though computed field strengths are of the right order for the Earth and irregular oscillatory behaviour is obtained.

The role of the inner core is particularly important. Firstly, the tangent cylinder that is parallel to the rotation axis and touches the inner core at the equator separates two quite different types of flow with the most vigorous motion lying within the tangent cylinder. Secondly and more importantly, because the inner core is a conductor with similar conductivity to that of the outer core any magnetic field lines that penetrate into the solid inner core can only evolve on the diffusion time scale. This pinning of the field lines helps to stabilise what would otherwise be wild oscillatory behaviour and to prevent reversals of the main dipole field on very short time scales.

The numerical solution of the coupled magnetic, flow and thermal equations (15.2.36)–(15.2.38) is commonly undertaken using a spectral technique coupled to a toroidal and poloidal decomposition. As in (15.1.4) the magnetic field is expressed as

$$\mathbf{B}(\mathbf{x}, t) = \nabla \times [T(r, t)\mathbf{x}] + \nabla \times \nabla \times [S(r, t)\mathbf{x}], \tag{15.3.1}$$

and so satisfies the condition $\nabla \cdot \mathbf{B} = 0$ (15.2.39). A similar form can be applied for an incompressible fluid so that

$$\mathbf{v}(\mathbf{x}, t) = \nabla \times [V(r, t)\mathbf{x}] + \nabla \times \nabla \times [U(r, t)\mathbf{x}], \tag{15.3.2}$$

and $\nabla \cdot \mathbf{v} = 0$ (15.2.40).

The defining scalars for the toroidal and poloidal vectors are then expanded in terms of the Y_l^m spherical harmonics (11.3.38), e.g.,

$$T(r, \theta, \phi, t) = \sum_{l=0}^{L_B} \sum_{m=-l}^{l} T_l^m(r, t) Y_l^m(\theta, \phi), \tag{15.3.3}$$

$$V(r, \theta, \phi, t) = \sum_{l=0}^{L_V} \sum_{m=-l}^{l} V_l^m(r, t) Y_l^m(\theta, \phi); \tag{15.3.4}$$

a comparable expansion can be made for the non-dimensional temperature

$$\Theta(r, \theta, \phi, t) = \sum_{l=0}^{L_T} \sum_{m=-l}^{l} \Theta_l^m(r, t) Y_l^m(\theta, \phi). \tag{15.3.5}$$

Hollerbach (2000) recommends that the truncations of the magnetic, velocity and temperature spherical harmonic expansions are distinct, because the various fields exhibit structure in rather different length scales. Thus, the velocity field $\mathbf{v}$ is likely to exhibit fine-scale structure when the Ekman number Ek is small, whereas the temperature field Θ will have fine structure when the Roberts number q is small. Independent truncation allows flexibility in the representation, with consequent savings in computational effort.

With the expansion in spherical harmonics we need to turn the original magnetohydrodynamic equations and boundary conditions for **v**, **B**, and Θ into equivalent differential equations and associated boundary conditions for U_l^m, T_l^m etc. that are suitable for numerical computation. Application of the curl $(\nabla\times)$ and the curl of the curl $(\nabla\times\nabla\times)$ to (15.2.37) removes the pressure term. The r components of these equations yield

$$\sum_l\sum_m \frac{\L^2}{r}\left[\mathrm{Ro_M}\frac{\partial}{\partial t} - \mathrm{Ek}\mathfrak{L}^2\right] V_l^m Y_l^m = \hat{\mathbf{x}}\cdot\nabla\times\mathbf{F}, \tag{15.3.6}$$

$$\sum_l\sum_m \frac{\L^2}{r}\left[\mathrm{Ro_M}\frac{\partial}{\partial t} - \mathrm{Ek}\mathfrak{L}^2\right] \mathfrak{L}^2 U_l^m Y_l^m = -\hat{\mathbf{x}}\cdot\nabla\times\nabla\times\mathbf{F}, \tag{15.3.7}$$

where, as in (15.1.8), $\L^2 = l(l+1)$ and $\mathfrak{L}^2$ is the angular momentum operator (15.1.6). The forcing vector for the flow equations

$$\mathbf{F} = -2\hat{\mathbf{e}}_z\times\mathbf{v} - \mathrm{Ro_M}(\mathbf{v}\cdot\nabla)\mathbf{v} + (\nabla\times\mathbf{B})\times\mathbf{B} + \widetilde{\mathrm{Ra}}\,q\theta\mathbf{x}. \tag{15.3.8}$$

A comparable development can be made for the induction equation (15.2.36),

$$\sum_l\sum_m \frac{\L^2}{r}\left[\frac{\partial}{\partial t} - \mathfrak{L}^2\right] S_l^m Y_l^m = \hat{\mathbf{x}}\cdot\nabla\times(\mathbf{v}\times\mathbf{B}), \tag{15.3.9}$$

$$\sum_l\sum_m \frac{\L^2}{r}\left[\frac{\partial}{\partial t} - \mathfrak{L}^2\right] T_l^m Y_l^m = \hat{\mathbf{x}}\cdot\nabla\times\nabla\times(\mathbf{v}\times\mathbf{B}). \tag{15.3.10}$$

For the temperature equation we can also make a spectral development

$$\sum_l\sum_m \frac{\L^2}{r}\left[\frac{\partial}{\partial t} - q\left(\mathfrak{L}^2 + \frac{2}{r}\frac{\partial}{\partial r}\right)\right] \Theta_l^m Y_l^m = -\mathbf{v}\cdot\nabla\Theta. \tag{15.3.11}$$

The orthogonality of the spherical harmonics allows us to extract equations for the individual spectral components so that, e.g., from (15.3.6) we find

$$\frac{\L^2}{r}\left[\mathrm{Ro_M}\frac{\partial}{\partial t} - \mathrm{Ek}\mathfrak{L}^2\right] V_l^m = \int_0^{2\pi} d\phi \int_0^{\pi} d\theta\, \sin\theta\, [Y_l^m]^*\hat{\mathbf{x}}\cdot\nabla\times\mathbf{F}. \tag{15.3.12}$$

The linear boundary conditions for the different fields apply directly to the individual spectral components.

The radial dependence of the various spectral components can be expanded in terms of Chebyshev polynomials, such as,

$$U_l^m(r,t) = \sum_{n=1}^{N_u+2} U_{l,n}^m(t) T_{n-1}(x) \quad \text{with} \quad x = \frac{2r - r_c - r_i}{r_c - r_i}, \tag{15.3.13}$$

and the variable x varies between -1 and 1. Radial derivatives can be evaluated using a recursion on n for the $U_{l,n}^m$ coefficients. The Chebyshev coefficients,

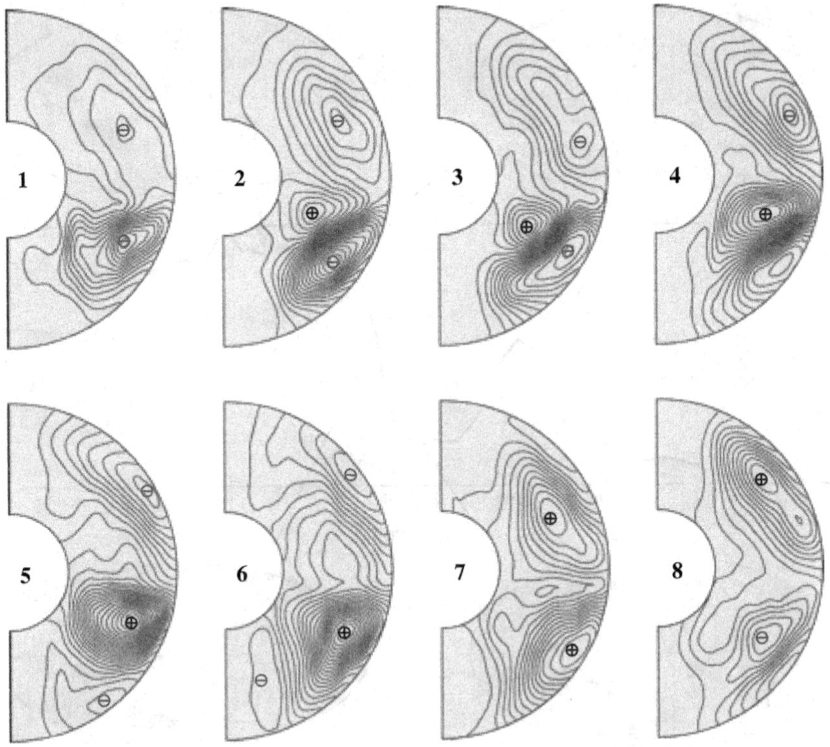

Figure 15.4. Evolution of the magnetic lines of force in a meridional plane averaged over longitude during a simulated polarity reversal. The simulated time sequence starts at the top left and ends at the bottom right after approximately 5200 years, which is consistent with the time scale of the geomagnetic field. Clockwise fields are shown in black (with a plus sign) and counterclockwise in grey (with a minus sign). [Courtesy of F. Takahashi.]

such as $U^m_{l,n}$, are to be found by evaluating (15.3.12) at the N_u collocation points representing the zeros of $T_{N_u}(x)$, together with the two boundary conditions to provide $N_u + 2$ equations. The Chebyshev representation automatically concentrates resolution close to the boundaries, which is of considerable value for resolving boundary layer features.

The differential equations for the spectral components are linear on the left-hand side of each equation, such as (15.3.12), but on the right-hand side we have spherical transforms of non-linear functions. Once the right-hand side terms are known, Runge–Kutta or similar systems can be used to solve the differential equations (see Hollerbach, 2000, for details of the computational scheme). The evaluation of the non-linear terms becomes rapidly infeasible as the number of spectral components included increases, because the computation cost is $O(L^4)$ where L is the truncation level of the spectral expansion. In consequence a 'pseudo-spectral' method is employed with repeated switches between real and spectral space. The field variables **v**, **B**, etc. are evaluated at appropriate points

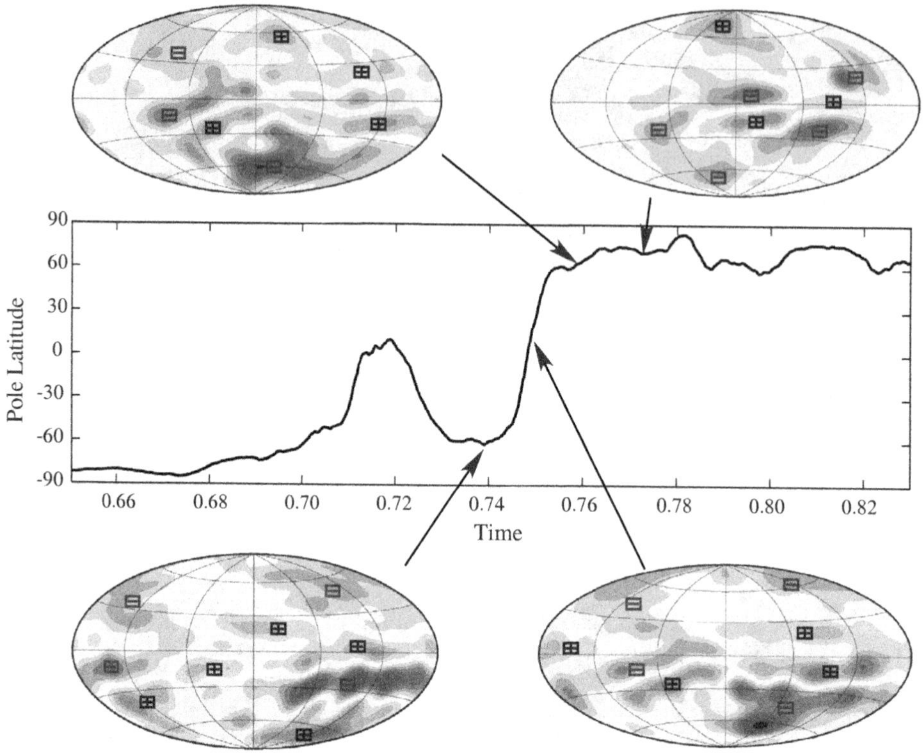

Figure 15.5. Evolution of the morphology of the radial component of the magnetic field at the core surface during a simulated polarity reversal. Positive signs indicate outward fields from the core and negative signs inward fields. The time scale is non-dimensionalised by the magnetic diffusion time of approximately 200,000 years. The maximum and minimum contour values are ± 0.5mT. The results are presented for spherical harmonic order less than 12 to allow comparison with the inferences from the observed geomagnetic field in Figure 15.1. [Courtesy of F. Takahashi.]

in real space, and the necessary operations performed to construct $\mathbf{F}$ and $\mathbf{v} \times \mathbf{B}$ in the spatial domain. The required derivatives are then constructed in the spectral domain. Hollerbach (2000) show how to make efficient choices for the spatial sampling to achieve accurate results without excessive computational effort.

In the presence of an inner core with finite conductivity, the evolution of the magnetic field is determined not only by changes at the boundary between the inner and outer core, but also by prior variations in the inner core. The associated magnetic diffusion time for the inner core imposes a new time scale on the whole dynamo system that has a significant effect on the behaviour (see, e.g., Glatzmeier & Roberts, 1996).

The complications in solving the coupled magnetohydrodynamic equations come from the very small values of the magnetic Rossby number Ro_M ($\sim 10^{-8}$) and the Ekman number Ek ($< 10^{-9}$) in equations such as (15.3.12). These terms are associated with very fine length scales (of the order of metres) and rapid time variations. Available computational resources have limited calculations to values several orders of magnitude larger than expected core conditions. Nevertheless dynamic behaviour with Earth-like states has been simulated.

Takahashi et al. (2005) have carried out a spectral approach using all spherical harmonics of degree and order up to 255, using the Earth Simulator computer in Japan, and have been able to reduce the Ekman number to 4×10^{-7}. Their dynamo models employ electrically insulating, no-slip and isothermal boundary conditions at the core–mantle boundary and the inner core boundary. Distinct reversals of the magnetic field are simulated with a relatively rapid transition between field states (Figure 15.4) In addition to full reversals this high resolution model shows excursions of the magnetic field that do not lead to a sustained polarity change (Figure 15.5).

Almost all numerical dynamo models produce a significant axial dipole component above the trend of the power law decay of the higher field harmonics, whereas the quadrupole term lies somewhat below the trend. A major limitation arises from limited knowledge of the physical properties of the core. Some aspects can be expected to be addressed by *ab initio* calculations as discussed in Chapter 9, but parameters such as the viscosity of the core are likely to remain elusive.

15.4 Evolution of the Earth's core

One of the intriguing questions about the Earth is the way that the core has evolved in time. When did the inner core form and how might it be growing? As we shall see the answers to such questions depend critically on the Earth's heat budget, and in particular the rate at which heat leaves the metallic core and enters the silicate mantle.

It is tempting to suggest that the onset of observable magnetic fields about 3.5 Ga ago marks the point at which the inner core started to form and the dynamo began to evolve towards its current configuration. However, dynamo action does appear to be possible in a fluid sphere and the presence of a small inner core should not have a profound effect on the flow field. As we shall see, it is likely that the growth of the inner core, once solidification occurs, is quite rapid. This change could well affect the amplitude of the magnetic field that could be sustained by dynamo action.

15.4.1 Energy balance

Following Buffet et al. (1996)we can use the energy balance equations to understand the evolution of the core. We assume that there is no mass transport across the core–mantle boundary S_c and then look at the energy budget in the core. We can integrate the expressions (15.2.24) and (15.2.24) over the whole volume of the core V_c to describe the rate of change of the kinetic and magnetic fields. The rate of change of mechanical energy for the core is thus

$$\frac{d}{dt}\int_{V_c} \tfrac{1}{2}\rho v^2 \, dV = \int_{V_c} \mathbf{v}\cdot(\mathbf{j}\times\mathbf{B})dV - \int_{V_c} \mathbf{v}\cdot(\rho\nabla\varphi + \nabla P)dV. \qquad (15.4.1)$$

With the magnetohydrodynamic assumption of the neglect of displacement currents, the rate of change of magnetic energy for the whole core is

$$\frac{d}{dt}\int_{V_c} \frac{1}{2}\frac{B^2}{\mu_0}\,dV = -\int_{V_c} \mathbf{v}\cdot(\mathbf{j}\times\mathbf{B})dV - \int_{V_c} \frac{J^2}{\breve{\sigma}}\,dV. \qquad (15.4.2)$$

We can express the internal heat budget in the form

$$\frac{d}{dt}\int_{V_c} \rho U\,dV = -\int_{S_c} \mathbf{q}\cdot d\mathbf{S} + \int_{V_c} \rho h\,dV + \int_{V_c} \frac{J^2}{\breve{\sigma}}\,dV - \int_{V_c} P(\boldsymbol{\nabla}\cdot\mathbf{v})dV, \qquad (15.4.3)$$

where U is the specific internal energy, h is the rate of radioactive heating per unit mass and $\mathbf{q}$ is the heat flux into the base of the mantle.

The contributions to the specific internal energy can be built up from thermodynamic considerations and so

$$dU = T\,dS + \frac{P}{\rho^2}\,d\rho + \mu_\zeta\,d\zeta, \qquad (15.4.4)$$

where ζ is the concentration of light elements in the core, and μ_ζ is the corresponding chemical potential. We also have the conservation of mass in terms of the continuity equation (8.1.1),

$$\frac{D\rho}{Dt} = -\rho(\boldsymbol{\nabla}\cdot\mathbf{v}), \qquad (15.4.5)$$

and can describe the diffusion of the lighter elements by

$$\rho\frac{D\zeta}{Dt} = -\boldsymbol{\nabla}\cdot\mathbf{i}, \qquad (15.4.6)$$

with $\mathbf{i}$ the flux of the lighter component.

With the aid of (15.4.4)–(15.4.6) we can rewrite the internal energy budget equation (15.4.3) as

$$\int_{V_c} \rho T\frac{DS}{Dt}\,dV = -\int_{S_c} \mathbf{q}\cdot d\mathbf{S} + \int_{V_c} \frac{J^2}{\breve{\sigma}}\,dV + \int_{V_c} \boldsymbol{\nabla}\cdot\mathbf{i}\,dV. \qquad (15.4.7)$$

The changes in the heat content of the core are due to the heat flow from the core, in the interior, and diffusive segregation of lighter elements. Gravitational energy is released by thermal contraction and compositional fractionation; there will be a modification of the compression, with a small amount of adiabatic heating, as the net radial mass distribution changes.

With respect to the thermal evolution of the core, the changes in the gravitational and internal energies will dominate those from the mechanical and magnetic energies so that from (15.4.1)–(15.4.2):

$$\frac{d}{dt}\int_{V_c} \left\{\tfrac{1}{2}\rho v^2 + \tfrac{1}{2}\frac{B^2}{\mu_0}\right\} dV = -\int_{V_c} \mathbf{v}\cdot(\rho\boldsymbol{\nabla}\varphi + \boldsymbol{\nabla}P)dV - \int_{V_c} \frac{J^2}{\breve{\sigma}}\,dV \approx 0. \qquad (15.4.8)$$

Thus, we can equate the total ohmic dissipation Φ to the difference between the change in gravitational energy and the work done against pressure gradients

$$\Phi = -\int_{V_c} \frac{J^2}{\breve{\sigma}} dV = \int_{V_c} \mathbf{v} \cdot (\rho \nabla \varphi + \nabla P) dV. \tag{15.4.9}$$

The second integral in (15.4.9) would vanish if the state of the core were hydrostatic, and so ohmic heating is maintained by the departures of the core from a hydrostatic state due to convection.

Under the same approximation of neglect of changes in magnetic, kinetic and nuclear energies, the rate of change $d\Sigma/dt$ of the total energy within the core obtained by summing (15.4.1)–(15.4.3) is

$$\begin{aligned} \frac{d}{dt}\Sigma &= \frac{d}{dt}\int_{V_c} \rho U \, dV + \int_{V_c} \rho \mathbf{v} \cdot \nabla \varphi \, dV, \\ &= -\int_{S_c} \mathbf{q} \cdot d\mathbf{S} - \int_{S_c} P\mathbf{v} \cdot d\mathbf{S}. \end{aligned} \tag{15.4.10}$$

The work done against the Lorentz forces and the work associated with $P\nabla \cdot \mathbf{v}$ do not appear in (15.4.10) because they represent transfer of energy between different forms within the core.

15.4.2 *Thermal and compositional effects*

Figure 15.6 illustrates the processes that are likely to contribute to the evolution of the core through the thermal and compositional release of gravitational energy. The processes are shown separately, but all four will in fact operate at the same time.

Because of the extraction of heat flux $\mathbf{q}$ from the core into the mantle above, there will be cooling of the core and a contraction in the core radius, $r_c \to r_c - \delta r_c$, with the generation of a thermal boundary layer below the cooler mantle. The heat flux can be expected to vary over the surface of the core because of the strong variation in the properties of the mantle just above the core–mantle boundary. Over time parts of the cold and dense thermal boundary layer will become unstable and sink to mix into the bulk of the outer core. At the inner core boundary, the lighter components are expelled as the inner core solidifies and the inner core radius will tend to grow, $r_i \to r_i + \delta r_i$. There will also be release of latent heat as the inner core material freezes. The light compositional boundary layer will be unstable and there will be overturn and mixing of the lighter components into the outer core.

The long term effect of inner core growth will be a redistribution of mass, with a concentration of the heavier iron in the inner core and a consequent reduction in density in the outer core, since the mass of the core must remain constant.

Because motions are rapid the mixing of cooler and lighter components will be efficient and the outer core temperature profile will be approximately adiabatic and the composition nearly uniform. Even though the fluctuations about the adiabatic, isentropic, well-mixed and hydrostatic state will be small, they are needed to produce the convective flows and buoyancy fluxes that sustain the dynamo.

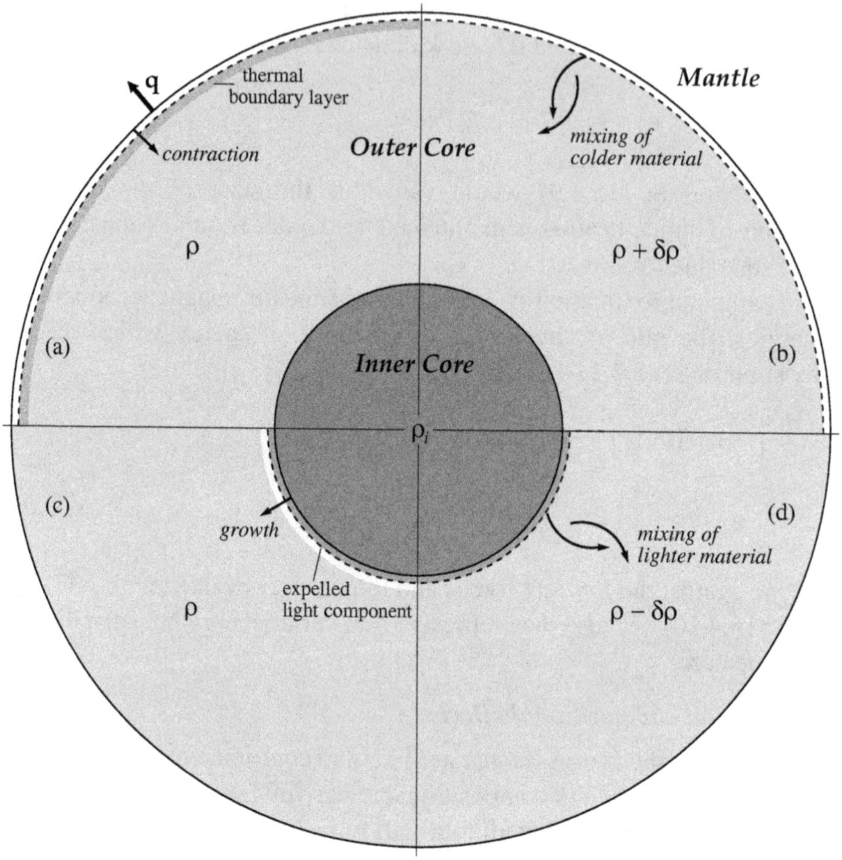

Figure 15.6. Core processes contributing to thermal and compositional release of gravitational energy: (a) heat flux extracted from the core leads to local contraction near the core mantle boundary, (b) the cold, dense thermal boundary layer becomes unstable and mixes deeper into the core, (c) lighter elements are expelled from the inner core in growth by solidification, (d) the light compositional boundary layer becomes unstable and mixes into the overlying fluid.

15.4.3 Inner core growth in a well-mixed core

The cooling of the core is controlled, in large part, by the much more massive mantle with relatively sluggish flows. The mantle regulates the total heat flux $\mathsf{Q}(\mathsf{t}) = \int_{S_c} (\mathsf{q}) \cdot \mathbf{dS}$ across the core-mantle boundary, and the magnitude and time dependence of $\mathsf{Q}(\mathsf{t})$ will depend on the details of mantle convection.

The part of the thermal heat flux that is significant is that part that is not associated with thermal conduction along the adiabat, i.e. $\mathsf{Q}(\mathsf{t})/\mathsf{Nu}$, where the Nusselt number Nu was introduced in (7.5.2). The effective heat flux

$$\int_{S_c} \mathbf{q}^* \cdot \mathbf{dS} = \left(1 - \frac{1}{\mathsf{Nu}}\right) \mathsf{Q}(\mathsf{t}). \tag{15.4.11}$$

Diffusive transport of light elements along the pressure gradient with radius is expected to be slower and less significant than thermal diffusion. So all the gravitational energy released by compositional segregation will contribute to Φ (15.4.9), but only the convective heat flux $\mathbf{q}^*$ contributes to the thermal part.

There is a density jump $(\Delta\rho)_\zeta$ across the boundary between the inner and outer cores $r = r_i$ associated with the compositional changes, which will be less than the seismological estimates that include the phase change to the solid state. The solidification at this boundary will release latent heat L per unit mass. It is also convenient to introduce the mass-averaged gravitational potential of the outer core,

$$\bar{\varphi} = \frac{1}{M_{V_{oc}}} \int_{oc} \rho\varphi \, dV. \tag{15.4.12}$$

Buffet et al. (1996) show that the rate of ohmic dissipation Φ under the assumption of efficient convective mixing of compositional and thermal anomalies can be expressed as

$$\Phi = 4\pi r_i^2 \frac{dr_i}{dt} \left[(\Delta\rho)_\zeta - \frac{\alpha_{th}\rho_0 L}{C_p} \right] [\bar{\varphi} - \varphi(r_i)] \\ - \frac{\alpha_{th}}{C_p} \int_{S_c} \mathbf{q}^* \cdot d\mathbf{S} \, [\bar{\varphi} - \varphi(r_c)], \tag{15.4.13}$$

where, as in (15.2.18), α_{th} is the coefficient of thermal expansion. If mixing is not efficient, diffusion would smooth out the effects before the motion had a chance to produce Joule heating. We can think of (15.4.13) as representing the effect of the redistribution of a compositional mass deficit from the inner core and a mass excess from the thermal boundary layer at the top of the core. The contribution $\Phi_\zeta = 4\pi r_i^2 (\Delta\rho)_\zeta \, dr_i/dt$ is purely compositional, and the remainder $\Phi_T = \Phi - \Phi_\zeta$ is thermal.

The contributions of compositional changes to the internal energy will be dominated by the other terms in the internal heat budget of the core, so that to an good approximation we can write (15.4.7) as

$$\int_{V_c} \rho T \frac{DS}{Dt} \, dV = -Q(t) + \int_{V_c} \frac{J^2}{\breve{\sigma}} \, dV, \tag{15.4.14}$$

$$= \int_{V_c} \rho C_p \frac{DT}{Dt} \, dV - \int_{V_c} \alpha_{th} T \frac{Dp}{Dt} \, dV - \int_{S_i} \rho L \frac{dr_c}{dt} \, dS, \tag{15.4.15}$$

where S_i is the surface of the inner core. Small effects, such as heat generated by mixing, are neglected in (15.4.15).

Buffet et al. (1996 - Appendix B) demonstrate that the advective term $\int_{V_c} \rho T(\mathbf{v} \cdot \boldsymbol{\nabla} S) dV$ in (15.4.15) cancels out the thermal contribution Φ_T to the convective energy release. As a result (15.4.15) can be rearranged to the form

$$Q(t) = -\int_{V_c} \rho C_p \frac{\partial T}{\partial t} \, dV + \int_{V_c} \alpha_{th} T \frac{\partial p}{\partial t} \, dV + \int_{S_i} \rho L \frac{dr_i}{dt} \, dS + \Phi_C. \tag{15.4.16}$$

The changes in the temperature T and the pressure P appearing in (15.4.16) can be estimated by looking at the changes in the average conditions in the core over time scales long compared with those of convective fluctuations, but short compared with the time over which the Earth cools. Thus

$$\frac{\partial p}{\partial r} = -\rho g, \qquad \frac{\partial T}{\partial r} = -\frac{\rho g \gamma_{th} T}{K_S}, \qquad r_i < r < r_c, \tag{15.4.17}$$

where γ_{th} is the Grüneisen parameter (6.1.13). The equations (15.4.17) need to be supplemented with the condition that the composition is uniform on these time scales in $r_i < r < r_c$. The acceleration due to gravity g, given by (6.1.1), will depend on P, T and composition through the density ρ. The pressure p must vanish at the free surface, and thus takes the form (6.1.2). The core liquid is regarded as an ideal mixture, with the light component present only in the fluid outer core. The mass fraction ζ is given by

$$\zeta(t) = M_l/M_{oc}(t), \quad r_i < r < r_c, \tag{15.4.18}$$

where M_l is the total mass of light element and $M_{oc}(t)$ is the slowly decreasing mass of the outer core as the heavy component freezes onto the inner core.

The temperature is controlled by the requirement that the surface of the inner core is in thermodynamic equilibrium with the surrounding fluid, i.e., $T(r_i, t)$ is equal to the liquidus temperature $T_L(p, \zeta)$. An approximate linear expansion for T_L can be made about the conditions where the inner core just begins to form, (p_0, ζ_0),

$$T_L(p,\zeta) = T_L(p_0,\zeta_0) + \frac{\partial T_L}{\partial p}(p - p_0) + \frac{\partial T_L}{\partial \zeta}(\zeta - \zeta_0). \tag{15.4.19}$$

The material properties ρ, K_S, γ_{th} will be controlled by the equation of state and will vary in time. The full set of equations describing the evolution of the core will therefore require numerical solution. However, as demonstrated by Buffet et al. (1996), the main properties can be found through analytical approximations based on power series that yield results within 8% of those from more detailed numerical models including the effect of self-gravitation.

The density and thermodynamic parameters are approximated by their values ρ_0, K_0, γ_0 at the centre of the Earth so that there are simple solutions for p and T as a function of radius:

$$p(r) = p_0 + Ar^2, \quad \text{with} \quad A = 2\pi G \rho_0^2/3, \tag{15.4.20}$$

and

$$T(r) = T_L(r_i) e^{-\phi(r^2 - r_i^2)/r_c^2}, \quad \text{with} \quad \phi = A r_c^2 \gamma_0 / K_0. \tag{15.4.21}$$

The liquidus temperature $T_L(r_i)$ as a function of inner core radius r_i takes the form

$$T_L(r_i) = \frac{A r_c^2}{\phi}\left(\frac{\partial T}{\partial p}\right)_0 - \frac{\partial T_L}{\partial p} A r_i^2 - \frac{\partial T_L}{\partial \zeta}\left(\frac{r_i^3}{r_c^3 - r_i^3}\right)\zeta_0, \tag{15.4.22}$$

where $(\partial T/\partial p)_0$ is the adiabatic gradient evaluated at $p = p_0$.

Because we have made the assumption of constant density ρ_0 the pressure does not change with time, $\partial p/\partial t = 0$, and the energy balance does not depend on pressure. This is an oversimplification, but the associated error is small. This means that once the total heat flux $Q(t)$ is prescribed, the energy balance only depends on the rates of change of temperature $\partial T/\partial t$ and the rate of change of the inner-core radius dr_i/dt. These quantities are related by

$$\frac{\partial T(r,t)}{\partial t} = \frac{d}{dr_i}\left(T_L(r_i)e^{\phi(r_i^2/r_c^2)}\right)e^{-\phi(r^2/r_c^2)}\frac{dr_i}{dt}, \tag{15.4.23}$$

where $T_L(r_i)$ is specified by (15.4.22). Because the temperature variation across the core is not large, $\phi = 0.26$, and we can make the approximation

$$e^{-\phi(r^2-r_i^2)/r_c^2} = 1 - \phi(r^2 - r_i^2)/r_c^2 + O(\phi^2). \tag{15.4.24}$$

We can now evaluate the volume integral over $\partial T/\partial t$ that is required in (15.4.16) as

$$\int_{V_c} \rho_0 C_p \frac{\partial T}{\partial t}\, dV = H\left(\frac{dT_L(r_i)}{dr_i} + 2\frac{r_i\phi}{r_c^2}T_L(r_i)\right)\frac{dr_i}{dt}, \tag{15.4.25}$$

where the scaling factor

$$H = \frac{2\pi}{3} r_c^3 \rho_0 C_p \left[1 - \phi\left(\frac{3}{5} - \frac{r_i^2}{r_c^2}\right)\right]. \tag{15.4.26}$$

The form (15.4.25) implicitly assumes that the inner core is adiabatic, which will be a reasonable form until the core is almost totally solid.

The gravitational potential for the pressure distribution (15.4.20) is $\varphi = Ar^2/\rho_0 + const$ and so the average potential in the outer core, which is needed in (15.4.13) , can be found from

$$\bar{\varphi} = \frac{3A}{5\rho_0}\frac{r_i^5 - r_c^5}{r_i^3 - r_c^3} + const; \tag{15.4.27}$$

the constant does not present any problem since we need differences in potential.

With the aid of the representation (15.4.25) we can express the heat balance equation (15.4.16) in terms of the radius of the inner core $r_i(t)$ by eliminating the dependence on temperature. In the level of approximation to which we are working we can neglect the small effect of thermal contraction at the top of the core, and introduce the fractional inner core radius

$$\xi_i = \frac{r_i(t)}{r_c}. \tag{15.4.28}$$

The heat flux equation (15.4.16) can then be rewritten as an ordinary differential equation for ξ_i of the form $f(\xi_i)d\xi_i/dt = Q(t)$, with the heat flux $Q(t)$ as a forcing term. The terms appearing in $f(\xi_i)$ are rational functions of ξ_i and so an analytical integral for $\xi_i(t)$ can be found once $Q(t)$ is specified.

For the current inner core $\xi_i = 0.349$ and so $\xi_i^4 = 0.015$, we can therefore make a further approximation of neglecting terms higher than cubic in ξ_i in the differential equation for the inner core radius, and then the solution can be expressed as

$$\mathcal{M}^{-1}\int_0^t Q(\tau)d\tau = \xi_i^2 + (\mathcal{G}_\zeta + \mathcal{L} - \mathcal{Z})\xi_i^3 + O(\xi_i^4). \qquad (15.4.29)$$

The time t is thus measured from the instant at which the temperature falls below the liquidus and the inner core begins to grow.

The factor $\mathcal{M}$ in (15.4.29) serves to make the heat flux dimensionless,

$$\mathcal{M} = 4\pi\left(\frac{1}{3} - \frac{\Phi}{5}\right)\rho_0 C_p A r_c^5\left[\frac{\partial T_L}{\partial p} - \left(\frac{\partial T}{\partial p}\right)_0\right], \qquad (15.4.30)$$

which is the heat that is required to cool the entire core to its solidification temperature.

The set of parameters on the right hand side of (15.4.29) describe the relative proportions of the various physical processes in the evolution of the core:

$$\mathcal{Z} = \frac{\zeta_0}{Ar_c^2}\frac{\partial T_L}{\partial \zeta}\left[\frac{\partial T_L}{\partial p} - \left(\frac{\partial T}{\partial p}\right)_0\right]^{-1},$$
$$\mathcal{G}_\zeta = \frac{4\pi}{5}\frac{Ar_c^5}{\mathcal{M}}\frac{(\Delta\rho)_\zeta}{\rho_0}, \qquad (15.4.31)$$
$$\mathcal{L} = \frac{4\pi}{3}\frac{\rho L r_c^3}{\mathcal{M}}.$$

The quantity $\mathcal{Z}$ represents the effect of composition on the liquidus temperature, $\mathcal{G}_\zeta$ the gravitational energy release and ohmic dissipation due to compositional segregation and $\mathcal{L}$ the effect of latent heat release.

Many of the quantities needed for evaluating (15.4.29) are not well known, but the greatest uncertainty is in the difference in temperature gradients appearing in $\mathcal{M}$, $\mathcal{Z}$. This difference can be cast into a slightly more helpful form by using Lindemann's relation to describe the melting curve

$$\frac{\partial T_L}{\partial p} = 2(\gamma_{th} - \tfrac{1}{3})\frac{T_L}{K_T}, \qquad (15.4.32)$$

where K_T is the isothermal bulk modulus (see, e.g., Stacey, 1992). This relation is based on the concept that melting will occur when the amplitude of atomic vibrations about their equilibrium position exceeds a certain fraction of the interatomic distance in the basic lattice.

The adiabatic temperature gradient takes the form $\gamma_{th}T/K_S$ in terms of the isentropic modulus K_S. The difference in the gradient at the centre of the Earth is thus approximately

$$\left[\frac{\partial T_L}{\partial p} - \left(\frac{\partial T}{\partial p}\right)_0\right] = \frac{T_L(0)}{K_T}\left[\left(2 - \frac{K_T}{K_S}\right)\gamma_{th} - \frac{2}{3}\right]; \qquad (15.4.33)$$

estimates of the central temperature show considerable variation in the range 4000–6000 K that will affect the size of the gradient difference.

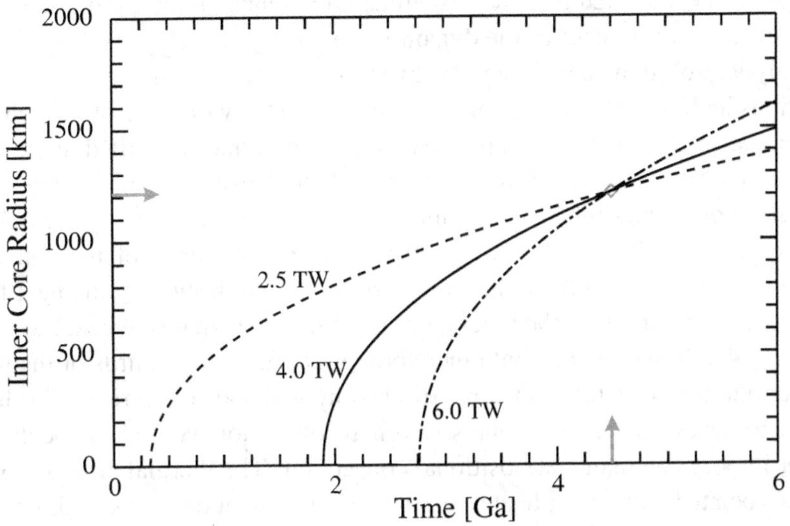

Figure 15.7. Evolution of the radius of the inner core to the present-day values and into the future for three estimates of the net flux across the core-mantle boundary (in TW = 10^{12} W) using the analytic approximation (15.4.34). The indicators mark the current Earth conditions.

The analytic equation for the growth of the inner core (15.4.29) can be used to develop an explicit expression for ξ_i to the same level of approximation as before:

$$\xi_i(t) = \xi_o(t) - \delta\xi_i(t). \tag{15.4.34}$$

The leading order approximation is controlled just by the heat loss into the mantle

$$\xi_o(t) = \left[\int_0^t Q(\tau)d\tau/\mathcal{M}\right]^{1/2}, \tag{15.4.35}$$

whereas the correction term

$$\delta\xi_i(t) = \frac{(\mathcal{G}_\zeta + \mathcal{L} - \mathcal{Z})\xi_o^2}{2 + 3(\mathcal{G}_\zeta + \mathcal{L} - \mathcal{Z})\xi_o} + O(\xi_o^4) \tag{15.4.36}$$

includes the influence of latent heat and gravitational energy associated with compositional change.

Figure 15.7 displays the results from (15.4.34) using the numerical values for the parameters suggested by Buffet et al. (1996), for different assumptions about the net heat flux into the mantle. The growth curves for the inner core have been adjusted so that they pass through the current Earth conditions. A low mantle heat flux means that an inner core needs to form early in Earth history, whereas more rapid cooling allows the process to be delayed. The solid curve for an average heat flux of 4.0 TW is representative of current estimates. A feature of all the solutions is the relatively rapid initial growth of the inner core; once solidification starts an inner core several hundred kilometres in radius is formed within 200 Ma.

This change can be expected to have a significant influence on the evolution of the magnetic field as the character of the dynamo changes.

The analytic solution has been obtained with the aid of a number of simplifications including neglect of self-gravitation and the variations of the density with radius. Despite the limitations the results agree with more detailed numerical solutions within 8%, which is less than the error in the density approximation! The numerical solutions give slightly more rapid growth for the same average heat flux.

From the analysis above, we see that the relative importance of thermal and compositional convection in the outer core is primarily controlled by the heat flux from the core and the radius of the inner core. For current estimates of heat fluxes of the order of 3–4 TW and the present inner core radius, about two thirds of the heat available for ohmic dissipation can come from compositional convection. This heat flux is close to the adiabatic value, and so thermal convection would be expected to be of lesser importance than compositional convection. The thermal effects would largely be associated with latent heat release. When the inner core was smaller, and the heat flux to the mantle most likely larger, thermal convection would have been the dominant energy source for the geodynamo.

Appendix: Table of Notation

Stress and Strain

$\mathbf{x}$	-	position vector
x_i	-	position coordinates
$\boldsymbol{\xi}$	-	initial position vector
ξ_i	-	initial position coordinates
$\mathbf{u}$	-	displacement vector
$\mathbf{v}$	-	velocity vector
$\mathbf{f}$	-	acceleration vector
$\mathbf{g}$	-	external force vector
$\boldsymbol{\nu}$	-	local body moment
$\mathbf{n}$	-	normal vector, orthogonal vector
F	-	deformation gradient
$J = \det \mathsf{F}$	-	Jacobian, determinant of deformation gradient
A	-	displacement gradient
$d\mathbf{S}, d\boldsymbol{\Sigma}$	-	surface elements
dV, dm	-	volume, mass elements
σ_{ij}	-	stress tensor
σ_r	-	principal stresses
σ^{PK}, σ^{SK}	-	first and second Piola–Kirchhoff tensors
$\boldsymbol{\sigma}, \mathbf{t}$	-	traction vectors
$\boldsymbol{\tau}$	-	traction vector at a surface
λ_r	-	stretches
U, V	-	left and right stretch tensors
$\mathsf{R}, \mathsf{Q}, \mathsf{P}$	-	rotation matrices
$\mathsf{N} \equiv \mathbf{n}\times$	-	skew matrix
$\mathsf{E}, \mathbf{e}$	-	Green strain tensor, Cauchy strain tensor
D/Dt	-	material derivative
p	-	pressure
Π_{ij}	-	momentum flux density tensor

Solids and Fluids

ϵ	-	strain
e_{ij}	-	strain tensor
D_{ij}	-	rate of deformation tensor
$\hat{D}_{ij}$	-	deviatoric strain rate tensor
ϖ_{ij}	-	spin tensor, vorticity tensor
c_{ijkl}	-	elastic modulus tensor
C_{ijkl}	-	anelastic relaxation tensor
κ, K	-	bulk modulus
μ, G	-	shear modulus
ρ	-	density
Φ	-	seismic parameter
λ	-	Lamé modulus
E	-	Young's modulus
υ	-	Poisson's ratio
φ	-	gravitational potential
p, p', p_f	-	pressure, incremental pressure, fluid pressure
g	-	acceleration due to gravity
η	-	shear viscosity
ζ	-	bulk viscosity
G_i^k	-	elastic Green's tensor
$\phi(t)$	-	relaxation function
$\psi(t)$	-	creep function, stream function
R_λ, R_μ	-	isotropic relaxation functions
τ_R, τ_C	-	relaxation times
$\mathbf{\Omega}$	-	angular velocity vector
Ω	-	angular velocity

Waves

ω	-	angular frequency
θ, ϕ	-	coordinate angles
α	-	*P* wavespeed
β	-	*S* wavespeed
a	-	*P* slowness $(1/\alpha)$
b	-	*S* slowness $(1/\beta)$
ϕ, c_0	-	wavespeed of sound waves in a fluid
p	-	horizontal slowness, phase slowness
$\mathbf{n}$	-	direction of travel
$\mathbf{s}$	-	ray direction
$\mathbf{p}$	-	slowness vector
Q^{-1}	-	loss factor

Thermodynamic Quantities

h	-	heat production
$\mathbf{q}$	-	heat flux vector
T	-	temperature
k	-	thermal conductivity
κ_H	-	thermal diffusivity
u	-	internal energy density
H	-	thermal energy, enthalpy
Y	-	work rate
W	-	work density, strain energy density
Q	-	thermal contribution to internal energy
$\mathcal{U}$	-	total energy
$\mathcal{F}, F$	-	Helmholtz free energy
$\mathcal{G}, G$	-	Gibbs free energy
$\mathcal{S}, S$	-	entropy
V	-	volume
R	-	gas constant
α_{th}	-	volume thermal expansion coefficient
γ_{th}	-	thermodynamic Grüneisen parameter
K_T, K_S	-	isothermal, adiabatic bulk modulus
C_P, C_V	-	specific heat at constant pressure, volume
p_H, e_H	-	pressure and energy on Hugoniot
L	-	latent heat
γ_c	-	Clapeyron slope

Non-dimensional Quantities for Fluids

Ek	-	Ekman number
Nu	-	Nusselt number
Pe	-	Péclet number
Pr	-	Prandtl number
Ra	-	Rayleigh number
$\widetilde{Ra}$	-	modified Rayleigh number for rotating system
Re	-	Reynolds number
q	-	Roberts number
Ro	-	Rossby number
Ro_M	-	magnetic Rossby number
θ	-	non-dimensional temperature

Electromagnetic

$\mathbf{E}$	-	electric vector
$\mathbf{B}$	-	magnetic induction
$\mathbf{D}$	-	electric displacement
$\mathbf{H}$	-	magnetic vector
$\mathbf{j}$	-	electric current density
$\breve{\rho}$	-	electric charge density
$\hat{\mathbf{j}}$	-	surface current density
$\hat{\rho}$	-	surface charge density
$\breve{\sigma}$	-	conductivity
μ_0, ϵ_0	-	magnetic permeability, permittivity of free space
μ, ϵ	-	magnetic permeability, permittivity
c	-	speed of light
$\gamma_{i,jk}$	-	piezoelectric tensor
λ	-	magnetic diffusivity $= 1/\mu_0\breve{\sigma}$
$\mathcal{W}$	-	stored energy in magnetic field
$\mathcal{Q}$	-	dissipation due to Joule heating
$\mathbf{S}$	-	Poynting vector
$\mathbf{Q}$	-	energy flux vector
d	-	penetration depth
n	-	complex refractive index
ϕ	-	phase difference

Quantum/Atomic

ψ	-	electron state
H	-	Hamiltonian
V	-	potential
T	-	kinetic energy
$\mathbf{D}$	-	dynamical matrix
$\mathbf{G}$	-	force tensor
μ	-	chemical potential
ρ_{ij}	-	interatomic interactions
$\omega_{\mathbf{q}}$	-	frequency of quantum state
$\mathcal{E}$	-	energy in lattice modes
H^*, E^*, V^*	-	activation quantities
$c, \mathbf{J}$	-	species concentration, flux
D	-	diffusivity
$\mathbf{b}$	-	Burgers vector
Θ_D	-	Debye temperature
k_B, R	-	Boltzmann constant, gas constant

Deformation

d	-	grain size
T_M	-	melting temperature
n, m	-	grain size exponent, stress exponent in creep dependence
σ_N	-	normal stress
σ_S	-	shear stress
$\Delta\sigma$	-	deviatoric stress
p_f	-	pore fluid pressure
λ_f	-	ratio between fluid and lithostatic pressure
$\sigma_1, \sigma_2, \sigma_3$	-	prevailing principal stresses
δ_s	-	slab dip angle
V_s	-	slab convergence rate
t_s	-	spreading time
$\mathfrak{R}$	-	thermal Reynolds number
Q_0	-	surface heat flow
T_M	-	mantle temperature
Δg	-	gravity anomaly
Z	-	gravitational admittance
D	-	flexural rigidity of a plate
T_e	-	effective elastic thickness
h, k, l	-	Love numbers
ζ	-	sea level function
b	-	bouyancy flux

Seismology

v	-	wavespeed
m_{ij}	-	moment tensor density
M_{ij}	-	moment tensor
$\mathfrak{M}_{ij}$	-	moment rate tensor
r	-	radial variable
$\mathfrak{H}$	-	modal operator
$\mathbf{u}_I^e$	-	modal eigendisplacement
c_I	-	modal coefficients
$W; T$	-	displacement; traction coefficients for toroidal modes
$U, V; P, S$	-	displacement; traction coefficients for spheroidal modes
Φ, Ψ	-	gravitational coefficients for spheroidal modes
$\mathfrak{b}$	-	radial column vector for spheroidal modes
$\mathcal{B}$	-	boundary term for modal estimate
$\mathcal{T}$	-	kinetic energy for modal estimate
$\mathfrak{T}$	-	kinetic energy per unit volume
$\mathfrak{V}$	-	potential energy per unit volume
G	-	gravitational constant

Core dynamics

P	-	augmented pressure
ζ	-	mass fraction of light constituent in core
$\mathbf{i}$	-	mass flux of light constituent
Θ	-	temperature deviation
θ	-	non-dimensional temperature perturbation
r_i, r_c	-	radii of inner and outer cores
ξ_i	-	fractional inner core radius

Mathematical

a, b, c, d	-	constants
$j_l(x), h_l^{(1),(2)}(x)$	-	spherical Bessel functions
$z_l(x)$	-	combination of spherical Bessel functions
$P_l(x), P_l^m(x)$	-	Legendre functions
$Q_l^{(1),(2)}$	-	travelling wave form of Legendre function
$\mathbf{P}_l^m, \mathbf{B}_l^m, \mathbf{C}_l^m$	-	vector surface harmonics (spherical)
Y_l^m	-	surface harmonics on a sphere
l	-	angular order of spherical harmonics
$\L$	-	$[l(l+1)]^{1/2}$

Bibliography

Reference Texts

General Continuum Mechanics

Hunter, S.C., 1976. - *Mechanics of a Continuous Medium*, Ellis Horwood.

Spencer, A.J.M., 1980. - *Continuum Mechanics*, Longman; reprinted by Dover.

Malvern, L.E., 1969. - *Introduction to the Mechanics of a Continuous Medium*, Prentice Hall.

Elasticity

Atkin, R.J. & Fox, N., 1980. - *An Introduction to the Theory of Elasticity*, Longman.

Sokolnikoff, I.S., 1956. - *Mathematical Theory of Elasticity*, McGraw-Hill.

Hudson, J.A., 1981. - *The Excitation and Propagation of Elastic Waves*, Cambridge University Press.

Kennett, B.L.N., 2001. *The Seismic Wavefield I: Introduction and Theoretical Development*, Cambridge University Press.

Nye, J.F., 1979. - *Physical Properties of Crystals*, Oxford University Press.

Fluid Mechanics

Batchelor, G.K., 1967. - *An Introduction to Fluid Mechanics*, Cambridge University Press.

Lautrop, B., 2005. - *Physics of Continuous Matter*, Institute of Physics Publishing.

Landau, L.D. & Lifshitz, E.M., 1987. - *Fluid Mechanics*, Second edition, Pergamon Press.

Geodynamic Applications

Turcotte, D.L. & Schubert, G., 2002. - *Geodynamics: Applications of Continuum Physics to Geological Problems*, second edition, Cambridge University Press.

Davies, G.F., 1998. - *Dynamic Earth: Plates, Plumes and Mantle Convection*, Cambridge University Press.

Continuum Mechanics for the Earth

Dalhen, F.A. & Tromp, J., 1998. - *Theoretical Global Seismology*, Princeton University Press.

Chapters 2 and 3 provide a detailed account of the application of continuum mechanics in a quasi-spherical body with self-gravitation. Particular attention is given to the description of the stress state in Lagrangian frames rather than the Eulerian frame.

Poirier, J.-P., 1991. - *Introduction to the Physics of the Earth's Interior*, Cambridge University Press

Ranalli, G., 1987. - *Rheology of the Earth*, Second edition, Allen & Unwin.

Seismology

Kennett, B.L.N., 2002. - *The Seismic Wavefield II: Interpretation of Seismograms on Regional and Global Scales*, Cambridge University Press.

Kasahara, K., 1981. - *Earthquake Mechanics*, Cambridge University Press.

Stein, S. & Wysession, M., 2003. - *An Introduction to Seismology, Earthquakes and Earth Structure*, Blackwell Publishing.

The Interior of the Earth

Fowler, C.M.R., 2005. - *The Solid Earth*, Second edition, Cambridge University Press.

A fine broad survey of geophysics.

Davies, G.F., 1998. - *Dynamic Earth: Plates, Plumes and Mantle Convection*, Cambridge University Press.

Karato, S.-I., 2003. - *The Dynamic Structure of the Deep Earth*, Cambridge University Press.

Both these books offer highly personal but engaging accounts of the interior of the Earth with links to mineral physics, geochemistry and geodynamics.

Geological deformation

Hobbs, B.E., Means, W.D. & Williams, P.F., 1976. - *An Outline of Structural Geology*, John Wiley.

Suppe, J., 1985. - *Principles of Structural Geology*, Prentice-Hall.

Leeder M. & Pérez-Arlucea, M., 2006. - *Physical Processes in Earth and Environmental Sciences*, Blackwell Publishing.

Heat Conduction

Carslaw, H.S. & Jaeger, J.C., 1959. - *Conduction of Heat in Solids*, Oxford University Press.

References

Ahrens, T.J., 1987. Shock wave techniques for geophysics and planetary physics, in *Methods of Experimental physics*, Eds. C.G. Samnmis and T.L. Henyey, Academic Press, **24A**, 185–235.

Ahrens, T.J. & Johnson, M.L., 1995a. Shock wave data for minerals, in *Mineral Physics and Crystallography: A Handbook of Physical Constants*, American Geophysical Union, Washington.

Ahrens, T.J. & Johnson, M.L., 1995b. Shock wave data for rocks, *Rock Physics and Phase Relations: a Handbook of Physical Constants*, American Geophysical Union, Washington.

Allen, M.P. & Tildesley, D.J., 1987. *Computer Simulation of Liquids*, Oxford University Press.

Ammon, C.J., Ji, C., Thio, H.-K., Robinson, D., Ni, S., Hjorleifsdottir, V., Kanamori, H., Lay, T., Das, S., Helmberger, D.V., Ichinose, G., Polet, J. & Wald, D., 2005. Rupture process of the 2004 Sumatra–Andaman earthquake, *Science*, **308**, 1133–1139.

Anderson, D.L., 1982. Hotspots, polar wander, Mesozoic convection and the geoid, *Nature*, **297**, 391–393.

Anderson, D.L. & Minster, J.B., 1979. The frequency dependence of Q in the Earth and implications for mantle rheology and Chandler wobble, *Geophys. J. R. Astr. Soc.*, **58**, 431–440.

Anderson, E.M., 1951. *The Dynamics of Faulting*, Oliver & Boyd, Edinburgh.

Barnhoorn, A., Bystricky, M. Burlini, L. & Kunze, K., 2004. The role of recrystallisation on the deformation behaviour of calcite rocks: large strain torsion experiments on Carrara marble. *J. Struct. Geol.*, **26**, 885–903.

Ben-Ismail, W. & Mainprice, D., 1998. A statistical view of the strength of seismic anisotropy in the upper mantle based on petrofabric studies of ophiolite and xenolith samples. *Tectonophysics*, **296**, 145–157.

Batt, G.E. & Braun, J., 1999. The tectonic evolution of the Southern Alps, New Zealand: insights from fully thermally coupled dynamical modeling, *Geophys. J. Int.*, **136**, 403–421.

Bercovici, D., Ricard, Y. & Richards, M.A., 2000. The relation between mantle dynamics and plate tectonics: a primer, in *The History and Dynamics of Global Plate Motions*, eds. M.A. Richards, R.G. Gordon & R.D. van der Hilst, AGU Geophysical Monograph, **121**, 113–137.

Bercovici, D. & Karato, S.-I., 2002. Theoretical analysis of shear localization in the lithosphere, *Rev. Mineral. Geochem.*, **51**, 387–420.

Beroza, G.C. & Zoback, M.D., 1993. Mechanism diversity of the Loma Prieta aftershocks and mechanisms for mainshock–aftershock interaction, *Science*, **259**, 210–213.

Besse, J., & Courtillot, V., 2002. Apparent and true polar wander and the geometry of the geomagnetic field over the last 200 Myr, *J. Geophys. Res*, **107**, B11:2300.

Birch, F., 1952. Elasticity and constitution of the Earth's interior, *J. Geophys. Res.*, **57**, 227–286.

Bloxham, J. & Gubbins, G., 1989. Geomagnetic secular variation, *Phil. Trans. R. Soc. Lond.*, **329A**, 415–502.

Born, M. & Huang, K., 1954. *Dynamical Theory of Crystal Lattices*, Oxford University Press, Oxford.

Brace, W.F. & Kohlstedt, D.L., 1980. Limits on lithospheric stress imposed by laboratory experiments, *J. Geophys. Res.*, **85**, 6248–6252.

Braginsky, S.L. & Roberts, P.H., 1995. Equations governing convection in Earth's core and the geodynamo, *Geophys. Astrophys. Fluid Dyn.*, **79**, 1–97.

Briggs W.L, 1987. *Multigrid Tutorial*, Society for Industrial & Applied Mathematics.

Braun, J. & Sambridge, M., 1994. Dynamical Lagrangian Remeshing (DLR): A new algorithm for solving large strain deformation problems and its application to fault-propagation folding, *Earth Planet. Sci. Lett.*, **124**, 211–220.

Buffet, B.A., 1998. Free oscillations in the length of the day: Inferences on physical processes near the core–mantle boundary, in *The Core–Mantle Boundary Region*, eds. M. Gurnis, M.E. Wysession, E. Knittle & B.A. Buffet, Geodynamics Monograph, **28**, American Geophysical Union.

Buffet, B.A., Huppert, H.E., Lister, J.R. & Woods, A., 1996. On the thermal evolution of the Earth's core, *J. Geophys. Res.*, **101**, 7989–8006.

Bullard, E., Everett, J.E., & Smith, A.G., 1965. The fit of the continents around the Atlantic, *Phil. Trans. R. Soc. Lond. A*, **222**, 41–51.

Bullen, K.E., 1975. *The Earth's Density*, Chapman & Hall, London.

Bunge, H.-P. & Baumgardner, J.R., 1995. Mantle convection modeling on parallel virtual machines, *Computers in Physics*, **9**, 207–215.

Bunge, H.-P. & Richards, M.A., 1996, The origin of long-wavelength structure in mantle convection: effects of plate motions and viscosity stratification, *Geophys. Res. Lett.*, **23**, 2987–2990.

Bunge, H.P., Richards, M.A. & Baumgardner, J.R., 2002. Mantle-circulation models with sequential data assimilation: inferring present-day mantle structure from plate-motion histories, *Phil Trans. R. Soc. Lond.*, A**360**, 2545–2567.

Bunge, H.-P., Hagelberg, C.R. & Travis, B.J., 2003. Mantle circulation models with variational data assimilation: inferring past mantle flow and structure from plate motion histories and seismic tomography, *Geophys. J. Int.*, **152**, 280–301.

Bunge, H.-P., Richards, M.A., Lithgow-Bertelloni, C., Baumgardner, J.R., Grand, S. & Romanowicz, B., 1998. Time scales and heterogeneous structure in geodynamic earth models, *Science*, **280**, 91–95.

Burov, E.B. & Watts, A.B., 2006. The long-term strength of the continental lithosphere: "jelly-sandwich" or "crème brûlée"?, *GSA Today*, **16**(1), 4–10.

Busse, F.H., Richards, M.A & Lenardic, A., 2006. A simple model of high Prandtl and high Rayleigh number convection bounded by thin low-viscosity layers, *Geophys. J. Int.*, **164**, 160–167.

Campbell, I.H., 2007. Testing the plume theory, *J. Chem. Geol.*, **241**, 153–176.

Catlow, C.R.A., 2003. Computer modelling of materials: an introduction, 1–29, in *Computational Materials Science*, Eds. C.R.A. Catlow & E.A. Kotomin, IOS Press, Amsterdam.

Chaikan, P.M. & Lubensky, T.C., 1995, *Principles of Condensed Matter Physics*, Cambridge University Press.

Chase C.G. & Sprowl, D.R., 1983. The modern geoid and ancient plate boundaries, *Earth Planet. Sci. Lett.*, **62**, 314–320.

Christensen, U.R., 1992. An Eulerian technique for thermomechanical modeling of lithosphere extension, *J. Geophys. Res.*, **97**, 2015–2036.

Courtillot, V.E., & Renne, P.R., On the ages of flood basalt events, *Compt. Rend. Geoscience*, **335**, 113–140.

Cox, S.F. & Ruming, K., 2004. The St Ives mesothermal gold system, Western Australia – a case of golden aftershocks?, *J. Structural Geology*, **26**, 1109–1125.

Creager, K.C., 1999. Large scale variations in inner core anisotropy, *J. Geophys. Res.*, **104**, 23 127–23 139.

Davies, G.F., Ocean bathymetry and mantle convection: 2. Small-scale flow, *J. Geophys. Res.*, **93**, 10 481–10 488.

Debayle, E., Kennett, B., & Priestley, K., 2005. Global azimuthal seismic anisotropy: the unique plate-motion deformation of Australia, *Nature*, **433**, 509–512.

Dahlen, F.A. & Tromp, J., 1998. *Theoretical Global Seismology*, Princeton University Press, Princeton.

Davies G.F., 1984. Lagging mantle convection, the geoid and mantle structure *Earth Planet. Sci. Lett.*, **69**, 187–194.

DeMets, C., Gordon, R.G., Argus, D.F., & Stein, S., 1990. Current plate motions, *Geophys. J. Int.*, **101**, 425–478.

DeMets, C., Gordon, R.G., Argus, D.F., & Stein, S., 1994. Effect of recentrevisions to the geomagnetic reversal time-scale on estimates of current plate motions, *Geophys. Res. Lett.*, **21**, 2191–2194.

Dixon, T.H., 1991. An introduction to the global positioning system and some geological applications *Rev. Geophys.*, **29**, 249–276.

de Wijs, G.A., Kresse, G. & Gillan, M.J., 1998. First order phase transitions by first principles free energy calculations: The melting of Al, *Phys. Rev. B*, **35**, 8233–8234.

Duess, A., Woodhouse, J.H., Paulssen, H. & Trampert, J., 2000. The observation of inner core shear phases, *Geophys. J. Int.*, **142**, 67–73.

Dziewonski, A.M. & Anderson D.L., 1981. Preliminary reference Earth model, *Phys. Earth Planet. Inter.*, **25**, 297–356.

England, P. & Wilkins, C., 2004. A simple analytic approximation to the temperature structure in subduction zones, *Geophys. J. Int.*, **159**, 1138–1154.

Engdahl, E.R., van der Hilst, R.D. & Buland, R., 1998. Global teleseismic earthquake relocation with improved travel times and procedures for depth determination, *Bull Seism. Soc. Am.*, **88**, 722–743.

Evans, B. & Kohlstedt, D.L., 1995, Rheology of rocks, in *Rock Physics and Phase Relations: A Handbook of Physical Constants*, American Geophysical Union, Washington.

Faul, U.H. & Jackson, I., 2005. The seismological signature of temperature and grain size variations in the upper mantle, *Earth Planet. Sci. Lett.*, **234**, 119–134.

Fishwick, S., Kennett, B.L.N. & Reading, A.M., 2005. Contrasts in lithospheric structure within the Australian Craton, *Earth Planet. Sci. Lett.*, **231**, 163–176.

Forsyth, D.W., 1985. Subsurface loading and estimates of the flexural rigidity of continental lithosphere, *J. Geophys. Res.*, **90**, 12 623–12 632.

Forsyth, D.W. & Uyeda, S., 1975. On the relative importance of the driving forces of plate motion, *Geophys. J. R. Astr. Soc.*, **43**, 103–162.

Fouch, M.J., Fischer, K.M., Parmentier, E.M., Wysession, M.E. & Clarke, T.J., 2000. Shear wave splitting, continental keels and patterns of mantle flow, *J. Geophys. Res.*, **105**, 6255–6275.

Frohlich, C., 2006. A simple analytic method to calculate the thermal parameter and temperature within subducted lithosphere, *Phys. Earth Planet. Inter.*, **155**, 281–285.

Funning, G.J., Parsons, B., Wright, T.J., Jackson, J.A. & Fielding, E.J., 2005. Surface displacements and source parameters of the 2003 Bam (Iran) earthquake from Envisat advanced synthetic aperture radar imagery, *J. Geophys. Res.*, **110**, B09406, doi:10.129/2004JB003338.

Garnero, E.J., Revenaugh, J., Williams, Q., Lay, T. & Kellogg, L.H., 1998. Ultralow velocity zone at the core mantle boundary, in *The Core-Mantle Boundary Region*, eds. M. Gurnis, M.E. Wysession, E. Knittle & B.A. Buffet, Geodynamics Monograph, **28**, American Geophysical Union.

Gerya, T.V., Stöckert, B. & Perchuk, A.L., 2002. Exhumation of high-pressure metamorphic rocks in a subduction channel: a numerical simulation, *Tectonics*, **21**, 1056, doi:10.129/2002TC001406.

Glatzmaier, G.A. & Roberts, P.H., 1996. Rotation and magnetism of earth's inner core, *Science*, **274**, 1887–1891.

Goleby, B.R., Shaw, R.S., Wright, C., Kennett, B.L.N. & Lambeck, K., 1989. Geophysical evidence for 'thick-skinned' crustal deformation in central Australia, *Nature*, **337**, 325–330.

Gorbatov, A., & Kennett, B.L.N., 2003. Joint bulk-sound and shear tomography for Western Pacific subduction zones, *Earth Planet. Sci. Lett.*, **210**, 527–543.

Gordon, R.G., & Jurdy, D.M., 1986. Cenozoic global plate motions, *J. Geophys. Res.*, **91**, 2389–2406.

Gudmundsson, O., Kennett, B.L.N. & Goody, A., 1994. Broadband observations of upper mantle seismic phases in northern Australia and the attenuation structure in the upper mantle. *Phys. Earth Planet Inter.*, **84**, 207–226.

Gung, Y.C., Panning, M. & Romanowicz, B., 2003. Anisotropy and thickness of the lithosphere, *Nature*, **422**, 707-711.

Gurnis, M., Wysession, M.E., Knittle, E. & Buffet, B.A. (eds.), 1998. *The Core–Mantle Boundary Region*, Geodynamics Monograph, **28**, American Geophysical Union.

Hager, B.H., 1984. Subducted slabs and the geoid, constraints on mantle rheology and flow, *J. Geophys. Res.*, **89**, 6003–6015.

Hager, B.H., & O'Connell, R.J., 1979. Kinematic models of large-scale flow in the mantle, *J. Geophys. Res.*, **84**, 1031–1048.

Harrison, N., 2003. An introduction to density functional theory, 45–70, in *Computational Materials Science*, Eds. C.R.A. Catlow & E.A. Kotomin, IOS Press, Amsterdam.

Hollerbach, R., 2000. A spectral solution of the magneto-convection equations in spherical geometry, *Int. J. Numeric. Meth. Fluids*, **32**, 773–797.

Holt, W.E., Shen-tu, B., Haines, J. & Jackson, J., 2000. On the determination of self-consistent strain rate fields within zones of distributed continental deformation, in *The History and Dynamics of Global Plate Motions*, eds. M.A. Richards, R.G. Gordon & R.D. van der Hilst, AGU Geophysical Monograph, **121**, 113–137.

Houseman, G. & England, P., 1986. A dynamical model of lithospheric extension and sedimentary basin formation, *J. Geophys. Res.*, **91**, 719–729.

Houseman, G. & Molnar, P., 1997. Gravitational (Rayleigh–Taylor) instability of a layer with non-linear viscosity and convective thinning of continental lithosphere, *Geophys. J. Int.*, **128**, 125–150

Husson, L., 2006. Dynamic topography above retreating subduction zones, *Geology*, **34**, 741–744.

Iaffaldano, G., Bunge H.P. & Dixon, T.H., 2006. Feedback between mountain belt growth and plate convergence, *Geology*, **34**, 893–896.

Ishii, M. & Dziewonski, A.M., 2003. Distinct seismic anisotropy at the centre of the Earth, *Phys. Earth Planet. Inter.*, **140**, 203–217.

Ishii, M., Shearer, P., Houston, H. & Vidale, J., 2005. Imaging the Sumatra rupture with Hi-Net seismic data, *Nature*, doi:10.1038/nature03675.

Jackson, A., 1997. Time-dependency of tangentially geostrophic core motions, *Phys. Earth Planet. Inter.*, **103**, 293–311.

Jackson, J., 2002. Strength of the continental lithosphere: Time to abandon the jelly sandwich?, *GSA Today*, **12**(9), 4–10.

Jackson, I. & Rigden, S.M., 1998. Composition and temperature of the Earth's mantle: seismological models interpreted through experimental studies of earth materials, in *The Earth's Mantle: Structure, Composition and Evolution*, 405–460, ed. I. Jackson, Cambridge University Press.

Jacobsen, S.D., Reichmann, H.J., Spetzler, H.A., Mackwell, S.J., Smyth, J.R., Angel R.A., & McCammon, C.A., 2002. Structure and elasticity of single-crystal (Mg,Fe)O and a new method of generating shear waves for gigahertz ultrasomic interferometry, *J. Geophys. Res.*, **107**(B2), 2037, doi:10.1029/2001JB000490.

Jaeger, J.C. & Cook, N.G.W., 1979. *Fundamentals of Rock Mechanics*, Third edition, Chapman & Hall, London.

Jeanloz, R., & Morris, S., 1986. Temperature distribution in the crust and mantle, *Ann. Rev. Earth Planet. Sci.*, **14**, 377–415.

Jeanloz, R., & Morris, S., 1987. Is the mantle geotherm subadiabatic?, *Geophys. Res. Lett.*, **14**, 335–338.

Jessell, M.W. & Bons, P.D., 2002. The numerical simulation of microstructure, *J. Geol. Soc. London*, **200**, DRT Conference Special Issue, 137–148.

Jochum, K.P., Hofmann, A.W., Ito, E., Seufert, H.M. & White W.M., 1983. K, U and Th in mid-ocean ridge basalt glass and heat-production, K/U and K/Rb in the mantle, *Nature*, **306**, 431–436.

Jordan, T.H., 1975. The continental tectosphere, *Rev. Geophys.*, **13**, 1–12.

Jordan, T.H., 1978. Composition and development of the continental tectosphere, *Nature*, **274**, 544–548.

Karato, S.-I., 2003. *The Dynamic Structure of the Deep Earth*, Princeton University Press.

Karato, S.-I. & Wu, P., 1993. Rheology of the upper mantle: a synthesis, *Science*, **260**, 771–778.

Kennett, B.L.N, 1998. On the density distribution within the Earth, *Geophys. J. Int.*, **132**, 374–382.

Kennett, B.L.N. & Engdahl, E.R., 1991. Traveltimes for global earthquake location and phase identification, *Geophys. J. Int.*, **105**, 429–465.

Kennett, B.L.N. & Gorbatov, A., 2004. Seismic heterogeneity in the mantle – strong shear wave signature of slabs from joint tomography, *Phys. Earth Planet. Inter.*, **146**, 88–100.

Kennett, B.L.N., Engdahl, E.R. & Buland, R., 1995. Constraints on seismic velocities in the Earth from travel times, *Geophys. J. Int.*, **122**, 108–124.

Kiefer, B., Stixrude, L., Hafner, J. & Kresse, G., 2001. Structure and elasticity of wadsleyite at high pressure, *Am. Mineral.*, **86**, 1387–1395.

King, G.C., Stein, R.S. & Lin, J., 1994. Static stress changes and the triggering of earthquakes, *Bull. Seism. Soc. Am.*, **84**, 567–585.

Kirby, S.H., Stein, S., Okal, E.A. & Rubie, D.C., 1996. Metastable phase transformations and deep earthquakes in subducting oceanic lithosphere, *Rev. Geophys.*, **34**, 261–306.

Kohn, W. & Sham, L.J., 1965, Self consistent equations including exchange and correlation effects, *Phys. Rev. B*, **140**, 1133–1138.

Kong, X. & Bird, P., 1995. SHELLS: a thin-shell program for modelling neotectonics of regional or global lithosphere with faults, *J. Geophys. Res.*, **100**, 22 129–22 132.

Kono, M. & Roberts, P.H., 2002, Recent geodynamo simulations and observations of the magnetic field, *Rev. Geophys.*, **40**, 4, doi:10.129/2000RG000102.

Kresse, G., Furthmüller, J. & Hafner, J., 1995. Ab-initio force-constant approach to phonon dispersion relations of diamond and graphite, *Europhys. Lett.*, **32**, 729–734.

Lambeck, K. & Johnston, P., 1998. Viscosity of the mantle: Evidence from analysis of glacial-rebound phenomena, in *The Earth's Mantle: Structure, Composition and Evolution*, 461–502, ed. I. Jackson, Cambridge University Press.

Lay, T., Garnero, E.J., Young, C.J. & Gaherty, J.B., 1997. Scale lengths of shear velocity heterogeneity at the base of the mantle from S wave differential times, *J. Geophys. Res.*, **102**, 9887–9910.

Lay, T., Williams, Q., Garnero, E.J., Kellogg, L.H. & Wysession, M.E., 1998. Seismic wave anisotropy in the D″ region and its implications, in *The Core–Mantle Boundary*

Region, eds. M. Gurnis, M.E. Wysession, E. Knittle & B.A. Buffet, Geodynamics Monograph, **28**, American Geophysical Union.

Li, L., Brodholt, P., Stackhouse, S., Weidner D.J., Alfredsson, M. & G.D. Price, 2005. The elasticity of (Mg, Fe)(Si, Al)O_3 perovskite at high pressure, *Earth Planet. Sci. Lett.*, **240**, 529–536.

Lister, G.S., Etheridge, M.A. & Symonds P.A., 1991. Detachment models for the formation of passive continental margins, *Tectonics*, **10**, 1038–1064.

Lithgow-Bertelloni, C., & Richards, M.A., 1998. The dynamics of Cenozoic and Mesozoic plate motions, *Rev. Geophys.*, **36**, 27–78.

Lithgow-Bertelloni, C. & Silver, P.G., 1998. Dynamic topography, plate driving forces and the African superswell *Nature*, **395**, 269–272.

Liu, J., Sieh, K., and Hauksson, E., 2003. A structural interpretation of the aftershock "cloud" of the 1992 Mw 7.3 Landers earthquake, *Bull. Seism. Soc. Am.*, **93**, 1333–1344.

McKenzie, D.P., 1969. Speculations on the consequences and causes of plate motion, *Geophys. J. R. Astr. Soc.*, **18**, 1–32.

McKenzie, D.P., 1978. Some remarks on the development of sedimentary basins, *Earth Planet. Sci. Lett.*, **40**, 25–32.

McKenzie, D.P., 2003. Estimating T_e in the presence of internal loads, *J. Geophys. Res.*, **108**, B9, 2438, doi:10.1029/2002JB001766.

McKenzie, D.P. & Fairhead, D., 1997. Estimates of the effective thickness of the continental lithosphere from Bouger and free-air gravity anomalies, *J. Geophys. Res.*, **102**, 27 523–27 552.

McKenzie, D.P., Jackson, J.A. & Priestley, K.F., 2005. The termal structure of oceanic and continental lithosphere, *Earth Planet. Sci. Lett.*, **233**, 337–349.

McNamara, A.K. & Zhong, S.J., 2005. Thermochemical structures beneath Africa and the Pacific Ocean, *Nature*, **437**, 1136–1139.

Masters, T.G. & Shearer, P.M., 1990. Summary of seismological constraints on the structure of the Earth's core, *J. Geophys. Res.*, **95**, 21 691–21 695.

Masters, G. & Widmer, R., 1995. Free oscillations: frequencies and attenuation, in *Global Earth Physics: A Handbook of Physical Constants*, 104–125, ed. Ahrens, T.J., American Geophysical Union.

Masters, G., Laske, G., Bolton, H. & Dziewonski, A., 2000. The relative behaviour of shear velocity, bulk sound speed, and compressional velocity in the mantle: implications for chemical and thermal structure, in *Earth's Deep Interior: Mineral Physics and Tomography from the Atomic to the Global Scale*, eds. S.I. Karato, A.M. Forte, R.C. Liebermann, G. Masters & L. Stixrude, AGU Geophysical Monograph, **117**, 63–87.

Megnin, C. & Romanowicz, B., 2000. The three-dimensional shear velocity structure of the mantle from the inversion of body, surface and high-mode waveforms, *Geophys. J. Int.*, **143**, 709–728.

Miller M.S., Gorbatov A. & Kennett B.L.N., 2005. Heterogeneity within the subducting Pacific plate beneath the Izu-Bonin-Mariana arc: evidence from tomography using 3D ray-tracing inversion techniques, *Earth Planet Sci. Lett.*, **235**, 331–342.

Mitrovica, J.X. & Forte, A.M., 2004. A new inference of mantle viscosity based upon joint inversion of convection and glacial isostatic adjustment data, *Earth Planet. Sci. Lett.*, **225**, 177–189.

Montagner, J-P & Kennett, B.L.N., 1996. How to reconcile body-wave and normal-mode reference Earth models?, *Geophys. J. Int.*, **125**, 229–248.

Micklethwaite, S. & Cox, S.F., 2004. Fault-segment rupture, aftershock-zone fluid flow, and mineralization, *Geology*, **32**, 813–816.

Morelli, A. & Dziewonski, A.M., 1993. Body wave traveltimes and a spherically symmetric *P*- and *S*-wave velocity model, *Geophys. J. Int.*, **112**, 178–194.

Moresi, L. & Solomatov, V., 1998. Mantle convection with a brittle lithosphere: Thoughts on the global tectonic styles of Earth and Venus, *Geophys. J. Int.*, **133**, 669–682.

Morgan, W.J., 1968 Rises, trenches, great faults and crustal blocks *J. Geophys. Res.*, **73** 1959–1982.

Murakami, M., Hirose, K., Sata, N., Ohishi, Y. & Kawamura, K., 2004. Phase transition of $MgSiO_3$ perovskite in the deep lower mantle, *Science*, **304**, 855-858.

Nishimura, C. & Forsyth, D., 1989. The anisotropic structure of the upper mantle in the Pacific, *Geophys. J. R. Astr. Soc.*, **96**, 203–226.

Nolet, G., Grand S. & Kennett B.L.N., 1994. Seismic heterogeneity in the Upper Mantle, *J. Geophys. Res.*, **99**, 23 753–23 766.

O'Neill, H.St.C. & Palme, H., 1998. Composition of the silicate Earth: implications for accretion and core formation, in *The Earth's Mantle: Structure, Composition and Evolution*, 3–126, ed. I. Jackson, Cambridge University Press.

Pieri, M., Burlini, L., Kunze, K., Stretton, I. & Olgaard, D.L., 2001. Rheological and microstructural evolution of Carrara marble with high shear strain: results from high temperature torsion experiments. *J. Struct. Geol.*, **23**, 1393–1413.

Pysklywec, R.N., 2006. Surface erosion control on the evolution of the deep lithosphere, *Geology*, **34**, 225–228.

Reynolds, S.D., Coblentz, D.D. & Hillis, R.R., 2003. Influences of plate-boundary forces on the regional intraplate stress field of continental Australia, *Geol. Soc. Australia Spec. Publ.* **22** and *Geol. Soc. America Spec. Pap.* **372**, 59–70.

Ricard, Y., Richards, M., Lithgow-Bertelloni, C., & Le Stunff, Y., 1993. A geodynamic model of mantle density heterogeneity, *J. Geophys. Res.*, **98**, 21 895–21 909.

Richards, M.A. & Hager, B.H., 1984. Geoid anomalies in a dynamic Earth, *J. Geophys. Earth*, **89**, 5987–6002.

Richards, M.A. & Engebretson, D.C., 1992. Large-scale mantle convection and the history of subduction, *Nature* **355**, 437–440.

Richards, M.A., Yong, W.S., Baumgardner, J.R. & Bunge, H.P., 2001. Role of a low viscosity zone in stabilizing plate tectonics: Implications for comparative terestrial planetology, *Geochem. Geophys. Geosyst.*, **2**, doi:10.129/2000GC00115.

Richter, F.M. & McKenzie, D.P., 1978. Simple models of plate convection, *J. Geophys.*, **44**, 441–471.

Rubatto, D. & Hermann, J., 2001. Exhumation as fast as subduction? *Geology*, **29**, 3–6.

Scholz, C., 1990. *The Mechanics of Earthquakes and Faulting*, Cambridge University Press, Cambridge.

Schilling, J.G., 1991. Fluxes and excess temperatures of mantle plumes inferred from their interaction with migrating midocean ridges, *Nature*, **352**, 397–403.

Sella, G.F., Dixon, T.H., & Mao, A.L., 2002. A model for recent plate velocities from space geodesy, *J. Geophys. Res.*, **107**, B4:Art. No. 2081.

Shearer, P.M., 1999. *An Introduction to Seismology*, Cambridge University Press, Cambridge.

Simons, F.J., Zuber, M.T. & Korenaga, J., 2000. Isostatic response of the Australian lithosphere: Estimates of effective elastic thickness and anisotropy using multitaper spectral analysis, *J. Geophys. Res.*, **105**, 19 163–19 184.

Sleep, N.H., 1990. Hotspots and mantle plumes – some phenomenology, *J. Geophys. Res.*, **95**, 6715–6736.

Song, X. & Helmberger, D.V., 1992. Velocity structure near the inner core boundary from waveform modelling, *J. Geophys. Res.*, **97**, 6573–6586.

Spakman, W. & Nyst, M.C.J., 2002. Inversion of relative motion data for estimates of the velocity gradient field and fault slip, *Earth Planet. Sci. Lett.*, **203**, 577–591.

Spera, F.J., Yuen, D.A. & Giles, G., 2006. Tradeoffs in chemical and thermal variations in the post-perovskite phase transition: Mixed phase regions in the lower mantle, *Phys. Earth Planet. Inter.*, **156**, 234–246.

Stacey, F., 1992. *Physics of the Earth*, Third edition, Brookfield Press, Brisbane, Australia.

Stein, S. & Rubie, D., 1999. Deep earthquakes in real slabs, *Science*, **286**, 909–910.

Steinberger, B., & O'Connell, R.J., 1997. Changes of the Earth's rotation axis owing to advection of mantle density heterogeneities, *Nature*, **387**, 169–173.

Steinberger, B. & O'Connell, R.J., 1998. Advection of plumes in mantle flow: implications for hotspot motion, mantle viscosity and plume distribution, *Geophys. J. Int.*, **132**, 412–434.

Steinle-Neumann, G., Stixrude, L. & Cohen, R., 1999. First principles elastic constants for the hcp transition metals Fe, Co and Re at high pressure, *Phys. Rev. B*, **60**, 791–799.

Tackley, P.J. & Xie, S., 2002. The thermochemical structure and evolution of Earth's mantle: constraints and numerical models, *Phil. Trans. R. Soc. Lond.*, A**360**, 2593–2609.

Takahashi, F., Matsushima, M. & Honkura, Y., 2005. Simulations of a quasi-Taylor state geomagnetic field including polarity reversals on the Earth Simulator, *Science*, **309**, 459–461.8:31 PM 11/19/2007

Takeuchi, H. & Saito, M., 1972. Seismic surface waves, *Methods of Computational Physics*, **11**, Academic Press.

Tarduno, J.A., Duncan, R.A., Scholl, D.W., Cottrell, R.D., Steinberger, B., Thordason, T., Kerr, B.C., Neal, C.R., Frey, F.A., Torii, M. & Carvallo, C., 2003. The Emperor Seamounts: southward motion of the Hawaiian hotspot plume in Earth's mantle, *Science*, **301**, 1064–1069.

Tozer, D.C., 1972. The present thermal state of the terrestrial planets, *Phys. Earth. Planet. Inter.*, **6**, 182–197.

Tregoning, P., Lambeck, K., Stolz, A., Morgan, P., McClusky, S.C., van der Beek, P., McQueen, H., Jackson, R.J., Little, R.P., Laing, A. & Murphy, B., 1998. Estimation of current plate motions in Papua New Guinea from GPS observations, *J. Geophys. Res.*, **103**, 12 181–12 203.

Turcotte, D.L., & Oxburgh, E.R., 1967. Finite amplitude convective cells and continental drift, *J. Fluid Mech.*, **28**, 29–42.

Van der Voo, R., Spakman, W., Bijwaard, H., 1999. Mesozoic subducted slabs under Siberia, *Nature*, **397**, 246–249.

Watts, A.B., 2001. *Isostasy and Flexure of the Lithosphere*, Cambridge University Press.

Whitehead, J.A., 1988. Fluid models of geological hotspots, *Ann. Rev. Fluid Mech.*, **20**, 61–87.

Yoshida, S., Koketsu, K., Shibazaki, B., Sagiya, T., Kato, T. & Yoshida Y., Joint inversion of near- and far-field waveforms and geodetic data for the rupture process of the 1995 Kobe earthquake, *J. Phys. Earth*, **44**, 437–454.

Young, C.J. & Lay, T., 1987. Evidence for a shear velocity discontinuity in the lower mantle beneath India and the Indian Ocean, *Phys. Earth Planet. Inter.*, **49**, 37–53.

Yue, L.-F., Suppe, J. & Hung, J.-H., 2005. Structural geology of a classic thrust belt earthquake: the 1999 Chi-Chi earthquake Taiwan (M_w 7.6), *J. Struct. Geol.*, **27**, 2058-2083.

Zhong, S.D. & Gurnis, M., 1996. Interaction of weak faults and non-Newtonian rheology produces plate tectonics in a 3D model of mantle flow, *Nature*, **383**, 245–247.

Zienkiewicz, O.C., Taylor, R.L. & Nithiarasu, P., 2005. *The Finite Element Method for Fluid Dynamics*, Sixth edition, Elsevier Butterworth-Heinemann.

Index